Asheville-Buncombe
Technical Community College
Learning Resources Center
340 Victoria Rd.
Asheville, NC 28801

D1218390

McGraw-Hill's

NATIONAL ELECTRICAL SAFETY CODE® (NESC®) 2007 HANDBOOK

About the Author

David J. Marne, P.E., B.S.E.E, is a consulting electrical engineer. He is company president and senior engineer for
Marne and Associates, Inc. in Missoula,
Montana, where he specializes in NESC®
training and engineering design. Mr.
Marne serves on NESC® Subcommittee 4,
Overhead Lines-Clearances, he is a senior
member the IEEE, and has presented sem-
inars on the NESC® to a variety of industry
professionals.

McGraw-Hill's

NATIONAL ELECTRICAL SAFETY CODE® (NESC®) 2007 HANDBOOK

Based on the Current 2007
National Electrical Safety Code® (NESC®)

David J. Marne

National Electrical Safety Code® and NESC®
are both registered trademarks and service marks
of the Institute of Electrical and Electronics Engineers, Inc. (IEEE)

New York Chicago San Francisco Lisbon London Madrid
Mexico City Milan New Delhi San Juan Seoul
Singapore Sydney Toronto

The McGraw·Hill Companies

Copyright © 2007 by The McGraw-Hill Companies, Inc. All rights reserved. Printed in the United States of America. Except as permitted under the United States Copyright Act of 1976, no part of this publication may be reproduced or distributed in any form or by any means, or stored in a data base or retrieval system, without the prior written permission of the publisher.

2 3 4 5 6 7 8 9 0 DOC/DOC 0 1 9 8 7

ISBN 13: 978-0-07-145367-7
ISBN 10: 0-07-145367-9

The sponsoring editor for this book was Steve Chapman, the editing supervisor was Mon Patty, the production supervisor was James Kussow, and the project manager was Samik Roy Chowdhury (Sam). It was set in Melior by International Typesetting and Composition. The art director for the cover was Anthony Landi.

Printed and bound by RR Donnelley.

This book was printed on acid-free paper.

McGraw-Hill books are available at special quantity discounts to use as premiums and sales promotions, or for use in corporate training programs. For more information, please write to the Director of Special Sales, Professional Publishing, McGraw-Hill, Two Penn Plaza, New York, NY 10121-2298. Or contact your local bookstore.

Contents

Part 1.   Rules for the Installation and Maintenance of Electric Supply Stations and Equipment

Part 3. Safety Rules for the Installation and Maintenance of Underground Electric Supply and Communication Lines

Preface

McGraw-Hill's National Electrical Safety Code® (NESC®) 2007 Handbook is written for the engineer, staking technician, power lineman, communications lineman, inspector, and safety administrator of an electric power or communication utility company, contracting company, or consulting firm. Employees involved with transmission, distribution, and substation design and construction should develop the highest possible understanding of the NESC to keep the public, utility workers, and themselves safe.

McGraw-Hill's NESC Handbook is designed to be used in conjunction with the National Electrical Safety Code (NESC) published by the Institute of Electrical and Electronic Engineers (IEEE). This Handbook presents hundreds of figures and photos, plus examples and discussions to explain and clarify the NESC rules. The straightforward and practical information in this Handbook is intended to aid the understanding of the sometimes confusing and complicated text in the NESC. This Handbook follows the order of parts, sections, and rules as presented in the NESC including the general rules, substation rules (Part 1), overhead line rules (Part 2), underground line rules (Part 3), and work rules (Part 4). This format assures quick and easy correlation between the NESC rules and the discussions and explanations in this Handbook. This Handbook is most effectively used by referring to the NESC for the precise wording of a rule and then referring to the corresponding rule number in this Handbook for a practical understanding of what the Code requires. Appendix A of this Handbook includes over one hundred photos of NESC applications and violations. Appendix B of this Handbook contains Occupational Safety and Health Administration (OSHA) standards related to the NESC work rules.

The **National Electrical Safety Code (NESC)** is the "bible" for developing power and communication utility standards. All utility construction has some relevance to the **NESC** including items such as line design, substation design, standard drawings, material purchases, pole placement, work practices, signage, and others. The **NESC** should not be confused with the National Electrical Code (NEC) published by the National Fire Protection Agency (NFPA). The NEC is used for residential, commercial, and industrial building wiring. It does not apply to utility systems although there is some overlap of the **NESC** and the NEC at a building service. The **NESC** is written as a voluntary standard. It can be adopted as law by individual states or other governmental authorities. To determine the legal status of the **NESC**, the state public service commission or public utility commission, or other governmental authority should be contacted. The **Code** is written by various **NESC** committees. The organizations represented, subcommittees, and committee members are listed in the front of the **Code** book. The procedure for revising the **NESC** is described in the back of the **Code** book. The **NESC** has an interpretation committee that issues formal interpretations. The procedure for obtaining a formal interpretation is outlined in the front of the **Code** book. The **NESC** is currently published on a five year cycle. Urgent safety matters that require a change in between **Code** documents are handled through a Tentative Interim Amendment (TIA) process. Original work on the **NESC** began in 1913. This Handbook is based on the 2007 edition of the **NESC**.

McGraw-Hill's NESC Handbook is the only **NESC** Handbook that focuses on the practical application of the current edition of the **NESC**. The numerous figures, photos, and examples make *McGraw-Hill's NESC Handbook* the most complete and useful handbook available.

Additional reference material for the interested reader includes, *The Lineman's and Cableman's Handbook* by Thomas M. Shoemaker and James E. Mack, and the *Standard Handbook for Electrical Engineers*, by Donald G. Fink and H. Wayne Beaty, both published by McGraw-Hill. The transmission line, distribution line, and substation standards published by the Rural Utility Service (RUS), formerly known as the Rural Electric Administration (REA), are also invaluable references.

This Handbook should be used as a reference for understanding and applying the rules in the **NESC**. This Handbook is not an official **Code** document and does not contain official **NESC** committee interpretations.

David J. Marne, P. E.

Acknowledgments

I wish to express my sincere appreciation to the many individuals who helped me bring the 2007 edition of this Handbook to print.

The following peer reviewers provided a technical review of my manuscript: Kirk P. Dewey, P.E., Grant D. Glaus, P.E., and Mathew C. Schwarz, P.E. The following people helped bring the text, figures, and photographs to life: Danielle "Nelly" Arnold and Grant D. Glaus. The following people at McGraw-Hill and their affiliates supported and guided my efforts: Katie Andersen, Steve Chapman, Patty Mon, Sam RC, and Paul Tyler.

Finally, I wish to thank my wife, Patty, and my children for their constant support.

David J. Marne, P.E.

General Sections

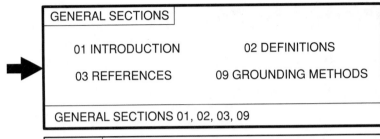

GENERAL SECTIONS

01 INTRODUCTION   02 DEFINITIONS

03 REFERENCES   09 GROUNDING METHODS

GENERAL SECTIONS 01, 02, 03, 09

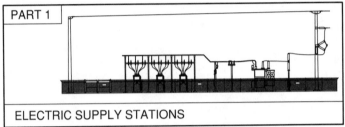

PART 1

ELECTRIC SUPPLY STATIONS

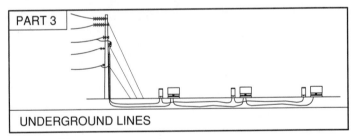

PART 2

OVERHEAD LINES

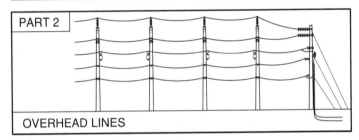

PART 3

UNDERGROUND LINES

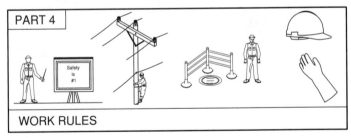

PART 4

WORK RULES

Introduction to the National Electrical Safety Code

010. PURPOSE

The very first rule in the NESC outlines the purpose of the entire book. The rules contained in the NESC are provided for the "practical safeguarding of persons during the installation, operation, or maintenance of electric supply and communication lines and associated equipment." The persons the Code is referring to are both utility employees and the general public. The Code uses the term electric supply for electric power. The Code covers electric supply lines and associated equipment. Examples of associated equipment are substations, transformers, reclosers, regulators, etc. The Code also covers communications lines and equipment. Communications utilities include, but are not limited to, telephone and cable TV.

The Code contains the basic provisions necessary for safety. The Code is not intended to be a design manual. The Code specifies what needs to be accomplished for safety, not how to accomplish it. Values for clearance and structure strength must not be less than the values indicated in the Code for safety purposes. Clearance and strength values can certainly be greater than the Code specifies, but greater values are not required for safety.

For example, if the Code clearance of a 12.47/7.2-kV phase conductor is 18.5 ft over the ground, the Code does not specify how high the poles have to be or where the wires are attached on the pole to obtain the 18.5-ft clearance. Selecting the proper pole heights and attachment heights is a design function. The Code does require that the phase wires be not less than 18.5 ft high (they can certainly be higher). Some utilities establish clearance values by using the Code clearance plus an adder. The adder can be thought of as a design or construction tolerance adder to maintain the required **Code** clearance. A line designed for exactly 18.5 ft of clearance could end up having problems meeting **Code**. There are several factors that could jeopardize the 18.5-ft clearance. One factor could be that the pole hole was dug 6 inches deeper than it should be dug. Another factor might be a slight rise in elevation at midspan that was not detected due to the fact that the ground line was not profiled or surveyed. Using a clearance adder (say 3 ft in this case) would require designing the line to 21.5 ft instead of 18.5 ft. The clearance adder could help meet and maintain the required **Code** clearance over time. See Rule 230I for an additional discussion of maintaining clearances over the life of an installation.

There are many design standards available for utility companies to reference. The Rural Utilities Service (RUS) publishes transmission, distribution, and substation design manuals. Many larger utilities develop their own design manuals. Each manual written must be in compliance with the **Code**, as the **Code** is the "bible" for all utility work.

The **Code** states that rules apply to specified conditions. One example of the specified conditions is overhead line clearance for open supply conductors, 750 V–22 kV, over a road subject to truck traffic. In **NESC** Appendix A (**NESC** Fig. A-1), the **Code** shows the reference component (of a truck) to be 14 ft. The mechanical and electrical clearance component shown in this figure is 4.5 ft, for a total of 18.5 ft of clearance when the conductors are at their maximum sag condition. These are the specified conditions. If an overhead line is being designed for a mining installation and the conditions are that 22-ft-high trucks are used, then this condition requires that the same line be designed with a 26.5-ft clearance (22' + 4.5' = 26.5').

011. SCOPE

This rule defines what is covered in the **National Electrical Safety Code** and what is not covered. The **NESC** covers supply and communication lines equipment and work practices carried out by a public or private utility company or a company functioning as a utility. The **NESC** basically applies to electric supply (power) and communications utilities. Communications utilities include, but are not limited to, telephone and cable TV. An example of an organization not normally thought of as a utility but functioning as a utility could be a university system campus or a large industrial complex that owns a high-voltage distribution system.

The **NESC** does not cover utilization wiring in buildings. The standard that does cover building wiring is the National Electrical Code (NEC). The NEC is the "bible" for the electrical building industry and is used primarily by engineers and electricians. The **NESC** is the "bible" used primarily by utility engineers and utility linemen. See Fig. 011-1.

In places the **NESC** and the NEC overlap. One location is at the service to a building. For an overhead electric service, the typical dividing point between

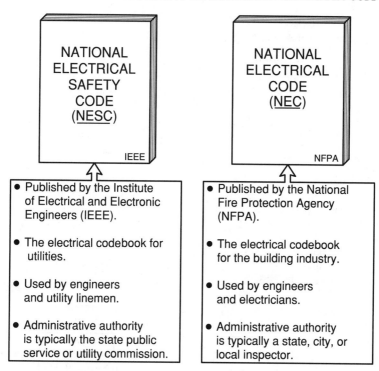

Fig. 011-1. Differences between the NESC and the NEC (Rule 011).

the NESC and the NEC is the conductor splice at the weatherhead. This is the dividing line between the electric utility function and the electric utilization function. For underground services the typical dividing point can vary. Some utilities provide service to the terminals on the meter base. Others provide service to the property line and the customer provides wiring after that point. Sometimes a utility will have the customer install secondary wiring from the pad mount transformer to the meter but then the utility will take ownership of this wiring. To determine whether the circuit is covered under the NESC or the NEC, the ownership of the circuit and who maintains and controls the circuit are important factors to consider. Some utilities use a direct buried splice or a junction box under the meter base to provide a clear transition point between the two codes. See Fig. 011-2.

Control of the street lighting is the important factor to consider when determining whether street lighting is covered under the NESC or the NEC. Street lighting that is metered usually falls under NEC requirements. Street lighting that is not metered and is owned, operated, and maintained by the utility typically is covered under the NESC. The main differences between street lighting covered under the NESC and street lighting covered under the NEC are grounding methods and overcurrent protection requirements.

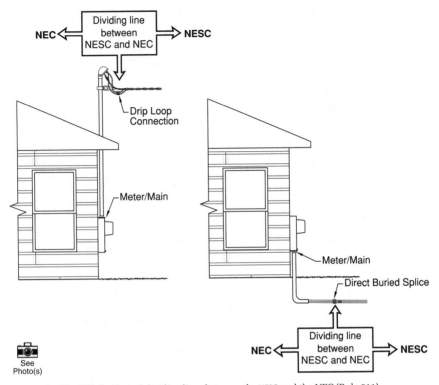

Fig. 011-2. Typical dividing lines between the NESC and the NEC (Rule 011).

The dividing line between the NESC and the NEC for a communications circuit is typically the building network interface device or demarcation point. This separates the utility communications function from the building communications function.

The NESC does not cover installations in mines, ships, railway loading equipment, aircraft, or automotive equipment. These industries have their own standards. See Rules 101 and 301 for a discussion of utilization wiring in Parts 1 and 3.

Although not specifically addressed in Rule 011, the NESC does not cover easements, environmental protection, raptor (bird) protection, FCC regulations, FAA regulations, electromagnetic fields (emf), or construction cost issues. These items are either covered by other standards or they do not present a safety concern and therefore they are not addressed in the NESC.

012. GENERAL RULES

This rule provides three general rules for applying the Code. Rule 012A requires that the design, construction, operation, and maintenance of electric supply and communication lines equipment must be in accordance with the NESC. Rule 012B requires that utilities or other organizations performing work for the utility, such as a contractor, are the responsible parties for meeting NESC requirements for design, construction, operation, and maintenance. Rule 012C acknowledges that the Code cannot cover every conceivable situation. Where the Code does not specify a rule to cover a particular installation, construction and maintenance should be done in accordance with "accepted good practice" for the local conditions known at the time. This does not allow the Code to be ignored for a condition that is covered in the Code. If "accepted good practice" is needed because a specific Code rule does not exist, a comparable NESC rule or the National Electrical Code (NEC) can be referenced to find an "accepted good practice." Other standards can also be referenced or "accepted good practice" may be determined by reviewing utility operating records to find out what practices work for local conditions. Rule 012 applies to both supply and communication utilities during the design, construction, operation, and maintenance of lines and equipment. The NESC applies during the initial design and construction and during the life of the installation (i.e., during operation and maintenance).

013. APPLICATION

New installations and extensions are covered under Rule 013A. This rule clearly states that all new installations and extensions must adhere to the provisions of the NESC. Rule 013A1 does allow the administrative authority (i.e., the public service or public utility commission) to waive or modify the NESC rules if safety is provided in other ways. An example is listed in the Code to clarify this statement. Rule 013A2 recognizes that new types of construction and methods may be used even if they are not covered in the Code. This must be done under qualified supervision, for example, under the direction of a registered professional engineer, who is using engineering judgment and collecting data on the new construction. Equivalent safety must be provided. On joint-use (power and communications) facilities, both the power and communication utilities must agree to the methods not covered in the Code. Since the Code is on a 5-year revision cycle, many times new construction methods arise during the 5-year period and then are included in a future edition of the Code.

Rule 013B discusses how the Code is applied to existing installations. Many people use the term "grandfather clause" when discussing this rule. The "grandfather clause" concept appears simple; however, caution must be used when applying this rule. Rule 013B1 states that if an existing installation meets or is altered to meet the current NESC, the installation is considered to apply to the current edition. Rule 013B2 states that existing installations, including maintenance replacements, that comply with a prior edition of the Code, need not be modified to comply with the current edition. Modifications may be required for safety reasons by the administrative authority (i.e., public service or public utility commission) or as required by Rule 202. Rule 202 requires use of the current edition of Rule 238C even for a pole maintenance replacement. See Rules 202 and 238C for additional information. Rule 013B2 can be applied to changes in the **Code**. Changes from the previous edition are indicated by a vertical black bar in the left margin of the **NESC** Codebook. Applying Rule 013B2 eliminates the need to fix existing installations that do not comply with a new Rule in the current edition. Rule 013B3 specifically discusses conductors and equipment that are added, altered, or replaced on an existing structure. Rule 013B3 states that the structure or the facilities on the structure need not be modified or replaced if the resulting installation will be in compliance with either:

- The **Code** rules that were in effect at the time of the original installation,
- The **Code** rules in effect in a subsequent edition to which the installation has been previously brought into compliance, or
- The rules of the current edition.

Choosing the third bullet in the list above (the rules of the current edition) results in using the best **Code** information available as the **NESC** has evolved and improved over time. Choosing the first or second bullet in the list above is acceptable as the **Code** provides a choice of any one of the three options. However, if the first or second bullet is chosen, caution needs to be taken to determine if the existing installation complies with a prior **Code**. Normally electric supply and communication engineers have the most current codebook sitting upon their desk and field workers do not have prior codebooks lying around in their trucks. Even if something as simple as adding a transformer to an existing installation or replacing an existing pole is being done, the utility performing the work must be careful not to blindly assume that the existing situation complies with a prior **Code**. Utilities may want to consult the advice of a legal expert when applying Rule 013B3 as interpreting this rule from a legal viewpoint may be more critical than from a technical viewpoint.

The application of a "grandfather clause" should not be confused with maintaining clearances and the change in land use under a supply or communications line. See Rule 230I for a discussion.

Rule 013C requires that inspections of new and existing facilities meet the inspection rules in the current edition of the **NESC**, not prior editions. Inspections of electric supply stations are covered in Rule 121, inspections of overhead lines are covered in Rule 214, and inspections of underground lines are covered in Rule 313. Rule 013C also requires that work performed on new and existing facilities meet the work rules in Part 4 of the current edition of the **NESC**, not prior editions.

014. WAIVER

In this rule the **Code** recognizes the need to waive or modify rules in cases of emergency or temporary installations. The **Code** grants the person responsible for the installation the ability to modify or waive rules with specific requirements. Both Rule 014A for emergency installations and Rule 014B for temporary overhead installations apply to clearance and strength of overhead lines. The **Code** does not specify a time length for temporary installations or define an emergency installation. The **Code** makes it very clear that temporary installations may not have reduced clearance. See Fig. 014-1 for a summary of the requirements of this rule.

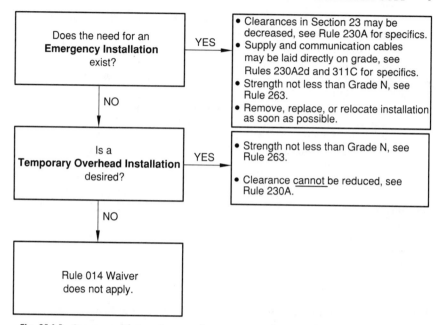

Fig. 014-1. Summary of the requirements for emergency and temporary installations (Rule 014).

015. INTENT

This rule defines three key **NESC** words: "shall," "should," and "RECOMMEN-DATION." **Code** rules containing the word "shall" indicate that the rule is mandatory and the rule must be met. Not complying with a rule that contains the word "shall" is a direct **Code** violation. **Code** rules containing the word "should" indicate requirements that are normally and generally practical for the specified conditions. The **Code** does recognize that under certain circumstances some rules may not be practical. Where this is the case, the word "should" is used. Under the definition of "should," the **Code** references Rule 012 that requires "accepted good practice" to be used. If at all possible the words "should" and "shall" should be considered the same. If it is not possible to treat "should" the same as "shall," it is prudent to document the specific conditions that prohibit a utility from applying the **Code** rule that contains the word "should" and it is prudent to apply an "accepted good practice" that provides an equivalent degree of safety for the specific conditions. The **Code** uses the word "RECOMMENDATION" for provisions that are considered desirable but not intended to be mandatory. "RECOMMENDATION" is the least stringent of the three terms "shall," "should," and "RECOMMENDATION." Since the **Code** does consider a "RECOMMENDATION" desirable, it seems prudent that the utility make some effort to comply with a "RECOMMENDATION" even though it is not mandatory.

Rule 015 also contains clarifications of the words "NOTE," "EXAMPLE," and "footnote." "NOTES" and "EXAMPLES" are not mandatory, and they are provided for information and illustrative purposes but are not considered part of the Code requirements. "Footnotes," however, are used for tables throughout the Code and they carry the full force and effect of the table or rule with which they are associated.

An exception to a rule has the same force as the rule itself. Exceptions are not reduced safety measures. For example, if a clearance value is reduced by an exception, some condition associated with the exception is provided to maintain safety. Typically, the Code provides a larger value in the rule and smaller value in the exception.

The physical location of the words "RECOMMENDATION," "EXCEPTION," and "NOTE," and how these words are indented with respect to other text, signifies to what rule the "RECOMMENDATION," "EXCEPTION," or "NOTE" applies to.

016. EFFECTIVE DATE

This rule states that the 2007 edition of the NESC may be used at any time on or after publication date. In addition, this edition shall become effective no later than 180 days (6 months) following the publication date. The 180-day grace period allows utilities and other agencies to acquire copies of the Code and revise regulations, standards, and procedures. The note to this rule clarifies the fact that this edition is not required to be used before the 180-day period; however, it is not prohibited to use it during this period.

The NESC is a standard published by the Institute of Electrical and Electronic Engineers (IEEE). For the NESC to become a legal requirement, it is typically adopted by a state authority having jurisdiction over utilities (i.e., a public service commission or public utility commission) or by some other authority. To determine the specific legal status of the NESC, the authority having jurisdiction must be contacted.

017. UNITS OF MEASURE

The Code uses the metric system as the primary unit for numerical values. The customary (English or inch-foot-pound) system is the secondary system. Metric values are always shown first with the customary inch-foot-pound system shown second and inside parentheses. Some tables in the Code have the metric and English system in the same table. Other more complex tables have separate tables. When separate tables are used, the first table in the Code will be the metric table and the second table will be the English table. Metric values are based on the modern version of the metric system titled, "The International System of

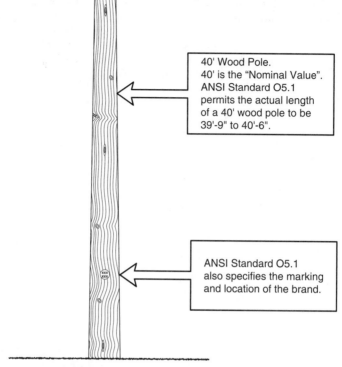

Fig. 017-1. Example of nominal values (Rule 017B).

Units" (or SI). The values in each system are rounded to convenient numbers. An exact conversion is not used so that the values appear functional for safety purposes. Units of measure discussed in this Handbook are based on the customary (English or inch-foot-pound) system.

Rule 017B states that physical items referenced in the **Code** are in "nominal values" unless specific dimensions are provided. Other standards may set tolerances for manufacturing. An example of nominal values is shown in Fig. 017-1.

018. METHOD OF CALCULATION

Rule 018 provides rounding requirements. In general, rounding "off" to the nearest significant digit is required unless otherwise specified in applicable rules. One rule that requires a different rounding method is Rule 230A4, which requires rounding "up" as Sec. 23 deals with overhead line clearances which are typically specified as "not less than" clearances. Rounding "off" follows the rules of traditional rounding learned in math class. An example of rounding "off" is rounding 20.02 down to 20.0 or rounding 20.66 up to 20.7. An example of rounding "up" for "not less than" clearance is rounding 20.02 to 20.1 because rounding "off" to 20.0 would not meet a clearance required to be "not less than" 20.02. See Rule 230A4 for additional information.

Section 02

Definitions of Special Terms

The **NESC** provides several terms and their definitions for use in the codebook. Section 02 is the first place to look for definitions of special terms. The **Code** text under the title of Section 02 references the Authoritative Dictionary of IEEE Standards Terms (IEEE 100) for definitions not contained in this section. IEEE 100 should be used as a second step if the definition is not provided in Sec. 02 of the **NESC**. The third and final step for definitions not provided in Sec. 02 of the **NESC** or in IEEE 100 is to look the word up in a standard dictionary. These three steps are outlined in Fig. 02-1.

Occasionally, terms are defined in individual rules instead of Sec. 02. One example of this is the words "shall," "should," and "RECOMMENDATION" that are defined in Rule 015. Another example is how the word "equipment" is defined relative to a specific application. Rule 238A defines equipment relative to clearance between communication and supply facilities located on the same overhead line structure. Rule 380A provides examples of equipment relative to underground construction.

Clarifications and drawings of key terms are provided in this handbook. They are provided in the individual rules in which the terms apply instead of this section with the exception of two terms, voltage and effectively grounded, which are discussed below. The term voltage has six definitions in Sec. 02 of the **NESC**.

In some code rules and tables, voltage is specifically stated as phase to phase or phase to ground. In some locations, a voltage is stated without a phase to phase or phase to ground reference. If a voltage is stated without a phase to phase or phase to ground reference, the voltage is dependent on the type of grounding system. For example, if the code states that a clearance adder is required for lines over 50 kV (without a phase to phase or phase to ground reference) and the line is effectively grounded (i.e., a WYE connected single

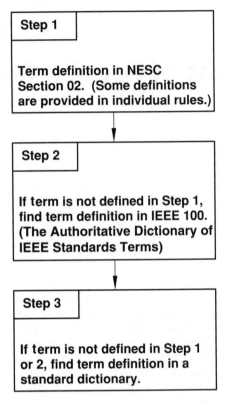

Step 1

Term definition in NESC Section 02. (Some definitions are provided in individual rules.)

Step 2

If term is not defined in Step 1, find term definition in IEEE 100. (The Authoritative Dictionary of IEEE Standards Terms)

Step 3

If term is not defined in Step 1 or 2, find term definition in a standard dictionary.

Fig. 02-1. Steps for finding definitions of terms (Sec. 02).

grounded system/unigrounded system with ground fault relaying), the clearance adder may be applied to lines over 50 kV to ground (or 86.6 kV phase to phase). If the line is ungrounded, the clearance adder must be applied to lines over 50 kV phase to phase.

The terms effectively grounded, multigrounded/multiple grounded system, and single grounded system/unigrounded system are all defined in Sec. 02 of the NESC. The important part of the definition of effectively grounded is what the definition does not say. The definition of effectively grounded does not provide a value in ohms (e.g., an effectively grounded system is 5 Ω or less). Multigrounded systems discussed in Rule 096C typically are considered effectively grounded, but in special cases (e.g., a very rocky 1-mile stretch of line), more than four grounds in each mile may be needed to make a multigrounded system an effectively grounded system.

Section 03

References

Section 03 provides a list of standards that are referenced in the **Code**. The standards listed form a part of the **Code** to the extent that they are referenced in the **Code** rules. If a standard is cited for information purposes only, it does not appear in Sec. 03; it appears in the Bibliography in Appendix E of the **NESC**. For example, ANSI O5.1, *American National Standard Specifications and Dimensions for Wood Poles,* appears in Sec. 03 as it is referenced in **Code** Rule 261A2b(1). However, IEEE Standard 80, *IEEE Guide for Safety in AC Substation Grounding,* is listed in Appendix E of the **NESC** (Bibliography) as it is cited for informational purposes in a "NOTE" in Rule 092E. Rule 015 states that a "NOTE" is provided for information purposes only. The **Code** recognizes that the standards listed in Sec. 03 provide information that does not need to be repeated in the **Code**. This helps keep the codebook from getting too wordy and utilizes standards that another agency has documented.

The **Code** acknowledges, with a footnote in Sec. 03, that current standards may be newer than ones listed, as the **Code** is updated on a 5-year revision cycle and some standards may be updated during the middle of this process. The standards listed in Sec. 03 are an important part of a technical library of reference material. Additional footnotes in Sec. 03 of the **NESC** provide mailing and web site addresses to order the various standards.

Section 09

Grounding Methods for Electric Supply and Communications Facilities

090. PURPOSE

The purpose of Sec. 09 is to provide practical methods of grounding. Grounding is one of the ways to protect people from hazardous voltages. Grounding also allows protective devices to operate during a fault condition. The basic theory behind grounding is to keep the voltage of a grounded part (e.g., equipment case, neutral conductor, communications messenger, etc.) as close as possible to the potential of the earth so that a voltage difference does not exist between a person and a grounded metal object. The Code states in this rule that grounding is used as one of the means of safeguarding employees and the public from injury. Other means include, but are not limited to, guarding, adequate clearance above ground, proper burial depth, etc.

091. SCOPE

The scope of Sec. 09 is to provide the *methods* of protective grounding for supply and communication conductors and equipment. The *requirements* for grounding are listed in the other parts of the Code. The requirements for grounding electric supply stations are predominantly in Rule 123. The requirements for grounding overhead lines are predominantly in Rule 215. The requirements for grounding underground lines are predominantly in Rule 314. The scope of Sec. 09 does not include the grounded return of electric railways or lightning protection not associated with supply and communication wires, for example, lightning protection wires connected to a lightning rod on top of a barn.

092. POINT OF CONNECTION OF GROUNDING CONDUCTOR

092A. Direct Current Systems That Are to Be Grounded.   This rule has basic connection requirements for direct-current (DC) systems. For 750 V and less, the grounding conductor connection must be made only at the supply station. For three-wire DC circuits, the connection must be made to the neutral. For DC systems over 750 V, the grounding conductor connection must be made at both the supply and load points. The connection must be made to the neutral of the system. The ground or grounding electrode can be external or remote from each of the stations. This permits separating the electrode from areas with ground currents that can cause electrolytic damage. The Code permits one of the two stations to have its grounding connection made through a surge arrester as long as the other station has the neutral effectively grounded. An exception is provided for the 750 V and greater category for back-to-back DC converter stations that are adjacent to each other. For this condition the neutral of the system should be connected to ground at one point only.

092B. Alternating-Current Systems That Are to Be Grounded

092B1. 750 V and Below.   The point of grounding connection on alternating-current (AC), wye-connected, three-phase, four-wire, and single-phase, three-wire systems operated at 750 V and below is shown in Fig. 092-1.

On other one-, two-, or three-phase systems feeding lighting circuits, a grounding connection must be made to a common circuit conductor. Common examples include a 120/240-V, three-phase, four-wire center tap delta service, a 120/208-V, single-phase, three-wire service fed from a 120/208-V, three-phase, four-wire service, or a 120-V, single-phase, two-wire service fed from a 120/240-V, single-phase, three-wire service.

Wye and delta circuits that are not grounded or do not use a common (neutral) conductor for grounding cannot be used to serve lighting loads. See Fig. 092-2.

Grounding connections must be made at the source and line side of a service as shown in Fig. 092-3.

092B2. Over 750 V.   Nonshielded conductors (e.g., bare neutral conductors) must be grounded as shown in Fig. 092-4.

The wording in Rule 092B2a requires unigrounding at the source (substation transformer) and permits, but does not require, multigrounding along the line. However, various rules in Part 2, Overhead Lines, and Part 3, Underground Lines, will require systems to be effectively grounded. Effectively grounded

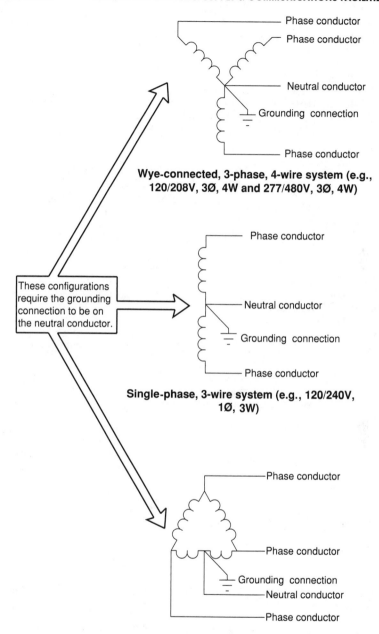

Wye-connected, 3-phase, 4-wire system (e.g., 120/208V, 3Ø, 4W and 277/480V, 3Ø, 4W)

These configurations require the grounding connection to be on the neutral conductor.

Single-phase, 3-wire system (e.g., 120/240V, 1Ø, 3W)

Single-phase, 3-wire system on a delta-connected, 3-phase, 4-wire system (e.g., 120/240V, 3Ø, 4W center tap delta)

Fig. 092-1. Grounding connection on wye-connected, three-phase, four-wire and single-phase, three-wire systems (Rule 092B1).

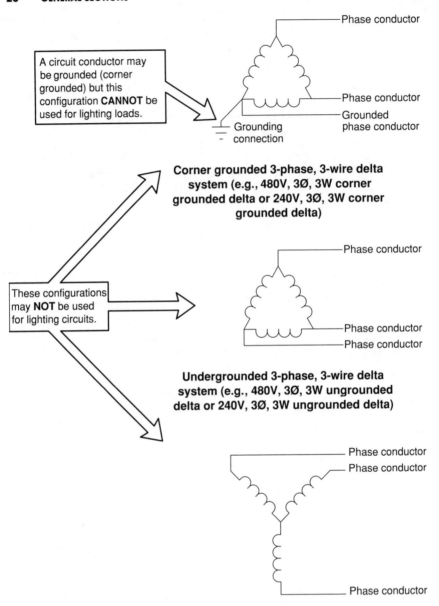

A circuit conductor may be grounded (corner grounded) but this configuration **CANNOT** be used for lighting loads.

Phase conductor

Phase conductor

Grounded phase conductor

Grounding connection

Corner grounded 3-phase, 3-wire delta system (e.g., 480V, 3Ø, 3W corner grounded delta or 240V, 3Ø, 3W corner grounded delta)

These configurations may **NOT** be used for lighting circuits.

Phase conductor

Phase conductor
Phase conductor

Undergrounded 3-phase, 3-wire delta system (e.g., 480V, 3Ø, 3W ungrounded delta or 240V, 3Ø, 3W ungrounded delta)

Phase conductor
Phase conductor

Phase conductor

Ungrounded 3-phase, 3-wire wye system (e.g., 480V, 3Ø, 3W ungrounded wye)

Fig. 092-2. Wye and delta systems not to be used for lighting loads (Rule 092B1).

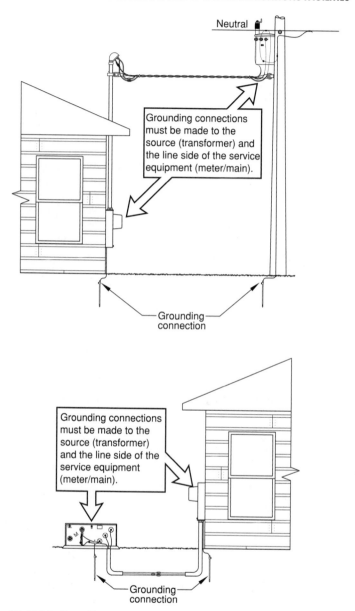

Fig. 092-3. Grounding connections at source and line side of a service (Rule 092B1).

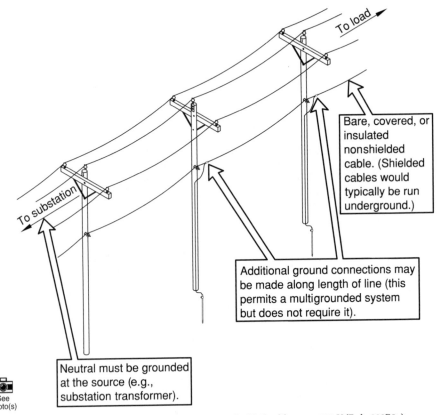

Bare, covered, or insulated nonshielded cable. (Shielded cables would typically be run underground.)

Additional ground connections may be made along length of line (this permits a multigrounded system but does not require it).

Neutral must be grounded at the source (e.g., substation transformer).

See Photo(s)

Fig. 092-4. Grounding connections for nonshielded cables over 750 V (Rule 092B2a).

systems typically need to be multigrounded to provide sufficiently low ground impedance. Multigrounded systems are discussed in Rule 096C.

Shielded conductors on riser poles must be grounded as shown in Fig. 092-5.

Shielded cables without an insulating jacket must be grounded as shown in Fig. 092-6.

Shielded cables with an insulating jacket must be grounded as shown in Fig. 092-7.

Shielded cable without an insulating jacket that is buried in direct contact with the earth has an advantage of being grounded all along its length. However, direct-buried shielded cable without an insulating jacket is susceptible to corrosion. The insulating jacket can prevent corrosion of the shield or concentric neutral, but grounding is not as effective. Shielded cables with an insulating jacket need to be terminated and grounded in pad-mounted enclosures or need to have a section of jacket removed and a ground attachment made to meet the multigrounding requirements of Rule 096C. The same applies to both bare and jacketed concentric neutral cables installed in conduit.

092B3. Separate Grounding Conductor.   If a separate grounding conductor is used on an AC system to be grounded as an adjunct (joined addition) to a cable

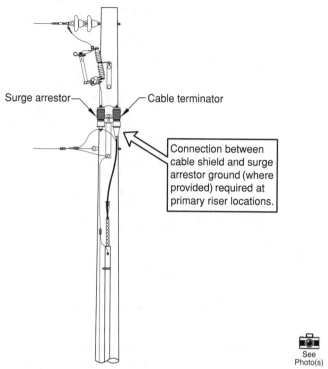

Surge arrestor ——

Cable terminator ——

Connection between cable shield and surge arrestor ground (where provided) required at primary riser locations.

See Photo(s)

Fig. 092-5. Surge arrester cable—shielding interconnection (Rule 092B2b(1)).

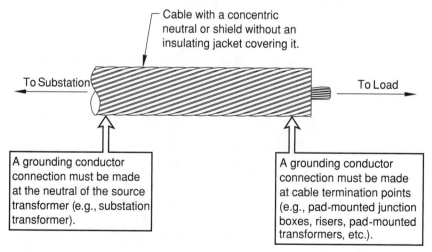

Cable with a concentric neutral or shield without an insulating jacket covering it.

To Substation ←

To Load →

A grounding conductor connection must be made at the neutral of the source transformer (e.g., substation transformer).

A grounding conductor connection must be made at cable termination points (e.g., pad-mounted junction boxes, risers, pad-mounted transformers, etc.).

See Photo(s)

Fig. 092-6. Grounding points for a shielded cable without an insulating jacket (Rule 092B2b(2)).

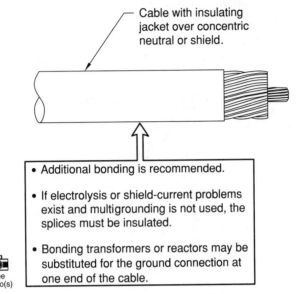

Cable with insulating jacket over concentric neutral or shield.

- Additional bonding is recommended.

- If electrolysis or shield-current problems exist and multigrounding is not used, the splices must be insulated.

- Bonding transformers or reactors may be substituted for the ground connection at one end of the cable.

See Photo(s)

Fig. 092-7. Grounding points for a shielded cable with an insulating jacket (Rule 092B2b(3)).

run underground, there are several conditions that apply. The separate grounding conductor must be connected directly or through the neutral to items that must be grounded. The conductor must be located as shown in Fig. 092-8.

Adjunct (joined addition) grounding conductors are typically used with shielded supply cables. If the shield on the supply cable is not a sufficient size to carry neutral current or fault current, an adjunct grounding cable can be used. An adjunct grounding conductor should not be used to replace a corroded concentric neutral conductor in a direct-buried cable. Rule 350B requires that a direct-buried cable operating above 600 V have a continuous metallic shield, sheath, or concentric neutral. The adjunct grounding conductor can be used to supplement the concentric neutral but not replace it if it has corroded away.

092C. Messenger Wires and Guys

092C1. Messenger Wires. The point of connection of the grounding conductor to messenger wires that are required to be grounded by other parts of the code is shown in Fig. 092-9.

Communications messenger wires on joint-use poles are required to be grounded in Part 2, Overhead Lines, Rule 215C and in Secs. 23 and 24 to meet certain clearance and grade of construction requirements. The messenger must meet certain ampacity and strength criteria defined in Rules 093C1, 093C2, and 093C5. For the messenger (on a joint-use structure) to meet the ampacity requirement in Rule 092C2, the messenger wire ampacity must be rated not less than one-fifth of the neutral wire ampacity. It is sometimes difficult to find the ampacity rating of messenger wires as many communications messenger wires are actually guy wires.

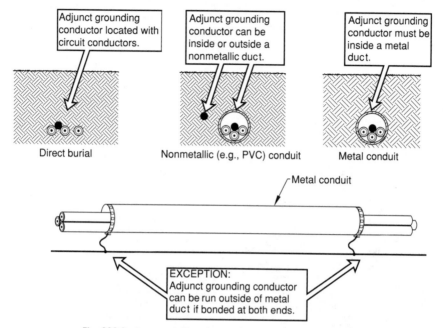

Fig. 092-8. Separate (adjunct) grounding conductor (Rule 092B3).

Manufacturers of guy wires typically provide mechanical strength ratings, not electrical ampacity ratings. The four grounds in each mile rule appears here for the first time in the **Code**. It is discussed in detail in Rule 096C.

092C2. Guys.   The point of connection of the grounding conductor to guys that are required to be grounded by other parts of the **Code** is shown in Fig. 092-10.

Guys must be either grounded (per Rule 215C2) or insulated (per the exception to Rule 215C2). If guys are grounded, they must be grounded using the methods in this rule.

092C3. Common Grounding of Messengers and Guys on the Same Supporting Structure.   When messengers and guys are on the same supporting structure and they are required to be grounded by other parts of the **Code**, they must be bonded together and grounded by the connection methods listed in this rule. The methods listed are a combination of the messenger and guy connection requirements provided in Rules 092C1 and 092C2.

092D. Current in Grounding Conductor.   This rule recognizes that multi-grounded systems, for example, a 12.47/7.2-kV, three-phase, four-wire circuit that has four or more grounds in each mile, may develop objectionable current flow on the grounding conductor (pole ground). This rule provides methods to alleviate the objectionable current flow.

Objectionable current flow may exist due to stray earth currents or other reasons. Fault currents and lightning discharge currents are not considered objectionable current flows when applying this rule and some amount of current will always be present on the grounding conductor (pole ground) during normal operation.

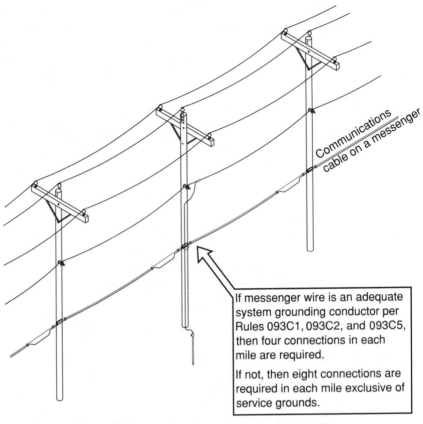

Communications cable on a messenger

If messenger wire is an adequate system grounding conductor per Rules 093C1, 093C2, and 093C5, then four connections in each mile are required.

If not, then eight connections are required in each mile exclusive of service grounds.

See Photo(s)

Fig. 092-9. Grounding of messenger wires (Rule 092C1).

092E. Fences.   When fences are required to be grounded by other parts of the Code (primarily in Rule 110A1), they must be connected to a grounding conductor as shown in Fig. 092-11.

This rule provides both specific requirements for fence grounding (Rules 092E1 through 092E6) and general requirements by noting IEEE Standard 80, which is the industry standard for substation grounding. Rule 093C6 also applies to fences. Fence mesh strands are only required to be bonded if the fence posts are nonconducting. For conducting (metal) fence posts, the fence mesh must be under tension and electrically connected to the post for the mesh to be grounded. A grounding conductor feed up to barbwire strands at the top of a fence can be woven through the chain-link mesh for added grounding continuity. A ground grid which is typically buried under an electric supply station and connected to the station fence is discussed in Rule 096B. An example of substation fence grounding is shown in Fig. 092-12.

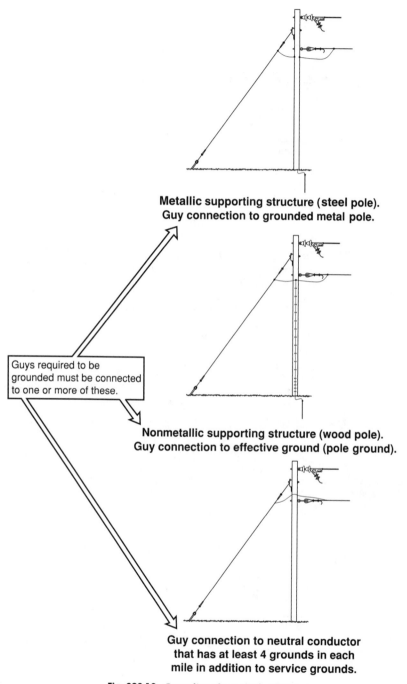

Metallic supporting structure (steel pole).
Guy connection to grounded metal pole.

Guys required to be
grounded must be connected
to one or more of these.

Nonmetallic supporting structure (wood pole).
Guy connection to effective ground (pole ground).

Guy connection to neutral conductor
that has at least 4 grounds in each
mile in addition to service grounds.

Fig. 092-10. Grounding of guys (Rule 092C2).

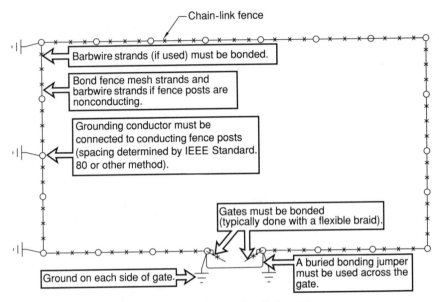

Chain-link fence

Barbwire strands (if used) must be bonded.

Bond fence mesh strands and barbwire strands if fence posts are nonconducting.

Grounding conductor must be connected to conducting fence posts (spacing determined by IEEE Standard. 80 or other method).

Gates must be bonded (typically done with a flexible braid).

Ground on each side of gate.

A buried bonding jumper must be used across the gate.

Fig. 092-11. Fence grounding (Rule 092E).

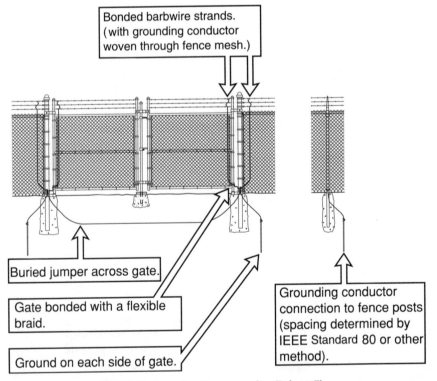

Bonded barbwire strands. (with grounding conductor woven through fence mesh.)

Buried jumper across gate.

Gate bonded with a flexible braid.

Ground on each side of gate.

Grounding conductor connection to fence posts (spacing determined by IEEE Standard 80 or other method).

See Photo(s)

Fig. 092-12. Example of fence grounding (Rule 092E).

093. GROUNDING CONDUCTOR
AND MEANS OF CONNECTION

093A. Composition of Grounding Conductors.   Grounding conductors can be copper or other metals or combinations of metals that will not corrode during their expected service life under the existing conditions. Surge arrester connections must be short, straight, and free from sharp bends. Metallic electrical equipment cases or the structural metal frame of a building can also be used as a grounding conductor. Many utilities use copper for the entire length of the grounding conductor (pole ground). Some utilities use aluminum or ACSR. If aluminum or ACSR is used above grade, it is typically spliced to copper, which then runs below grade (see Rule 093E5). Some utilities utilize copper-coated steel. Copper substitutes have become popular due to copper theft.

The grounding conductor must not have a switching device connected to it. Some exceptions apply, including high-voltage DC systems, testing under competent supervision, and surge arrester operation. This rule provides an important note stating that the normally grounded base of the surge arrester may be at line potential (fully energized) following the operation of the disconnector.

093B. Connection of Grounding Conductors.   The connection between the grounding conductor (pole ground) and grounded conductor (neutral) must be made considering the metals involved and exposure to the environment. The connector must not corrode and must be rated for the type of metals it is connecting. Dissimilar metals connected together with an improper connector will set up a battery action which will accelerate corrosion. Soldering is not acceptable, except on lead sheath cable, as fault currents will produce enough heat to melt the solder. Suitable connection methods and clarification of the terms grounded and grounding are shown in Fig. 093-1.

093C. Ampacity and Strength.   This rule defines short-time ampacity requirements for bare and insulated grounding conductors. A bare conductor can carry a larger fault current than an insulated conductor of the same size because the bare conductor is only limited by melting or damaging the conductor material. The insulated grounding conductor has the additional constraint of not damaging the insulation. See Fig. 093-2.

Short-time ampacity of both bare and insulated conductors can be obtained from conductor manufacturers. This information is typically referred to as a conductor short-circuit withstand chart or a conductor damage curve.

Short-time ampacity for a single-grounded system is shown in Fig. 093-3.

Short-time ampacity for a multigrounded AC system is shown in Fig. 093-4.

Rule 093C2 references Rule 093C8, which also specifies ampacity limits based on the ampacity of phase conductors and grounding electrode resistance. The one-fifth ampacity requirement applies to the normal operating current, not to the short-time fault ampacity. An example of pole ground sizing is shown in Fig. 093-5.

In addition to checking the pole ground to the primary neutral, the service transformer neutral should also be considered. A bare AWG No. 6 copper pole ground connected to the neutral of a large secondary service may not have the required one-fifth ampacity of the secondary neutral. Large secondary services require careful application of Rules 093C2 and 093C8.

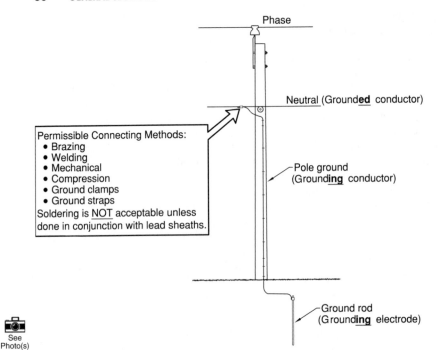

Fig. 093-1. Connection of grounding conductor to grounded conductor (Rule 093B).

In addition to single-grounded and multigrounded system requirements, Rule 093C requires AWG No. 12 copper or larger conductors to ground instrument transformers and AWG No. 6 copper or AWG No. 4 aluminum or larger conductors to ground primary surge arresters. The primary surge arrester rule has an exception permitting use of copper-clad or aluminum-clad steel wires.

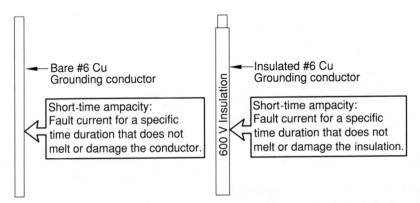

Fig. 093-2. Short-time ampacity of bare and insulated grounding conductors (Rule 093C).

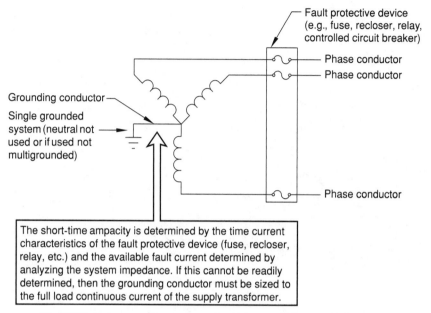

Fig. 093-3. System grounding conductor for single-grounded systems (Rule 093C1).

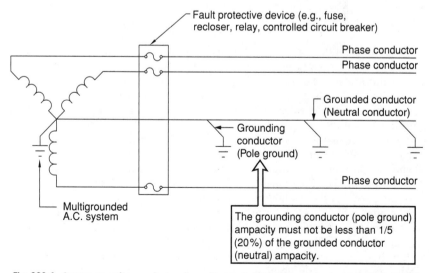

Fig. 093-4. System grounding conductors for multigrounded alternating current systems (Rule 093C2).

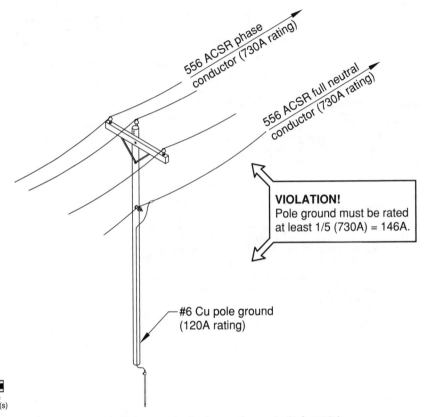

Fig. 093-5. Example of pole ground ampacity (Rule 093C2).

Grounding conductors for equipment, messenger wires, and guys must have a short-time ampacity based on the available fault current and operating time of the circuit protective device. If the circuit does not have an overcurrent or fault protection device (e.g., fuse, recloser, relay-controlled circuit breaker, etc.), then the design and operating conditions of the circuit must be analyzed and the grounding conductor cannot be smaller than AWG No. 8 copper. If a conductor enclosure (i.e., rigid steel conduit) is connected to a metal equipment enclosure with suitable lugs, bushings, etc., the metallic conduit and metallic equipment path can be used as an equipment-grounding conductor. When grounding conductors are used, they shall be connected to a suitable lug, terminal, or other device without disruption.

The ampacity and strength of the grounding conductor used for grounding fences must also have adequate short-time ampacities or must be Stl WG No. 5 or larger.

Bonding of equipment frames and enclosures must consist of a metallic path back to the grounded terminal of the local supply. If the supply is remote, metallic parts within reach must be bonded and connected to ground.

Rule 093C8 specifies an ampacity limit such that no grounding conductor needs to have an ampacity greater than either:

- The phase conductor that would supply the ground fault, or
- The maximum current in the grounding conductor calculated by dividing the supply voltage by the electrode resistance

Consider an example related to Rule 093C8b. Assuming a 7200-V phase to ground circuit and assuming a 25-Ω ground rod resistance, 7200 V divided by 25 Ω = 288 A. For a 120/240-V secondary, 120 V to ground divided by a 25-Ω ground rod resistance would be 4.8 A. Rule 093C8 may limit the size of the ground wire specified in other parts of Rule 093C based on required ampacity. Secondary services may have large grounded (neutral) conductors; however, the grounding (pole ground) conductor size may be limited by applying Rule 093C8. In this example, the assumption of a 25-Ω ground rod resistance is just that, an assumption. Ground rod resistance will vary by type of soil, moisture in the soil, length of rod, etc. Field measurements must be taken to determine actual ground rod resistance.

The mechanical strength of grounding conductors must be suitable to the conditions they are exposed to (i.e., lawn mowers, weed eaters, car bumpers, etc.). Unguarded grounding conductors must have a tensile strength equal to or greater than AWG No. 8 soft-drawn copper except for conductors noted in Rule 093C3 (i.e., AWG No. 12 copper for instrument transformers).

093D. Guarding and Protection.   Guards over grounding conductors are only required for single-grounded systems that are exposed to the public. If the grounding electrode is on a single-grounded system that is not exposed to the public (e.g., in a fenced substation), it does not have to be guarded. Grounding electrodes on multigrounded systems are not required to be guarded even if they are exposed to mechanical damages. A multigrounded system requires at least four grounds in each mile, and Rule 214 requires inspection of overhead lines. These two requirements provide a method to assure safe grounding on multigrounded systems; therefore, guards on multigrounded systems are not required. If guards are not required but they are installed, they should be installed in a manner as if they were required.

Rules 239D and 360A provide additional guarding requirements for various types of conductors. If guarding of the grounding conductor is required, guards must be suitable for the damage to which they will be exposed. If guarding of the grounding conductor is not required, a typical installation method is stapling the grounding conductor to a wood pole. The requirements for grounding conductors with or without guards are outlined in Fig. 093-6.

Rule 093D4 recognizes that an inductive choke is created when a conductor is run through a metallic raceway. This can create a hazardous voltage during a lightning strike (or even during a fault condition). The **Code** requires a nonmetallic guard (e.g., conduit) to avoid this condition. The strength of nonmetallic materials (i.e., plastics) has increased to the point where they can be used for protection without cracking or breaking. A U-shaped metallic raceway is acceptable, as it does not completely enclose the grounding conductor. If a metallic guard similar to a steel

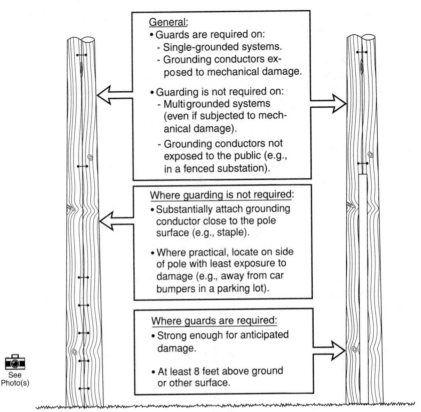

Fig. 093-6. Requirements for grounding conductors with or without guards (Rules 093D1, 093D2, and 093D3).

pipe or rigid metal conduit is used, it must be bonded to the grounding conductor at both ends, as shown in Fig. 093-7.

093E. Underground. Grounding conductors laid underground require slack due to the settling of the earth. Direct-buried joints or splices must be made with corrosion resistance in mind. Corrosion must be kept to a minimum. A cable insulation shield (e.g., concentric neutral, metallic foil, braid, etc.) must be connected to other grounded equipment in underground enclosures. Looped magnetic elements must not be positioned between the grounding conductor and the phase conductors.

The metals used for grounding in earth, concrete, or masonry must not corrode. This rule specifically notes that aluminum is not generally acceptable when used underground. An example of an aluminum ground wire that transitions to copper for underground burial is shown in Fig. 093-8.

093F. Common Grounding Conductor for Circuits, Metal Raceways, and Equipment. This rule allows one common grounding conductor for both the supply system (neutral) and equipment (e.g., a recloser) where the ampacity of the

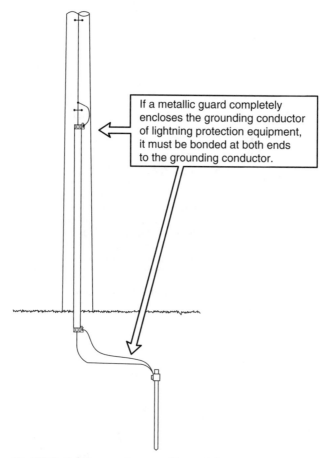

Fig. 093-7. Requirements for a metallic guard that completely encloses the grounding conductor of lightning protection equipment (Rule 093D4).

grounding conductor is adequate for both. Ampacity for the system grounding conductor and equipment grounding conductor is discussed in Rule 093C. Rule 097 addresses a common grounding conductor for primary and secondary neutrals at transformer locations. An example of one common grounding conductor for the circuit and equipment is shown in Fig. 093-9.

094. GROUNDING ELECTRODES

Grounding electrodes can be existing electrodes or made electrodes. Existing electrodes are existing conductive items buried in the earth for a purpose other than grounding but can also serve as a grounding electrode. Most utilities use

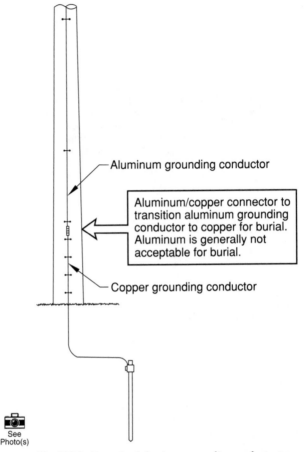

Aluminum grounding conductor

Aluminum/copper connector to transition aluminum grounding conductor to copper for burial. Aluminum is generally not acceptable for burial.

Copper grounding conductor

See Photo(s)

Fig. 093-8. Example of aluminum grounding conductor transitioning to copper for burial (Rule 093E5).

made electrodes, which are purposely constructed and buried to serve as grounding electrodes. Requirements for existing electrodes are outlined in Figs. 094-1 through 094-3.

Made electrodes must penetrate the moisture level and be below the frost line. They must be metal or combined metals that do not corrode and they must not be painted, enameled, or covered in any way with an insulating material. The driven ground rod is the most commonly used made electrode. Many utilities require the ground rod to be located in undisturbed earth a fixed distance away from the pole hole, although no such requirement is provided in the Code. The requirements for various types of made electrodes listed in the Code are outlined in Figs. 094-4 through 094-12.

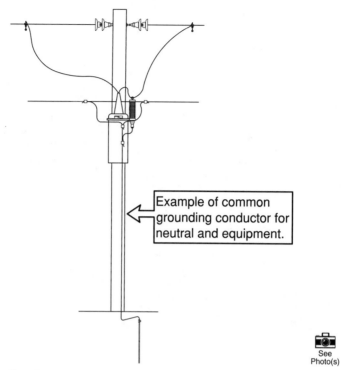

See
Photo(s)

Fig. 093-9. Example of common grounding conductor for neutral
and equipment (Rule 093F).

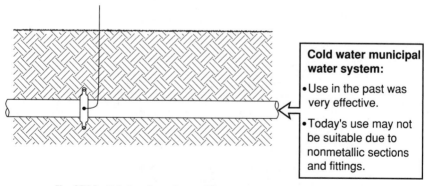

Fig. 094-1. Existing electrode—metallic water piping system (Rule 094A1).

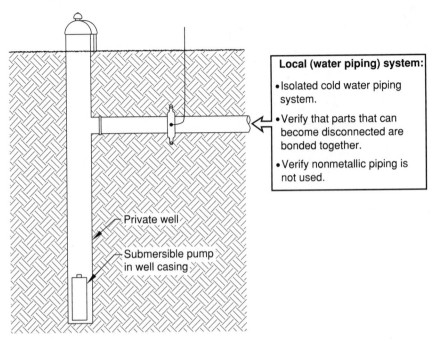

Local (water piping) system:

- Isolated cold water piping system.

- Verify that parts that can become disconnected are bonded together.

- Verify nonmetallic piping is not used.

Private well

Submersible pump in well casing

Fig. 094-2. Existing electrode—local (water piping) system (Rule 094A2).

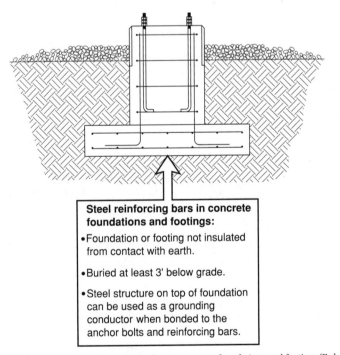

Steel reinforcing bars in concrete foundations and footings:

- Foundation or footing not insulated from contact with earth.

- Buried at least 3' below grade.

- Steel structure on top of foundation can be used as a grounding conductor when bonded to the anchor bolts and reinforcing bars.

Fig. 094-3. Existing electrode—steel reinforcing bars in concrete foundations and footings (Rule 094A3).

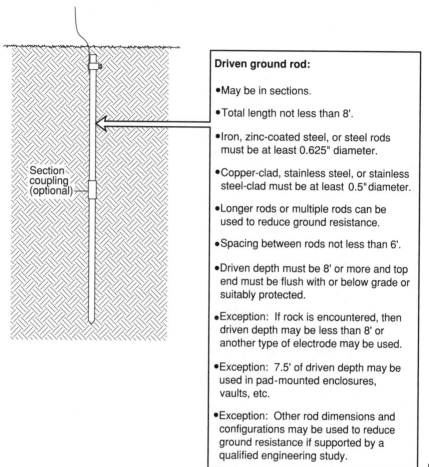

Driven ground rod:

- •May be in sections.

- •Total length not less than 8'.

- •Iron, zinc-coated steel, or steel rods must be at least 0.625" diameter.

- •Copper-clad, stainless steel, or stainless steel-clad must be at least 0.5" diameter.

- •Longer rods or multiple rods can be used to reduce ground resistance.

- •Spacing between rods not less than 6'.

- •Driven depth must be 8' or more and top end must be flush with or below grade or suitably protected.

- •Exception: If rock is encountered, then driven depth may be less than 8' or another type of electrode may be used.

- •Exception: 7.5' of driven depth may be used in pad-mounted enclosures, vaults, etc.

- •Exception: Other rod dimensions and configurations may be used to reduce ground resistance if supported by a qualified engineering study.

Section coupling (optional)

See Photo(s)

Fig. 094-4. Made electrodes—driven ground rods (Rule 094B2).

095. METHOD OF CONNECTION TO ELECTRODE

The connection to the grounding electrode must be permanent (except for removal due to inspection or maintenance) and be mechanically sound, corrosion-resistant, and have the required ampacity for the fault current to which it will be subjected. Suitable connection methods are shown in Fig. 095-1.

The Code also has specific rules for connecting to steel framed and non-steel-framed structures. The connection to water piping systems is also outlined. When water piping is used as the grounding electrode, bonds must be made around meters or other removable fittings.

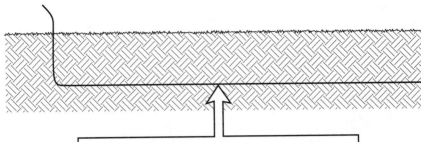

Buried wire (counterpoise):

- Used in areas of high soil resistivity, or shallow bedrock, or where lower resistance is required than obtainable with rods.
- Material must be suitable for direct burial.
- Must be at least 0.162" in diameter and at least 100' long.
- Must be buried at least 18" deep.
- Must be laid as straight as possible.
- May be arranged in a grid.
- Exception: 18" burial depth may be reduced for rock.
- Exception: Other lengths or configurations may be used per a qualified engineering study.

Fig. 094-5. Made electrodes—buried wire (counterpoise) (Rule 094B3a).

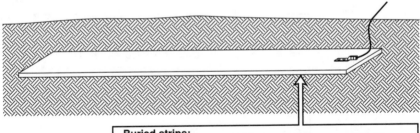

Buried strips:

- Used in areas of high soil resistivity, or shallow bedrock, or where lower resistance is required than obtainable with rods.
- Must be at least 10' in length.
- Must have total (two sides) surface not less than 5 sq. ft. (e.g., 10' long by 0.25' wide).
- Must be buried at least 18" deep.
- Ferrous metal must be at least 0.25" thick.
- Nonferrous metal must be at least 0.06" thick.
- Used for rocky areas with irregular-shaped pits of excavation.

Fig. 094-6. Made electrodes—buried strips (Rule 094B3b).

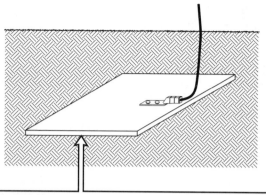

Buried plates or sheets:

• Used in areas of high soil resistivity, or shallow bedrock, or where lower resistance is required than obtainable with rods.

• Must have at least 2 sq. ft. of surface exposed to soil. Therefore, 1' × 1' if both top and bottom are exposed to soil. If the top was exposed to soil and the bottom was exposed to rock or if the bottom was exposed to soil and the top was exposed to the bottom and sides of a pole, then 1' × 2' would be required.

• Must be buried at least 5' deep.

• Ferrous metal must be at least 0.25" thick.

• Nonferrous metal must be at least 0.06" thick.

Fig. 094-7. Made electrodes—buried plates or sheets (Rule 094B3c).

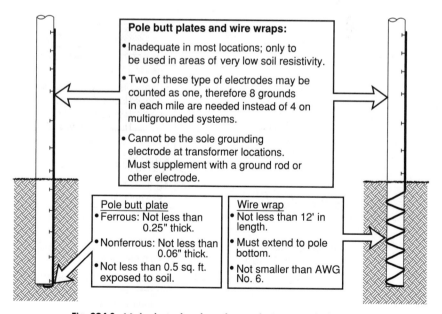

Pole butt plates and wire wraps:

• Inadequate in most locations; only to be used in areas of very low soil resistivity.

• Two of these type of electrodes may be counted as one, therefore 8 grounds in each mile are needed instead of 4 on multigrounded systems.

• Cannot be the sole grounding electrode at transformer locations. Must supplement with a ground rod or other electrode.

Pole butt plate
• Ferrous: Not less than 0.25" thick.
• Nonferrous: Not less than 0.06" thick.
• Not less than 0.5 sq. ft. exposed to soil.

Wire wrap
• Not less than 12' in length.
• Must extend to pole bottom.
• Not smaller than AWG No. 6.

Fig. 094-8. Made electrodes—butt plates and wire wraps (Rule 094B4).

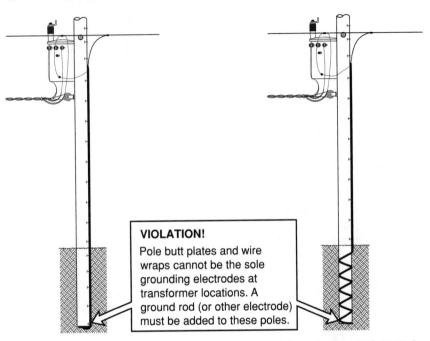

Fig. 094-9. Made electrodes—butt plates and wire wraps at transformer locations (Rule 094B4a).

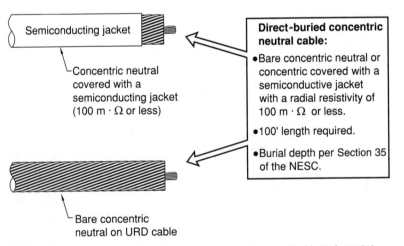

Fig. 094-10. Made electrodes—direct-buried concentric neutral cable (Rule 094B5).

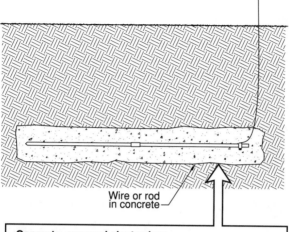

Wire or rod
in concrete

Concrete-encased electrodes:

- Wire or rod which is suitable for burial and concrete encasement and not insulated from contact with the earth.
- Top of concrete not less than 1' below grade, 2.5' is recommended.
- Wire not smaller than AWG No. 4 if copper, or 3/8" diameter or AWG No. 1/0 if steel.
- Not less than 20' long inside the concrete except for the external connection.
- Run as straight as practical.
- Shorter pieces arrayed similar to a structural footing is acceptable.
- May be more practical or effective than driven rods, strips, or plates.
- Exception: Other lengths or configurations may be used per a qualified engineering study.

Fig. 094-11. Made electrodes—concrete-encased electrodes (Rule 094B6).

The Code (in Sec. 094, "Grounding Electrodes") does not list gas piping as an acceptable electrode. Made electrodes or grounded structures should be separated from high-pressure (150 lb/in^2 or greater) pipelines containing flammable liquids or gases by a distance of 10 ft or more. No distances are specified for separating grounding electrodes from low-pressure gas lines. High-pressure pipelines are used as transmission facilities. Low-pressure gas lines are most commonly used to supply natural gas to homes. The requirements for separating grounding electrodes from high-pressure pipelines are shown in Fig. 095-2.

Rule 095C requires that the connection to the grounding electrode be free from rust, enamel, or scale. This may be done by cleaning or using fittings that penetrate such coatings.

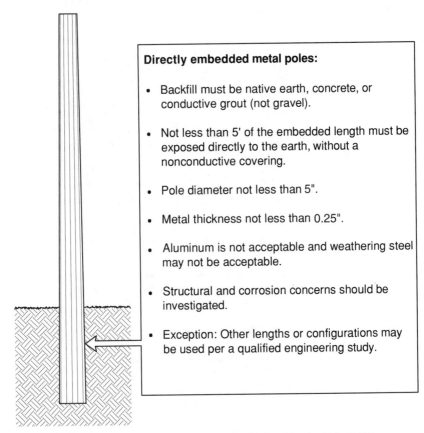

Directly embedded metal poles:

- Backfill must be native earth, concrete, or conductive grout (not gravel).

- Not less than 5' of the embedded length must be exposed directly to the earth, without a nonconductive covering.

- Pole diameter not less than 5".

- Metal thickness not less than 0.25".

- Aluminum is not acceptable and weathering steel may not be acceptable.

- Structural and corrosion concerns should be investigated.

- Exception: Other lengths or configurations may be used per a qualified engineering study.

Fig. 094-12. Made electrodes—directly embedded metal poles (Rule 094B7).

096. GROUND RESISTANCE REQUIREMENTS

096A. General.   The main intent of Rule 096 is to assure a grounding resistance low enough to permit prompt operation of circuit protective devices (e.g., fuses, reclosers, relay, controlled circuit breakers, etc.).

096B. Supply Stations.   Supply stations typically require extensive grounding systems consisting of a ground grid or mat combined with grounding electrodes. They are designed to limit touch, step, mesh, and transferred potentials. The Code notes IEEE Standard 80 as a reference for substation grounding. Rules 092E and 093C6 apply to grounding the fence enclosing the electric supply station. The requirements of Rule 096B are outlined in Fig. 096-1.

096C. Multigrounded Systems.   Multigrounded systems are the most common type of distribution system. A typical 12.47/7.2-kV, three-phase, four-wire grounded-wye distribution system is multigrounded. For a system to be multigrounded, the following must occur:

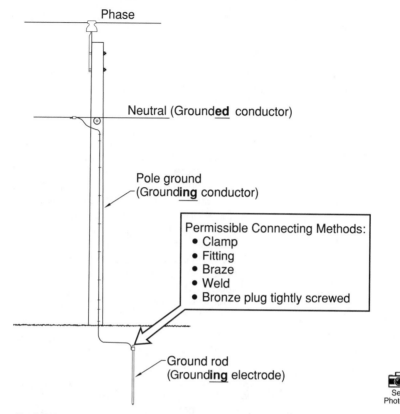

Phase

Neutral (Grounded conductor)

Pole ground
(Grounding conductor)

Permissible Connecting Methods:
- Clamp
- Fitting
- Braze
- Weld
- Bronze plug tightly screwed

Ground rod
(Grounding electrode)

See
Photo(s)

Fig. 095-1. Connection of grounding conductor to grounding electrode (Rule 095A).

- The circuit must have a neutral of sufficient size and ampacity.
- The neutral must be connected to a grounding electrode at each transformer location.
- The neutral must be connected to a grounding electrode not less than four times in each mile of the entire line. The grounds at transformers can be counted in the four grounds in each mile, but the grounds at individual services (i.e., meters) cannot be counted.

The intent of a multigrounded system is to always carry a neutral and to have not less than four grounds in each mile of the entire line. This results in grounds being placed approximately 1/4 mile or shorter apart, although some intervals may be longer. To check the four grounds in each mile requirement, a "one-mile window" can be used. Examples are shown in Fig. 096-2.

The **Code** does not specify a ground resistance for multigrounded systems. The **Code** notes that multigrounded systems are dependent on the multiplicity of grounding electrodes, not the ground resistance of any individual electrode. See Sec. 02, Definitions, for additional information on multigrounded and effectively grounded systems.

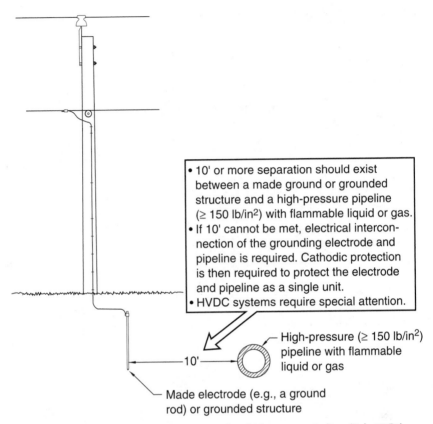

- 10' or more separation should exist between a made ground or grounded structure and a high-pressure pipeline ($\geq$ 150 lb/in^2) with flammable liquid or gas.
- If 10' cannot be met, electrical interconnection of the grounding electrode and pipeline is required. Cathodic protection is then required to protect the electrode and pipeline as a single unit.
- HVDC systems require special attention.

High-pressure ($\geq$ 150 lb/in^2) pipeline with flammable liquid or gas

10'

Made electrode (e.g., a ground rod) or grounded structure

Fig. 095-2. Grounding electrode separation from high-pressure pipelines (Rule 095B2).

For underground installations where the supply cable has an insulating jacket over the concentric neutral or the supply cable is in conduit, the cable must be terminated and grounded so that there are not less than four grounds in each mile along the line. If an express direct-buried underground feeder is constructed with an insulating jacket but without frequent termination points, the cable jacket must be stripped back and a suitable grounding electrode must be connected not less than four times in each mile along the line. If a supply cable has a semiconducting jacket, the cable can be treated similar to a bare concentric neutral cable and the jacket does not need to be stripped back for grounding. The semiconducting jacket must not exceed 100 m · Ω radial resistivity per Rule 094B5. Use of semiconducting jacketed cable is not very common due to the fact that these cables are higher in cost than insulated jacketed cable.

Rule 096C provides an exception to the four grounds in every mile for underwater crossings. Grounding on each side of the underwater crossing should be given special attention to make up for any lack of grounding in the underwater portion of the cable.

- Supply stations may require extensive grounding systems consisting of buried conductors, grounding electrodes, or interconnected combinations of both.

- The grounding system must be designed to limit touch, step, mesh, and transferred potentials. (No specifics are provided for conductor sizes, conductor spacing, or number of ground rods. IEEE Standard 80 is noted as a reference.)

- The fence grounding requirements of Rules 092E and 093C6 also apply.

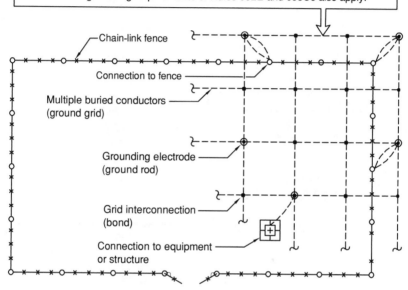

Chain-link fence

Connection to fence

Multiple buried conductors
(ground grid)

Grounding electrode
(ground rod)

Grid interconnection
(bond)

Connection to equipment
or structure

Fig. 096-1. Ground resistance requirements for supply stations (Rule 096B).

Rule 096C provides a recommendation that permits using this rule for shield wires (also referred to as overhead ground wires, static wires, and surge-protection wires) which are typically located at the top of transmission lines.

There are instances in the Code where eight grounds in each mile of line are required instead of four grounds in each mile. Rule 092C1 requires eight grounds in each mile for a communications messenger that is not an adequate grounding conductor. Rule 094B4 requires eight grounds in each mile for pole butt plates and wire wraps. Rule 354D3 requires eight grounds in each mile for direct-buried power and communications conductors in random separation (less than 12 inches apart).

096D. Single-Grounded (Unigrounded or Delta) Systems. Single-grounded systems, typically grounded wye transmission systems, that do not carry a neutral and are grounded only at the source transformer must have a ground resistance not exceeding 25 Ω. This rule states that if a single electrode exceeds 25 Ω, then other grounding methods must be used. If the single-grounded system originates in a substation, Rule 096B also applies.

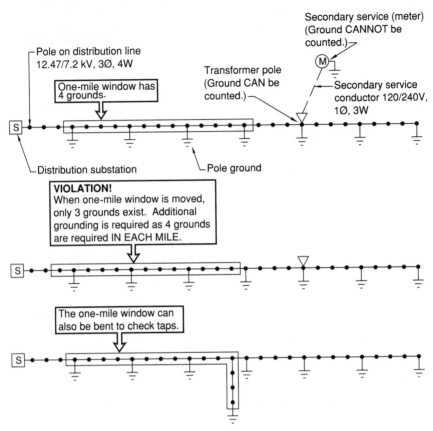

Fig. 096-2. Example of checking "four grounds in each mile" (Rule 096C).

097. SEPARATION OF GROUNDING CONDUCTORS

Rule 097A requires that separate grounding conductors be run for primary surge arresters over 750 V, secondary circuits under 750 V, and shield wires. But, Rule 097B allows a single grounding conductor and single grounding electrode if a ground connection exists at each surge arrester location and the primary neutral or shield wire and secondary neutral are connected together. When the primary and secondary neutrals are connected, Rule 097C requires the common neutral to be multigrounded (see Rule 096C). Rule 097A is typically applied in conjunction with Rule 097D1. An example of this application is a delta–grounded-wye transformer bank fed from an ungrounded primary system as shown in Fig. 097-1.

Rules 097B and 097C are typically applied to grounded-wye–grounded-wye three-phase systems and grounded-wye single-phase systems fed from a multigrounded primary system as shown in Fig. 097-2.

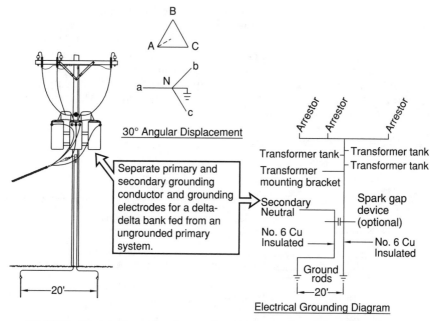

Fig. 097-1. Example of separate primary and secondary grounding (Rules 097A and 097D1)

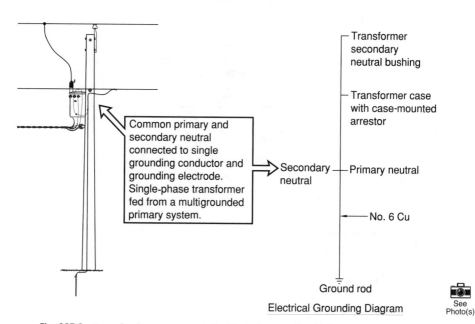

Fig. 097-2. Example of a common neutral with single grounding (Rules 097B and 097C).

On multigrounded systems the primary and secondary neutrals should be interconnected. The **NESC** uses the word "should" in this case, not "shall," as there are times when separation of primary and secondary neutrals on a multigrounded system is applicable. The most common reason for separating primary and secondary neutrals on a multigrounded system is to minimize stray voltage on the secondary neutral imposed by the primary neutral. The requirements separating primary and secondary neutrals for stray voltage or other valid reasons are outlined in Fig. 097-3.

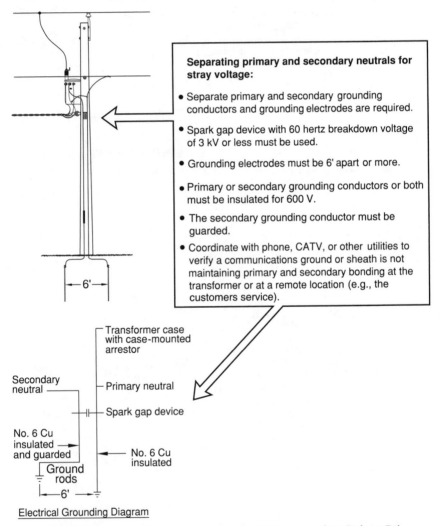

Separating primary and secondary neutrals for stray voltage:

- Separate primary and secondary grounding conductors and grounding electrodes are required.

- Spark gap device with 60 hertz breakdown voltage of 3 kV or less must be used.

- Grounding electrodes must be 6' apart or more.

- Primary or secondary grounding conductors or both must be insulated for 600 V.

- The secondary grounding conductor must be guarded.

- Coordinate with phone, CATV, or other utilities to verify a communications ground or sheath is not maintaining primary and secondary bonding at the transformer or at a remote location (e.g., the customers service).

Transformer case with case-mounted arrestor

Secondary neutral

Primary neutral

Spark gap device

No. 6 Cu insulated and guarded

Ground rods

No. 6 Cu insulated

6'

Electrical Grounding Diagram

Fig. 097-3. Separating primary and secondary neutrals for stray voltage (Rule 097D2).

If a made electrode is used to ground surge arresters on an ungrounded system exceeding 15 kV phase to phase, the NESC requires that the ground rod(s) be at least 20 ft from buried communication cables.

Rule 097G focuses on grounding requirements for joint-use poles. If separate grounding conductors (pole grounds) are run to the supply neutral and the communications messenger, a bond between the pole grounds must be added. If a single grounding conductor (pole ground) is used on a joint-use pole, it must be connected to both the supply neutral and the communications messenger. Most utilities use a single-pole ground for grounding both power and communications. The single-pole ground method will require a review for special cases like the delta–grounded-wye transformation or the stray voltage application discussed in this rule. See Fig. 097-4.

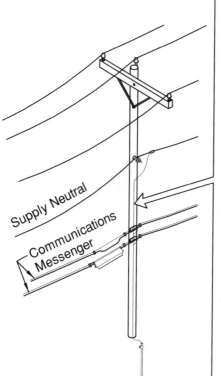

- If a single grounding conductor (pole ground) is used, it must be connected to both the supply neutral and the communications messenger.

- If separate grounding conductors (pole grounds) are run to the supply neutral and the communications messenger, they must be bonded together.

- Exceptions apply when separate grounding conductors are required by other rules (e.g., delta primary systems).

- If isolation is being maintained between primary and secondary neutrals (e.g., for stray voltage), the communications messenger must be connected only to the primary grounding conductor.

See Photo(s)

Fig. 097-4. Bonding of communication systems to electric supply systems on a joint-use structure (Rule 097G).

098. NUMBER 098 NOT USED IN THIS EDITION.

099. ADDITIONAL REQUIREMENTS FOR GROUNDING AND BONDING OF COMMUNICATION APPARATUS

This rule outlines how to ground communication apparatus when grounding is required in other parts of the **Code**. This rule references Note 2 of Rule 097D2, which discusses cooperation between supply and communications employees to isolate primary and secondary neutrals (typically for resolving stray-voltage problems).

A communications grounding conductor shall preferably be made of copper or other material that will not corrode and shall not be less than AWG No. 14. The communications grounding conductor must be connected as shown in Fig. 099-1.

A separate communications ground rod is not required per Rule 099A. If a communications ground rod is used because a supply service does not exist, the communications ground rod may be smaller in diameter and length per

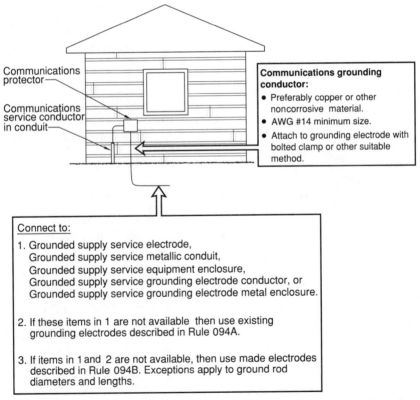

Fig. 099-1. Additional requirements for communications grounding (Rules 099A and 099B).

the exception to Rule 099A3. However, if a supply service does exist and a communications ground rod is used to supplement the supply grounding system, the exception to Rule 099A3 permitting smaller rods does not apply. Rule 099A does not prohibit a supplemental communications ground rod, but only if the supply service does not exist can the smaller communications-size ground rod be used. If a standard-size ground rod (per Rule 094B2) is used for communications grounding to supplement the supply ground rod, an AWG No. 6 copper or equivalent jumper must bond the two ground rods together as shown in Fig. 099-2.

The requirements in this rule overlap the requirements in the National Electrical Code (NEC). The NEC should be reviewed to resolve any service entrance issues with the local building inspection authority.

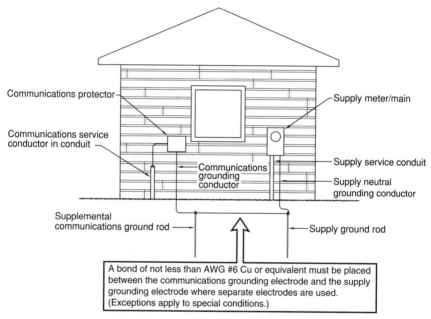

Fig. 099-2. Bonding of communications and supply electrodes (Rule 099C).

Part 1

Rules for the Installation and Maintenance of Electric Supply Stations and Equipment

GENERAL SECTIONS

01 INTRODUCTION   02 DEFINITIONS

03 REFERENCES   09 GROUNDING METHODS

GENERAL SECTIONS 01, 02, 03, 09

PART 1

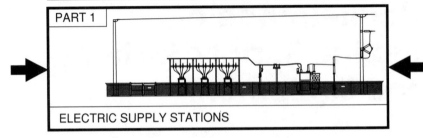

ELECTRIC SUPPLY STATIONS

PART 2

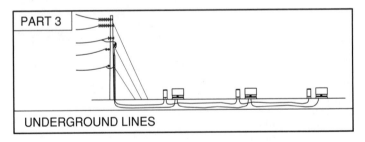

OVERHEAD LINES

PART 3

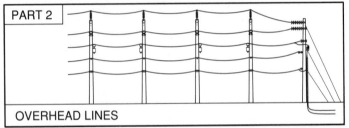

UNDERGROUND LINES

PART 4

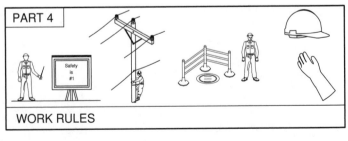

WORK RULES

Section 10

Purpose and Scope of Rules

100. PURPOSE

The purpose of Part 1, Electric Supply Stations, is similar to the purpose of the entire NESC outlined in Rule 010, except Rule 100 is specific to electric supply stations and equipment. Part 1 of the NESC focuses on the practical safeguarding of persons during the installation, operation, and maintenance of electric supply stations and equipment.

101. SCOPE

The scope of Part 1, Electric Supply Stations, includes electric supply conductors and equipment (in electric supply stations), and associated structural arrangements (in electric supply stations). Electric supply stations can consist of generating stations, substations, and switching stations. The term arrangements is important, as Part 1, Electric Supply Stations, provides rules for arranging items in electric supply stations for clearance purposes, but it does not provide strength and loading factors for the structural components as Part 2, Overhead Lines, does.

A key phrase in this rule is, "accessible only to qualified personnel." The rules of Part 1, Electric Supply Stations, assume that the general public is not exposed to the conductors and equipment located in the electric supply stations. For example, in Part 2, Overhead Lines, NESC Table 232-2 specifies a vertical clearance of 18.0 ft for a 12.47/7.2-kV, three-phase, four-wire rigid live part above a roadway, driveway, parking lot, or alley, and 14.0 ft when it is located in a pedestrian-only area. When this same 12.47/7.2-kV rigid live part is located inside a substation fence (accessible only to qualified personnel), NESC Table 124-1 specifies a vertical clearance of 9.0 ft for a 15-kV phase to phase, 110-kV BIL rigid live part above the ground or other accessible surface. This example shows that Part 1, Electric Supply Stations, is applicable when the supply facilities are accessible to qualified personnel only, via a fence, locked room, or other method (see Rule 110A), and that the clearance values are lower in electric supply stations than in Part 2, Overhead Lines, which are accessible to the general public.

The last sentence of Rule 101 clarifies the application of the NESC versus the National Electrical Code (NEC) to supply substations. Rule 011 discusses the scope of the NESC and the NEC. The NESC covers conductors and equipment in an electric supply station when they are serving a utility function (not an office building wiring function). The NESC electric supply station rules cover utility functions. Generation stations in particular and even the control buildings of substations and switching stations involve utilization wiring for lighting, ventilation, and controls. The NESC does not provide specific rules for utilization wiring. Rule 012C, which requires accepted good practice, must be applied when specific conditions are not covered. The NEC is an excellent reference for accepted good practice in this case.

As evidenced by reading all the rules in Part 1, Electric Supply Stations, the scope of Part 1 applies to both indoor and outdoor substations.

Communications utility personnel can skip over Part 1, Electric Supply Stations, as it does not apply to them. The definition of electric supply station in Sec. 02 includes generating stations and substations, but not communications central offices.

102. APPLICATIONS OF RULES

Rule 102 references Rule 013 for the general application of Code rules; see Rule 013 for a discussion.

103. REFERENCED SECTIONS

This rule references four sections related to Part 1, Electric Supply Stations, so that rules do not have to be duplicated and the reader of the Code realizes that other sections are related to the information provided in Part 1. The related sections are:

- Introduction—Sec. 01
- Definitions—Sec. 02
- References—Sec. 03
- Grounding Methods—Sec. 09

The rules in Part 1, predominantly Rules 110A and 123, will provide the requirements for grounding electric supply stations. The grounding methods are provided in Sec. 09.

Protective Arrangements in Electric Supply Stations

110. GENERAL REQUIREMENTS

110A. Enclosure of Equipment. Rule 110A is the defining rule of Part 1, Electric Supply Stations. If Rule 110A is met, the rules in Part 1, Electric Supply Stations, can be used. If Rule 110A is not met, then the rules in Part 2, Overhead Lines, or Part 3, Underground Lines, apply instead of Part 1, Electric Supply Stations. An example of how Rule 110A applies to Part 1, Electric Supply Stations, or Part 2, Overhead Lines, is shown in Fig. 110-1.

An example of how Rule 110A applies to Part 1, Electric Supply Stations, or Part 3, Underground Lines, is shown in Fig. 110-2.

The NESC discusses enclosures of rooms (for indoor applications) and spaces (for outdoor applications). The following are required to enclose the room or space:

- Fences,
- Screens,
- Partitions, or
- Walls

The enclosure formed is required to "limit the likelihood" of entrance by unauthorized people (i.e., the general public) or unauthorized workers. Even the best prison system cannot avoid an escape, therefore, "limit the likelihood" is used rather than "prevent entry." The entrance to the room or space must be locked or under observation by an authorized attendant. This wording can become critical when utility employees are working inside a substation with

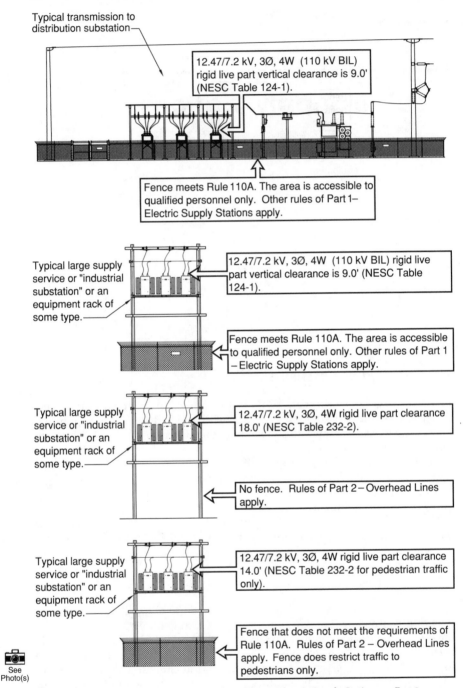

Typical transmission to distribution substation—

12.47/7.2 kV, 3Ø, 4W (110 kV BIL) rigid live part vertical clearance is 9.0' (NESC Table 124-1).

Fence meets Rule 110A. The area is accessible to qualified personnel only. Other rules of Part 1– Electric Supply Stations apply.

Typical large supply service or "industrial substation" or an equipment rack of some type.—

12.47/7.2 kV, 3Ø, 4W (110 kV BIL) rigid live part vertical clearance is 9.0' (NESC Table 124-1).

Fence meets Rule 110A. The area is accessible to qualified personnel only. Other rules of Part 1 – Electric Supply Stations apply.

Typical large supply service or "industrial substation" or an equipment rack of some type.—

12.47/7.2 kV, 3Ø, 4W rigid live part clearance 18.0' (NESC Table 232-2).

No fence. Rules of Part 2–Overhead Lines apply.

Typical large supply service or "industrial substation" or an equipment rack of some type.—

12.47/7.2 kV, 3Ø, 4W rigid live part clearance 14.0' (NESC Table 232-2 for pedestrian traffic only).

Fence that does not meet the requirements of Rule 110A. Rules of Part 2 – Overhead Lines apply. Fence does restrict traffic to pedestrians only.

See Photo(s)

Fig. 110-1. Example of how Rule 110A applies to Part 1, Electric Supply Stations, or Part 2, Overhead Lines (Rule 110A).

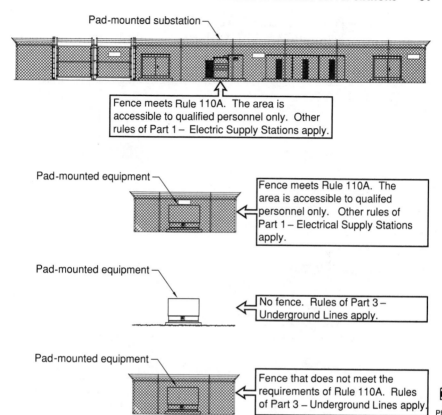

Pad-mounted substation

Fence meets Rule 110A. The area is accessible to qualified personnel only. Other rules of Part 1 – Electric Supply Stations apply.

Pad-mounted equipment

Fence meets Rule 110A. The area is accessible to qualifed personnel only. Other rules of Part 1 – Electrical Supply Stations apply.

Pad-mounted equipment

No fence. Rules of Part 3 – Underground Lines apply.

Pad-mounted equipment

Fence that does not meet the requirements of Rule 110A. Rules of Part 3 – Underground Lines apply.

See Photo(s)

Fig. 110-2. Example of how Rule 110A applies to Part 1, Electric Supply Stations, or Part 3, Underground Lines (Rule 110A).

the gate open. At least one employee must observe the unlocked gate, or the gate must be locked after the employees enter the substation.

The **Code** requires a safety sign at each entrance (i.e., gate or door). In addition, a fenced substation must have a safety sign located on each side of the substation. See Fig. 110-3.

The **NESC** notes ANSI Z535 series documents for sign applications. Substation fences and pad-mounted transformers and enclosures are two of the most common signage applications for electric supply utilities. The ANSI Z535 approach to signage uses the philosophy that a warning is appropriate on the outer barrier (i.e., fence or enclosure), and if that barrier is breached, a "danger" sign is then appropriate. Traditionally, the "danger" sign was the most common choice for substation fence applications with little or no signage used inside the substation. Using the ANSI Z535 signage philosophy, "warning" signs would be placed on the substation fence and "danger" signs would be placed inside the substation on structures that support energized parts. This same philosophy can be applied to a pad-mounted transformer. A "warning" sign is placed on the outside of the

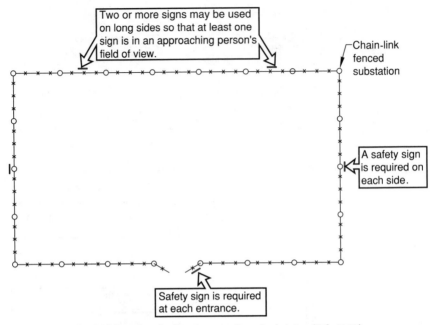

Fig. 110-3. Safety sign locations on a fenced substation (Rule 110A).

enclosure and a "danger" sign on the inside. Utilities should consult the ANSI Z535 documents, federal or state regulatory agencies, and the utility's insurance company for signage application recommendations. It is important to note that **NESC** Rule 110A does not require safety signs inside the substation. Safety signs are only required at each entrance and for fenced substations, on each side. Furthermore, **NESC** Rule 381G does not require a safety sign on the outside of pad-mounted equipment. Safety signs are only RECOMMENDATION (see Rule 015) for the inside of the equipment. Examples of safety signs are shown in Fig. 110-4.

When a metal (i.e., chain-link) fence is used to enclose a substation, the **NESC** details the height and type of construction. Other types of construction must present an equivalent barrier to climbing and unauthorized entry as the metal fence. To provide an equivalent barrier to climbing, fences should not have handholes or footholes more predominant than the mesh on a chain-link fence. Pad-mounted equipment, park benches, parked vehicles, adjacent fences, etc., should not be placed near a substation fence, as they can create "steps" for climbing the fence. Metal fences must be grounded in accordance with the grounding methods in Sec. 09 (see Rule 092E).

The requirements for metal fence heights are outlined in Fig. 110-5.

Many utilities establish substation fence height values by using the **code** requirement plus an adder. The adder (1 ft for example) can be thought of as a design or construction tolerance adder to maintain the required fence height over time. There are several factors that can jeopardize the fence height. Factors could include the addition of gravel or some other type of fill outside

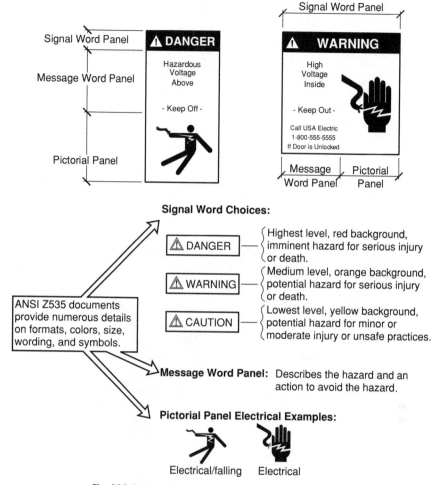

Fig. 110-4. Examples of ANSI Z535 safety signs (Rule 110A).

the substation. Installing a substation fence with an overall height of 8 ft can help maintain the Code-required 7-ft height over the life of the installation.

In addition to the fence height requirements, Rule 110A specifies a safety clearance zone from the fence to live (energized) parts inside the substation. NESC Fig. 110-1 and NESC Table 110-1 convey the required values. The distances in NESC Table 110-1 are required to place live parts far enough back from the fence so that a person poking an object through the fence or swinging an object over the fence will not contact energized parts. An example of applying NESC Fig. 110-1 and NESC Table 110-1 is shown in Fig. 110-6.

Rule 110A2 provides two exceptions to using the safety clearance zone requirements. The first exception involves a solid fence or wall that will not

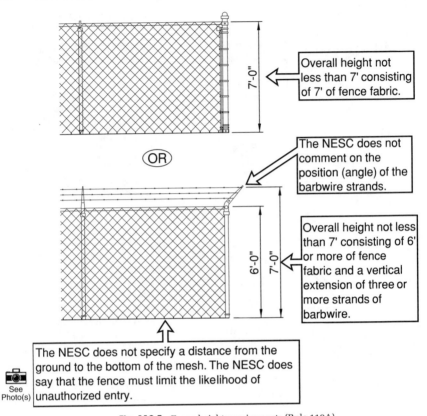

Overall height not less than 7' consisting of 7' of fence fabric.

The NESC does not comment on the position (angle) of the barbwire strands.

Overall height not less than 7' consisting of 6' or more of fence fabric and a vertical extension of three or more strands of barbwire.

The NESC does not specify a distance from the ground to the bottom of the mesh. The NESC does say that the fence must limit the likelihood of unauthorized entry.

See Photo(s)

Fig. 110-5. Fence height requirements (Rule 110A).

permit a person to poke an object through the fence. If Exception 1 is applied, Rule 125, Working Space about Electric Equipment, would apply to the space between the wall and the energized parts. The second exception involves interior fences, which do not need to comply with the safety clearance zone as the interior fence is accessible to qualified employees only. The exceptions are outlined in Fig. 110-7.

110B. Rooms and Spaces. The rooms (i.e., interior) or spaces (i.e., exterior) that comprise an electric supply station must be noncombustible. The Code uses the phrase, "as much as practical noncombustible." This wording recognizes the oil-filled equipment may be combustible; however, the substation structure should not be. Rules 152 and 172 provide additional requirements related to oil-filled equipment located in electric supply stations. Steel is the most common choice for modern outdoor substation construction but the Code recognizes that wood poles are still commonly located within the substation fence. Dry grass or weeds should be removed from an outdoor substation to maintain the noncombustible requirement.

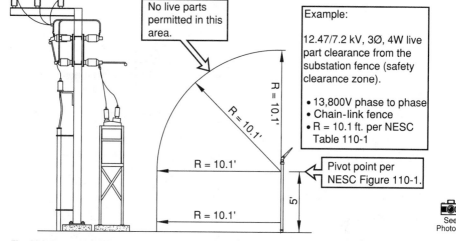

No live parts permitted in this area.

Example:

12.47/7.2 kV, 3Ø, 4W live part clearance from the substation fence (safety clearance zone).

- 13,800V phase to phase
- Chain-link fence
- R = 10.1 ft. per NESC Table 110-1

R = 10.1'

R = 10.1'

R = 10.1'

R = 10.1'

Pivot point per NESC Figure 110-1.

See Photo(s)

Fig. 110-6. Example of how to apply the safety clearance zone to fences per NESC Table 110-1 and NESC Fig. 110-1 (Rule 110A2).

The substation room or space must not contain combustible materials or fumes and must not be used for manufacturing or storage. Three exceptions to Rule 110B2 apply to storage of materials in an electric supply station.

The first exception permits storage of equipment or material used for maintenance of the electric supply station, for example, a spare substation transformer, spare fuses, etc. This equipment or material must be guarded (e.g., stored in a shed) or separated from live parts per Rule 124.

The second exception permits storage of virtually any type of electric construction and maintenance material if it is fenced separately from the electric supply substation equipment. The fence separating the electric supply equipment and the storage materials must meet the requirements of Rule 110A. If this exception is applied, the storage ends up not being in the electric supply station per se, but in a separate space adjacent to it.

The third exception permits storage on a temporary basis (no time period is specified) for electric construction and maintenance work in progress. For example, if a new distribution line is being built near a substation, or a maintenance pole replacement project exists near a substation, the substation may temporarily be used to store materials for the project. This exception requires the stored materials to be associated with work in progress. In other words, the substation site cannot be used as a storage yard for warehousing future construction or maintenance materials. To apply exception three, the Code lists five conditions that must be met to maintain a safe working area.

The ventilation in the room or space must be adequate and the room or space, if indoors, should be dry. If the electric supply station space is located outdoors, the equipment in the space must be designed for the atmospheric conditions. The general requirements for rooms and spaces are outlined in Fig. 110-8.

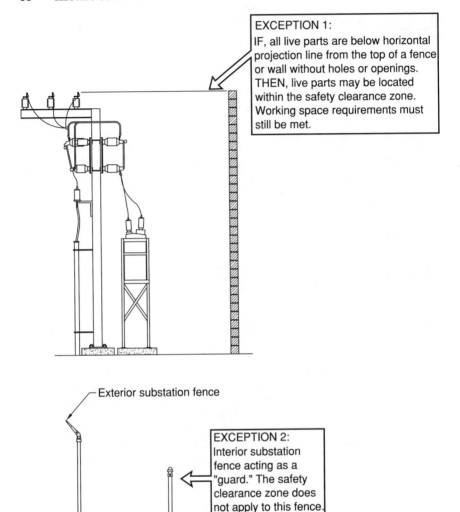

EXCEPTION 1:
IF, all live parts are below horizontal projection line from the top of a fence or wall without holes or openings. THEN, live parts may be located within the safety clearance zone. Working space requirements must still be met.

Exterior substation fence

EXCEPTION 2:
Interior substation fence acting as a "guard." The safety clearance zone does not apply to this fence.

Fig. 110-7. Exceptions to safety clearance zone requirements (Rule 110A2).

110C. Electric Equipment.   Electric equipment in the supply station must be supported and secured. The **Code** does not specifically address any seismic (earthquake) construction requirements. Rule 012C, which requires accepted good practice, must be applied in this case. Heavy equipment such as a substation transformer may be secured in place by its own weight. Heavy equipment such as an electric generator that has dynamic (rotating movement) forces will require supporting measures in addition to its own weight. See Fig. 110-9.

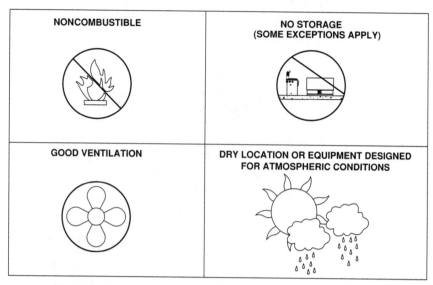

Fig. 110-8. General requirements for substation rooms and spaces (Rule 110B).

111. ILLUMINATION

111A. Under Normal Conditions. This rule provides illumination (lighting) levels for electric supply station rooms and spaces. NESC Table 111-1 provides illumination values in lux (metric) and foot-candles (English) for generating station areas, both interior and exterior. The table includes indoor switchgear and outdoor substation areas. Various light fixture manufacturers publish calculation aids for determining lighting levels. The Illuminating Engineering Society (IES) publishes books on lighting design applications. Rule 111A does not require that the lighting be permanently installed; therefore, portable lighting can be used to meet the rule. Rule 111C discusses receptacles for portable cords.

111B. Emergency Lighting. Attended electric supply stations must have automatically initiated emergency lighting for power failure. The exit paths in attended stations must have 1 foot-candle of lighting at all times. This can be done using an emergency generator or storage batteries. The duration of backup lighting should be evaluated, but in no case should the duration be less than 90 min (1.5 h). It is recommended that the wiring for the emergency lighting fixtures be kept independent from the normal wiring.

111C. Fixtures. The lighting fixtures should be permanent, or receptacles for portable lighting must be located to minimize cord lengths. Switches for lighting must be in a safely accessible location.

111D. Attachment Plugs and Receptacles for General Use. Plugs and receptacles used in electric supply stations must disconnect all poles by one operation and must be of the grounding type. Special voltages, amperages, or frequencies

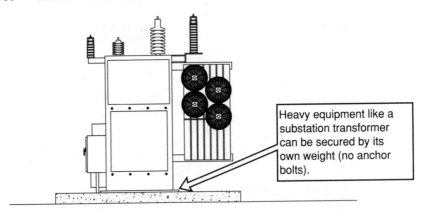

Heavy equipment like a substation transformer can be secured by its own weight (no anchor bolts).

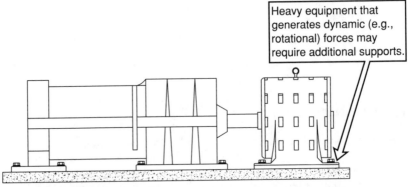

Heavy equipment that generates dynamic (e.g., rotational) forces may require additional supports.

Fig. 110-9. Supporting and securing heavy equipment (Rule 110C).

must have plugs and receptacles that are not interchangeable. Manufacturers of wiring devices (e.g., receptacles, plugs, switches, etc.) use National Electrical Manufacturers Association (NEMA) standard configurations for various voltage, phase, and current ratings.

111E. Receptacles in Damp or Wet Locations.   If the receptacle is in a damp or wet location, it must have ground-fault interruption (GFI) as part of either the receptacle or the circuit breaker feeding the receptacle. As an alternative to using GFI protection, the NESC allows testing of a grounded circuit (e.g., using a Megger to verify adequate insulation resistance) as often as experience has shown necessary.

112. FLOORS, FLOOR OPENINGS, PASSAGEWAYS, AND STAIRS

This rule gives special attention to floors, floor openings, passageways, and stairs, as they are accident-prone areas. The requirements for floors and passageways are outlined in Fig. 112-1.

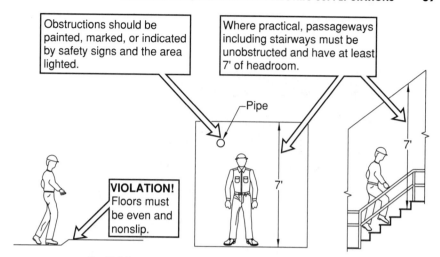

Fig. 112-1. Requirements for floors and passageways (Rule 112).

Railings are required for floor openings and raised platforms or walkways in excess of 1 ft in height. Handrails are required for stairways with four or more risers. A 3-in unobstructed clearance is required around handrails to assure an adequate grip. The rule requiring handrails for stairways consisting of four or more risers is outlined in Fig. 112-2.

113. EXITS

Exits in spaces and rooms must be kept clear of obstructions. Double exits must be provided if the arrangement of equipment and an accident can make a single exit inaccessible. The exit doors must swing out and have some type of panic hardware (e.g., a push bar, not a door knob) except for fence gates in outdoor

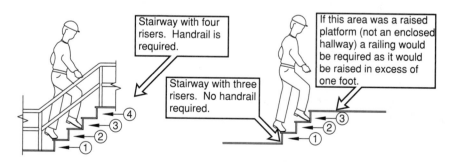

Fig. 112-2. Requirements for stair handrails (Rule 112).

substations and doors in rooms containing only low-voltage nonexplosive equipment. See Fig. 113-1.

114. FIRE-EXTINGUISHING EQUIPMENT

This rule requires fire-extinguishing equipment in the electric supply station. Typically a portable fire extinguisher will meet the approval, location, and marking requirements. A portable extinguisher will not be sufficient to extinguish a major transformer fire; therefore, a separate plan for this type of failure should be made.

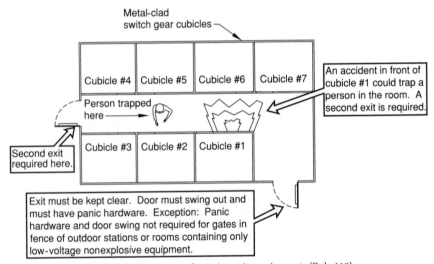

Fig. 113-1. Electric supply station exit requirements (Rule 113).

Installation and Maintenance of Equipment

120. GENERAL REQUIREMENTS

This rule discusses installation and maintenance of electric supply station equipment. Safeguarding personnel during installation, construction, and maintenance is the primary concern of Sec. 12. The rules of Sec. 12 apply to both alternating-current (AC) and direct-current (DC) electric supply stations.

121. INSPECTIONS

The inspections discussed in this rule and the inspections discussed in Rule 214, Part 2, Overhead Lines, and in Rule 313, Part 3, Underground Lines, form the basic requirements for inspecting electric supply stations and supply and communication lines. The inspections required in this rule for electric supply stations are outlined in Fig. 121-1.

Any equipment or wiring found defective must be permanently disconnected or promptly corrected. New equipment must be tested in accordance with industry practice. The Code does not specify the details of the inspection program; it simply states that it be regular and scheduled. Accepted good practice, which is discussed in Rule 012C, would imply keeping accurate records or logs of the inspections completed.

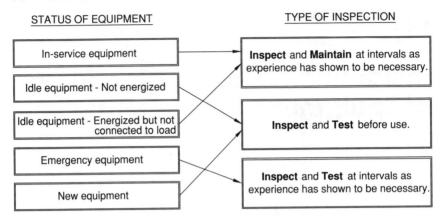

Fig. 121-1. Inspection requirements (Rule 121).

122. GUARDING SHAFT ENDS, PULLEYS, BELTS, AND SUDDENLY MOVING PARTS

Mechanical parts located in an electric supply station must be safeguarded. Mechanical transmission machinery is abundant in generating stations. Substations typically do not have much mechanical machinery. This rule requires the use of ANSI/ASME B15.1. Many times the Code makes note of a standard, but in this case, the standard is required, as it is part of the Code text. Suddenly moving parts must also be guarded or isolated.

123. PROTECTIVE GROUNDING

This rule states the requirements for electric supply station grounding. The methods of protective grounding are found in Sec. 09, "Grounding Methods for Electric Supply and Communications Facilities."

Non-current-carrying metal parts in the electric supply station must be grounded or isolated. Metallic fences must be grounded. IEEE Standard 80 is noted in this rule and in Sec. 09, as it is the standard for electric supply station grounding.

Provisions must also exist for grounding during maintenance. When a conductor, bus section, or piece of equipment is disconnected for maintenance, it must be grounded. The grounding can be done with permanent grounding switches or a readily accessible means for connecting portable grounding jumpers. The Part 4 Work Rules are referenced for proper procedures.

Direct-current (DC) systems have unique rules for grounding, which are discussed in Sec. 09.

124. GUARDING LIVE PARTS

124A. Where Required. Live parts in an electric supply station over 300 V phase to phase must be guarded, isolated by location (using vertical and horizontal clearance), or insulated to avoid inadvertent contact by qualified utility personnel in the substation. The basic intent of this rule is that utility personnel who are qualified to be in the substation (but not necessarily working on the substation) can walk around without accidentally contacting energized parts. The clearances in this section do not apply to the general public. Rule 110A, "General Requirements, Enclosure of Equipment," must be met before the clearances of Rule 124 apply. Otherwise Part 2, Overhead Lines, is applicable. See Rule 110A for an example.

Rule 124 constantly uses the word "guard," which implies a physical barrier between the utility employee and the live part. In most applications, physical guards are not nearly as common as the alternative to providing a guard, which is providing adequate clearance.

Rule 124A1 references **NESC** Table 124-1 and **NESC** Fig. 124-1. **NESC** Table 124-1 and **NESC** Fig. 124-1 are used to determine clearance to live (i.e., energized) parts in the electric supply station. The footnotes to **NESC** Table 124-1 provide additional information on Basic Impulse Insulation Levels (BIL) selection methods. Appendix D in the **NESC** provides additional information on per-unit overvoltage factors.

In addition to clearance to live parts, Rule 124A3 specifies an 8-ft, 6-in clearance to parts on indeterminate potential. A bushing or insulator has a surface along it of indeterminate potential. The top of the bushing has a known voltage, for example, 7.2 kV to ground. The bottom of the bushing has a known voltage, 0 V if it is grounded. The space between the top of the bushing and the bottom is of unknown voltage, or as the **Code** calls it, indeterminate potential. It is somewhere between 7.2 kV and 0 V. The 8-ft, 6-in value is an important number when designing and inspecting substation facilities. If the substation equipment purchased is not tall enough to meet the 8-ft, 6-in dimension at the bottom of a part of indeterminate potential (e.g., the bottom of a transformer bushing), concrete pad thickness and/or equipment stands must be used to provide the required height for both the live part clearance (top of the bushing) and the indeterminate voltage clearance (bottom of the bushing). Both clearances must be met or exceeded to avoid a **Code** violation. An example of how to apply the vertical clearances of **NESC** Table 124-1 and **NESC** Fig. 124-1 is shown in Fig. 124-1.

Rule 124A1 specifies that the clearance to live parts may be obtained by using vertical clearance, horizontal clearance, or a combination of both. Rule 124A3 specifies a vertical clearance only for parts of indeterminate potential. These rules are outlined in Fig. 124-2.

If a concrete pad under substation equipment is large enough to stand on, then the measurement needs to be taken from the top of the concrete pad. The interpretation of what size concrete pad is a permanent supporting surface for workers is not well defined in the **Code**. If the concrete pad is oversized for easy maneuverability of a worker, the clearance measurement must be made from the top of the concrete pad. If the pad is oversized for the equipment but

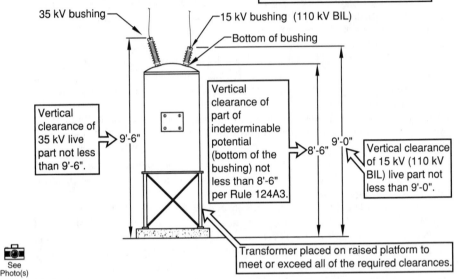

Example:
34.5/19.4 kV, 3Ø, 4W primary
12.47/7.2 kV, 3Ø, 4W, secondary
Vertical clearance:
35 kV between phases for primary.
15 kV (110 kV BIL) between phases for secondary.
Vertical clearance per NESC Table 124-1, Column 2.

35 kV bushing

15 kV bushing (110 kV BIL)

Bottom of bushing

Vertical clearance of 35 kV live part not less than 9'-6".

9'-6"

Vertical clearance of part of indeterminable potential (bottom of the bushing) not less than 8'-6" per Rule 124A3.

8'-6"   9'-0"

Vertical clearance of 15 kV (110 kV BIL) live part not less than 9'-0".

See Photo(s)

Transformer placed on raised platform to meet or exceed all of the required clearances.

Fig. 124-1. Example of how to apply vertical clearances per NESC Table 124-1 and NESC Fig. 124-1 (Rules 124A1 and 124A3).

a worker has to "hug the equipment" to stay standing on the concrete pad, the clearance measurement can be made from the substation surface (e.g., gravel) instead of the top of the concrete pad. This rule is outlined in Fig. 124-3.

Rule 124A2 recognizes that additional clearances or guarding may be needed where material may be carried such as passageways, corridors, storage areas, etc. (primarily indoor areas). Additional clearance values are not specified. If physical guards are used for these areas, they must be removed with tools or keys.

The Code does not specify bus to bus clearances, conductor to bus clearances, or conductor to conductor clearances in Part 1, Electric Supply Stations. Rule 012C, which requires accepted good practice, must be applied. The conductor to conductor clearances given in Part 2, Overhead Lines, Sec. 23, are not required to be used in Part 1, Electric Supply Stations, but they are a reference for accepted good practice. Other common references for accepted good practice for bus to bus clearance in outdoor electrical substations are ANSI C37.32 and NEMA SG6. The absence of a Code rule related to bus to bus clearance is outlined in Fig. 124-4.

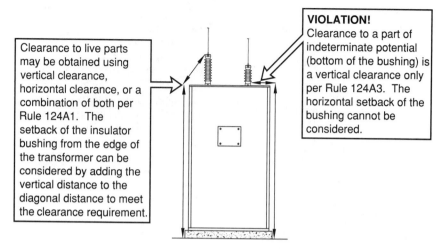

Clearance to live parts may be obtained using vertical clearance, horizontal clearance, or a combination of both per Rule 124A1. The setback of the insulator bushing from the edge of the transformer can be considered by adding the vertical distance to the diagonal distance to meet the clearance requirement.

VIOLATION!
Clearance to a part of indeterminate potential (bottom of the bushing) is a vertical clearance only per Rule 124A3. The horizontal setback of the bushing cannot be considered.

Fig. 124-2. Clearance measurements are specified as vertical, horizontal, or a combination of both (Rules 124A1 and 124A3).

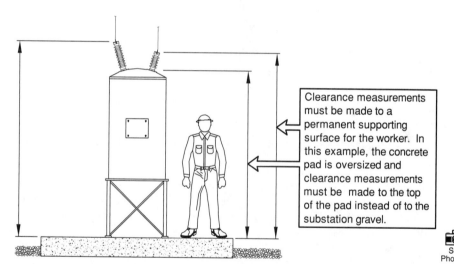

Clearance measurements must be made to a permanent supporting surface for the worker. In this example, the concrete pad is oversized and clearance measurements must be made to the top of the pad instead of to the substation gravel.

See Photo(s)

Fig. 124-3. Clearance measurements made to a permanent supporting surface (Rule 124A1).

124B. Strength of Guards. Physical guards, when used instead of clearance, must be rigid and secure such that a person falling or slipping will not displace or deflect the guard.

124C. Types of Guards. The first sentence of this rule points out that meeting the clearances in NESC Table 124-1 permits guarding by location.

Providing adequate clearance is the most common form of guarding by location. When guarding by isolation is used, entrances to the guarded space must be locked, barricaded, or roped off, and safety signs must be posted at entrances.

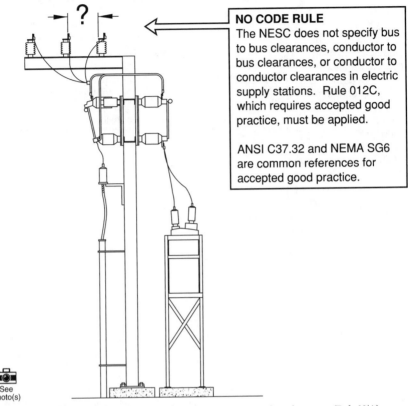

NO CODE RULE
The NESC does not specify bus to bus clearances, conductor to bus clearances, or conductor to conductor clearances in electric supply stations. Rule 012C, which requires accepted good practice, must be applied.

ANSI C37.32 and NEMA SG6 are common references for accepted good practice.

See Photo(s)

Fig. 124-4. Absence of Code rule related to bus to bus clearances (Rule N/A).

Rules 124C2 through 124C6 discuss various types of physical guards including shields, enclosures, barriers, mats, supporting surfaces for persons above live parts, and insulating covering. When railings or fences are used as guards, NESC Fig. 124-2 applies. An example of a railing or fence used as a guard is shown in Fig. 124-5.

The requirement in Rule 124C3 to locate the guard railing or fence "preferably not more than 4 ft" from the nearest point in the guard zone may not be practical in some cases. A NOTE to the rule indicates that additional working space may be required (more than 4 ft) when Rules 125 and 441 are considered.

Using conductor insulation as a guard is discussed in Rule 163.

125. WORKING SPACE ABOUT ELECTRIC EQUIPMENT

125A. Working Space (600 V or Less).   Working space is required around electrical equipment for inspection or servicing. Adequate working space avoids equipment crowding and provides a safe working environment. The working

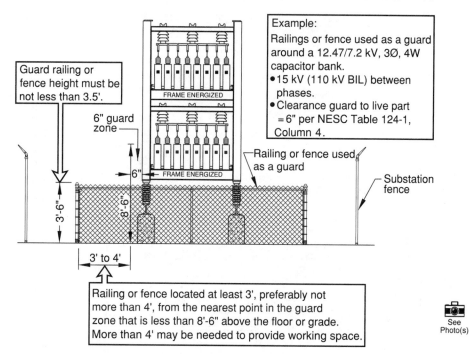

Fig. 124-5. Example of how to apply NESC Table 124-1 and NESC Fig. 124-2 (Rule 124C3).

space required for equipment operated at 600 V or less is outlined in NESC Table 125-1. In addition to the horizontal distances shown in the table, a minimum of 7 ft of headroom is required and a width of not less than 30 in is required. If the equipment is wider than 30 in, then the working space must be available for the full width of the equipment.

The Code specifically states that concrete, brick, or tile walls are considered grounded. A sheetrock wall is not referenced.

The back of a switchboard is assumed to be nonaccessible if all parts replacements, wire connections, etc., can be done from the front. If this were not the case, working space would also be needed behind a switchboard. The distance must be measured from the front of the enclosure if the energized parts are normally enclosed.

The conditions in NESC Table 125-1 are for exposed energized parts. If the parts are always de-energized during inspection, servicing, etc., then the NESC does not specify a workspace dimension. See Figs. 125-1 and 125-2.

The working space in Fig. 125-1 must be guarded to avoid encroachment by others into the equipment or into the service person when the working space is in an open area or passageway. There must be at least one entrance for access into the working space. See Rule 113 for exit requirements. The working space must not be used for storage. See Fig. 125-3.

125B. Working Space over 600 V. For voltages above 600 V, the working space is provided in accordance with NESC Table 124-1. This working space is for

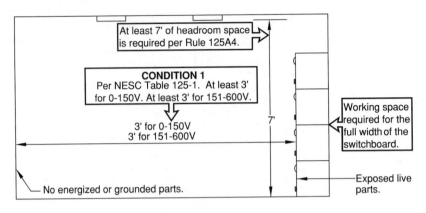

At least 7' of headroom space is required per Rule 125A4.

CONDITION 1
Per NESC Table 125-1. At least 3' for 0-150V. At least 3' for 151-600V.

3' for 0-150V
3' for 151-600V

7'

Working space required for the full width of the switchboard.

No energized or grounded parts.

Exposed live parts.

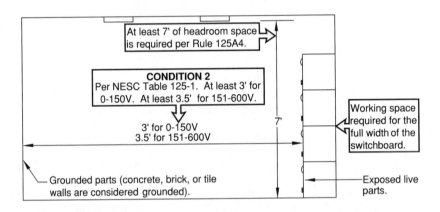

At least 7' of headroom space is required per Rule 125A4.

CONDITION 2
Per NESC Table 125-1. At least 3' for 0-150V. At least 3.5' for 151-600V.

3' for 0-150V
3.5' for 151-600V

7'

Working space required for the full width of the switchboard.

Grounded parts (concrete, brick, or tile walls are considered grounded).

Exposed live parts.

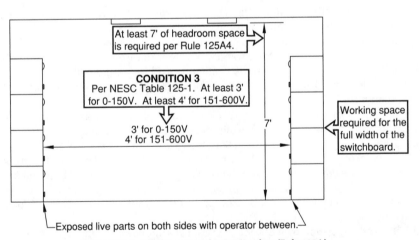

At least 7' of headroom space is required per Rule 125A4.

CONDITION 3
Per NESC Table 125-1. At least 3' for 0-150V. At least 4' for 151-600V.

3' for 0-150V
4' for 151-600V

7'

Working space required for the full width of the switchboard.

Exposed live parts on both sides with operator between.

Fig. 125-1. Working space for 600 V or less (Rule 125A).

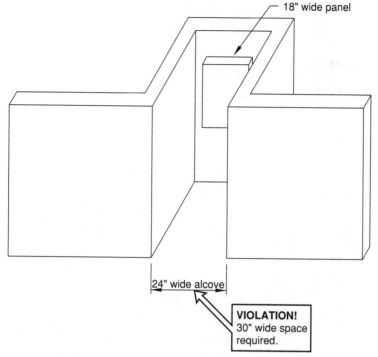

Fig. 125-2. Minimum width of working space (Rule 125A3).

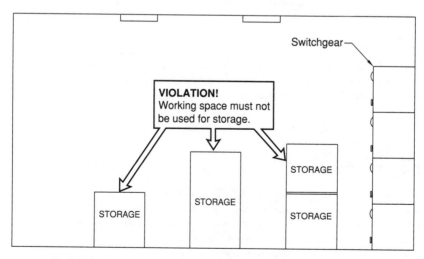

Fig. 125-3. Storage materials must not be in the working space (Rule 125A1).

See
Photo(s)

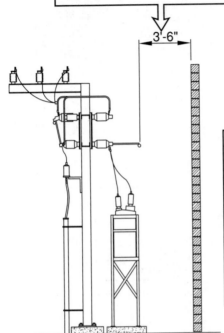

Working space not less than 3'-6".
In this example, the safety clearance
zone to the fence or wall does not
apply per Rule 110A2, Exception 1.

3'-6"

Example: 12.47/7.2 kV, 3∅, 4W
electric supply station working
space.
• 15 kV (110 kV BIL) between
  phases.
• Fence or wall without holes or
  openings.
• Working space is horizontal
  clearance per NESC Table 124-1,
  Column 3.

Fig. 125-4. Example of working space over 600 V (Rule 125B).

electrical equipment in substations. It does not apply to Part 2, Overhead Lines,
or Part 3, Underground Lines. Examples of working space in areas with equipment over 600 V are shown in Figs. 125-4 and 125-5.

126. EQUIPMENT FOR WORK
ON ENERGIZED PARTS

This rule indirectly ties the work rules of Part 4 into the working space of
Rule 125. If a worker is within the guard zone of NESC Table 124-1, Column 4,
then the worker must utilize protective equipment that is properly tested and
rated for the voltage involved. The Work rules in Part 4, specifically Rule 441,
contain minimum approach distances to live parts.

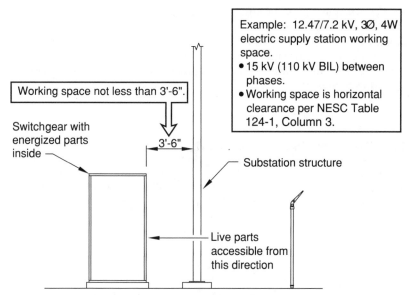

Example: 12.47/7.2 kV, 3∅, 4W electric supply station working space.
• 15 kV (110 kV BIL) between phases.
• Working space is horizontal clearance per NESC Table 124-1, Column 3.

Working space not less than 3'-6".

Switchgear with energized parts inside

3'-6"

Substation structure

Live parts accessible from this direction

Fig. 125-5. Example of working space over 600 V (Rule 125B).

127. CLASSIFIED LOCATIONS

Classified locations are locations where fire or explosion hazards may exist due to flammable gases, vapors, liquids, dust, or fibers. The NESC requires that classified locations in the vicinity of electric supply stations meet the National Electrical Code (NEC) hazardous (classified) locations requirements. The NEC has very detailed requirements related to classified locations, and rather than repeating them, the NESC requires that the NEC rules be met. In addition to the NEC requirements, the NESC provides requirements for coal-handling areas and various other hazardous locations primarily related to electric supply generating stations. Several National Fire Protection Association (NFPA) documents are required to be referenced for specific types of installations.

Separation from or ventilation of a hazardous area can reduce the classified area requirements. High-voltage facilities are typically located away from classified areas. Low-voltage equipment (i.e., under 600 V) can be installed in explosionproof enclosures and located in a classified location in accordance with the NEC rules.

128. IDENTIFICATION

This rule requires identification or labeling of equipment and devices in the electric supply station. This includes indoor switchboards and outdoor equipment.

The identification must be uniform throughout any one station and not be placed on removable covers or doors that could be interchanged, therefore creating a labeling error. The Code is not specific as to the type of identification, but equipment must be sufficiently labeled for safe use and operation. Identification may include voltage levels, nameplate information, phase color coding, feeder numbering, etc.

129. MOBILE HYDROGEN EQUIPMENT

Hydrogen gas is highly flammable. Bonding of mobile hydrogen equipment will reduce the chance for a difference in potential, which will reduce the chance of sparks that may cause an explosion.

Section 13

Rotating Equipment

130. SPEED CONTROL AND STOPPING DEVICES

Section 13 deals with generators, motors, motor generators, and rotary converters in electric supply stations. NESC Sec. 02, "Definitions of Special Terms," defines an electric supply station as a generating station or substation. Rotating equipment is more applicable to generating stations, as substations typically do not involve rotating equipment.

Sec. 13 only provides general rules related to basic safety features. Other IEEE standards are available, but not referenced in the Code, for generation station design. Using IEEE Standards and the National Electrical Code (NEC) for specific design features is a good application of Rule 012C, which requires using accepted good practice for particulars not specified in the NESC.

Rule 130 requires overspeed trip of prime movers (i.e., diesel engines, turbines, etc.) in addition to governors. Manual stopping devices are also required. Speed-limiting devices are required for separately excited AC motors and series motors, as they typically have problems with overspeed or runaway. Adjustable-speed motors must also have a speed-limiting device and must be equipped or connected to avoid weak fields that produce overspeed. Mechanical protection of the control circuits for stopping and speed limiting is also required.

131. MOTOR CONTROL

Rule 131 requires that motors do not restart automatically after a power outage if the unexpected starting could create injury to personnel. Two types of basic motor starting circuits are used in the motor control industry, low-voltage protection and low-voltage release.

Low-voltage protection (three-wire control) requires manual restarting by an equipment operator. This rule does provide an option to use low-voltage release (two-wire control with automatic restarting) if warning signals and time-delay features are designed into the control scheme. Examples of simple two-wire and three-wire motor control schemes are shown in Fig. 131-1.

132. NUMBER 132 NOT USED IN THIS EDITION

133. SHORT-CIRCUIT PROTECTION

This rule requires short-circuit protection for electric motors. Only a simple requirement is stated. Rule 012C, which requires accepted good practice, should

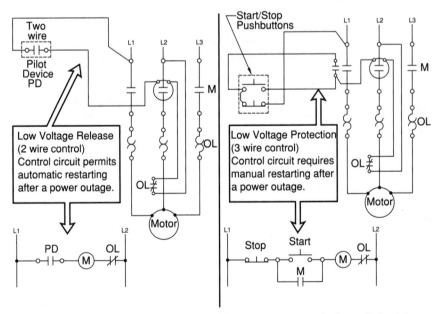

Fig. 131-1. Example of simple two-wire and three-wire motor control schemes (Rule 131).

be applied for specific details. The National Electrical Code (NEC) specifies maximum fuse and circuit breaker sizes for various types and sizes of motors. The NEC is an excellent reference for accepted good practice in this case.

An example of motor protection components and terminology is shown in Fig. 133-1.

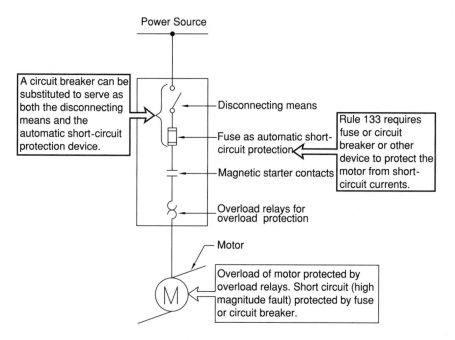

Fig. 133-1. Example of motor protection components and terminology (Rule 133).

Section 14

Storage Batteries

140. GENERAL

This section outlines the requirements for storage batteries used in electric supply stations. The battery system is typically used for DC control and DC operation of circuit breakers, circuit switchers, reclosers, etc. Batteries produce hydrogen gas, which is explosive, and electrolyte, which is acidic and corrosive. Sealed batteries can minimize these concerns. Typical battery systems used in substation applications include lead-acid and nickel-cadmium (NiCa). Batteries can be located with other equipment or in a dedicated battery room. The number of batteries and the amount of hydrogen produced by the batteries affect this decision. Rule 142 states that the natural or powered ventilation must limit the hydrogen accumulation to less than an explosive mixture. The Code does not provide a percent concentration that is considered safe. Battery manufacturers can provide guidelines for calculating hydrogen gas emissions and ventilation recommendations. The NESC outlines general, location, ventilation, rack, floor, illumination, and service facility requirements in Rules 140 through 146 (Rule 147 is not used). Code Rules 140 through 146 are outlined in Fig. 140-1.

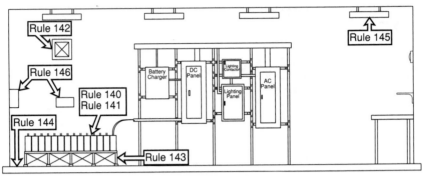

CONTROL BUILDING

Rule 140. General	Rule 144. Floors in Battery Areas
• Adequate space for operation and batteries.	• Acid-resistive floor. • Provisions to contain spilled electrolyte.
Rule 141. Location	**Rule 145. Illumination for Battery Areas**
• Batteries located in a protective enclosure or area accessible only to qualified personnel.	• Guard or isolate lighting fixtures (e.g., lenses or wire guards). • No receptacles or light switches in battery area.
Rule 142. Ventilation	**Rule 146. Service Facilities**
• Natural or powered ventilation. • Enunciate ventilation failure if required to keep hydrogen level below an explosive mixture.	• Goggles or face shield. • Acid-resistant gloves. • Protective aprons and overshoes. • Portable or stationary wash for eyes and skin. • Safety sign prohibiting smoking, sparks, or flames.
Rule 143. Racks	
• Racks firmly anchored, preferably to the floor. • Metal racks must be grounded.	

See
Photo(s)

Fig. 140-1. Storage batteries (Rules 140 through 146).

141. LOCATION

See discussion in Rule 140 and Fig. 140-1.

142. VENTILATION

See discussion in Rule 140 and Fig. 140-1.

143. RACKS

See discussion in Rule 140 and Fig. 140-1.

144. FLOORS IN BATTERY AREAS

See discussion in Rule 140 and Fig. 140-1.

145. ILLUMINATION FOR BATTERY AREAS

See discussion in Rule 140 and Fig. 140-1.

146. SERVICE FACILITIES

See discussion in Rule 140 and Fig. 140-1.

Section 15

Transformers and Regulators

150. CURRENT-TRANSFORMER SECONDARY CIRCUITS PROTECTION WHEN EXCEEDING 600 V

This rule requires secondary circuits of current transformers (with primary circuits over 600 V) to be protected by conduit, covering, or some other protection. If the conduit or covering is metal, it must be effectively grounded and consideration must be given to circulating currents. Current transformers (CTs) must also have a provision for shorting the secondary wiring. These requirements are needed, as open or damaged current-transformer secondary circuits may cause a hazardous high voltage and arcing.

An example of protecting current-transformer secondary circuits is shown in Fig. 150-1.

151. GROUNDING SECONDARY CIRCUITS OF INSTRUMENT TRANSFORMERS

Instrument transformers consist of both voltage transformers (VTs) and current transformers (CTs). Voltage transformers are sometimes referred to as potential transformers (PTs). The secondaries of voltage transformers and current transformers must be effectively grounded. An example of basic VT and CT connections is shown in Fig. 151-1.

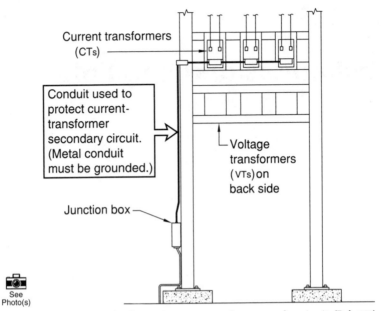

Current transformers (CTs)

Conduit used to protect current-transformer secondary circuit. (Metal conduit must be grounded.)

Voltage transformers (VTs) on back side

Junction box

See Photo(s)

Fig. 150-1. Example of protecting current-transformer secondary circuits (Rule 150).

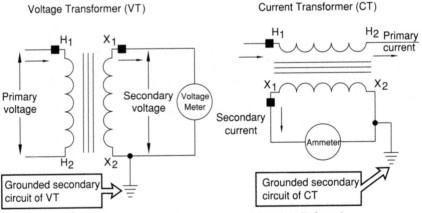

Voltage Transformer (VT)

Current Transformer (CT)

H_1 X_1

H_1 H_2 Primary current

Primary voltage

Secondary voltage   Voltage Meter

X_1 X_2

Secondary current

H_2 X_2

Ammeter

Grounded secondary circuit of VT

Grounded secondary circuit of CT

Fig. 151-1. Example of basic VT and CT connections (Rule 151).

152. LOCATION AND ARRANGEMENT OF POWER TRANSFORMERS AND REGULATORS

152A. Outdoor Installations. This rule requires that energized parts of power transformers be enclosed, guarded, or physically isolated as discussed in Rule 124. The case (enclosure) of a substation transformer and regulator must be effectively grounded or guarded.

The **Code** methods required for minimizing fire hazards related to liquid-filled power transformers and regulators installed in outdoor substations are outlined below (one or more methods must be used):
- Less flammable liquids
- Space separation
- Fire-resistant barriers
- Automatic extinguishing systems
- Absorption beds
- Enclosures

The **Code** requires that the degree of the fire hazard and the amount and characteristics of the liquid be considered when selecting a method, but no specifics are provided for the amount or type of oil, separation distance, etc. Rule 012C, which requires accepted good practice, must be applied. IEEE Standard 979, *IEEE Guide for Substation Fire Protection,* is an excellent reference for accepted good practice in this case.

The **NESC** does not address Spill Prevention Control & Countermeasure (SPCC) plans. SPCC plans are required by the Environmental Protection Agency (EPA). The amount of oil contained in the substation and the location of the substation relative to water are two of the many considerations that trigger the need for a substation SPCC plan.

Rule 012C is sometimes referenced when an engineer or designer is trying to determine how close an oil-filled service transformer can be placed to a building. Rule 152 is in Part 1 and Part 1 only applies to electric supply stations, not to overhead or pad-mounted service transformers outside the substation fence. Part 2, Overhead Lines, and Part 3, Underground Lines, do not provide any additional rules related to locating oil-filled transformers near buildings. Part 2 does provide clearances to buildings for live parts and equipment cases, but these clearances are based on electrical safety, not fire safety. Rule 012C, which requires accepted good practice, must be applied in this instance. The National Electrical Code (NEC), Underwriters Laboratories (UL), and Factory Mutual (FM) are all good references for determining service transformer location standards.

152B. Indoor Installations. Transformers and regulators installed indoors require even greater consideration of fire hazards. Rule 152B is divided into three parts. The first paragraph covers traditional oil-filled transformers. The second paragraph covers dry-type transformers. The third paragraph covers oil-filled transformers with less flammable oil. Traditional oil-filled transformers use mineral oil, which is considered flammable. Less flammable oil is typically a fluid consisting of fire-resistant hydrocarbon. Dry-type transformers use no fluid oil for cooling. They are vented for air cooling. Dry-type transformers usually do not have the overload capability that oil-filled transformers have. The term "fire walls" is used but no specifics as to the fire wall ratings are provided (i.e., 1 hour rating, 2 hour rating, etc.). In addition to the accepted good practice documents discussed in Rule 152A, the Uniform Building Code (UBC) or other applicable document must be referenced for fire wall construction details. NFPA 850, *Recommended Practice for Fire Protection for Electric Generating Plants and High Voltage Direct Current Converter Stations,* can be referenced when designing generating stations.

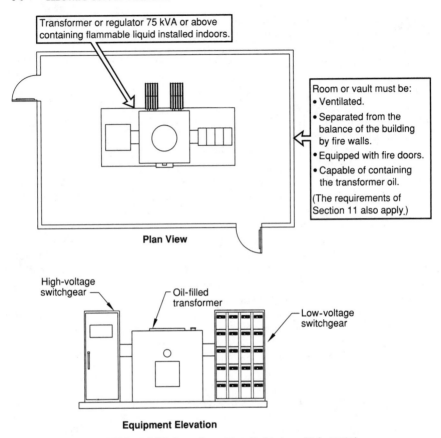

Fig. 152-1. Oil-filled transformer installed indoors (Rule 152B1).

The requirements for locating traditional oil-filled transformers indoors are outlined in Fig. 152-1.

The requirements for locating dry-type transformers indoors are outlined in Fig. 152-2.

The requirements for locating less flammable liquid transformers indoors are outlined in Fig. 152-3.

153. SHORT-CIRCUIT PROTECTION OF POWER TRANSFORMERS

Electric supply station power transformers are required to have short-circuit protection. A short circuit within the transformer provides a high-magnitude fault current. The short-circuit protection device typically will not be sized to

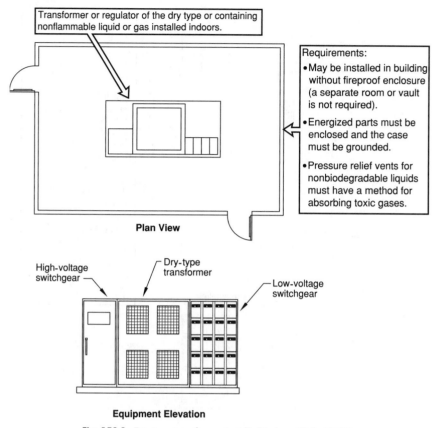

Plan View

Transformer or regulator of the dry type or containing nonflammable liquid or gas installed indoors.

Requirements:
- May be installed in building without fireproof enclosure (a separate room or vault is not required).
- Energized parts must be enclosed and the case must be grounded.
- Pressure relief vents for nonbiodegradable liquids must have a method for absorbing toxic gases.

High-voltage switchgear

Dry-type transformer

Low-voltage switchgear

Equipment Elevation

Fig. 152-2. Dry-type transformer installed indoors (Rule 152B2).

protect transformer overload. Transformer overload is commonly monitored by metering the load and comparing the metered load to the transformer nameplate capacity.

This rule provides a list of automatically disconnecting short-circuit protection devices. Circuit breakers typically open all three phases of a three-phase line simultaneously. Fuses, on the other hand, open just the faulted phase. Removing just the faulted phase is acceptable. This rule does not apply to every transformer in the station, only to power transformers. For example, current transformers (CTs) are not typically fused. Voltage transformers (VTs) may be fused, but the Code does not require them to be fused. The choices for short-circuit protection of power transformers are outlined in Fig. 153-1.

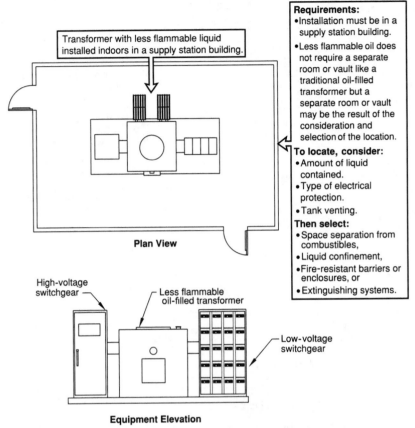

Transformer with less flammable liquid installed indoors in a supply station building.

Plan View

High-voltage switchgear

Less flammable oil-filled transformer

Low-voltage switchgear

Equipment Elevation

Requirements:
- •Installation must be in a supply station building.
- •Less flammable oil does not require a separate room or vault like a traditional oil-filled transformer but a separate room or vault may be the result of the consideration and selection of the location.

To locate, consider:
- • Amount of liquid contained.
- • Type of electrical protection.
- • Tank venting.

Then select:
- • Space separation from combustibles,
- • Liquid confinement,
- •Fire-resistant barriers or enclosures, or
- • Extinguishing systems.

Fig. 152-3. Less flammable oil-filled transformer installed indoors (Rule 152B3).

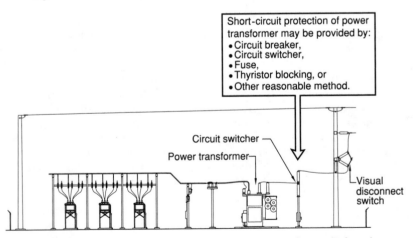

Short-circuit protection of power transformer may be provided by:
- • Circuit breaker,
- • Circuit switcher,
- • Fuse,
- • Thyristor blocking, or
- • Other reasonable method.

Circuit switcher

Power transformer

Visual disconnect switch

See Photo(s)

Fig. 153-1. Choices for short-circuit protection of power transformers (Rule 153).

Section 16

Conductors

160. APPLICATION

Selecting conductors may appear to be a simple task but there are literally thousands to pick from. This rule requires that the following conductor features must be considered when choosing a conductor.

- Suitable for the location
- Suitable for the use
- Suitable for the voltage (insulation rating)
- Adequate ampacity (conductor size)

This section only applies to conductors in an electric supply station. Part 2 applies to overhead line conductors outside the substation and Part 3 applies to underground conductors outside the substation. Conductors in this rule, in Part 2, Overhead Lines, and in Part 3, Underground Lines, are discussed in general terms. The NESC does not specify conductor ampacity for various conductor sizes and types. To determine a conductor's ampacity, Rule 012C, which requires accepted good practice, must be applied. Ampacity tables can be found in the National Electrical Code (NEC) and in conductor manufacturers' literature. Both of these sources are excellent applications of Rule 012C. The NEC also specifies fuse and circuit breaker sizes for protecting 600 V conductors.

Conductors, by the definition in Sec. 02, include substation bus bars, as well as wires, which are typically bare, and cables, which are typically insulated.

Typical examples of conductors found in substations are shown in Fig. 160-1.

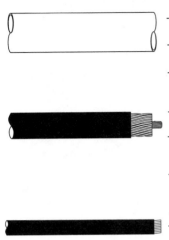

Bare 3" aluminum tube (bus bar)
- Voltage is dependent on the insulators that the bus bar is supported on.
- Ampacity (normal load current) is dependent on conductor material, size, and ambient temperature.
- Location: Above ground with proper clearance.

Insulated 1/0 aluminum URD cable (> 600 V)
- Voltage is dependent on insulation type and thickness.
- Ampacity (normal load current limit) is dependent on conductor material, size, insulation type, and conduit or burial method.
- Location: In conduit or direct buried at proper depths.

Insulated 1/0 copper (600 V or less)
- Voltage is dependent on insulation type and thickness.
- Ampacity (normal load current limit) is dependent on conductor material, size, insulation type, and conduit or burial method.
- Location: In conduit or direct buried at proper depths.

Fig. 160-1. Typical examples of conductors in substations (Rule 160).

161. ELECTRICAL PROTECTION

After the conductors are properly selected and applied, electrical protection of conductors is required. Overcurrent protection of conductors consists of two components, overload and short-circuit protection. Overload protection keeps the load current from exceeding the conductor's normal load current ampacity. Short-circuit protection keeps the fault current from exceeding the conductor's short-time fault current ampacity.

This rule requires overcurrent (both overload and short-circuit) protection by the following methods:
- Design of the system *and*
- Overcurrent devices *or*
- Alarm devices *or*
- Indication devices *or*
- Trip devices

A typical 120-V, 20-A, single-pole, thermal-magnetic circuit breaker protecting a No. 12 AWG copper, 600-V conductor provides both overload protection via the circuit breaker thermal element and short-circuit protection via the circuit breaker magnetic element. A typical 15-kV recloser protecting an overhead or underground distribution feeder may be set to only provide short-circuit protection. The overload protection of the feeder is monitored by checking percent conductor loading of the conductor during a system load study.

The electrical protection device may protect the conductor and a piece of equipment. For example, the high-side fuse of a substation transformer can provide short-circuit protection for the substation transformer and the bus bars.

Overload protection of the transformer and bus bars can be provided by monitoring the load and comparing the load to the transformer and bus bar ratings.

Grounded conductors must be configured without overcurrent protection so there is not an interruption in the grounding protection. When a single-phase or three-phase circuit is fused or connected to a circuit breaker or recloser, the grounded conductor (neutral) must not be fused or connected to the circuit breaker or recloser contacts so that the continuity to ground is maintained.

162. MECHANICAL PROTECTION AND SUPPORT

Conductors contained within the electric supply station must be adequately supported to withstand forces caused by the maximum short-circuit current that the conductor may experience. This is a very general requirement, unlike the conductor strength and loading requirements of Part 2, Overhead Lines. Sections 24, 25, 26, and 27 in Part 2 provide very specific loading and strength requirements for conductor supports. The strength and loading rules of Part 2 are required to be applied to conductors that extend outside the electric supply station. Calculating magnetic forces caused by short-circuit currents and applying the overload and strength factors for conductors from Part 2 to conductors inside the substation is one application of Rule 012C, which requires using accepted good practice when specifics are not provided in the Code. Other commonly referenced design standards for substation conductor supports are IEEE Standard 605, *IEEE Guide for Design of Substation Rigid Bus Structures*, and ASCE 7, *Minimum Design Loads for Buildings* and *Other Structures*. The Uniform Building Code (UBC) is commonly referenced for seismic (earthquake) zones. The Rural Utilities Service (RUS) publishes Bulletin 1724E-300, *Design Guide for Rural Substations*, which is also useful in substation design.

If a conductor, its insulation, or support is subject to mechanical damage, the following are required to limit the likelihood of damage or disturbance:

- Casing,
- Armor, or
- Other means

Typically in a substation, galvanized rigid steel conduit is used to protect and support insulated conductors subject to mechanical damage.

The mechanical protection and support requirements of this rule are outlined in Fig. 162-1.

163. ISOLATION

Conductors inside the substation fence, when bare above 150 V to ground or insulated without a shield above 2500 V to ground, must be treated like the live parts in Rule 124. Nonshielded, insulated, and jacketed conductors may be installed in accordance with Rule 124C6, which discusses conductor guarding using insulation. The rules related to guarding a conductor with insulation are outlined in Fig. 163-1.

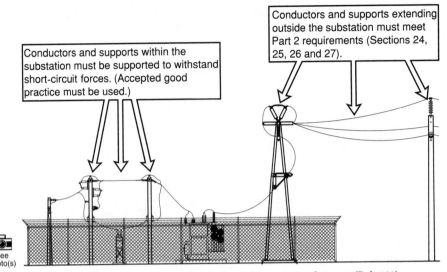

Conductors and supports within the substation must be supported to withstand short-circuit forces. (Accepted good practice must be used.)

Conductors and supports extending outside the substation must meet Part 2 requirements (Sections 24, 25, 26 and 27).

See Photo(s)

Fig. 162-1. Substation conductor mechanical protection and support (Rule 162).

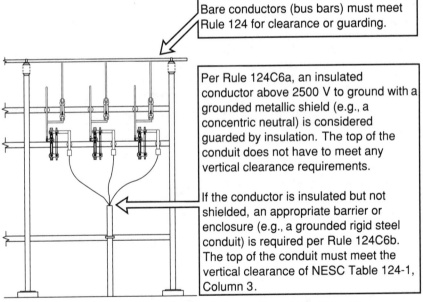

Bare conductors (bus bars) must meet Rule 124 for clearance or guarding.

Per Rule 124C6a, an insulated conductor above 2500 V to ground with a grounded metallic shield (e.g., a concentric neutral) is considered guarded by insulation. The top of the conduit does not have to meet any vertical clearance requirements.

If the conductor is insulated but not shielded, an appropriate barrier or enclosure (e.g., a grounded rigid steel conduit) is required per Rule 124C6b. The top of the conduit must meet the vertical clearance of NESC Table 124-1, Column 3.

See Photo(s)

Fig. 163-1. Guarding a substation conductor with insulation (Rule 163).

164. CONDUCTOR TERMINATIONS

The rules for conductor terminations in an electric supply substation are outlined in Fig. 164-1.

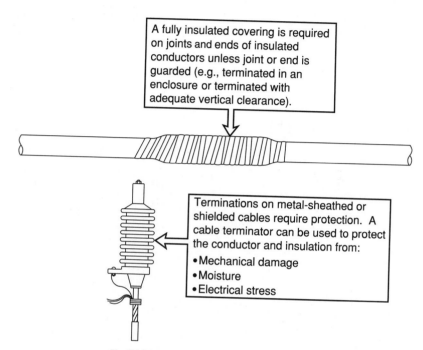

A fully insulated covering is required on joints and ends of insulated conductors unless joint or end is guarded (e.g., terminated in an enclosure or terminated with adequate vertical clearance).

Terminations on metal-sheathed or shielded cables require protection. A cable terminator can be used to protect the conductor and insulation from:
- Mechanical damage
- Moisture
- Electrical stress

Fig. 164-1. Conductor terminations (Rule 164).

Circuit Breakers, Reclosers, Switches, and Fuses

170. ARRANGEMENT

This rule requires circuit breakers, reclosers, switches, and fuses be accessible only to qualified persons. Section 17 is part of Part 1, Electric Supply Stations; therefore, the rules of this section apply only to circuit breakers, reclosers, switches, and fuses located in electric supply stations. Circuit breakers, reclosers, switches, and fuses in electric supply stations are accessible only to qualified persons when Rule 110A is met.

To protect persons from energized parts or arcing, Rule 170 requires the following:

- Walls,
- Barriers,
- Latched doors,
- Location,
- Isolation, or
- Other means

The requirements of Rule 124, Guarding Live Parts, also apply.

Conspicuous and unique markings must be provided at the switching device or at any remote operating points to identify the equipment (e.g., circuit breaker, recloser, switch) that is being controlled.

When the switch contacts are not normally visible (e.g., under oil, contained in a vacuum bottle, etc.), the switching device must be equipped with an operating

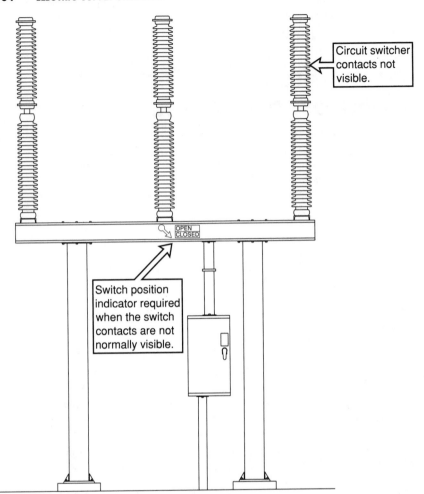

Fig. 170-1. Example of a switch position indicator (Rule 170-1).

position indicator. The work rules in Part 4 of the **NESC** provide switching control procedures and rules for de-energizing equipment or lines to protect employees. An example of a switch position indicator is shown in Fig. 170-1.

171. APPLICATION

The following ratings must be considered when applying a circuit breaker, circuit switcher, recloser, switch, or fuse:

- Voltage
- Continuous current
- Momentary current
- Short-circuit current interrupt rating

The maximum short-circuit current interrupt rating must be considered if the device is used to interrupt fault current.

When to apply ratings for momentary currents and interrupt currents is dependent on how the switch is used. If a switch is used as a disconnect only (i.e., it does not open under a fault condition), then it must be able to withstand a fault current flowing through it but it does not need to be rated to interrupt the fault current. The fault current it must withstand is termed the momentary fault current rating. If a switch is used to interrupt a fault, it must be rated to withstand the momentary fault current and interrupt the fault current without damage to the switch itself. A device that interrupts fault current must have a fault current interrupt rating.

It is important to note that the **NESC** does not address the selection of fuse, relay, or recloser settings such as time-current curves, delay intervals, number of operations to lockout, etc. These ratings are typically addressed in a system sectionalizing or coordination study. Careful selection of these and other settings is required for the safe and reliable operation of the electric system and accepted good practice must be applied in this case. The work rules in Part 4 of the **NESC** do address disabling a reclosing feature during work activities.

The interrupting capacity should be reviewed prior to each significant system change. For example, if a substation recloser is rated to interrupt 1250 A of short-circuit current and the substation transformer is replaced, the available fault current may increase to a value larger than 1250 A, which could cause damage to the recloser and personnel when the recloser operates to interrupt a fault. An example is given in Fig. 171-1.

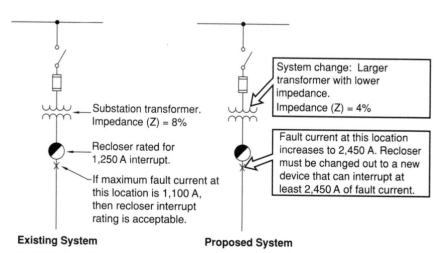

Existing System   **Proposed System**

Fig. 171-1. Example of checking interrupting capacity prior to a significant system change (Rule 171).

172. CIRCUIT BREAKERS, RECLOSERS, AND SWITCHES CONTAINING OIL

Circuit breakers, reclosers, and switches containing oil receive special attention due to the flammability of the oil. Similar requirements are outlined in Rule 152 for oil-filled power transformers and regulators.

Segregation of oil-filled circuit-interrupting devices is required in the electric supply station. Segregation of oil-filled equipment minimizes fire damage to adjacent equipment or buildings. Segregation may be provided by the following methods:

- Spacing,
- Fire-resistant barrier walls, or
- Metal cubicles

Gas release vents require oil-separating devices or piping to a safe location. Means to control oil discharges from vents or tank rupture are required. Methods to control oil are outlined below:

- Absorption beds,
- Pits,
- Drains, or
- Any combination of the above

This list is slightly different from the requirements for oil-filled transformers and regulators in Rule 152 but the general intent is the same. IEEE Standard 979, *IEEE Guide for Substation Fire Protection,* is an excellent reference for fire protection requirements.

Buildings or rooms housing circuit breakers, reclosers, and switches containing flammable oil must be of fire-resistant construction (see Rule 152 for additional information). Not all circuit breakers, reclosers, and switches contain oil. Some are air break, some are vacuum break, and some are SF_6 gas insulated. This rule applies only to circuit breakers, reclosers, and switches that contain oil and are located inside the electric supply station.

173. SWITCHES AND DISCONNECTING DEVICES

Switches and disconnecting devices must have capacity for the following system ratings:

- Voltage
- Current
- Load break current (if required)

Switches can be used to break load currents or open under no-load conditions. If required to break load current, the load current they are rated to interrupt must be marked on the switch. This value should not be confused with the short-circuit (fault) current-interrupt rating discussed in Rule 171.

Switches and disconnectors must be able to be locked open and locked closed, or plainly tagged where locks are not practical. Part 4 of the NESC specifies the work rules applicable to operating, locking, and tagging switches. Switches that are operated remotely and automatically must have a disconnecting means for the control circuit near the disconnecting apparatus to limit the likelihood of accidental operation of the switch.

174. DISCONNECTION OF FUSES

Disconnecting an energized fuse can be dangerous at any voltage. This rule requires fuses in circuits of more than 150 V to ground or more than 60 A (at any voltage) to be classified as disconnecting, or arranged to be disconnected from the source of power, or removed with insulating handles.

In most cases, fused cutout-type switches are configured such that the switchblade or fuse is "dead" when the switch is in the open position. Loop feeds can create energized blades when the switch is in the open position. The NESC does not have a requirement for how the switch blades or fuses are energized but does require the proper handling of the fuse. Proper consideration must be given to clearance requirements when fuse and switch blades are in the open position.

Section 18

Switchgear and Metal-Enclosed Bus

180. SWITCHGEAR ASSEMBLIES

180A. General Requirements for All Switchgear. This rule covers general requirements for all switchgear. Examples of switchgear found in electric supply stations are shown in Fig. 180-1.

The general requirements for all switchgear are outlined below:
- Secure to minimize movement.
- Support cables to minimize force on terminals.
- Locate away from liquid piping unless protected.
- Locate away from flammable gases or liquids.
- Install after general construction or provide temporary protection.
- Protect when doing maintenance in the area.
- Do not use as a physical support unless so designed.
- Interiors shall not be used for storage unless so designed.
- Metal instrument cases are to be grounded, enclosed in grounded metal covers, or made of insulating materials.

180B. Metal-Enclosed Power Switchgear. In addition to the general requirements for all switchgear, requirements exist for metal-enclosed power switchgear. The rules for metal-enclosed power switchgear are outlined in Fig. 180-2.

180C. Dead-Front Power Switchboards. Dead-front power switchboards with uninsulated rear connections must be installed in rooms capable of being locked with access limited to qualified personnel only. The uninsulated rear connections

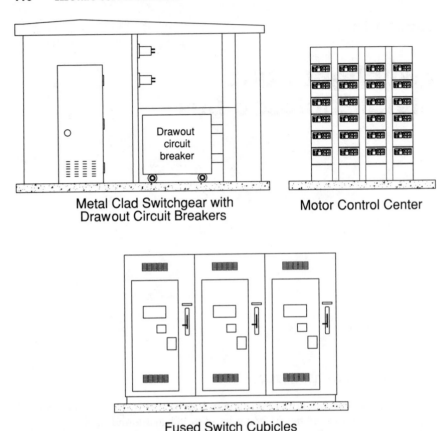

Metal Clad Switchgear with
Drawout Circuit Breakers

Motor Control Center

Fused Switch Cubicles

Fig. 180-1. Examples of switchgear assemblies (Rule 180A).

typically will not meet the vertical and horizontal clearances of Rule 124; therefore, a separate locked room is required for these energized parts.

180D. Motor Control Centers.   The fault current within the electric supply station can be very high due to low source impedance and little or no distribution-line impedance. The bus withstand ratings of motor control centers must therefore be given proper consideration. Bus bracing in low-voltage (e.g., 480-V) motor control centers commonly come in 50,000-, 100,000-, and 200,000-A values. This is a withstand rating, not an interrupt rating, as the bus must withstand the force of a momentary fault but not interrupt it. A fuse or circuit breaker in the motor control center will interrupt the fault. Current-limiting fuses can reduce the available fault current to which the motor control center bus and motor control center circuit breakers are exposed. Peak let-through currents for current-limiting fuses can be obtained from fuse manufacturer's literature. This rule also requires a safety sign for a motor control center cubicle having more than one voltage source. The rules regarding motor control center short-circuit ratings are outlined in Fig. 180-3.

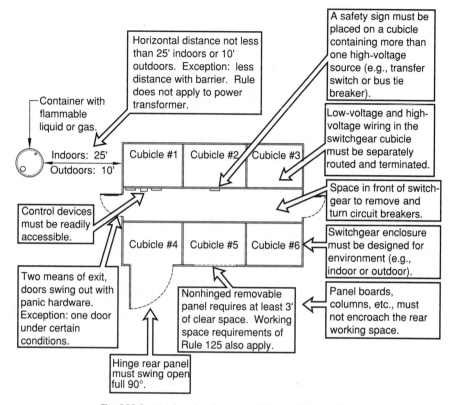

Fig. 180-2. Metal-enclosed power switchgear (Rule 180B).

180E. Control Switchboards. The requirements for control switchboards are outlined below:

- Control switchboards include cabinets with meters, relays, annunciators, computers, etc., for substation and generating station controls.
- Carpeting in control switchboard rooms must be antistatic and minimize toxic gas emissions under any conditions (e.g., water damage, fire damage, etc.).
- Adequate clearance in front and rear to read meters without stools.
- Personnel openings must be covered when not in use and must not be used for cable routing.

Reading control switchboard meters from stools produces a tripping or falling hazard. The rule related to adequate clearance for reading meters without stools is outlined in Fig. 180-4.

181. METAL-ENCLOSED BUS

181A. General Requirements for All Types of Bus. The rules for metal-enclosed bus are outlined in Fig. 181-1.

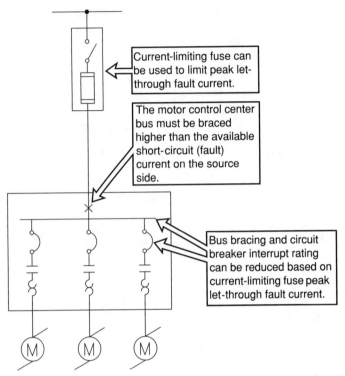

Current-limiting fuse can be used to limit peak let-through fault current.

The motor control center bus must be braced higher than the available short-circuit (fault) current on the source side.

Bus bracing and circuit breaker interrupt rating can be reduced based on current-limiting fuse peak let-through fault current.

Fig. 180-3. Motor control center short-circuit ratings (Rule 180D).

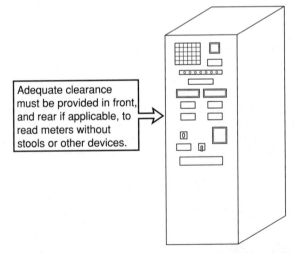

Adequate clearance must be provided in front, and rear if applicable, to read meters without stools or other devices.

See Photo(s)

Fig. 180-4. Adequate clearance for reading meters on control switchboards (Rule 180E3).

181B. Isolated-Phase Bus. Metal-enclosed bus is available in three common designs; segregated (isolated) phase bus bar, nonsegregated phase bus bar, and cable bus.

The use of isolated-phase bus requires the following special conditions:
- Clearance to magnetic material per manufacturer's recommendations
- Nonmagnetic conduit for alarm circuits
- Piping if enclosure drains are used
- Nonmagnetic wall plates
- Nonmagnetic conduit for grounding conductors

Examples of metal-enclosed bus are shown in Fig. 181-2.

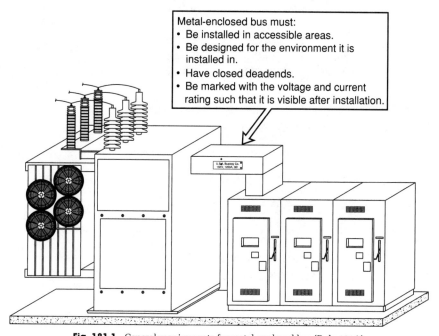

Metal-enclosed bus must:
- Be installed in accessible areas.
- Be designed for the environment it is installed in.
- Have closed deadends.
- Be marked with the voltage and current rating such that it is visible after installation.

Fig. 181-1. General requirements for metal-enclosed bus (Rule 181A).

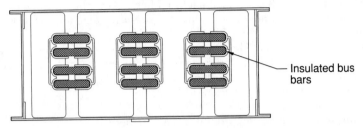

Nonsegregated phase metal-enclosed bus

Insulated bus bars

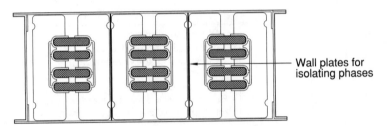

Segregated (isolated) phase metal-enclosed bus

Wall plates for isolating phases

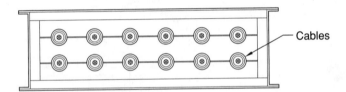

Cables

Cable type metal-enclosed bus

Fig. 181-2. Examples of metal-enclosed bus (Rule 181).

Section 19

Surge Arresters

190. GENERAL REQUIREMENTS

Surge arresters are used to protect equipment from overvoltage primarily due to switching surges and lighting. They are used throughout the electric utility system. The rules for surge arresters in Sec. 19 are for surge arresters installed in an electric supply station. Critical equipment is located in the electric supply station. The careful application of surge arresters is used to limit damage to expensive equipment such as the substation power transformer. Locating a surge arrester as close as practical to the equipment it protects limits voltage buildup across the equipment insulation. IEEE Standard C62.1 and IEEE Standard C62.11 are noted in the NESC. They are excellent references for surge arrester applications.

191. INDOOR LOCATIONS

Arresters can discharge hot gases and produce electric arcs. To avoid damage to its surroundings, if an arrester is located inside a building, it must be enclosed or located far away from passageways and combustible parts.

192. GROUNDING CONDUCTORS

Short conductive grounding leads are required to keep the voltage buildup across equipment insulation as low as possible. The grounding methods of Sec. 09 are referenced for additional information.

193. INSTALLATION

Surge arresters can be installed in various locations throughout the electric supply station. The rules regarding surge arrester installation in electric supply stations are outlined in Fig. 193-1.

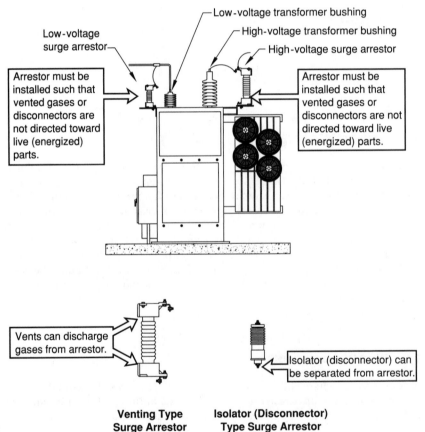

See
Photo(s)

Venting Type   **Isolator (Disconnector)**
Surge Arrestor   **Type Surge Arrestor**

Fig. 193-1. Arrester installation (Rule 193).

Safety Rules for the Installation and Maintenance of Overhead Electric Supply and Communication Lines

GENERAL SECTIONS

01 INTRODUCTION   02 DEFINITIONS

03 REFERENCES   09 GROUNDING METHODS

GENERAL SECTIONS 01, 02, 03, 09

PART 1

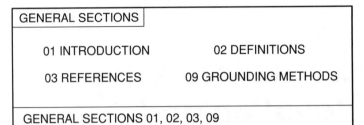

ELECTRIC SUPPLY STATIONS

PART 2

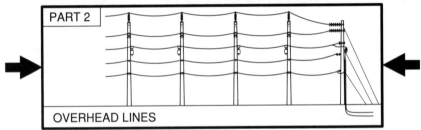

OVERHEAD LINES

PART 3

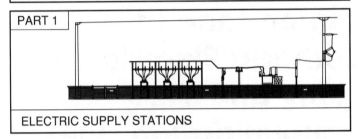

UNDERGROUND LINES

PART 4

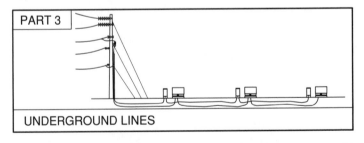

WORK RULES

Section 20

Purpose, Scope, and Application of Rules

200. PURPOSE

The purpose of Part 2, Overhead Lines, is similar to the purpose of the entire NESC outlined in Rule 010, except Rule 200 is specific to overhead supply and communication lines and equipment. Part 2 of the NESC focuses on the practical safeguarding of persons during the installation, operation, and maintenance of overhead supply (power) and communication lines and equipment.

201. SCOPE

The scope of Part 2, Overhead Lines, includes supply (power) and communication (phone, cable TV, etc.) conductors and equipment in overhead line applications. Separate rules apply to Electric Supply Stations (Part 1) and Underground Lines (Part 3). Part 2 covers overhead structural arrangements and extensions into buildings. Requirements are provided in Part 2 for:
- Spacing (center to center measurement)
- Clearances (surface to surface measurement)
- Strength of construction

There is some overlap between Overhead Lines (Part 2) and Underground Lines (Part 3). The overlap is at the underground riser on the overhead pole. Risers are covered in Sec. 23, Rule 239 of Part 2, and in Sec. 36 of Part 3. See Fig. 201-1.

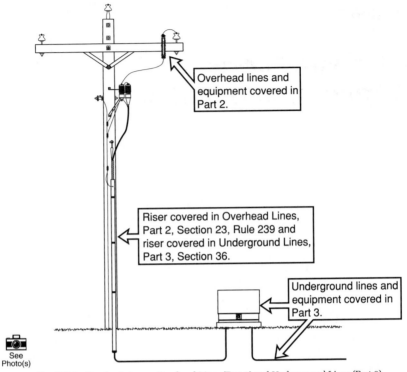

Overhead lines and
equipment covered in
Part 2.

Riser covered in Overhead Lines,
Part 2, Section 23, Rule 239 and
riser covered in Underground Lines,
Part 3, Section 36.

Underground lines and
equipment covered in
Part 3.

See
Photo(s)

Fig. 201-1. Overlap between Overhead Lines (Part 2) and Underground Lines (Part 3)
(Rule 201).

There is some overlap between Overhead Lines (Part 2) and Electric Supply
Stations (Part 1). The overlap occurs when an overhead conductor extends outside the electric supply station fence. The conductors and supports (e.g., poles,
deadend towers, etc.) extending outside the substation must meet the Part 2
requirements. Conductors and supports within the substation must meet the
Part 1 requirements. See Fig. 201-2.

When overhead supply equipment is fenced, the rules of Part 1, Electric Supply Stations, may apply. See Rule 110A for a discussion.

The notes at the end of Rule 201 remind the reader that Part 4 of the NESC and
the Occupational Health and Safety Administration (OSHA) regulations cover
the work rules related to overhead lines. See Part 4, Work Rules, for a discussion.

See Rule 011, Scope, for additional information.

202. APPLICATION OF RULES

Rule 202 references Rule 013 for the general application of Code rules. See
Rule 013 for a discussion. Rule 202 has a "however" statement that is essentially an exception to Rule 013. If a structure (e.g., pole) is being replaced,

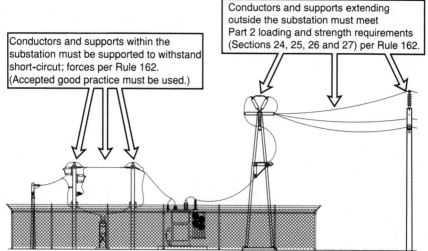

Conductors and supports within the substation must be supported to withstand short-circuit; forces per Rule 162. (Accepted good practice must be used.)

Conductors and supports extending outside the substation must meet Part 2 loading and strength requirements (Sections 24, 25, 26 and 27) per Rule 162.

See Photo(s)

Fig. 201-2. Overlap between Overhead Lines (Part 2) and Electric Supply Stations (Part 1) (Rule 201).

the arrangement of equipment must conform to the current NESC require-ments for vertical clearance between span wires or brackets carrying lumi-naires, traffic signals, or trolley conductors, and communications equipment covered in Rule 238C. No note or other explanation is provided in the Code as to why Rule 238C was singled out amongst all the overhead line rules.

Section 21

General Requirements

210. REFERENCED SECTIONS

This rule references four sections related to Part 2, Overhead Lines, so that rules do not have to be duplicated and the reader of the Code realizes that other sections are related to the information provided in Part 2. The related sections are:
- Introduction—Sec. 01
- Definitions—Sec. 02
- References—Sec. 03
- Grounding Methods—Sec. 09

The rules in Part 2, predominantly Rule 215, will provide the requirements for grounding overhead lines and equipment. The grounding methods are provided in Sec. 09.

211. NUMBER 211 NOT USED IN THIS EDITION.

212. INDUCED VOLTAGES

Rule 212 recognizes that voltages induced by the supply line may affect the communication line in some manner. Noise on the communication line due to the supply line is one example of induced voltage problems. Specific requirements to deal with induced voltage are not specified other than cooperation and advance notice. The clearances provided in Part 2 between supply and communication lines and equipment are for safety purposes. They were not developed to mitigate induced voltage.

213. ACCESSIBILITY

This rule generally states the requirement that overhead parts that need to be examined or adjusted during operation must be accessible to authorized supply and communication workers. This rule requires the following, all of which are discussed in more detail throughout Part 2.

- Climbing space
- Working space
- Working facilities
- Clearances between conductors

Climbing space is specifically discussed in Rule 236. Working space is covered in Rule 237. Clearances between conductors are covered in various rules throughout Sec. 23.

214. INSPECTION AND TESTS OF LINES AND EQUIPMENT

The rules for inspection and testing on lines and equipment are broken down into two parts, when in-service and when out-of-service. The open or closed status of a switch, fused cutout, recloser, etc., will determine if a line is in- or out-of-service. A line that is permanently abandoned must be removed or maintained in a safe condition. Opening the cutouts on a line and grounding the line may make it electrically safe but not structurally safe. If the permanently abandoned line is leaning or falling over, it must be removed or maintained in a safe condition. The rules for inspecting and testing overhead lines and equipment are outlined in Fig. 214-1.

Rule 214 applies to overhead supply and communications lines. Similar inspection rules are outlined in Rule 121 for electric supply stations and Rule 313 for underground lines.

Rule 214 does not provide a specific requirement as to how often to inspect overhead lines. The inspection frequency must be determined by the utility "at such intervals as experience has shown to be necessary." This statement should consider what the local conditions are; for example, do poles rot faster in some areas due to acidic soil conditions, do insulators fail more in some areas due to salt fog, pollution from a nearby factory, lightning, etc. A utility must have an inspection program that has proper record keeping and inspection frequency; however, the frequency of the inspection is not specified. Finding out what other utilities use for an inspection frequency and making proper adjustments may be a starting point for a utility without an inspection program.

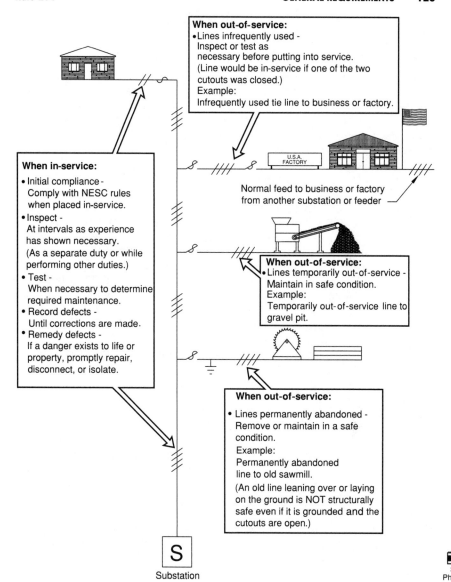

When out-of-service:
- Lines infrequently used -
Inspect or test as
necessary before putting into service.
(Line would be in-service if one of the two
cutouts was closed.)
Example:
Infrequently used tie line to business or factory.

When in-service:

- Initial compliance -
Comply with NESC rules
when placed in-service.
- Inspect -
At intervals as experience
has shown necessary.
(As a separate duty or while
performing other duties.)
- Test -
When necessary to determine
required maintenance.
- Record defects -
Until corrections are made.
- Remedy defects -
If a danger exists to life or
property, promptly repair,
disconnect, or isolate.

Normal feed to business or factory
from another substation or feeder

When out-of-service:
- Lines temporarily out-of-service -
Maintain in safe condition.
Example:
Temporarily out-of-service line to
gravel pit.

When out-of-service:

- Lines permanently abandoned -
Remove or maintain in a safe
condition.
Example:
Permanently abandoned
line to old sawmill.
(An old line leaning over or laying
on the ground is NOT structurally
safe even if it is grounded and the
cutouts are open.)

S

Substation

See
Photo(s)

Fig. 214-1. Inspection and tests of lines and equipment when in- and out-of-service (Rule 214).

A note in Rule 214A2 recognizes that inspections may be performed as a separate duty or performed while performing other duties. This statement does not eliminate the need for an inspection program or permit an "inspect it only if I see it" program. If an employee notices a defect while performing another task, the defect must be addressed, not ignored.

215. GROUNDING OF CIRCUITS, SUPPORTING STRUCTURES, AND EQUIPMENT

Rule 215 is broken into three main paragraphs. Rule 215A reminds us that the methods of grounding are specified in Sec. 09 and the requirements of grounding are provided in this rule for circuits and non-current-carrying parts. Rule 215B focuses on the requirements for grounding circuits. Rule 215C focuses on the requirements for grounding non-current-carrying parts. See Sec. 02, Definitions, for a discussion of the term "effectively grounded." The circuit grounding requirements of Rule 215B are outlined in Fig. 215-1.

Rule 215C uses the term "non-current-carrying parts." When a communication messenger is grounded and bonded to the supply neutral conductor, it can carry current. When a transformer is grounded and bonded to the supply neutral conductor, it can carry current. But, by definition of a current-carrying part in Sec. 02 of the NESC, messengers and transformer cases are not considered current-carrying parts as they are not connected to a source of voltage. Rule 314, which discusses grounding of underground circuits and equipment, uses the term "conductive parts to be grounded" instead of "non-current-carrying parts."

Guys must be effectively grounded or insulated per Rule 215C2. If guys are grounded, Rule 092C2 provides the methods for grounding of guys. If guys are insulated, Rule 215C5 provides requirements for the location of the guy insulator. The material properties and the electrical and mechanical strength of the guy insulator is specified in Rule 279A1.

The non-current-carrying parts grounding requirements of Rule 215C are outlined in Figs. 215-2 through 215-5.

216. ARRANGEMENT OF SWITCHES

This rule addresses overhead line switches. Rule 381C covers underground or pad-mounted switches and Sec. 17 covers switches in electric supply stations. Transmission and distribution line switches are the most common application of Rule 216. The Code rules related to the arrangement of overhead line switches are outlined in Fig. 216-1.

The intent of the requirements in Rule 216 is to minimize switching errors. Switching is sometimes done during poor weather conditions or low visibility. Accessibility, marking, uniform positioning, etc., help minimize switching errors.

217. GENERAL

Rule 217 provides general information regarding supporting structures (i.e., poles) except for clearance to other objects. Clearance of a supporting structure to other objects is covered in the overhead line clearance rules in Sec. 23, Rule 231. Supporting structures includes poles, lattice towers, etc. The rules related to

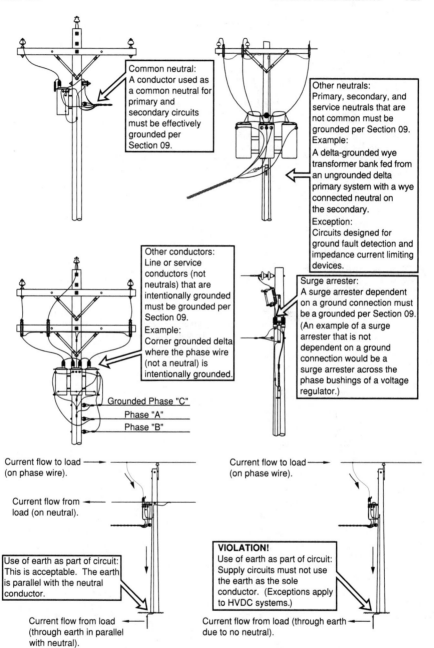

Common neutral:
A conductor used as a common neutral for primary and secondary circuits must be effectively grounded per Section 09.

Other neutrals:
Primary, secondary, and service neutrals that are not common must be grounded per Section 09.
Example:
A delta-grounded wye transformer bank fed from an ungrounded delta primary system with a wye connected neutral on the secondary.
Exception:
Circuits designed for ground fault detection and impedance current limiting devices.

Other conductors:
Line or service conductors (not neutrals) that are intentionally grounded must be grounded per Section 09.
Example:
Corner grounded delta where the phase wire (not a neutral) is intentionally grounded.

Surge arrester:
A surge arrester dependent on a ground connection must be a grounded per Section 09. (An example of a surge arrester that is not dependent on a ground connection would be a surge arrester across the phase bushings of a voltage regulator.)

Grounded Phase "C"
Phase "A"
Phase "B"

Current flow to load ⟶
(on phase wire).

Current flow to load ⟶
(on phase wire).

Current flow from ⟵
load (on neutral).

Use of earth as part of circuit:
This is acceptable. The earth is parallel with the neutral conductor.

VIOLATION!
Use of earth as part of circuit:
Supply circuits must not use the earth as the sole conductor. (Exceptions apply to HVDC systems.)

Current flow from load ⟵
(through earth in parallel with neutral).

Current flow from load (through earth ⟵
due to no neutral).

Fig. 215-1. Grounding of circuits (Rule 215B).

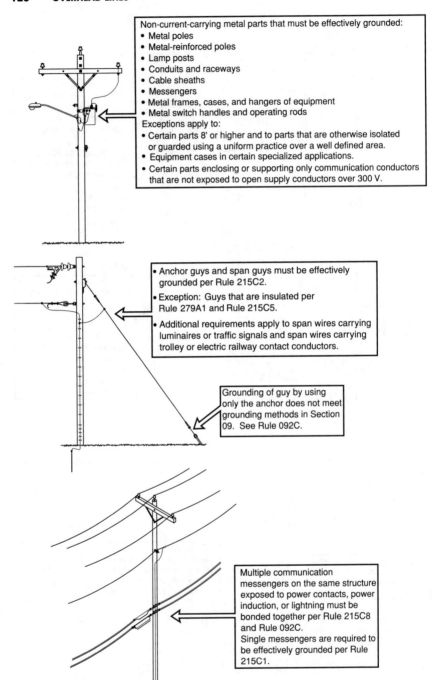

Non-current-carrying metal parts that must be effectively grounded:
• Metal poles
• Metal-reinforced poles
• Lamp posts
• Conduits and raceways
• Cable sheaths
• Messengers
• Metal frames, cases, and hangers of equipment
• Metal switch handles and operating rods
Exceptions apply to:
• Certain parts 8' or higher and to parts that are otherwise isolated
  or guarded using a uniform practice over a well defined area.
• Equipment cases in certain specialized applications.
• Certain parts enclosing or supporting only communication conductors
  that are not exposed to open supply conductors over 300 V.

• Anchor guys and span guys must be effectively
  grounded per Rule 215C2.
• Exception: Guys that are insulated per
  Rule 279A1 and Rule 215C5.
• Additional requirements apply to span wires carrying
  luminaires or traffic signals and span wires carrying
  trolley or electric railway contact conductors.

Grounding of guy by using
only the anchor does not meet
grounding methods in Section
09. See Rule 092C.

Multiple communication
messengers on the same structure
exposed to power contacts, power
induction, or lightning must be
bonded together per Rule 215C8
and Rule 092C.
Single messengers are required to
be effectively grounded per Rule
215C1.

Fig. 215-2. Non-current-carrying parts to be grounded (Rule 215C).

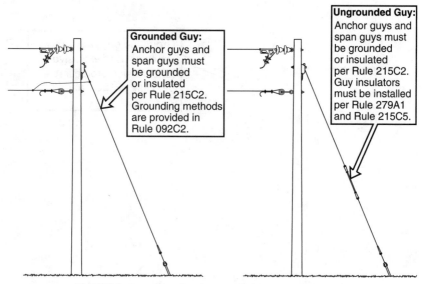

Grounded Guy:
Anchor guys and
span guys must
be grounded
or insulated
per Rule 215C2.
Grounding methods
are provided in
Rule 092C2.

Ungrounded Guy:
Anchor guys and
span guys must
be grounded
or insulated
per Rule 215C2.
Guy insulators
must be installed
per Rule 279A1
and Rule 215C5.

Fig. 215-3. Grounding of anchor guys and span guys (Rule 215C2).

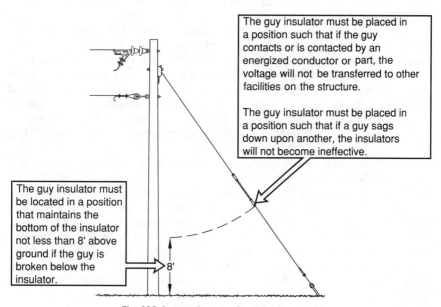

The guy insulator must be placed in
a position such that if the guy
contacts or is contacted by an
energized conductor or part, the
voltage will not be transferred to other
facilities on the structure.

The guy insulator must be placed in
a position such that if a guy sags
down upon another, the insulators
will not become ineffective.

The guy insulator must
be located in a position
that maintains the
bottom of the insulator
not less than 8' above
ground if the guy is
broken below the
insulator.

8'

Fig. 215-4. Use of guy insulators (Rule 215C5).

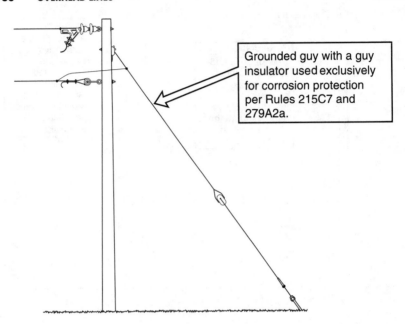

Grounded guy with a guy insulator used exclusively for corrosion protection per Rules 215C7 and 279A2a.

See Photo(s)

Fig. 215-5. Example of a guy insulator used to limit galvanic corrosion (Rule 215C7).

protection of supporting structures covered in Rule 217A1 are outlined in Fig. 217-1.

Rule 217A2 focuses on unauthorized climbing of supporting structures. The **Code** uses the term "readily climbable" in Rule 217A2a when discussing lattice towers. A definition of "supporting structure" (readily climbable and not readily climbable) is provided in Sec. 02 of the **NESC**. The **Code** uses 8-ft spacing in Rules 217A2b and 217A2c between pole steps and standoff brackets to limit climbing. The concern of readily climbable structures, steps, and standoff brackets is to keep unauthorized people from climbing the pole or structure.

The rules related to climbing of supporting structures covered in Rule 217A2 are outlined in Figs. 217-2, 217-3, and 217-4.

Rule 217A3 provides requirements for identifying supporting structures (i.e., poles) on overhead lines. Pole identification is required but not specifically limited to pole numbers. Location, construction, or marking can also be used to identify poles for employees. Identification of switch poles (or the switches themselves) is especially critical so that switching operations discussed in Part 4, Work Rules, are done correctly. Some utilities number every pole and some only number corner, tap, or other selected poles. A small rural utility can use construction and location for identification (e.g., the transformer pole at the Smith Ranch). The method used to identify poles should consider the size and complexity of the utility's system. Rule 411A2 in Part 4, Work Rules, requires that diagrams (i.e., maps) be maintained on file for employees to use.

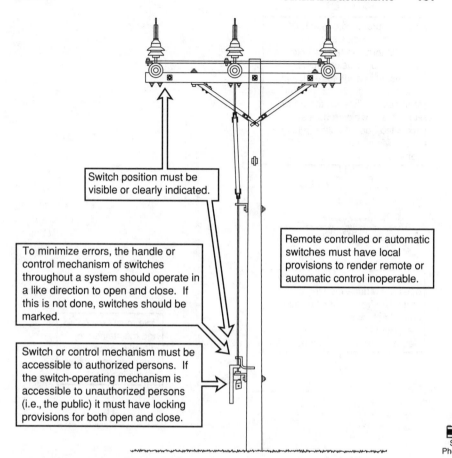

Switch position must be visible or clearly indicated.

To minimize errors, the handle or control mechanism of switches throughout a system should operate in a like direction to open and close. If this is not done, switches should be marked.

Remote controlled or automatic switches must have local provisions to render remote or automatic control inoperable.

Switch or control mechanism must be accessible to authorized persons. If the switch-operating mechanism is accessible to unauthorized persons (i.e., the public) it must have locking provisions for both open and close.

See Photo(s)

Fig. 216-1. Arrangement of overhead line switches (Rule 216).

The rules related to identification of supporting structures covered in Rule 217A3 are outlined in Fig. 217-5.

Obstructions can produce a climbing hazard such as a cut or fall. The rules related to obstructions on supporting structures covered in Rule 217A4 are outlined in Fig. 217-6.

Decorative lighting can produce climbing hazards, clearance problems, and pole strength problems when wind load is considered. Decorative lighting must meet approval of the owner and occupants on the pole.

The rules related to decorative lighting on supporting structures covered in Rule 217A5 are outlined in Fig. 217-7.

Rule 217B discusses unusual conductor supports. The NESC does not prohibit unusual conductor supports. It does require applying Code rules and additional precautions to assure safety. The Code specifically comments on trees and roofs. The Code states that attaching overhead conductors to trees or roofs "should

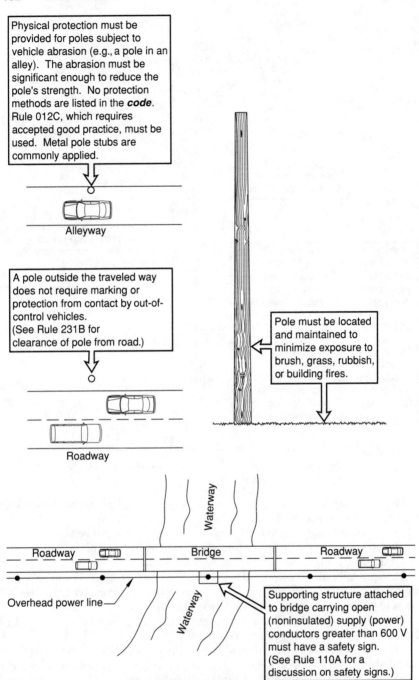

Physical protection must be provided for poles subject to vehicle abrasion (e.g., a pole in an alley). The abrasion must be significant enough to reduce the pole's strength. No protection methods are listed in the *code*. Rule 012C, which requires accepted good practice, must be used. Metal pole stubs are commonly applied.

Alleyway

A pole outside the traveled way does not require marking or protection from contact by out-of-control vehicles.
(See Rule 231B for clearance of pole from road.)

Roadway

Pole must be located and maintained to minimize exposure to brush, grass, rubbish, or building fires.

Waterway

Roadway   Bridge   Roadway

Overhead power line

Waterway

Supporting structure attached to bridge carrying open (noninsulated) supply (power) conductors greater than 600 V must have a safety sign.
(See Rule 110A for a discussion on safety signs.)

Fig. 217-1. Protection of supporting structures (Rule 217A1).

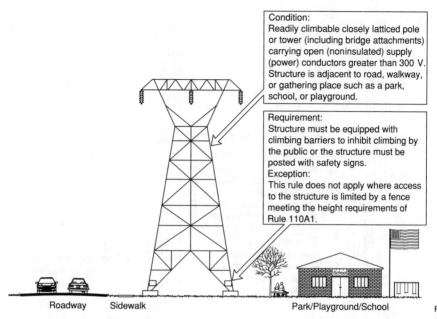

Condition:
Readily climbable closely latticed pole or tower (including bridge attachments) carrying open (noninsulated) supply (power) conductors greater than 300 V. Structure is adjacent to road, walkway, or gathering place such as a park, school, or playground.

Requirement:
Structure must be equipped with climbing barriers to inhibit climbing by the public or the structure must be posted with safety signs.
Exception:
This rule does not apply where access to the structure is limited by a fence meeting the height requirements of Rule 110A1.

Roadway   Sidewalk   Park/Playground/School   See Photo(s)

Fig. 217-2. Readily climbable supporting structures (Rule 217A2a).

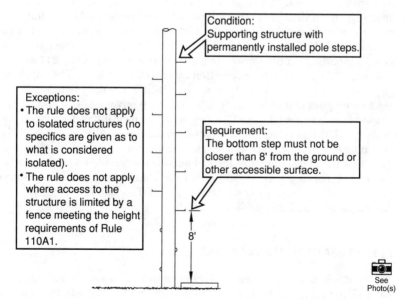

Condition:
Supporting structure with permanently installed pole steps.

Exceptions:
• The rule does not apply to isolated structures (no specifics are given as to what is considered isolated).
• The rule does not apply where access to the structure is limited by a fence meeting the height requirements of Rule 110A1.

Requirement:
The bottom step must not be closer than 8' from the ground or other accessible surface.

8'

See Photo(s)

Fig. 217-3. Permanently mounted steps on supporting structures (Rule 217A2b).

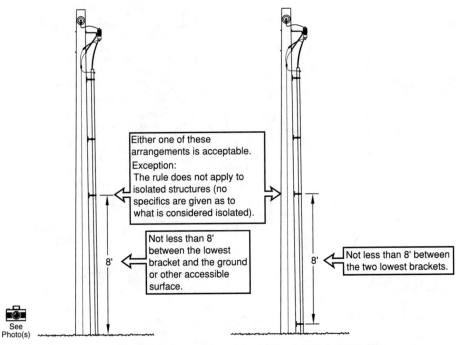

Either one of these arrangements is acceptable.
Exception:
The rule does not apply to isolated structures (no specifics are given as to what is considered isolated).

Not less than 8' between the lowest bracket and the ground or other accessible surface.

8'

8'

Not less than 8' between the two lowest brackets.

See Photo(s)

Fig. 217-4. Arrangement of standoff brackets (Rule 217A2c).

be avoided." The **Code** does not use stronger language like "shall not be attached." Even though "should" is used instead of "shall," everything possible must be done to avoid attaching conductors to trees or roofs. See Rule 015 for a discussion of the intent of the words "should" and "shall." The requirement to avoid supporting conductors on trees and roofs is shown in Fig. 217-8.

Rule 217C applies to the protection and marking of guys. The application of guy markers varies from utility to utility. Utilities are only required to use guy markers on guys exposed to pedestrian traffic. A guy marker or protection must be used for guys in established parking areas. Some utilities use a guy marker on every guy but this is not a **Code** requirement. Some utilities use bold fluorescent multicolored markers; others use conspicuous single color markers. Rule 217C3 clearly states that guys located outside the traveled way of a roadway or established parking areas do not require marking. Similar wording appears in Rule 217A1a for supporting structures (poles). The **Code** rules for guy markers and protection are outlined in Fig. 217-9.

218. VEGETATION MANAGEMENT

The **NESC** does not provide a specific clearance from a conductor (of any voltage) to vegetation (e.g., trees). Figure 218-1 shows the items that must be considered.

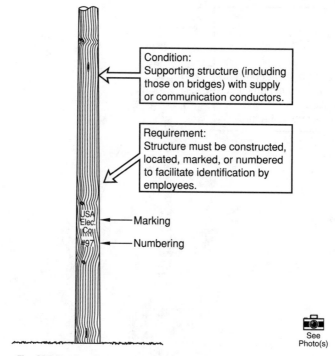

Fig. 217-5. Identification of supporting structures (Rule 217A3).

Right-of-way tree clearing and trimming for overhead lines can be an ongoing project in heavy tree areas. Workers who trim trees in the vicinity of energized conductors must be qualified to do so.

The **Code** recognizes that at times tree trimming or removal may not be practical. If tree trimming or removal is not practical, the **Code** requires using methods to separate the conductor from the tree to avoid damage by abrasion and grounding. This requirement is very general. No specifics are provided as to how to accomplish the separation. A supply utility should review the available options before assuming that tree trimming or removal is not practical. Unprotected conductors through trees can start forest fires and kill or severely injure anyone climbing the tree.

Line crossings, railroad crossings, and limited access highway crossings receive special considerations to minimize downed lines. Tree trimming for these areas should consider decaying trees or limbs and trees that are overhanging the line. See Fig. 218-2.

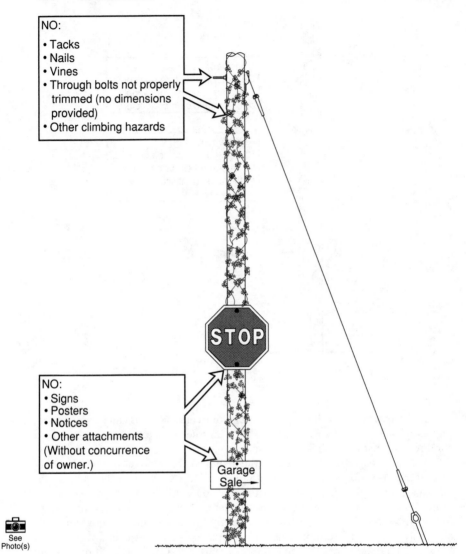

Fig. 217-6. Obstructions on supporting structures (Rule 217A4).

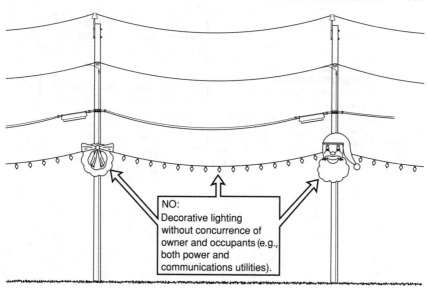

Fig. 217-7. Decorative lighting on supporting structures (Rule 217A5).

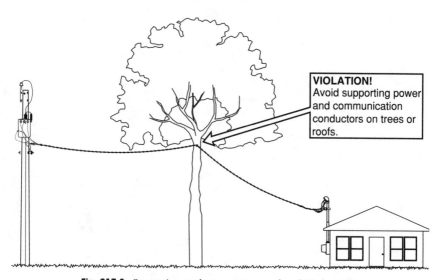

Fig. 217-8. Supporting conductors on trees and roofs (Rule 217B).

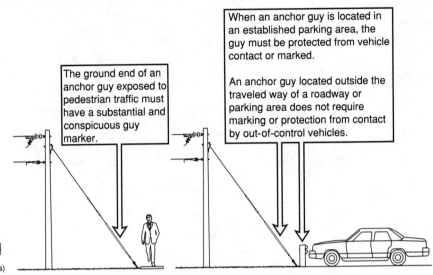

The ground end of an anchor guy exposed to pedestrian traffic must have a substantial and conspicuous guy marker.

When an anchor guy is located in an established parking area, the guy must be protected from vehicle contact or marked.

An anchor guy located outside the traveled way of a roadway or parking area does not require marking or protection from contact by out-of-control vehicles.

See Photo(s)

Fig. 217-9. Protection and marking of guys (Rule 217C).

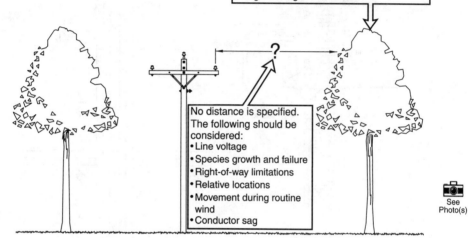

Vegetation that may damage ungrounded supply conductors should be pruned or removed. (Grounded neutral conductor and communication conductors are not addressed.)

Vegetation management should be performed as experience has shown necessary.

Where pruning or removal is not practical, the conductor should be separated from the tree with suitable materials or devices to avoid abrasion and grounding of the circuit.

?

No distance is specified. The following should be considered:
• Line voltage
• Species growth and failure
• Right-of-way limitations
• Relative locations
• Movement during routine wind
• Conductor sag

See Photo(s)

Fig. 218-1. General vegetation management requirements (Rule 218A).

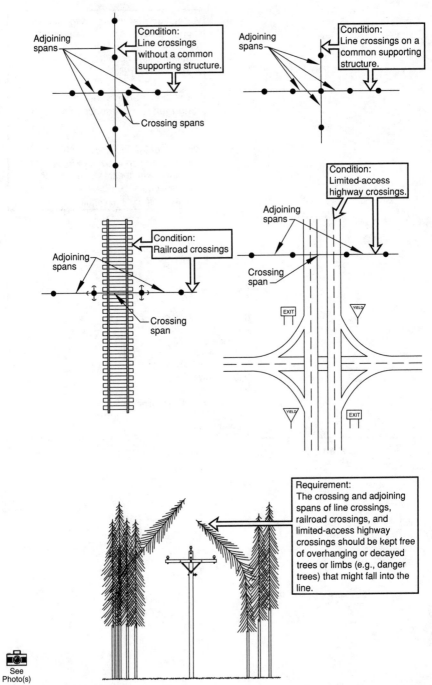

Fig. 218-2. Vegetation management at line, railroad, and limited-access highway crossings (Rule 218B).

Relations between Various Classes of Lines and Equipment

220. RELATIVE LEVELS

Rule 220A requires that the levels of different classes of conductors should be standardized by agreement of the utilities concerned. A joint use agreement between supply and communications utilities is one type of agreement that can be used to standardize utility locations on a pole. Rule 220B1 states that it is preferred that supply (power) conductors be located above communications conductors at crossings and on the same structures. See Fig. 220-1.

Rule 220B2, Special Construction for Supply Circuits, the Voltage of Which Is 600 Volts or Less and Carrying Power Not in Excess of 5 kW, is a rule for special construction related to railroad signal circuits. There are seven conditions (paragraphs a through g) that must be met to apply this rule. Rule 220B2 does not apply to modern cable television and telephone circuits.

Rule 220C discusses where to position supply (power) lines of different voltages on overhead structures. The terms "crossings" and "conflicts" are used in Rule 220C1. Crossings are discussed in detail in Rule 241C. Conflicts are discussed in Rule 221 and a definition of "structure conflict" is provided in Sec. 02 of the NESC. Relative levels of supply lines at crossings and conflicts are shown in Figs. 220-2 and 220-3.

Rule 220C2 has requirements for structures used only for supply (power) conductors. Rule 220C2a covers structures with circuits owned by one utility and Rule 220C2b covers structures with circuits owned by separate supply utilities. See Figs. 220-4 and 220-5.

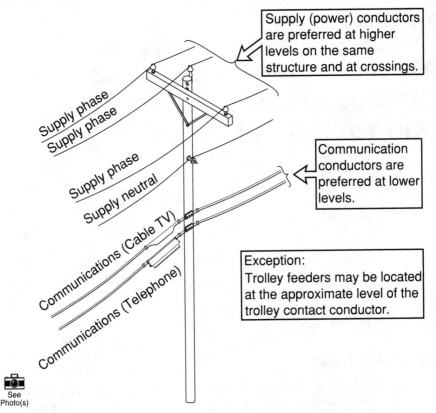

Supply (power) conductors are preferred at higher levels on the same structure and at crossings.

Supply phase
Supply phase
Supply phase
Supply neutral
Communications (Cable TV)
Communications (Telephone)

Communication conductors are preferred at lower levels.

Exception:
Trolley feeders may be located at the approximate level of the trolley contact conductor.

See Photo(s)

Fig. 220-1. Preferred supply and communication conductor levels (Rule 220B1).

Positioning a higher voltage line above a lower voltage line is a common construction practice. Clearance above ground is greater for higher voltage circuits, so having them at higher positions permits greater clearance. Rule 220C states the relative levels of conductors, not the clearance between them. NESC Table 235-5 is referenced in Rule 220C as this table provides vertical clearances between conductors on a common supporting structure. Lower voltage circuits are typically worked on more than higher voltage circuits, so having them at lower positions makes them easier to access.

Rules 220D and 220E require uniform positions of supply and communication conductors and equipment or constructing, locating, marking, or numbering to facilitate identification by authorized employees who have to work on them. This rule is similar to Rule 217A3, Identification of Supporting Structures. Identifying overhead conductors by attachment to distinctive insulators or crossarms is also acceptable. Using uniform positions of conductors does not prohibit changing locations systematically.

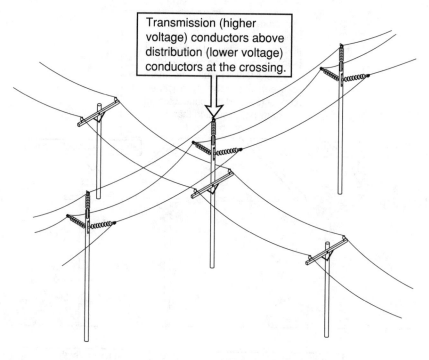

Transmission (higher voltage) conductors above distribution (lower voltage) conductors at the crossing.

See Photo(s)

Fig. 220-2. Relative levels of supply lines of different voltages at crossings (Rule 220C1).

A neutral conductor, when on a crossarm with the phase conductors, is commonly identified with a different color or style insulator or by labeling the crossarm with a letter "N" below the neutral insulator. A pole with crossarms owned by multiple utilities can have the crossarms labeled (i.e., marked or numbered) with the utility name, or the crossarms can be constructed or located such that the employees authorized to work on them can recognize them as their own. Similarly, a pole with multiple communication cable attachments can have the communication cables labeled (i.e., marked or numbered) with the utility name, or the cables can be constructed or located such that the employees authorized to work on them can recognize them as their own.

221. AVOIDANCE OF CONFLICT

The term "conflict" was introduced in Rule 220C. Avoidance of conflict can be accomplished by sufficient separation of lines, or by structure strength, or by combining the lines on the same structure. See Fig. 221-1.

Many times, right-of-way constraints will prohibit two separate lines and collinear or joint use construction will be required. The term "collinear construction" applies to two or more lines on the same structure. The term "joint

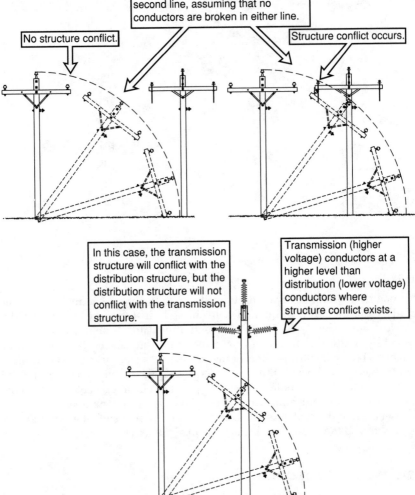

Structure conflict is defined in Section 02. Structure conflict exists if a line is so situated with respect to a second line that the overturning of the first line will result in contact between its supporting structures or conductors and the conductors of the second line, assuming that no conductors are broken in either line.

No structure conflict.

Structure conflict occurs.

In this case, the transmission structure will conflict with the distribution structure, but the distribution structure will not conflict with the transmission structure.

Transmission (higher voltage) conductors at a higher level than distribution (lower voltage) conductors where structure conflict exists.

See Photo(s)

Fig. 220-3. Relative levels of supply lines of different voltages at structure conflict locations (Rule 220C1).

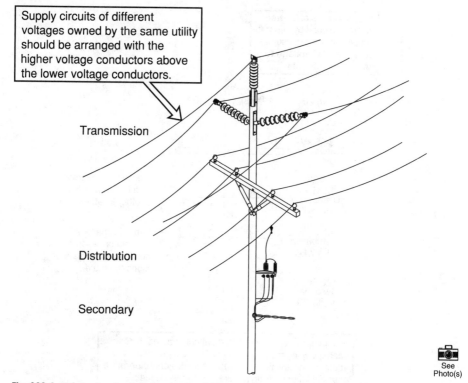

Supply circuits of different voltages owned by the same utility should be arranged with the higher voltage conductors above the lower voltage conductors.

Transmission

Distribution

Secondary

See Photo(s)

Fig. 220-4. Relative levels of supply circuits of different voltages owned by one utility (Rule 220C2a).

use construction," per the definitions in Sec. 02 of the **NESC**, applies to two or more kinds of utilities on the same structure (i.e., power and communication).

222. JOINT USE OF STRUCTURES

Per the definition of "joint use" in Sec. 02 of the **NESC**, joint use refers to two or more kinds of utilities on the same structure (i.e., power and communication). The **NESC** encourages the use of joint use construction along highways, roads, streets, and alleys. See Fig. 222-1.

To decide between joint use and separate lines along highways, roads, streets, and alleys, the following must be considered by the power and communication utilities considering the joint occupancy:

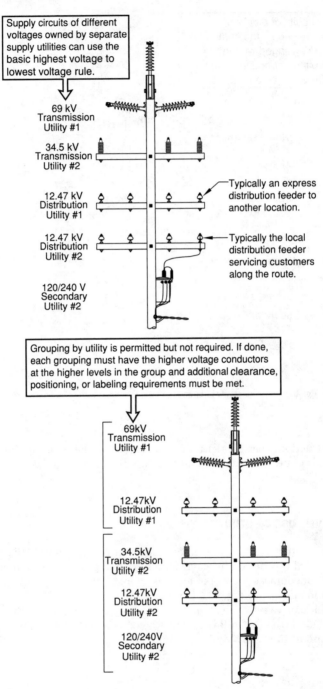

Supply circuits of different voltages owned by separate supply utilities can use the basic highest voltage to lowest voltage rule.

69 kV
Transmission
Utility #1

34.5 kV
Transmission
Utility #2

12.47 kV
Distribution
Utility #1

Typically an express distribution feeder to another location.

12.47 kV
Distribution
Utility #2

Typically the local distribution feeder servicing customers along the route.

120/240 V
Secondary
Utility #2

Grouping by utility is permitted but not required. If done, each grouping must have the higher voltage conductors at the higher levels in the group and additional clearance, positioning, or labeling requirements must be met.

69kV
Transmission
Utility #1

12.47kV
Distribution
Utility #1

34.5kV
Transmission
Utility #2

12.47kV
Distribution
Utility #2

120/240V
Secondary
Utility #2

Fig. 220-5. Relative levels of supply circuits of different voltages owned by separate utilities (Rule 220C2b).

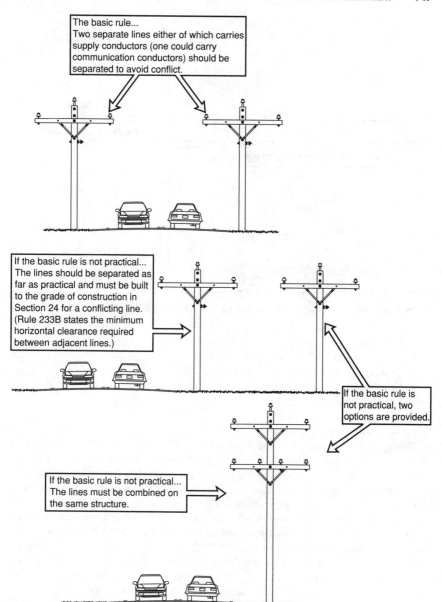

The basic rule...
Two separate lines either of which carries supply conductors (one could carry communication conductors) should be separated to avoid conflict.

If the basic rule is not practical...
The lines should be separated as far as practical and must be built to the grade of construction in Section 24 for a conflicting line. (Rule 233B states the minimum horizontal clearance required between adjacent lines.)

If the basic rule is not practical, two options are provided.

If the basic rule is not practical...
The lines must be combined on the same structure.

See Photo(s)

Fig. 221-1. Avoiding conflict between two separate lines (Rule 221).

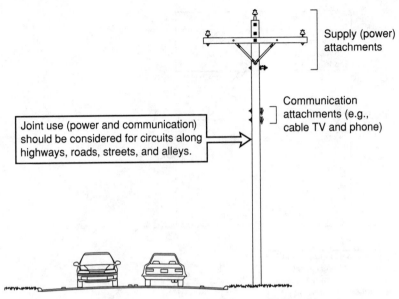

Fig. 222-1. Joint use (power and communication) of structures (Rule 222).

- Character of circuits
- Total number and weight of conductors
- Tree conditions
- Number of branches (taps)
- Number of service drops
- Structure conflict
- Availability of right of way

When joint use is used, it must meet the grade of construction specified in Sec. 24. A joint use agreement is appropriate when joint use construction is used. Joint use agreements typically include wording regarding an attachment fee that is paid by the utility that is attaching to a pole owned by a different utility. Joint use agreements also commonly include references to **NESC** requirements, methods to address payment for structures that must be replaced to meet joint use requirements, and attachment locations. Some supply utilities reserve the top 10 or 12 ft of a distribution pole for supply attachments. This method assures supply space will be available for future distribution service transformers, primary taps, secondary services, etc. Other supply utilities permit communication attachments as long as space is available. This method requires that the communications utility vacate the pole or pay for a pole upgrade if distribution service transformers, primary taps, secondary services, etc. need to be added in the future and space is not available for these items. Similar approaches may be applied to underground ducts. Federal Communications Commission (FCC) or state regulations may specify pole attachment or underground duct procedures and rental fee methodologies. A joint use agreement should be used for collecting pole rental fees, resolving make-ready issues, and assuring joint use attachments are in accordance with the NESC.

223. COMMUNICATIONS PROTECTIVE REQUIREMENTS

Rule 223 requires that a communication apparatus that is subjected to lightning, contact with supply conductors exceeding 300 V to ground, a ground potential rise greater than 300 V, or a steady-state induced voltage of a hazardous level be protected by insulation and, where necessary, surge arresters in conjunction with fusible elements. Additional communication protective devices are also listed for severe conditions.

A typical joint use (power and communication) overhead distribution line is subjected to the conditions listed in Rule 223A. The most common method used for the means of protection required in Rule 223B is insulating the communication conductors (in the form of a communication cable) and grounding the communication messenger and bonding it to the grounded supply neutral.

This measure must be taken to satisfy Footnote 7 of NESC Table 242-1 and to meet the messenger grounding requirements in Rules 215C1 and 215C8.

The additional communication protective devices for severe conditions are typically applied to a communication line entering an electric supply station. The electric supply station typically has large fault current duties, which can severely damage a metallic communication cable. Commonly, isolation equipment or a fiber-optic communication cable is used to serve substations to mitigate this concern.

224. COMMUNICATION CIRCUITS LOCATED WITHIN THE SUPPLY SPACE AND SUPPLY CIRCUITS LOCATED WITHIN THE COMMUNICATION SPACE

A communication circuit may be operated by a supply utility and used for communicating between supply stations or used as a line of business to provide Internet, data, television, or telephone services. Rule 224A applies to communication circuits located in the supply space.

If a communication circuit is located in the supply space (not the communication space), it must be installed and maintained by an employee qualified to work in the supply space per the work rules (Secs. 42 and 44) of the NESC. If a communication circuit is located in the communication space (not in the supply space), it can be installed and maintained by an employee qualified to work in the communication space per the work rules (Secs. 42 and 43) of the NESC.

Rule 224A2 states that an insulated communication cable supported by an effectively grounded messenger and located in the supply space must have the same clearance as neutrals meeting Rule 230E1 from communication circuits in the communication space and from supply conductors in the supply space.

This requirement can be applied to **NESC** Table 235-5, Footnote 5 and Footnote 9. See Rules 235C and 238 for examples and additional information. Fiber-optic cables located in the supply space must meet Rule 230F.

Examples of communication cables located in the supply space and the communication space are shown in Fig. 224-1.

Rule 224A3 provides conditions for locating a communication circuit in the supply space in one portion of the system and in the communication space in

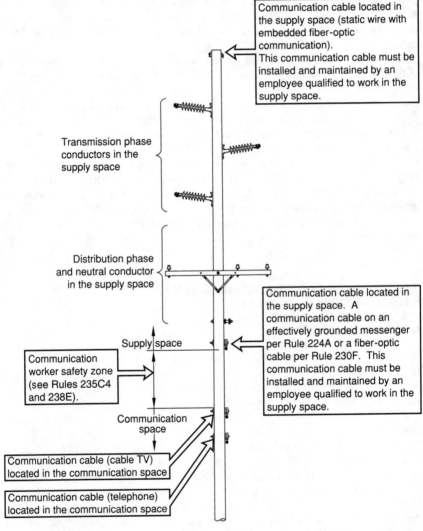

Fig. 224-1. Examples of communication cables located in the supply space and the communication space (Rule 224A).

another portion of the system. The transition of the communication cable from the supply space to the communication space must occur on a single structure. See Fig. 224-2.

Rule 224B applies to special supply circuits used exclusively in the operation of communication circuits. Rule 224B1 applies to open wire (noninsulated) circuits. Rule 224B2 applies to a communication cable with a supply circuit embedded in it.

225. ELECTRIC RAILWAY CONSTRUCTION

Electric railways can be in the form of electric locomotives on railroad tracks or electric trolleys on streetcar tracks. Rule 225 provides general information related to these types of installations. The NESC addresses clearances of electric railway conductors in Rule 225 and throughout Sec. 23. Size, strength, and loading issues related to electric railway and trolley contact conductors are addressed in Rule 261H3 and throughout Secs. 24, 25, 26, and 27.

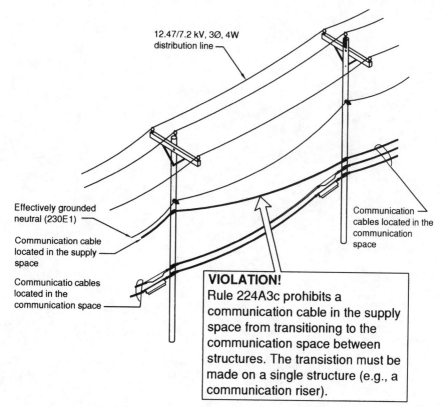

Fig. 224-2. Example of a communication cable transition from the supply space to the communication space (Rule 224A3c).

Section 23

Clearances

Section 23 is probably the most referenced section in the **NESC**. Rule 232, Vertical Clearance of Wires, Conductors, Cables, and Equipment above Ground, Roadway, Rail, or Water Surfaces, is probably the most referenced rule. Chances are that if a person only uses the **NESC** once a year, he or she will be using it to find a clearance value somewhere in this section. When an overhead line clearance question comes up, one of the first problems a person has is determining which rule applies to the **Code** clearance question at hand. Figure 23-1 outlines each rule in Sec. 23 with an icon providing a graphical representation of what the rule covers and a column indicating if a sag chart is needed to determine clearance values.

Many of the clearances found in the **NESC** tables in this section are based on conductor sags. The **NESC** specifies maximum sags, minimum sags, sags with wind, sags with ice, sags at specified temperatures, initial sags, and final sags. It is very important that the person using the **Code** to determine overhead line clearances first understands how to use a sag and tension chart. Important concepts that the user of a sag and tension chart must understand are outlined below:

- Sag and tension charts are created for a conductor type (e.g., 1/0 ACSR) at a ruling span (e.g., 300 ft) in a loading district (e.g., Medium Loading) and clearance zone (e.g., Zone 2). This information typically appears at the heading of the sag chart.
- A sag and tension chart describes what position a conductor will be in at various temperature and physical loading conditions. Examples of physical loading conditions are ice and wind.
- A ruling span is the span that "rules" or "governs" the behavior of all the spans between two conductor deadends.

INFO	RULE 230	General	Sag chart needed to check clearance? ☐YES ☒NO
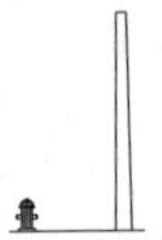	RULE 231	Clearances of supporting structures from other objects	☐YES ☒NO
	RULE 232	Vertical clearances of wires, conductors, cables, and equipment above ground, roadway, rail, or water surfaces	☒YES ☐NO
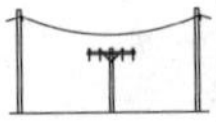	RULE 233	Clearances between wires, conductors, and cables carried on different supporting structures	☒YES ☐NO
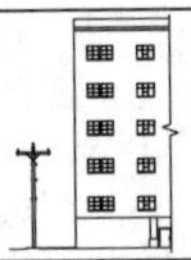	RULE 234	Clearances of wires, conductors, cables, and equipment from buildings, bridges, rail cars, swimming pools, and other installations	☒YES ☐NO
	RULE 235	Clearance for wires, conductors, or cables carried on the same supporting structure	☒YES ☐NO
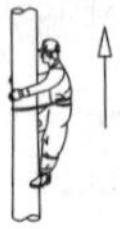	RULE 236	Climbing space	☐YES ☒NO
	RULE 237	Working space	☐YES ☒NO
	RULE 238	Vertical clearance between certain communications and supply facilities located on the same structure	☐YES ☒NO
	RULE 239	Clearance of vertical and lateral facilities from other facilities and surfaces on the same supporting structure	☐YES ☒NO

Fig. 23-1. Summary of overhead line clearance rules (Sec. 23).

- A ruling span is calculated between deadends by using the following formula:

$$RS = \sqrt{\frac{S_1^3 + S_2^3 + S_3^3 + \cdots + S_n^3}{S_1 + S_2 + S_3 + \cdots + S_n}}$$

Where RS = Ruling Span
$S_1, S_2, S_3 \cdots S_n$ = 1st, 2nd, 3rd $\cdots$ nth span lengths between the conductor deadends.

- The sags on a sag and tension chart are for a span length equal to the ruling span. To estimate sags for spans longer or shorter than the ruling span, the following formula can be used:

$$\text{Sag in feet} = \left(\frac{\text{span length}}{\text{ruling span length}}\right)^2 \times \text{ruling span sag in feet}$$

- The temperature on a sag chart is the conductor temperature, not the ambient air temperature. The conductor temperature is based on the ambient air temperature, the cooling effect of the wind, the radiant heating effect of the sun, and the amount of electrical current flowing through the conductor. IEEE Standard 738 provides methods to calculate the current-temperature relationship of bare overhead conductors. Below are approximate examples for a 1/0 ACSR conductor:
  - ✓ 1/0 ACSR bare conductor has an approximate current-carrying capacity (ampacity) of 230 A. The 230 A is based on a conductor temperature of 167°F.
  - ✓ 1/0 ACSR bare conductor:
    Assume the conductor carries a small electrical load (e.g., 11.5 A, which is approximately 5 percent of the 230-A ampacity).
    During summer conditions: Assume a 104°F ambient temperature combined with the heating effect of the electrical current yields a 110°F conductor temperature.
    During winter conditions: Assume a −20°F ambient temperature combined with the heating effect of the electrical current yields a −15°F conductor temperature.
    The maximum sag conditions per Rule 232A would be:
    - 32°F with ice (e.g., 0.25 in of ice for Zone 2)
    - 120°F (since the maximum operating temperature from above is less than 120°F)
  - ✓ 1/0 ACSR bare conductor:
    Assume the conductor carries a large electrical load (e.g., 175 A, which is approximately 75 percent of the 230-A ampacity).
    During summer conditions: Assume a 104°F ambient temperature combined with the heating effect of the electrical current yields a 167°F conductor temperature.
    During winter conditions: Assume a −20°F ambient temperature combined with the heating effect of the electrical current yields a 32°F conductor temperature.

The maximum sag conditions per Rule 232A would be:

- 32°F with ice (e.g., 0.25 in of ice for Zone 2)
- 120°F
- 167°F (since the maximum operating temperature from above is greater than 120°F)

✓ 1/0 ACSR bare conductor:

Assume the conductor carries an emergency electrical load (e.g., 230 A, which is 100 percent of the 230-A ampacity).

During summer conditions: Assume a 104°F ambient temperature combined with the heating effect of the electrical current yields a 212°F conductor temperature.

During winter conditions: Assume a −20°F ambient temperature combined with the heating effect of the electrical current yields a 60°F conductor temperature.

The maximum sag conditions per Rule 232A would be:

- 32°F with ice (e.g., 0.25 in of ice for Zone 2)
- 120°F
- 212°F (since the maximum operating temperature from above is greater than 120°F)

- It is common to see conductor temperatures of 104°F, 167°F, and 212°F on a sag chart as they represent 40°C, 75°C, and 100°C, respectively.
- Electrical loads can peak in the summer or winter or both. Areas with a large amount of electric heating load tend to peak in the winter. Areas with a large amount of air conditioning load tend to peak in the summer.
- Sag and tension charts provide both initial and final sags and tensions. Initial values apply to the day a new conductor is strung up. Final values can occur anywhere in time after that point. For example, if the conductor is exposed to an ice storm the first week it is installed, the conductor could be at final sag. If the conductor was never exposed to ice and was exposed to very little wind, it would take many years of hanging under its own weight for the conductor to reach final sag.
- The maximum sag on a sag chart typically occurs at one of the following conditions:

  ✓ 32°F with ice (e.g., 0.25 in of ice for Zone 2), final
  ✓ 120°F, final
  ✓ Greater than 120°F, final (e.g., 167°F, 212°F, etc.)

Larger sags may be observed at the following conditions, but these conditions are not required for checking vertical clearance:

  ✓ 15°F with heavy, medium, or light loading (e.g., 0.25 in of ice, 4-lb/ft² wind, 0.20 K factor)

  - The heavy, medium, and light conditions (Rule 250B) are also referred to as the Zone 1, 2, and 3 conditions (Rule 230B).
  - One of the Zone 1, 2, and 3 conditions (e.g., Zone 2, 15°F, 0.25 in of ice, 4 lb/ft² wind, 0.20 K factor) must be applied to the conductor for physical loading purposes before the final sag is checked at 32°F with ice (e.g., 0.25 in of ice for Zone 2), 120°F, and greater than 120°F if so designed. Note that sag is not required to be checked at 15°F, 0.25 in

of ice, 4 lb/ft² wind, 0.20 K factor. The heavy, medium, or light (Zone 1, 2, or 3) conditions must be applied to the conductor for physical loading purposes, but sag is checked at other conditions.

✓ 60°F with a 6-lb/ft² wind (this condition is required for conductor horizontal blowout, not vertical sag)

✓ 60°F with extreme wind (e.g., 17.76 lb/ft²) (this condition is required for conductor tension limits, not vertical sag)

✓ 15°F with extreme ice with concurrent wind (e.g., 0.25 in of ice with 2.30 lb/ft²) (this condition is required for conductor tension limits, not vertical sag)

- The minimum sag on a sag chart usually occurs at the initial sag of the coldest temperature without ice or wind loading (e.g., −20°F, initial).

- The maximum tension must be checked at the heavy, medium, or light loading condition (Rule 250B), the extreme wind condition (Rule 250C), if applicable, and the extreme ice with concurrent wind condition (Rule 250D), if applicable. It is possible for the maximum tension to occur at the minimum temperature condition.

- The minimum tension usually occurs at the maximum sag condition due to high temperatures at final tension.

- The sag and tensions on a sag chart can be varied by increasing tension which will decrease sag or by decreasing tension which will increase sag. Too much tension is not good for guying and structure strength and loading issues. Too much sag is not good for clearance issues. A balance must be reached between the two.

- The sag values on a sag and tension chart are used to calculate clearance. The tensions on a sag and tension chart are used to calculate strength and loading of structures and conductor supports.

- The following headings and abbreviations are used on a typical sag and tension chart:

  ✓ Conductor: Raven (this is the nickname for the conductor).

  ✓ 1/0 AWG, 6/1 stranding, ACSR: 1/0 AWG (American Wire Gauge) is the conductor size. ACSR stands for Aluminum Conductor, Steel Reinforced. The 6/1 represents six strands of aluminum twisted around one strand of steel.

  ✓ Area = cross-sectional area of the conductor

  ✓ Dia = diameter of the conductor

  ✓ Wt = weight of the conductor

  ✓ RTS = rated tensile strength of the conductor

  ✓ Span = ruling span

  ✓ The design condition is the limiting design condition entered into the program. It may be an **NESC** tension limit, a conductor manufacturer tension limit, or a user-defined sag or tension limit.

- **NESC**-defined tension limits (percentages) must be entered into the sag and tension program.

- User-defined sag and tension limits can be entered into a sag and tension program in addition to **NESC** design conditions. An example of a user-defined tension limit is 20 percent tension at the initial condition at the average annual minimum temperature (e.g., 20 percent tension at 0°F, initial). This limit is used to control Aeolian vibration. Another example of a user-defined limit is specifying a conductor sag for a lower circuit to match the sag of a higher circuit. The sags of two different conductor sizes or types cannot perfectly match throughout the entire temperature range. Matching sags at 60°F at the final tension typically produces the closest matching sags throughout the entire temperature range. Other examples of user-defined sag and tension limits are 3 ft of sag at 60°F final, a sag of 1 percent of the ruling span at 60°F final, a 2000-lb maximum tension limit, etc. The conductor or cable manufacturer may also specify a sag or tension limit. When a user-defined tension is entered into a sag and tension program, the NESC design conditions must also be entered to verify that none of the NESC design conditions are exceeded.

- The phase and neutral conductor of the same circuit can carry different amounts of current and therefore operate at different temperatures. The neutral conductor may carry less current due to phase balance and neutral current cancellation or due to the fact that the earth and a communication messenger on a joint use pole are in parallel with a multigrounded neutral.

- If a line is existing, a sag chart can be created for the existing line by measuring the sag at a known conductor temperature and span length. This information is then entered into a program to create a sag chart. Another method is to find the 3rd or 5th return wave for a known conductor temperature and span length. The return wave time can be converted to sag using the following formula:

$$\text{Sag in feet} = 4.025\left(\frac{\text{time in seconds}}{2 \times \text{number of return waves}}\right)^2$$

In both methods described above, the line must be assumed to be in its initial or final sag condition. The return wave method is not accurate when used on deadend spans as the deadend insulator string absorbs the conductor wave and distorts the wave timing.

- Sag and tension charts must be run for secondary and communication conductors and cables to accurately check clearance and strength issues for these circuits.

Required sag and tension values in Rules 230, 232, 233, 234, 235, 250B, 250C, 250D, 251, and 261H are noted in the sample sag and tension chart provided in Fig. 23-2.

The sag and tension chart provides a sag value at the midspan or center of the span. There are times when sag needs to be checked somewhere else in the span, for example, if a rise in the ground line occurs around the quarter span instead of the midspan. Another example is if two lines cross in the span, one at 10% of the span distance, and one at 33% of the span distance. A graph can be used to estimate the percentage of midspan sag at various distances along the span length. See Fig. 23-3.

Sample Sag and Tension Chart

Conductor RAVEN   #1/0 AWG 6/ 1 Stranding ACSR

Area=   .0968 Sq. In   Dia= .398 In Wt= .145 Lb/F RTS= 4380 Lb
Span=   300.0 Feet (NESC Medium Load Zone/NESC Clearance Zone 2)

	Design Points					Final			Initial		
	Temp F	Ice In	Wind Psf	K Lb/F	Weight Lb/F	Sag Ft	Tension Lb	RTS %	Sag Ft	Tension Lb	RTS %
Note 1	15.	.25	4.00	.20	.658	5.43	1366.	31.2	5.43	1366.	31.2
Note 2	15.	.25	2.30	.00	.387	4.31	1012.	23.1	4.07	1070.	24.4
Note 3	32.	.25	.00	.00	.346	4.49	868.	19.8	4.13	945.	21.6
Note 4	60.	.00	17.76	.00	.607	6.14	1114.	25.4	5.92	1154.	26.4
Note 5	60.	.00	6.00	.00	.246	4.66	595.	13.6	4.03	688.	15.7
Note 6	-20.	.00	.00	.00	.145	1.78	916.	20.9	1.61	1015.	23.2
	-10.	.00	.00	.00	.145	1.98	825.	18.8	1.73	945.	21.6
Note 11	0.	.00	.00	.00	.145	2.20	740.	16.9	1.86	876.	20.0*
	10.	.00	.00	.00	.145	2.46	663.	15.1	2.02	808.	18.5
	20.	.00	.00	.00	.145	2.74	595.	13.6	2.20	743.	17.0
	30.	.00	.00	.00	.145	3.04	536.	12.2	2.40	680.	15.5
	32.	.00	.00	.00	.145	3.11	526.	12.0	2.44	668.	15.3
	40.	.00	.00	.00	.145	3.36	486.	11.1	2.62	622.	14.2
	50.	.00	.00	.00	.145	3.68	444.	10.1	2.87	569.	13.0
Note 7	60.	.00	.00	.00	.145	4.00	408.	9.3	3.14	520.	11.9
	70.	.00	.00	.00	.145	4.32	378.	8.6	3.42	478.	10.9
	80.	.00	.00	.00	.145	4.63	353.	8.1	3.70	441.	10.1
	90.	.00	.00	.00	.145	4.93	331.	7.6	4.00	408.	9.3
	100.	.00	.00	.00	.145	5.22	313.	7.1	4.29	380.	8.7
	104.	.00	.00	.00	.145	5.34	306.	7.0	4.41	370.	8.5
	110.	.00	.00	.00	.145	5.47	299.	6.8	4.58	356.	8.1
Note 8	120.	.00	.00	.00	.145	5.61	291.	6.7	4.87	335.	7.7
Note 9	167.	.00	.00	.00	.145	6.25	262.	6.0	6.14	266.	6.1
Note 10	212.	.00	.00	.00	.145	6.84	239.	5.5	6.76	242.	5.5

* Design Condition

Note 1:
15°F, 0.25" ice, 4 lb/ft^2 wind, 0.20 lb/ft K factor.
This condition (Medium Loading) is specified in Rules 250B and 251.
This condition (Zone 2) is specified in Rule 230B.
The tension from this line, 1366 lb initial, is used for
guying and other structure strength and loading calculations.
This tension must be compared to the tension for extreme wind loading (Rule 250C), if
applicable, and the extreme ice with concurrent wind loading (Rule 250D), if applicable,
to determine the maximum design tension. The maximum tension on the chart is commonly
referred to as the "design tension".
The percent rated tensile strength from this line, 31.2%, is used in Rule 261H1a.
Rule 261H1a specifies a 60% limit.

Note 2:
15°F, 0.25" ice, 2.30 lb/ft^2 wind.
This condition is specified in Rule 250D.
A 2.30 lb/ft^2 wind corresponds to a 30 mph wind in west-central Washington
State. 0.25" of ice is also required at this location.
The tension from this line, 1070 lb initial, is used for guying and
other structure strength and loading calculations when Rule 250D applies
(for structures and supported facilities over 60' above ground).
The percent rated tensile strength from this line, 24.4%, is used in Rule 260B2.
Rule 260B2 specifies an 80% limit. This tension must be compared
to the tension for heavy, medium, and light loading (Rule 250B) and the extreme
wind loading (Rule 250C), if applicable, to determine the maximum design tension.

Fig. 23-2. Sample sag and tension chart (Sec. 23).

Note 3:
32°F, 0.25" ice.
This condition is specified in Rules 232A, 233A, 234A, and 235C2b(1)(c)(ii).
The sag from this line, 4.49' final, is used for clearance calculations.
This sag must be compared to the sag at 120°F, final, and greater than 120°F, final, if so designed, to determine the largest final sag.

Note 4:
60°F, 17.76 lb/ft^2 wind.
This condition is specified in Rule 250C. A 17.76 lb/ft^2 wind corresponds to an 85 mph wind in central Washington State.
The tension from this line, 1154 lb initial, is used for guying and other structure strength and loading calculations when Rule 250C applies (for structures and supported facilities over 60' above ground). The percent rated tensile strength from this line, 26.4%, is used in Rule 260B2. Rule 260B2 specifies an 80% limit. This tension must be compared to the tension for heavy, medium, and light loading (Rule 250B) and the extreme ice with concurrent wind loading (Rule 250D), if applicable, to determine the maximum design tension.

Note 5:
60°F, 6 lb/ft^2 wind.
This condition is specified in Rules 233A and 234A.
The sags from this line, 4.66' final and 4.03' initial, are used for horizontal clearance calculations.

Note 6:
–20°F.
This condition is required in Rule 234A.
The sag from this line, 1.61' initial, is used for vertical clearance under signs or buildings.

Note 7:
60°F.
This condition is required in Rules 233A, 235B, and 261H1b.
The sags from this line, 4.00' final and 3.14' initial, are used for vertical and horizontal clearance calculations. The percent rated tensile strengths from this line, 9.3% final, and 11.9% initial, are used in Rule 261H1b. Rule 261H1b specifies a 35% initial tension limit and a 25% final tension limit.

Note 8:
120°F.
This condition is specified in Rules 232A, 233A, 234A, and 235C2b(1)(c)(i).
The sag from this line, 5.61' final, is used for clearance calculations.
The sag must be compared to the sag at 32°F, 0.25" ice, final, and greater than 120°F, final, if so designed, to determine the largest final sag.

Fig. 23-2 (*Continued*). Sample sag and tension chart (Sec. 23).

Note 9:
167°F.
This condition is specified in Rules 232A, 233A, 234A, and 235C2b(1)(c)(i).
The sag from this line, 6.25' final, is used for clearance calculations.
This condition is only required if the line is designed to operate at a temperature greater than 120°F.

Note 10:
212°F.
This condition is specified in Rules 232A, 233A, 234A, and 235C2b(1)(c)(i).
The sag from this line, 6.82' final, is used for clearance calculations.
This condition is only required if the line is designed to operate at a temperature greater than 120°F.

Note 11:
0°F initial, 20% tension limit
This condition is not specified in the **Code**. It is a user-defined condition. For example, a 20% tension limit at the average minimum cold temperature may help prevent aeolian vibration. (See the note in Rule 261H1b.) Other user-defined conditions can also be entered into a sag and tension program. The program will determine the most limiting condition of all the conditions entered. In this case the 20% tension limit at 0°F initial was the most limiting condition and the program flagged this value as the " * Design Condition."

Fig. 23-2 (*Continued*). Sample sag and tension chart (Sec. 23).

230. GENERAL

230A. Application. Rule 230A provides an introduction to Sec. 23. Sec. 23 covers clearances for overhead supply and communication lines. Burial depths for underground lines are covered in Part 3 of the NESC and clearances in electric supply stations are covered in Part 1 of the NESC.

Rule 230A provides a note regarding the development of clearance values and makes reference to Appendix A of the NESC, which outlines how NESC clearances are calculated starting with the 1990 edition of the NESC. Prior to 1990, clearance was calculated using a 60°F sag condition with a clearance adder for long spans instead of using a maximum sag condition. Therefore, clearance values prior to 1990 cannot be directly compared to today's clearances. The examples and discussion presented in this Handbook will all focus on using today's **Code** and today's clearance calculations. Reading and understanding prior methods may be needed if a person is trying to apply the "grandfather" requirements of Rule 013.

Temporary clearances are required to always be the same as permanent clearances. In other words, no clearance reductions exist for a temporary situation. This rule cannot be understated and is conveyed in Fig. 230-1.

Temporary installations are permitted to have lower grades of construction. See Rules 014 and 241.

Emergency installations do permit some slight decreases in clearance if certain conditions are met. Rule 014 is referenced as it provides general waiver information. Rule 230A2 provides the specifics, the most common of which are outlined in Fig. 230-2.

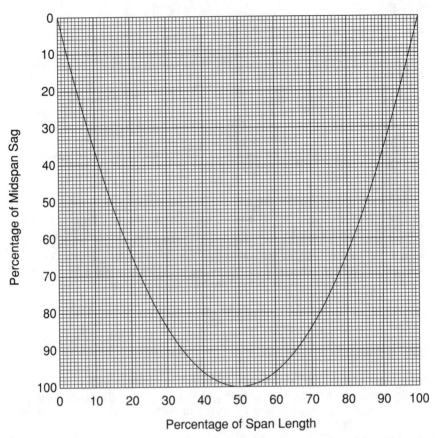

Fig. 23-3. Estimate of the percentage of midspan sag at various distances along the span length (Sec. 23).

Emergency installations permit laying certain supply (power) and communication cables directly on the ground. This same permission can be found in Rule 311C. An example of supply and communication cables laid on the ground is shown in Fig. 230-3.

Where access is limited to qualified personnel only, as inside a properly fenced electric substation, no emergency clearance is specified.

The general requirements in Rule 014 require emergency installations to be removed, replaced, or relocated as soon as practical.

Rule 230A3 describes how clearance and spacing are measured. See Fig. 230-4.

Throughout Sec. 23 the term clearance is used more often than spacing.

Rule 230A3 also states how to classify line hardware. See Fig. 230-5.

Rule 230A4 provides rounding requirements specific to the clearance calculations in Sec. 23. In general, Rule 018 permits rounding "off" to the nearest significant digit unless otherwise specified in applicable rules. One rule where other rounding is specified is Rule 230A4 which requires rounding "up" as Sec. 23 deals with overhead line clearances which are typically specified as "not less

TEMPORARY CLEARANCE = PERMANENT CLEARANCE

Fig. 230-1. Relationship between temporary and permanent clearance (Rule 230A1).

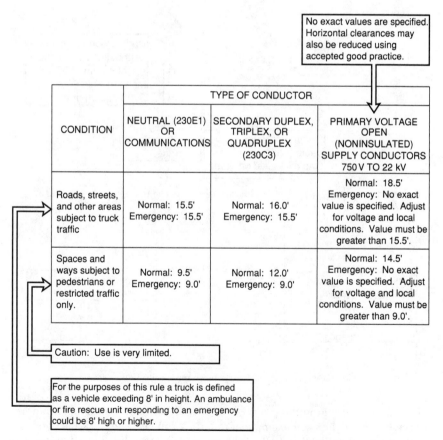

No exact values are specified. Horizontal clearances may also be reduced using accepted good practice.

CONDITION	TYPE OF CONDUCTOR		
	NEUTRAL (230E1) OR COMMUNICATIONS	SECONDARY DUPLEX, TRIPLEX, OR QUADRUPLEX (230C3)	PRIMARY VOLTAGE OPEN (NONINSULATED) SUPPLY CONDUCTORS 750 V TO 22 kV
Roads, streets, and other areas subject to truck traffic	Normal: 15.5' Emergency: 15.5'	Normal: 16.0' Emergency: 15.5'	Normal: 18.5' Emergency: No exact value is specified. Adjust for voltage and local conditions. Value must be greater than 15.5'.
Spaces and ways subject to pedestrians or restricted traffic only.	Normal: 9.5' Emergency: 9.0'	Normal: 12.0' Emergency: 9.0'	Normal: 14.5' Emergency: No exact value is specified. Adjust for voltage and local conditions. Value must be greater than 9.0'.

Caution: Use is very limited.

For the purposes of this rule a truck is defined as a vehicle exceeding 8' in height. An ambulance or fire rescue unit responding to an emergency could be 8' high or higher.

Fig. 230-2. Reductions of overhead clearances for emergency conditions (Rule 230A2).

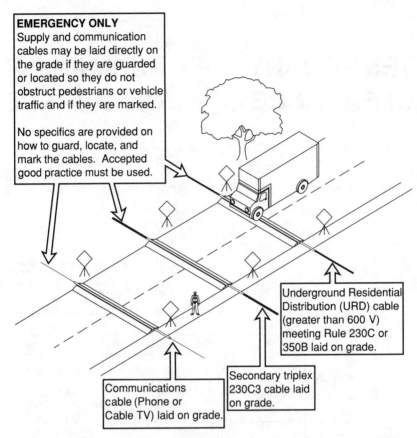

EMERGENCY ONLY
Supply and communication cables may be laid directly on the grade if they are guarded or located so they do not obstruct pedestrians or vehicle traffic and if they are marked.

No specifics are provided on how to guard, locate, and mark the cables. Accepted good practice must be used.

Underground Residential Distribution (URD) cable (greater than 600 V) meeting Rule 230C or 350B laid on grade.

Secondary triplex 230C3 cable laid on grade.

Communications cable (Phone or Cable TV) laid on grade.

Fig. 230-3. Example of cables laid on grade for emergency conditions (Rules 230A2d and 311C).

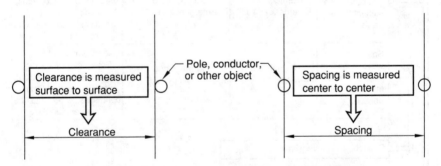

Clearance is measured surface to surface

Pole, conductor, or other object

Spacing is measured center to center

Clearance

Spacing

Fig. 230-4. Measurement of clearance and spacing (Rule 230A3).

than" clearances. Rounding "off" follows the rules of traditional rounding learned in math class. An example of rounding "off" is rounding 20.02 down to 20.0 or rounding 20.66 up to 20.7. An example of rounding "up" for "not less than" clearance is rounding 20.02 to 20.1 because rounding "off" to 20.0 would

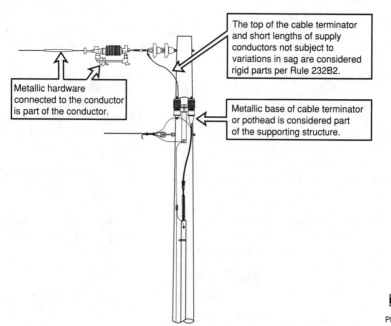

The top of the cable terminator and short lengths of supply conductors not subject to variations in sag are considered rigid parts per Rule 232B2.

Metallic hardware connected to the conductor is part of the conductor.

Metallic base of cable terminator or pothead is considered part of the supporting structure.

See Photo(s)

Fig. 230-5. Clearance measurements for parts of conductors and parts of supporting structure (Rule 230A3).

not meet a clearance required to be "not less than" 20.02. The number of decimal places for the final rounded value depends on the number of decimal places for the original starting value from the **Code** table or rule. An exception is provided for millimeters which are rounded to increments of 5mm. For example, the clearance values in **NESC** Table 232-1 have one decimal place (e.g., 18.5 ft). Therefore, a calculated clearance for a voltage greater than 22 kV per Rule 232C would be rounded "up" to the tenths place (e.g., 20.02 would be rounded up to 20.1).

230B. Ice and Wind Loading for Clearances. Rule 230B specifies ice and wind loading for clearance purposes **NESC** Fig. 230-1 and **NESC** Tables 230-1 and 230-2 are the three basic references to determine ice and wind loading for Zone 1, Zone 2, and Zone 3. **NESC** Fig. 230-1 is nearly identical to **NESC** Fig. 250-1. The only difference is Zone 1 in **NESC** Fig. 230-1 is labeled "Heavy" in **NESC** Fig. 250-1, Zone 2 is labeled "Medium," and Zone 3 is labeled "Light." The values specified in **NESC** Tables 230-1 and 230-2 are comparable to the values specified in **NESC** Tables 250-1 and 251-1. The values are identical; however, the presentation format varies slightly. The requirements in Sec. 25 (**NESC** Fig. 250-1, and **NESC** Tables 250-1 and 251-1) are for determining conductor tension and calculating strength and loading issues. The requirements in Sec. 23 (**NESC** Fig. 230-1, and **NESC** Tables 230-1 and 230-2) are for determining conductor sag and therefore calculating clearance. The requirements in **NESC** Table 230-2 (e.g., 15°F, 0.25 in of ice, 4 lb/ft² wind, 0.20 K factor) must be applied to the conductor for physical loading purposes before the final sag is checked at 32°F with ice (e.g., 0.25 in of ice for Zone 2), 120°F, and greater than 120°F if so designed (see Rule 232A). Note that sag is not required to be checked at 15°F, 0.25 in of ice, 4 lb/ft² wind, 0.20 K factor. The Zone 1, 2, or 3 condition from Table 230-2 must be applied to the conductor for physical loading purposes, but sag is checked at the ice condition in **NESC** Table 230-1. Rule 250C (extreme wind) and 250D (extreme ice with concurrent wind) can impose greater loads (and therefore more sag) than the loads in Zones 1, 2, and 3. However, only the loads of Zones 1, 2, and 3 are required for clearance purposes. Rule 230I, Maintenance of Clearances and Spacings, would require that conductors be resagged if an excessive ice or wind storm stretched conductors to the point of a clearance violation. Rules 230B3 and 230B4 are similar to Rules 251A and 251B. See Rules 251A and 251B for a discussion and related figures. The requirements for Zones 1, 2, and 3 are outlined in Fig. 230-6.

230C. Supply Cables. The terms 230C1, 230C2, and 230C3 cables will be used over and over throughout the rules of Sec. 23 and throughout the clearance tables in Sec. 23. The most common of these three cables used in construction today is the 230C3 cable, which is an overhead secondary duplex, triplex, or quadruplex cable. The rules defining the construction of 230C1, 230C2, and 230C3 supply cables are outlined in Fig. 230-7.

Clearance Zone	Ice and Wind Condition at a Specified Temperature (Clearance Loading)	Location
Zone 1	1/2" of radial ice Conductor at 0°F 4 lb/ft² wind pressure on conductor with ice Ice weight: 57 lb/ft³ Constant: 0.30 lb/ft	North Central and Northeastern portions of the United States and Alaska. See NESC Figure 230-1.
Zone 2	1/4" of radial ice Conductor at 15°F 4 lb/ft² wind pressure on conductor with ice Ice weight: 57 lb/ft³ Constant: 0.20 lb/ft	Northwestern, Mid Central, and Mid Eastern portions of the United States. See NESC Figure 230-1.
Zone 3	Conductor at 30°F 9 lb/ft² wind pressure on conductor without ice no ice Constant: 0.05 lb/ft	Southwestern, South Central, Southeastern portions of the United States and Hawaii. See NESC Figure 230-1.

Note: Values shown above are for "clearance loading" in Section 23. Clearance measurement conditions in Section 23 are covered in Rules 232A, 233A, 234A, 235B, and 235C2b(1)(c).

Fig. 230-6. Zone 1, 2, and 3 ice and wind loading for clearances (Rule 230B).

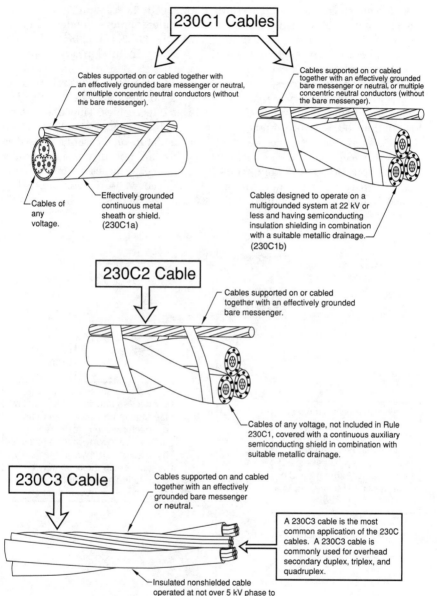

230C1 Cables

Cables supported on or cabled together with an effectively grounded bare messenger or neutral, or multiple concentric neutral conductors (without the bare messenger).

Cables supported on or cabled together with an effectively grounded bare messenger or neutral, or multiple concentric neutral conductors (without the bare messenger).

Cables of any voltage.

Effectively grounded continuous metal sheath or shield. (230C1a)

Cables designed to operate on a multigrounded system at 22 kV or less and having semiconducting insulation shielding in combination with a suitable metallic drainage. (230C1b)

230C2 Cable

Cables supported on or cabled together with an effectively grounded bare messenger.

Cables of any voltage, not included in Rule 230C1, covered with a continuous auxiliary semiconducting shield in combination with suitable metallic drainage.

230C3 Cable

Cables supported on and cabled together with an effectively grounded bare messenger or neutral.

A 230C3 cable is the most common application of the 230C cables. A 230C3 cable is commonly used for overhead secondary duplex, triplex, and quadruplex.

Insulated nonshielded cable operated at not over 5 kV phase to phase or 2.9 kV phase to ground.

See Photo(s)

Fig. 230-7. Construction of 230C1, 230C2, and 230C3 supply cables (Rule 230C).

A bare messenger or neutral is required for 230C1, 230C2, and 230C3 cables. An insulated neutral is more common for underground secondary duplex, triplex, and quadruplex construction. The 230C1, 230C2, and 230C3 cables are relying on the effectively grounded bare messenger or neutral to carry fault current if an insulation failure occurs.

230D. Covered Conductors.   The covered conductors Rule 230D refers to are commonly called tree wire. Tree wire cables are not fully insulated like underground residential distribution (URD) cables or 230C cables. They are covered to limit the likelihood of a short circuit in case of momentary contact with a tree limb. Covered conductors can be attached to insulators on crossarms or used as part of a spacer cable system. Since 230D cables are not fully insulated, they must be considered bare conductors for clearance purposes except that the clearance between the conductors may be reduced when the conductors are owned, operated, or maintained by the same utility. See Fig. 230-8.

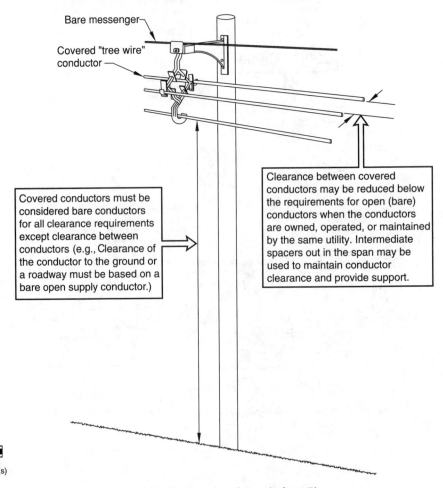

Bare messenger

Covered "tree wire" conductor

Clearance between covered conductors may be reduced below the requirements for open (bare) conductors when the conductors are owned, operated, or maintained by the same utility. Intermediate spacers out in the span may be used to maintain conductor clearance and provide support.

Covered conductors must be considered bare conductors for all clearance requirements except clearance between conductors (e.g., Clearance of the conductor to the ground or a roadway must be based on a bare open supply conductor.)

See Photo(s)

Fig. 230-8. Covered conductors (Rule 230D).

230E. Neutral Conductors. The phrase "neutral conductors meeting 230E1" will be used over and over throughout the rules of Sec. 23 and throughout the clearance tables in Sec. 23. See Sec. 02, Definitions, for a discussion of the term "effectively grounded." The rules for a 230E1 neutral are outlined in Fig. 230-9.

230F. Fiber-Optic Cable. Fiber-optic supply cable and fiber-optic communication cable are NESC terms to categorize where a fiber-optic cable is located. Examples of communication cables located in the supply space and in the communication space are discussed in Rule 224. Rule 224A2a applies to insulated communication circuits supported on an effectively grounded messenger. Rule 224A2b references Rule 230F for fiber-optic cable requirements.

Fiber-optic cables that are strung on overhead pole lines are supported on a messenger or they are contained in an All-Dielectric Self-Supporting (ADSS) cable. Fiber-optic cables can also be embedded in a metallic messenger or static wire.

Fiber-optic cables that are located in the supply space and embedded in a messenger or conductor must have the same clearance from communication facilities as the messenger or conductor the fiber-optic cables are embedded in. Fiber-optic cables that are supported on an effectively grounded messenger or contained in an all-dielectric self-supporting cable are required to have the same clearance as neutrals meeting Rule 230E1 from communication facilities. This requirement can be applied to NESC Table 235-5, Footnotes 5 and 9. See Rules 235C and 238 for examples and additional information.

Fiber-optic cables located in the communication space, whether or not supported by a messenger, must have the clearance from supply facilities as required for a communication messenger.

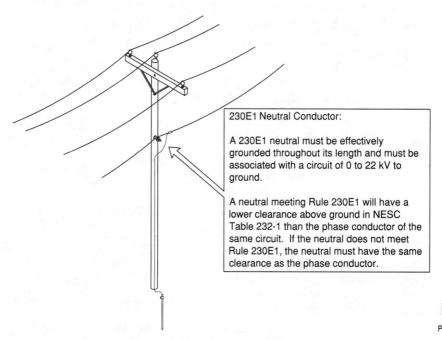

230E1 Neutral Conductor:

A 230E1 neutral must be effectively grounded throughout its length and must be associated with a circuit of 0 to 22 kV to ground.

A neutral meeting Rule 230E1 will have a lower clearance above ground in NESC Table 232-1 than the phase conductor of the same circuit. If the neutral does not meet Rule 230E1, the neutral must have the same clearance as the phase conductor.

See Photo(s)

Fig. 230-9. Neutral conductors (Rule 230E).

230G. Alternating- and Direct-Current Circuits.   The clearances in Sec. 23 apply to AC and DC circuits. The voltage used in a typical American home is 120-V alternating current at 60 Hz (60 cycles per second). The 120-V value is a root mean square (rms) measurement. Another term used for the rms voltage is the effective voltage. If an oscilloscope were plugged into a home receptacle, the scope would show that the top of the sine wave for the 120-V circuit actually crests at $120 \times \sqrt{2}$ or 169.7 V. This same discussion is true for 7.2 kV, 115 kV, 500 kV, etc. The clearance tables in Sec. 23 apply to the rms voltage (e.g., 120 V) not the crest voltage (e.g., 169.7 V). The NESC does use the crest value when applying the formula for alternate clearances in Rule 232D and in other alternate clearance calculations in Sec. 23.

DC, AC crest, and AC rms values are outlined in Fig. 230-10.

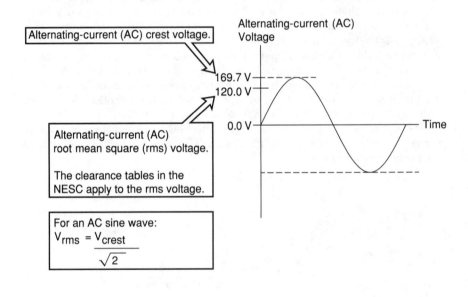

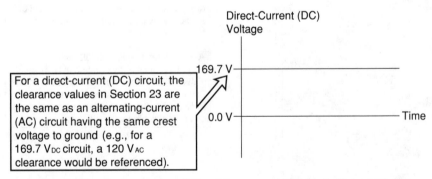

Fig. 230-10. Alternating-current (AC) and direct-current (DC) circuits (Rule 230G).

For three-phase circuits, it is important to note the difference between phase-to-phase and phase-to-ground voltage. The phase-to-ground voltage is equal to the phase-to-phase voltage divided by the square root of 3. An example of phase-to-phase and phase-to-ground voltages is shown in Fig. 230-11.

Many of the clearance tables in Sec. 23 use phase-to-ground values; however, some tables that involve voltage between conductors will use phase-to-phase values.

230H. Constant-Current Circuits.   Normal utility circuits are constant voltage and the current varies with the load connected to the circuit. Constant-current circuits are just the opposite. They operate with a constant current and the voltage varies with the load connected to the circuit. Constant-current circuits were once very common for street lighting and can still be found in use today. The voltage used to determine the clearance of constant-current circuit is the normal full-load voltage of the circuit. See Fig. 230-12.

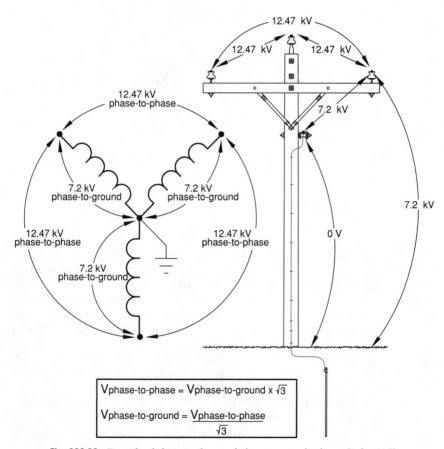

$$V_{\text{phase-to-phase}} = V_{\text{phase-to-ground}} \times \sqrt{3}$$

$$V_{\text{phase-to-ground}} = \frac{V_{\text{phase-to-phase}}}{\sqrt{3}}$$

Fig. 230-11. Example of phase-to-phase and phase-to-ground voltages (Rule 230G).

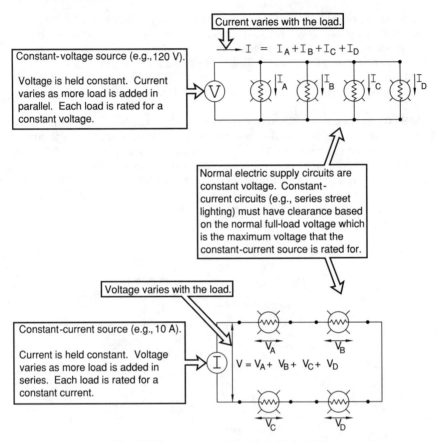

Current varies with the load.

$$I = I_A + I_B + I_C + I_D$$

Constant-voltage source (e.g.,120 V).

Voltage is held constant. Current varies as more load is added in parallel. Each load is rated for a constant voltage.

Normal electric supply circuits are constant voltage. Constant-current circuits (e.g., series street lighting) must have clearance based on the normal full-load voltage which is the maximum voltage that the constant-current source is rated for.

Voltage varies with the load.

Constant-current source (e.g., 10 A).

Current is held constant. Voltage varies as more load is added in series. Each load is rated for a constant current.

$$V = V_A + V_B + V_C + V_D$$

Fig. 230-12. Constant-current circuits (Rule 230H).

230I. Maintenance of Clearances and Spacings.   This is a small rule with a big implication. Clearances and spacing must be maintained. Forever. Per Rule 010, the NESC applies during installation, operation, and maintenance of supply and communication lines, not just during the initial installation. Several changes can occur that force a utility to be active in maintaining clearance. One is a change in land use under the power line. Two is a change in structures under or adjacent to the power line. Another is excessive sag due to a major ice- or wind-storm. In each of these cases it is the responsibility of the utility to maintain clearance. Examples of maintenance of clearances and spacings are shown in Fig. 230-13.

In the case of the excessive sag due to a major ice or wind storm, Rule 230I recognizes that clearance cannot be maintained during or after an abnormal event of this nature. There is some reasonable time period for storm damage

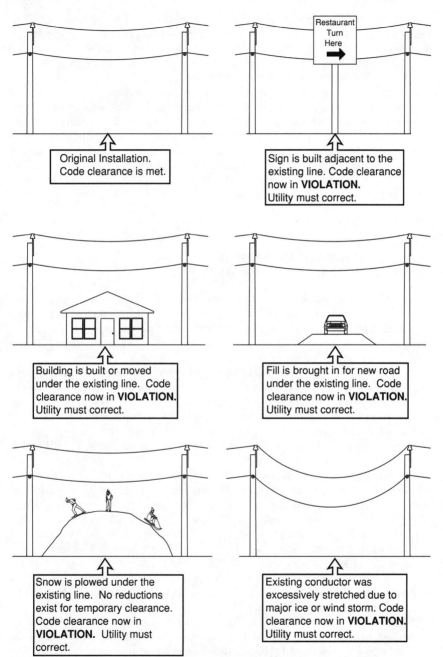

Fig. 230-13. Examples of maintenance of clearances and spacings (Rule 230I).

work to be repaired; however, the NESC does not address the time period. This same discussion applies to an abnormal event such as a car hitting a power pole and knocking down the wires or reducing the clearance of the wires. In this instance, clearance cannot be maintained during or sometime after the accident; however, time frames are not addressed.

A utility must correct a Code clearance violation, even if the line was originally built with the proper Code clearance. The NESC does not address who pays for the correction. The utility may consider billing a third party for the correction, but the utility is responsible for maintaining clearance or, in other words, fixing the problem. Some utilities establish clearance values by using the Code clearance plus an adder. See Rule 010 for a discussion of clearance adders. The line inspection requirements of Rule 214 should bring clearance issues to the utility's attention.

231. CLEARANCES OF SUPPORTING STRUCTURES FROM OTHER OBJECTS

Rule 231 applies to supporting structures. The most common supporting structure is a pole but since other structures exist (e.g., lattice towers) the NESC uses the term supporting structure, not pole. Rule 231 does not apply to conductors, only to the supporting structure. A simple title for this section could be "Where can I set my pole?"

Included with the supporting structure are the support arms, anchor guys, equipment, and braces. The clearance requirements of this section are between the nearest parts of the objects concerned.

231A. From Fire Hydrants. Rule 231A is needed to provide the fire department crews adequate space to connect hoses and equipment to the fire hydrant. The rule for clearance of a supporting structure from a fire hydrant is outlined in Fig. 231-1.

231B. From Streets, Roads, and Highways. Rule 231B provides clearances between supporting structures (poles) and roads. Avoiding vehicle contact with poles is just as important as maintaining clearance to energized conductors. Although not referenced, Rule 217A relates to Rule 231B. The note in Rule 217A1a regarding out-of-control vehicles helps distinguish between an ordinary vehicle using the road and an out-of-control vehicle that cannot be planned for.

Rule 231B1 applies to roads with curbs. The Code references two types of curbs, a redirectional curb and a swale-type curb. The rules for clearance of supporting structures from roads that have curbs are outlined in Figs. 231-2 and 231-3.

Rule 231B2 applies to roads without curbs. In this case no dimensions are specified, just the requirement to locate the pole "a sufficient distance from the roadway to avoid contact by ordinary vehicles using and located on the traveled way." The NESC has definitions in Sec. 02 for the terms shoulder, roadway, and traveled way to clarify the requirements in this rule. The rules for clearance

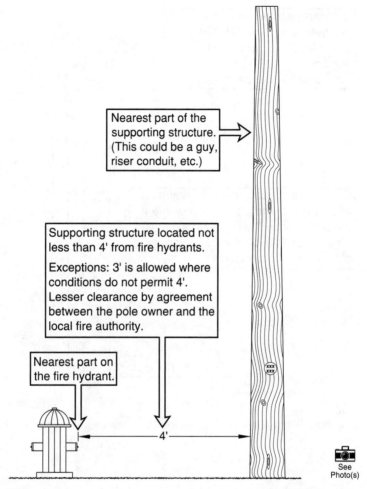

Nearest part of the supporting structure. (This could be a guy, riser conduit, etc.)

Supporting structure located not less than 4' from fire hydrants.

Exceptions: 3' is allowed where conditions do not permit 4'. Lesser clearance by agreement between the pole owner and the local fire authority.

Nearest part on the fire hydrant.

4'

See Photo(s)

Fig. 231-1. Clearance of a supporting structure from a fire hydrant (Rule 231A).

of supporting structures from roads that do not have a curb are outlined in Fig. 231-4.

Rule 231B3 recognizes various utilities compete for narrow rights-of-way (overhead and underground) along roads, streets, and highways. Special cases may exist and must be resolved using accepted good practice for the conditions at hand.

Rule 231B4 recognizes that in many cases a state highway department, county road department, city, or other governmental authority may require a permit to place poles in a right of way. The permit may have special

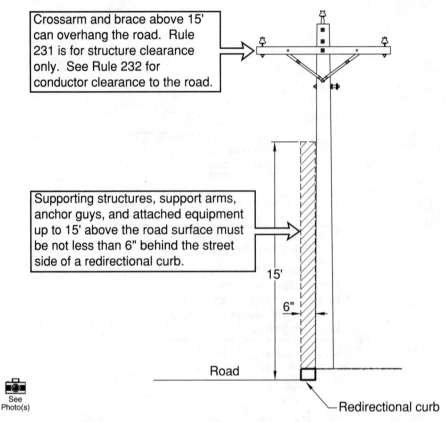

Crossarm and brace above 15' can overhang the road. Rule 231 is for structure clearance only. See Rule 232 for conductor clearance to the road.

Supporting structures, support arms, anchor guys, and attached equipment up to 15' above the road surface must be not less than 6" behind the street side of a redirectional curb.

15'

6"

Road

See Photo(s)

Redirectional curb

Fig. 231-2. Clearance of a supporting structure from a redirectional curb (Rule 231B1).

requirements for pole placement. Many times this requirement is more restrictive than the **NESC**. In these cases the permit or approval process will take precedence over the **NESC** rules.

231C. From Railroad Tracks. Rule 231C applies to supporting structures (poles) on lines paralleling or crossing railroad tracks. Four exceptions to Rule 231C1 and Rule 231C2 permit reductions to the basic clearance provided in Rule 231C1. Many times when applying for a permit to parallel or cross a railroad right-of-way, the railroad company will require greater clearances than listed in this rule. The **Code** makes a reference to Rule 234I, which covers clearance of wires, conductors, and cables to rail cars. Rule 232, which applies to conductor clearance over railroad tracks, should also be checked. Rule 231C only applies to structures, not conductors. The rules for clearance of supporting structures from railroad tracks are outlined in Fig. 231-5.

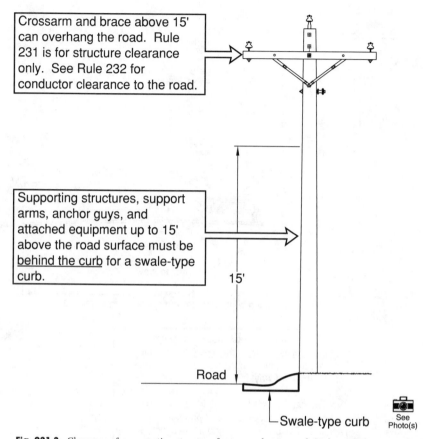

Crossarm and brace above 15'
can overhang the road. Rule
231 is for structure clearance
only. See Rule 232 for
conductor clearance to the road.

Supporting structures, support
arms, anchor guys, and
attached equipment up to 15'
above the road surface must be
<u>behind the curb</u> for a swale-type
curb.

15'

Road

Swale-type curb

See
Photo(s)

Fig. 231-3. Clearance of a supporting structure from a swale-type curb (Rule 231B1).

232. VERTICAL CLEARANCES OF WIRES, CONDUCTORS, CABLES, AND EQUIPMENT ABOVE GROUND, ROADWAY, RAIL, OR WATER SURFACES

Rule 232 is probably referenced more than any other rule in the NESC. NESC Table 232-1 is probably referenced more than any other table. One problem that can occur is referencing NESC Table 232-1 without reading and understanding all the related rules that apply to the table. For example, Rules 230A through 230I define general clearance requirements and types of conductors used in NESC Table 232-1. Rule 232A describes the conductor temperature and loading (sag) conditions that apply to NESC Table 232-1. Just as important as the related rules is the use and understanding of a sag and tension chart. The basics of how to use a sag and tension chart and a sample sag and tension chart are provided at the beginning of Sec. 23 of this Handbook.

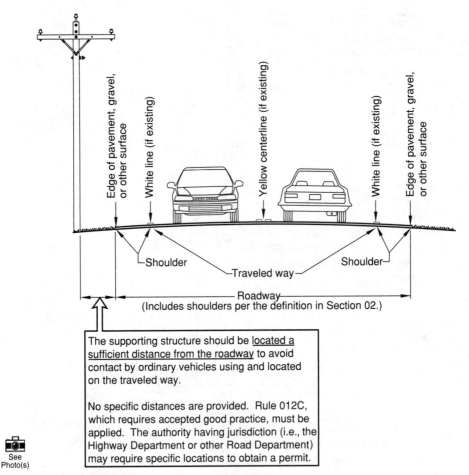

Fig. 231-4. Clearance of a supporting structure from a roadway without a curb (Rule 231B2).

232A. Application. Rule 232A states the temperature and loading conditions that apply to the clearances in **NESC** Table 232-1. Three conditions are specified, and the condition that produces the largest final sag must be used. The conditions are:

- 120°F, no wind displacement, final sag.
- Greater than 120°F, if so designed, no wind displacement, final sag.
- 32°F, no wind displacement, final sag, with ice from the Zone specified in Rule 230B (e.g., Zone 2 would require 0.25 in of ice).

The conditions above are outlined on the sample sag and tension chart at the beginning of Sec. 23.

The conductor conditions specified in Rule 232A are used to check clearance, not conductor tension. Conductor tension limits for open supply conductors are provided in Rule 261H.

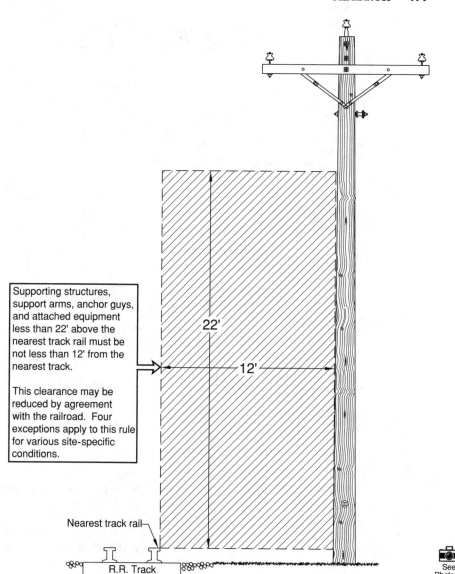

Supporting structures, support arms, anchor guys, and attached equipment less than 22' above the nearest track rail must be not less than 12' from the nearest track.

This clearance may be reduced by agreement with the railroad. Four exceptions apply to this rule for various site-specific conditions.

22'

12'

Nearest track rail

R.R. Track

See Photo(s)

Fig. 231-5. Clearance of a supporting structure from railroad tracks (Rule 231C).

The temperatures listed in Rule 232A are the conductor temperatures, not the ambient air temperature. See the beginning of Sec. 23 for additional information on conductor versus air temperature. The term loading in Rule 232A refers to the physical loads on the conductors in the form of ice and wind, not electrical loads in kilowatts or amperes. The conductor temperature and loading conditions for measuring vertical clearance are outlined in Fig. 232-1.

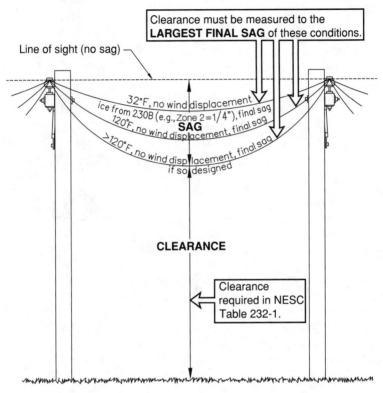

Fig. 232-1. Conductor temperature and loading conditions for measuring vertical clearance (Rule 232A).

Using the sample sag and tension chart at the beginning of Sec. 23, 212°F, no wind, no ice, final tension, produces the largest sag (6.84 ft) of the conditions that need to be checked in this rule. If the line was not designed to operate at temperatures above 120°F, then the 120°F, no wind, no ice, final value (5.61 ft) would be used as it is larger than the 32°F, no wind, 0.25 in ice, final value (4.49 ft). This example is not true for every case. Some conductors will have larger sags with ice, others will have larger sags with high temperatures. The wire size, span length, amount of ice, and design tension will all affect which condition produces the largest final sag. Examples that reinforce the fact that **Code** clearance for overhead lines is based on the largest final sag condition are shown in Figs. 232-2 and 232-3.

The majority of the discussion and examples in Sec. 23 revolve around the sample 1/0 ACSR sag and tension chart at the beginning of Sec. 23. Sag and tension charts can also be produced for secondary duplex, triplex, and quadruplex, communication cables on messengers (including phone, cable TV, and fiber-optic), self-supporting all-dielectric fiber-optic cables, overhead ground

Example:
- 300' span of 1/0 ACSR conductor.
- Assume the calculated ruling span is 300'.
- The clearance measurement is taken on a 50°F day; the conductor has a light electrical load all year long.
- Assume the conductor temperature is 60°F.
- Assume the maximum temperature the line is designed to operate at is 120°F.
- The midspan clearance measurement to a phase wire of 12.47/7.2kV, 3Ø 4W effectively grounded circuit is 20'.
- Assume the line was strung using the sag and tension chart at the beginning of Section 23.
- Assume the line is at the final sag condition.

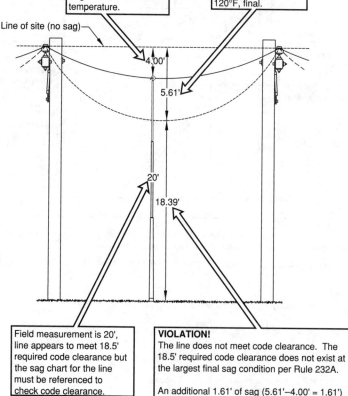

Sag for a 60° conductor temperature.

Largest final sag the line is designed to operate. In this example, the sag at 120°F, final.

Line of site (no sag)

4.00'

5.61'

20'

18.39'

Field measurement is 20', line appears to meet 18.5' required code clearance but the sag chart for the line must be referenced to check code clearance.

VIOLATION!
The line does not meet code clearance. The 18.5' required code clearance does not exist at the largest final sag condition per Rule 232A.

An additional 1.61' of sag (5.61'–4.00' = 1.61') can occur between the measured sag on a 50°F day (60°F conductor temperature) and the largest final sag (120°F in this example). 20.00'–1.61' = 18.39'

Fig. 232-2. Example of overhead clearance based on the largest final sag condition (Rule 232A).

Example:
- 400' span of 1/0 ACSR conductor.
- The 400' span is one span in a series of five spans from deadend to deadend.
- Assume the calculated ruling span is 300'.
- The clearance measurement is taken on a 50°F day; the conductor has a light electrical load all year long.
- Assume conductor temperature is 60°F.
- Assume the maximum temperature the line is designed to operate at is 120°F.
- The quarter span clearance measurement to phase wire of a 12.47/7.2 kV, 3Ø 4W effectively grounded circuit is 20'.
- The quarter span rise is 5'.
- Assume the line was strung using the sag and tension chart at the beginning of Section 23.
- Assume the line is at the final sag conditions.

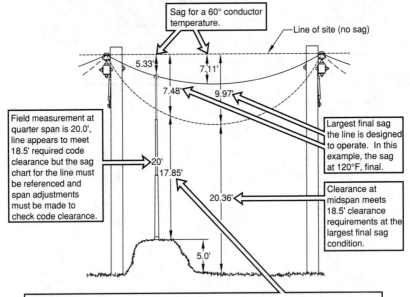

Sag for a 60° conductor temperature.

Line of site (no sag)

5.33'

7.11'

7.48' 9.97'

Field measurement at quarter span is 20.0', line appears to meet 18.5' required code clearance but the sag chart for the line must be referenced and span adjustments must be made to check code clearance.

Largest final sag the line is designed to operate. In this example, the sag at 120°F, final.

20'
17.85'

20.36'

Clearance at midspan meets 18.5' clearance requirements at the largest final sag condition.

5.0'

VIOLATION!
The line does not meet code clearance. The 18.5' required code clearance does not exist at the quarter span location at the largest final condition per Rule 232A.

The sag on a 50° day (60° conductor temperature) for a 400' span at midspan is:

$$\left(\frac{400}{300}\right)^2 \times 4' = 7.11'$$

The sag at the quarter span is 75% of the midspan sag per Figure 23-3.
7.11' x 0.75 = 5.33'

The largest final sag (at 120°F) in this example for a 400' span at midspan is:

$$\left(\frac{400}{300}\right)^2 \times 5.61' = 9.97'$$

The sag at the quarter span is 75% of the midspan sag per Figure 23-3.
9.97 x 0.75 = 7.48'

An additional 2.15' of sag (7.48' – 5.33') can occur between the measured sag on a 50°F day (60°F conductor temperature) and the largest final sag (120°F in this example) at the quarter span location. 20.00' – 2.15' = 17.85'

Fig. 232-3. Example of overhead clearance based on the largest final sag condition (Rule 232A).

wires (static wires), and virtually any other type of overhead power and communication cables.

The exception to Rule 232A recognizes that electric railroad and trolley car conductors experience different conditions than typical power and communication conductors. Finally, a note to Rule 232A reminds us that the phase and neutral conductor may not operate at the same temperatures. The neutral conductor may carry less current due to phase balance and neutral current cancellation or due to the fact that the earth and a communication messenger on a joint use pole are in parallel with a multigrounded neutral.

Prior to 1990, a 60°F sag was used to determine clearance plus additional clearance was required for long spans. In 1990, the **Code** clearance rules experienced major revisions and the method to determine **Code** clearance was changed from using the 60°F sag condition to using the maximum sag conditions in Rule 232A. This change partly came about due to the fact that computer programs became available to generate detailed sag and tension charts like the sample at the beginning of Sec. 23 of this Handbook.

232B. Clearance of Wires, Conductors, Cables, Equipment, and Support Arms Mounted on Supporting Structures.   Rule 232B references the most commonly used table in the NESC, Table 232-1. Table 232-1 cannot be used without first reading and understanding the temperature and loading conditions in Rule 232A and without understanding how to read and interpret a sag and tension chart.

NESC Table 232-1 covers vertical clearance of wires, conductors, and cables above ground, roadway, rail, and water surfaces. NESC Table 232-2 covers unguarded rigid live parts of equipment. The basic difference between these two tables is that Table 232-1 addresses items that vary due to sag and Table 232-2 addresses items that are fixed or rigid and do not vary with sag. Equipment cases fall under Table 232-2. Secondary drip loops are included in Table 232-1 even though they do not vary much with sag.

An example of how clearance values are determined in NESC Tables 232-1 and 232-2 is shown in Fig. 232-4.

One of the main differences between NESC Table 232-1 (conductors) and NESC Table 232-2 (rigid parts) is that the mechanical and electrical component of clearance is reduced by 0.5 ft for rigid parts as they are not subject to sag variations. This is evident when examining similar clearance conditions in NESC Tables 232-1 and 232-2.

NESC Tables 232-1 and 232-2 cover multiple conditions under the line, but the tables cannot cover every specific condition. For conditions not covered, Rule 012C, which requires accepted good practice, must be applied. Footnote 25 of NESC Table 232-1 explains how the clearances on the table were determined. Footnote 26 of NESC Table 232-1 clarifies how to increase **Code** clearance for oversized vehicles. For example, what if the truck in Fig. 232-4 was a 22-ft-high mining truck on private property? The engineer or designer of the line exposed to this type of vehicle can still use the 4.5-ft (conductor) and 4.0-ft (rigid part) mechanical and electrical component of clearance if the line is still 12.47/7.2 kV (Rules 232C and 232D discuss when additional clearance is required for higher voltages). A 22-ft height must be substituted for the 14-ft reference component. See Fig. 232-5.

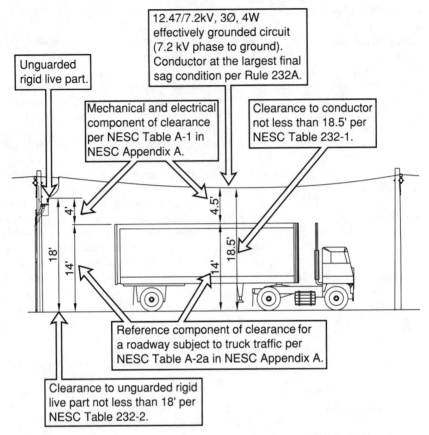

Fig. 232-4. Example of how clearance values are determined (Rule 232B).

The **NESC** clearance tables provide a wealth of information. They must be carefully reviewed before selecting the proper value from the table. Many of the clearances in **NESC** Table 232-1 are the same except that different footnotes apply to each. An explanation of how **NESC** Table 232-1 is formatted is shown in Fig. 232-6.

All of the clearances in **NESC** Table 232-1 are "not less than" values. See Rule 010 for a discussion of using **Code** clearance values plus an adder. Using the **Code** clearance plus an adder can help meet and maintain clearance over the life of an installation as required in Rule 230I. If the authority issuing a occupancy permit (e.g., a railroad company, highway department, etc.) requires greater clearance than the **NESC**, the greater clearance required in the occupancy permit will take precedence over the **NESC** clearance values. The phase conductor, neutral conductor, secondary cable (0 to 750 V), and insulated communication cable

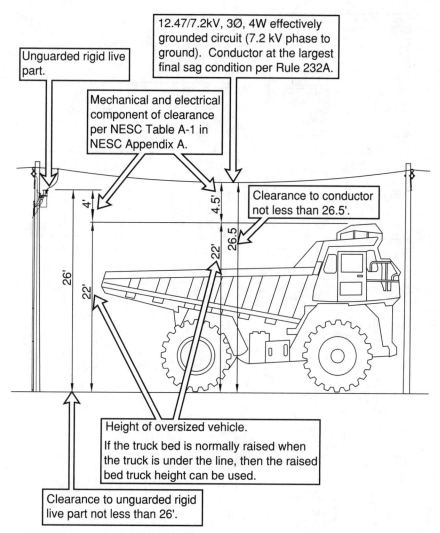

Unguarded rigid live part.

12.47/7.2kV, 3Ø, 4W effectively grounded circuit (7.2 kV phase to ground). Conductor at the largest final sag condition per Rule 232A.

Mechanical and electrical component of clearance per NESC Table A-1 in NESC Appendix A.

Clearance to conductor not less than 26.5'.

Height of oversized vehicle.

If the truck bed is normally raised when the truck is under the line, then the raised bed truck height can be used.

Clearance to unguarded rigid live part not less than 26'.

Fig. 232-5. Example of how clearance values can be determined for conditions not specified in the Code (Rule 232).

clearances for the 10 land use categories in **NESC** Table 232-1 are shown in Figs. 232-7 through 232-16.

   NESC Table 232-2 covers vertical clearance to equipment cases and unguarded rigid live parts above ground, roadway, and water surfaces. The land use categories are similar to **NESC** Table 232-1. The clearances in **NESC** Table 232-2 are commonly $1/2$ ft less than the corresponding clearance in **NESC** Table 232-1 as the rigid parts are not subject to variations in sag. Rules 232B2

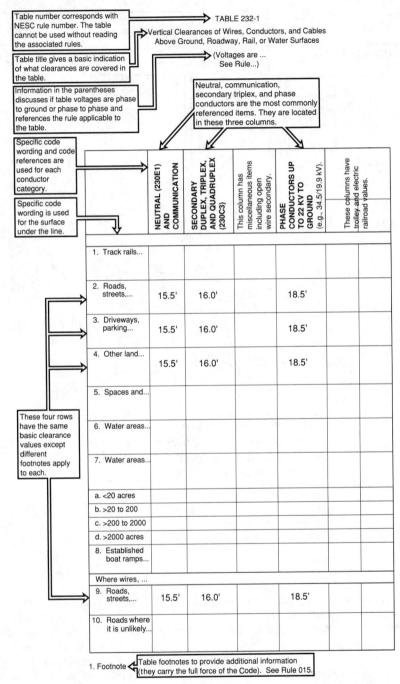

Fig. 232-6. Explanation of how NESC Table 231-1 is formatted (Rule 232).

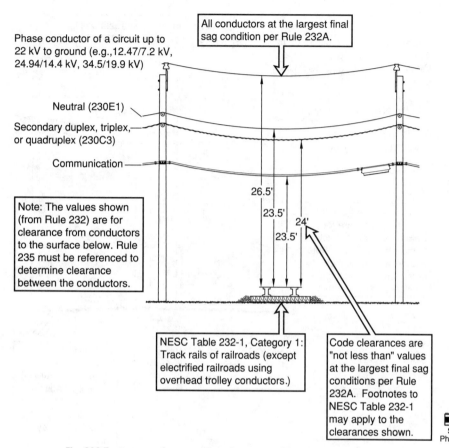

Fig. 232-7. Common clearance values from NESC Table 232-1 (Rule 232B1).

and 232B3 both reference **NESC** Table 232-2. An example of clearance to equipment cases and ungrounded rigid live parts is shown in Fig. 232-17.

Rule 232B4 includes vertical clearance requirements for effectively grounded and ungrounded conductive parts of luminaires (light fixtures). Modern luminaires are effectively grounded and operate on a constant voltage system. An example of clearance to an effectively grounded street light is shown in Fig. 232-18.

The clearances specified in Rule 232 are vertical clearances. Rule 232 does not provide horizontal or diagonal clearances to land surfaces. Rule 234 provides horizontal clearances to buildings and other structures but not to a land surface. To determine the horizontal or diagonal clearance to a land surface, Rule 012C, which requires accepted good practice, must be applied. See Fig. 232-19.

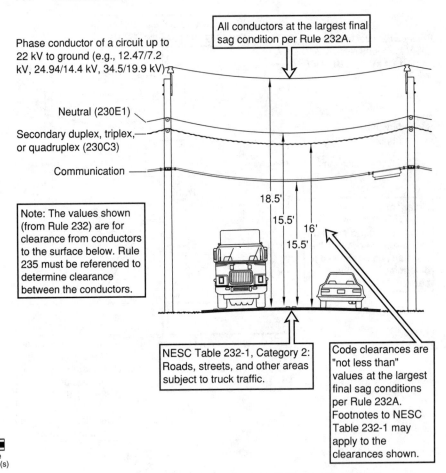

Phase conductor of a circuit up to 22 kV to ground (e.g., 12.47/7.2 kV, 24.94/14.4 kV, 34.5/19.9 kV)

All conductors at the largest final sag condition per Rule 232A.

Neutral (230E1)

Secondary duplex, triplex, or quadruplex (230C3)

Communication

18.5'

15.5'

16'

15.5'

Note: The values shown (from Rule 232) are for clearance from conductors to the surface below. Rule 235 must be referenced to determine clearance between the conductors.

NESC Table 232-1, Category 2: Roads, streets, and other areas subject to truck traffic.

Code clearances are "not less than" values at the largest final sag conditions per Rule 232A. Footnotes to NESC Table 232-1 may apply to the clearances shown.

See Photo(s)

Fig. 232-8. Common clearance values from NESC Table 232-1 (Rule 232B1).

The **Code** clearances in Rule 232 and clearance values elsewhere in the **Code** are "not less than" values. Clearance greater than the **Code** requires is certainly acceptable, but not required. Certain agencies like the State Highway Department, railroad company, or other agency issuing a crossing or occupancy permit may require more clearance than the **Code**. There is one agency that gets involved if too much clearance exists above ground or near an airport. That agency is the Federal Aviation Administration (FAA). The **NESC** does not address placing marker balls on conductors or marking of tall structures. FAA Advisory Circular AC 70/7460-1K and AC 70/7460-2K address obstruction marking and lighting for structures including overhead power lines. See Fig. 232-20.

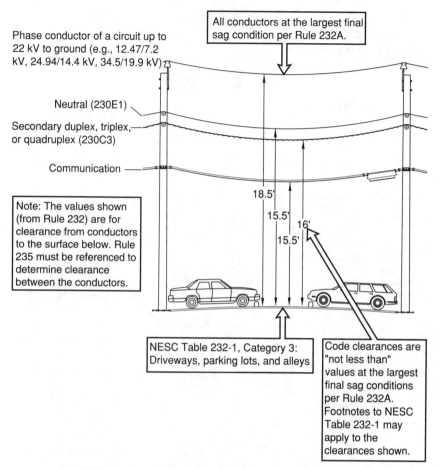

Phase conductor of a circuit up to 22 kV to ground (e.g., 12.47/7.2 kV, 24.94/14.4 kV, 34.5/19.9 kV)

All conductors at the largest final sag condition per Rule 232A.

Neutral (230E1)

Secondary duplex, triplex, or quadruplex (230C3)

Communication

Note: The values shown (from Rule 232) are for clearance from conductors to the surface below. Rule 235 must be referenced to determine clearance between the conductors.

18.5'
15.5'
16'
15.5'

NESC Table 232-1, Category 3: Driveways, parking lots, and alleys

Code clearances are "not less than" values at the largest final sag conditions per Rule 232A. Footnotes to NESC Table 232-1 may apply to the clearances shown.

Fig. 232-9. Common clearance values from NESC Table 232-1 (Rule 232B1).

232C. Additional Clearances for Wires, Conductors, Cables, and Unguarded Rigid Live Parts of Equipment. The clearance values in Tables 232-1 and 232-2 must be increased for the following reasons:

- The voltage of the line exceeds 22 kV.
  - ✓ 22 kV to ground for an effectively grounded circuit. The wording in parentheses under the titles of NESC Tables 232-1 and 232-2 permits effectively grounded circuits and those other circuits where all ground faults are cleared by promptly de-energizing the faulted section, both initially and following subsequent breaker operations.
  - ✓ 22 kV phase-to-phase for an ungrounded (e.g., delta) circuit.
  - ✓ The maximum operating voltage, typically 1.05 times the nominal operating voltage, must be used for lines over 50 kV.
  - ✓ The clearance adder is 0.4 in per kilovolt in excess of 22 kV.

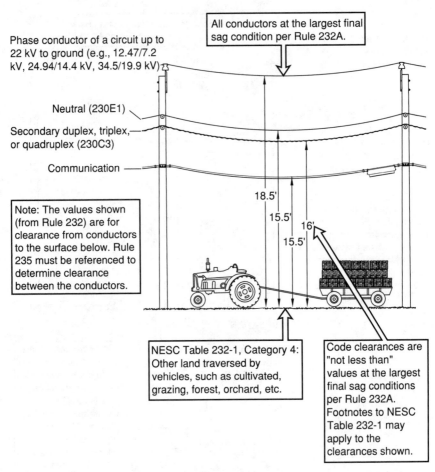

Phase conductor of a circuit up to 22 kV to ground (e.g., 12.47/7.2 kV, 24.94/14.4 kV, 34.5/19.9 kV)

All conductors at the largest final sag condition per Rule 232A.

Neutral (230E1)

Secondary duplex, triplex, or quadruplex (230C3)

Communication

Note: The values shown (from Rule 232) are for clearance from conductors to the surface below. Rule 235 must be referenced to determine clearance between the conductors.

18.5'
15.5'
16'
15.5'

NESC Table 232-1, Category 4: Other land traversed by vehicles, such as cultivated, grazing, forest, orchard, etc.

Code clearances are "not less than" values at the largest final sag conditions per Rule 232A. Footnotes to NESC Table 232-1 may apply to the clearances shown.

Fig. 232-10. Common clearance values from NESC Table 232-1 (Rule 232B1).

- The voltage of the line exceeds 50 kV and the line is above an elevation of 3300 ft.
  - ✓ 50 kV to ground for an effectively grounded circuit. The wording in parentheses under the titles of NESC Tables 232-1 and 232-2 permits effectively grounded circuits and those other circuits where all ground faults are cleared by promptly de-energizing the faulted section, both initially and following subsequent breaker operations.
  - ✓ 50 kV phase-to-phase for an ungrounded (e.g., delta) circuit.
  - ✓ A clearance adder of 3 percent (.03) for each 1000 ft in excess of 3300 ft of elevation. The percent increase is applied to the additional clearance required for lines above 22 kV, not to the total clearance.

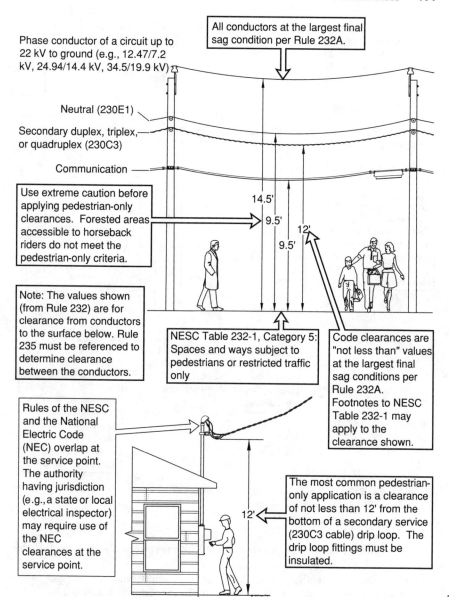

Phase conductor of a circuit up to 22 kV to ground (e.g., 12.47/7.2 kV, 24.94/14.4 kV, 34.5/19.9 kV)

All conductors at the largest final sag condition per Rule 232A.

Neutral (230E1)

Secondary duplex, triplex, or quadruplex (230C3)

Communication

Use extreme caution before applying pedestrian-only clearances. Forested areas accessible to horseback riders do not meet the pedestrian-only criteria.

14.5'
9.5'
12'
9.5'

Note: The values shown (from Rule 232) are for clearance from conductors to the surface below. Rule 235 must be referenced to determine clearance between the conductors.

NESC Table 232-1, Category 5: Spaces and ways subject to pedestrians or restricted traffic only

Code clearances are "not less than" values at the largest final sag conditions per Rule 232A. Footnotes to NESC Table 232-1 may apply to the clearance shown.

Rules of the NESC and the National Electric Code (NEC) overlap at the service point. The authority having jurisdiction (e.g., a state or local electrical inspector) may require use of the NEC clearances at the service point.

The most common pedestrian-only application is a clearance of not less than 12' from the bottom of a secondary service (230C3 cable) drip loop. The drip loop fittings must be insulated.

12'

See Photo(s)

Fig. 232-11. Common clearance values from NESC Table 232-1 (Rule 232B1).

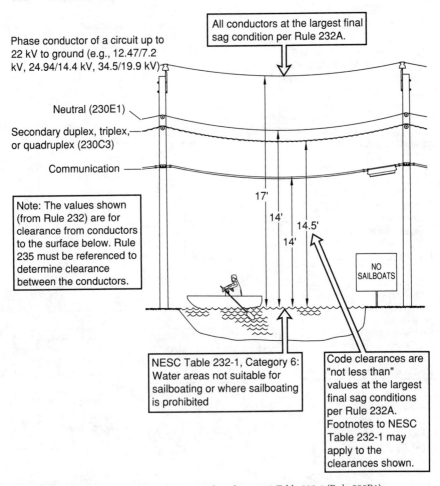

Phase conductor of a circuit up to 22 kV to ground (e.g., 12.47/7.2 kV, 24.94/14.4 kV, 34.5/19.9 kV)

All conductors at the largest final sag condition per Rule 232A.

Neutral (230E1)

Secondary duplex, triplex, or quadruplex (230C3)

Communication

Note: The values shown (from Rule 232) are for clearance from conductors to the surface below. Rule 235 must be referenced to determine clearance between the conductors.

17'

14'

14.5'

14'

NO SAILBOATS

NESC Table 232-1, Category 6: Water areas not suitable for sailboating or where sailboating is prohibited

Code clearances are "not less than" values at the largest final sag conditions per Rule 232A. Footnotes to NESC Table 232-1 may apply to the clearances shown.

Fig. 232-12. Common clearance values from NESC Table 232-1 (Rule 232B1).

✓ No increase in clearance is required for elevations above 3300 ft if the line is rated 50 kV or less.

• For voltages exceeding 98 kV AC to ground, additional clearance or other means are required to limit electrostatic field effects to 5 mA or less.

• For voltages over 470 kV, the clearance must be determined using the formulas in Rule 232D. Rule 232D can also be used as an alternate clearance method for lines over 98 kV AC to ground or 189 kV DC to ground.

• The "not less than" rounding requirements of Rule 230A4 apply.

An example of an additional clearance calculation is shown in Fig. 232-21.

Rule 232C1c requires that lines exceeding 98 kV AC to ground have additional clearance or other means to reduce electrostatic field effects to 5 mA or less.

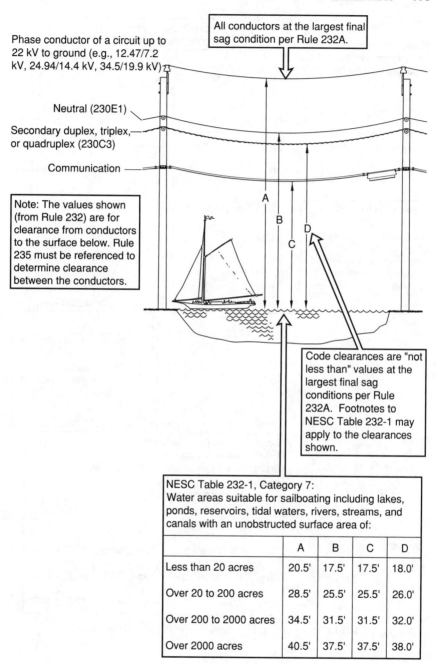

Phase conductor of a circuit up to 22 kV to ground (e.g., 12.47/7.2 kV, 24.94/14.4 kV, 34.5/19.9 kV)

All conductors at the largest final sag condition per Rule 232A.

Neutral (230E1)

Secondary duplex, triplex, or quadruplex (230C3)

Communication

Note: The values shown (from Rule 232) are for clearance from conductors to the surface below. Rule 235 must be referenced to determine clearance between the conductors.

A
B
C
D

Code clearances are "not less than" values at the largest final sag conditions per Rule 232A. Footnotes to NESC Table 232-1 may apply to the clearances shown.

NESC Table 232-1, Category 7:
Water areas suitable for sailboating including lakes, ponds, reservoirs, tidal waters, rivers, streams, and canals with an unobstructed surface area of:

	A	B	C	D
Less than 20 acres	20.5'	17.5'	17.5'	18.0'
Over 20 to 200 acres	28.5'	25.5'	25.5'	26.0'
Over 200 to 2000 acres	34.5'	31.5'	31.5'	32.0'
Over 2000 acres	40.5'	37.5'	37.5'	38.0'

Fig. 232-13. Common clearance values from NESC Table 232-1 (Rule 232B1).

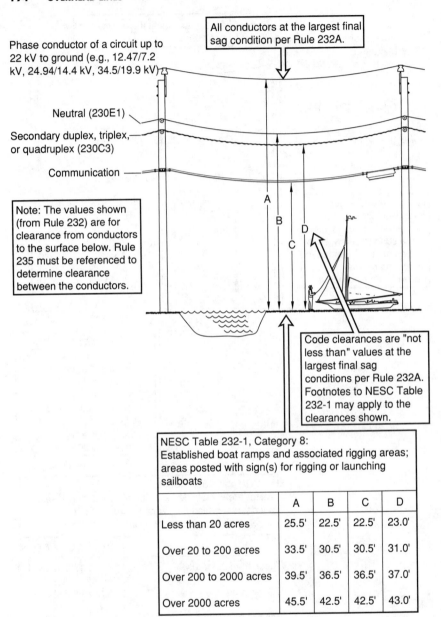

All conductors at the largest final sag condition per Rule 232A.

Phase conductor of a circuit up to 22 kV to ground (e.g., 12.47/7.2 kV, 24.94/14.4 kV, 34.5/19.9 kV)

Neutral (230E1)

Secondary duplex, triplex, or quadruplex (230C3)

Communication

Note: The values shown (from Rule 232) are for clearance from conductors to the surface below. Rule 235 must be referenced to determine clearance between the conductors.

Code clearances are "not less than" values at the largest final sag conditions per Rule 232A. Footnotes to NESC Table 232-1 may apply to the clearances shown.

NESC Table 232-1, Category 8:
Established boat ramps and associated rigging areas; areas posted with sign(s) for rigging or launching sailboats

	A	B	C	D
Less than 20 acres	25.5'	22.5'	22.5'	23.0'
Over 20 to 200 acres	33.5'	30.5'	30.5'	31.0'
Over 200 to 2000 acres	39.5'	36.5'	36.5'	37.0'
Over 2000 acres	45.5'	42.5'	42.5'	43.0'

Fig. 232-14. Common clearance values from NESC Table 232-1 (Rule 232B1).

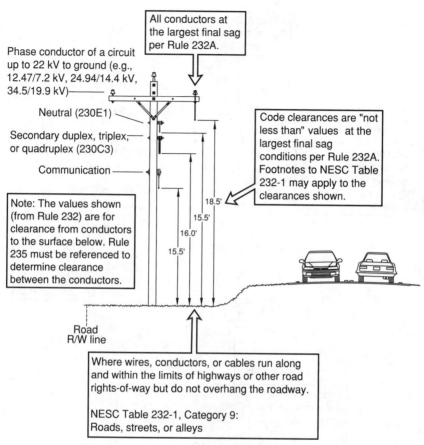

Fig. 232-15. Common clearance values from NESC Table 232-1 (Rule 232B1).

High-voltage transmission lines can induce a voltage and therefore induce currents into metal objects like a truck under the line. The average adult human body can detect an electrical shock at currents around 2 mA. Currents above 5 mA can cause pain and can be harmful to the body. Since no specific clearance adders or other methods are provided, Rule 012C, which requires accepted good practice, must be applied. Accepted good practice in this case may be an engineering analysis or field testing. The rules related to limiting electrostatic field effects to 5 mA are outlined in Fig. 232-22.

232D. Alternate Clearances for Voltages Exceeding 98 kV AC to Ground or 139 kV DC to Ground.   The use of Rule 232D will permit reduced clearances for some voltages exceeding 98 kV to ground alternating current (AC) and 138 kV to ground direct current (DC). See Rule 230G for a discussion of AC and DC circuits. The best method for checking the application of Rule 232D

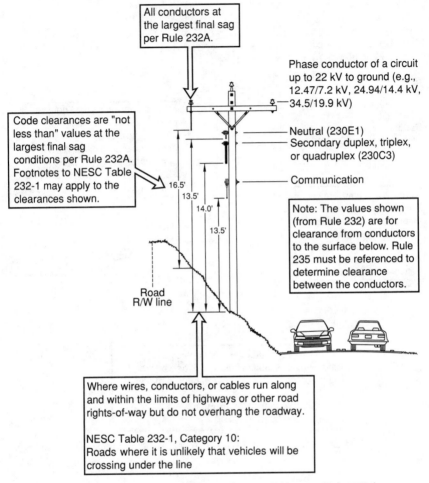

All conductors at the largest final sag per Rule 232A.

Phase conductor of a circuit up to 22 kV to ground (e.g., 12.47/7.2 kV, 24.94/14.4 kV, 34.5/19.9 kV)

Code clearances are "not less than" values at the largest final sag conditions per Rule 232A. Footnotes to NESC Table 232-1 may apply to the clearances shown.

Neutral (230E1)
Secondary duplex, triplex, or quadruplex (230C3)

Communication

16.5'
13.5'
14.0'
13.5'

Note: The values shown (from Rule 232) are for clearance from conductors to the surface below. Rule 235 must be referenced to determine clearance between the conductors.

Road R/W line

Where wires, conductors, or cables run along and within the limits of highways or other road rights-of-way but do not overhang the roadway.

NESC Table 232-1, Category 10:
Roads where it is unlikely that vehicles will be crossing under the line

Fig. 232-16. Common clearance values from NESC Table 232-1 (Rule 232B1).

is to use the formulas in Rule 232D and check the calculated results to **NESC** Table 232-4. If the calculated result does not match the examples provided in Table 232-4 for the same input conditions, a value was incorrectly entered or the formula was misapplied. Rule 232D also requires application of **NESC** Table 232-3, which lists reference heights similar to **NESC** Table A-2a in Appendix A of the **NESC**.

A list of important factors to consider when using Rule 232D is outlined below:
- The alternate clearances are for lines exceeding 98 kV (AC) to ground and 139 kV (DC) to ground only.
- The alternate clearance method must be used for voltages over 470 kV.
- **NESC** Table 232-3 reference heights are to be used.
- Crest voltage (not rms voltage) must be used. See Rule 230G for a discussion.

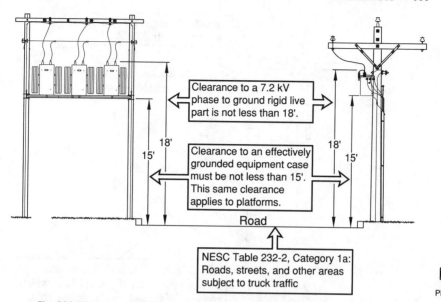

Fig. 232-17. Example of clearance to equipment cases and unguarded rigid live parts (Rules 232B2 and 232B3).

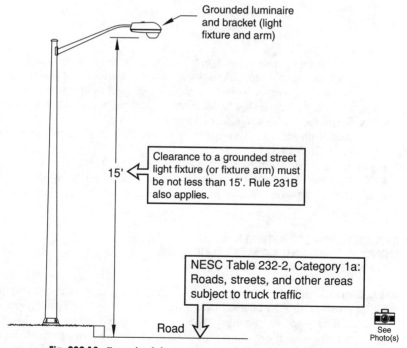

Fig. 232-18. Example of clearance to street lighting (Rule 232B4).

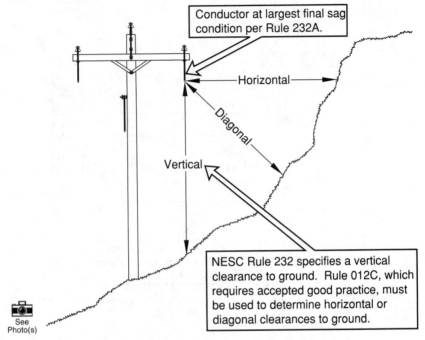

Conductor at largest final sag condition per Rule 232A.

Horizontal

Diagonal

Vertical

NESC Rule 232 specifies a vertical clearance to ground. Rule 012C, which requires accepted good practice, must be used to determine horizontal or diagonal clearances to ground.

See Photo(s)

Fig. 232-19. Vertical clearance to land surface (Rule 232).

- The maximum operating voltage, typically 1.05 times the nominal voltage, must be used. ANSI C84.1, American National Standard for Electric Power Systems and Equipment—Voltage Ratings (60 Hertz), provides a table listing nominal voltage levels and the corresponding maximum voltage level.
- Maximum switching surge factors must be known.
- Additional clearances are required for elevation and electrostatic effects similar to Rule 232C.
- There is a lower limit for clearance based on Rule 232C.
- The "not less than" rounding requirements of Rule 230A4 apply.
- NESC Table 232-4 can be used to verify that calculations were done correctly.

233. CLEARANCES BETWEEN WIRES, CONDUCTORS, AND CABLES CARRIED ON DIFFERENT SUPPORTING STRUCTURES

233A. General.   The first sentence of Rule 233 is very important and very powerful. "Crossings should be made on a common supporting structure, where practical." The wording includes "should" and "where practical." There are times when crossing on a common supporting structure is not practical. Turning lanes at roadway intersections can interfere with the location of a common

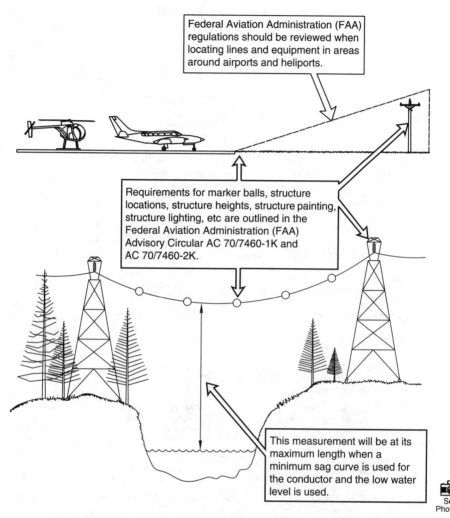

Federal Aviation Administration (FAA) regulations should be reviewed when locating lines and equipment in areas around airports and heliports.

Requirements for marker balls, structure locations, structure heights, structure painting, structure lighting, etc are outlined in the Federal Aviation Administration (FAA) Advisory Circular AC 70/7460-1K and AC 70/7460-2K.

This measurement will be at its maximum length when a minimum sag curve is used for the conductor and the low water level is used.

See Photo(s)

Fig. 232-20. Federal Aviation Administration (FAA) requirements (Rule N/A).

crossing pole. Transmission lattice towers do not provide a practical means of attaching a distribution line crossing. If a line crossing can be attached to a common supporting structure, Rule 235 applies instead of Rule 233. Rule 235, Clearance for Wires, Conductors, or Cables Carried on the Same Supporting Structure, is simpler to apply because the conductors are tied onto the structure, not moving around in a span as they are in Rule 233. Examples of line crossings with and without a common supporting structure are shown in Fig. 233-1.

Two "envelopes" are discussed in this rule, the conductor movement envelope in Rule 233A1 that outlines an area of where the conductor can be positioned at

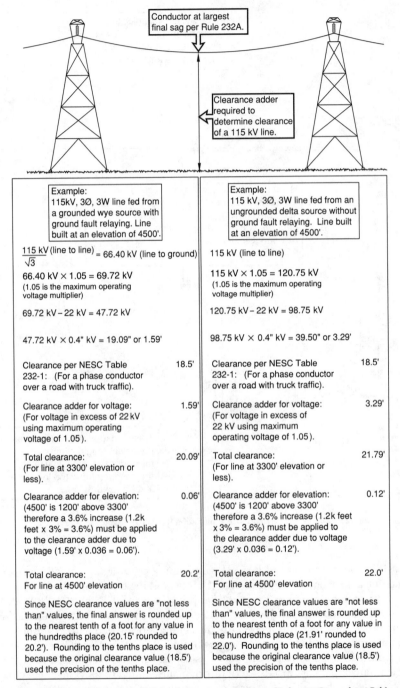

Conductor at largest
final sag per Rule 232A.

Clearance adder
required to
determine clearance
of a 115 kV line.

Example:
115kV, 3Ø, 3W line fed from
a grounded wye source with
ground fault relaying. Line
built at an elevation of 4500'.

$\dfrac{115\ kV\ \text{(line to line)}}{\sqrt{3}} = 66.40\ kV\ \text{(line to ground)}$

66.40 kV × 1.05 = 69.72 kV
(1.05 is the maximum operating
voltage multiplier)

69.72 kV – 22 kV = 47.72 kV

47.72 kV × 0.4" kV = 19.09" or 1.59'

Clearance per NESC Table
232-1: (For a phase conductor
over a road with truck traffic).	18.5'
Clearance adder for voltage:
(For voltage in excess of 22 kV
using maximum operating
voltage of 1.05).	1.59'
Total clearance:
(For line at 3300' elevation or
less).	20.09'
Clearance adder for elevation:
(4500' is 1200' above 3300'
therefore a 3.6% increase (1.2k
feet x 3% = 3.6%) must be applied
to the clearance adder due to
voltage (1.59' x 0.036 = 0.06').	0.06'
Total clearance:
For line at 4500' elevation	20.2'

Since NESC clearance values are "not less
than" values, the final answer is rounded up
to the nearest tenth of a foot for any value in
the hundredths place (20.15' rounded to
20.2'). Rounding to the tenths place is used
because the original clearance value (18.5')
used the precision of the tenths place.

Example:
115 kV, 3Ø, 3W line fed from an
ungrounded delta source without
ground fault relaying. Line built
at an elevation of 4500'.

115 kV (line to line)

115 kV × 1.05 = 120.75 kV
(1.05 is the maximum operating
voltage multiplier)

120.75 kV – 22 kV = 98.75 kV

98.75 kV × 0.4" kV = 39.50" or 3.29'

Clearance per NESC Table
232-1: (For a phase conductor
over a road with truck traffic).	18.5'
Clearance adder for voltage:
(For voltage in excess of
22 kV using maximum
operating voltage of 1.05).	3.29'
Total clearance:
(For line at 3300' elevation or
less).	21.79'
Clearance adder for elevation:
(4500' is 1200' above 3300'
therefore a 3.6% increase (1.2k feet
x 3% = 3.6%) must be applied to
the clearance adder due to voltage
(3.29' x 0.036 = 0.12').	0.12'
Total clearance:
For line at 4500' elevation	22.0'

Since NESC clearance values are "not less
than" values, the final answer is rounded up
to the nearest tenth of a foot for any value in
the hundredths place (21.91' rounded to
22.0'). Rounding to the tenths place is used
because the original clearance value (18.5')
used the precision of the tenths place.

Fig. 232-21. Example of an additional clearance calculation (Rules 232C1a and 232C1b).

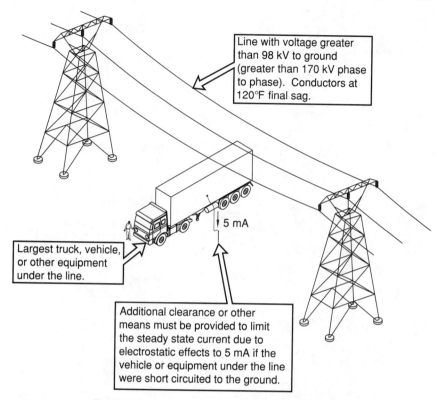

Line with voltage greater than 98 kV to ground (greater than 170 kV phase to phase). Conductors at 120°F final sag.

Largest truck, vehicle, or other equipment under the line.

↑ 5 mA

Additional clearance or other means must be provided to limit the steady state current due to electrostatic effects to 5 mA if the vehicle or equipment under the line were short circuited to the ground.

Fig. 232-22. Electrostatic field requirements (Rule 232C1c).

various temperature and loading conditions; and the clearance envelope in Rule 233A2 that outlines the horizontal and vertical clearance area that must be maintained between the two conductors.

The conductor movement envelope in Rule 233A1a(3) has the same largest final sag conditions described in Rule 232A. See Rule 232A for a discussion. In addition to the largest final sag condition, Rule 233A1a(1) requires a 60°F, no wind displacement, initial and final sag condition and Rule 233A1a(2) requires a 60°F, 6-lb/ft² wind, initial and final sag condition. These values are outlined on the sample sag and tension chart at the beginning of Sec. 23.

The temperatures listed in Rule 233A are the conductor temperatures, not the ambient air temperature. See the beginning of Sec. 23 for additional information on conductor versus air temperature. The term loading in Rule 233A refers to the physical loads on the conductors in the form of ice and wind, not electrical loads in kW or amps. The conductor temperature and loading conditions for measuring the conductor movement envelope are outlined in Fig. 233-2.

Using the sample sag and tension chart at the beginning of Sec. 23, 212°F, no wind, no ice, final produces the largest sag (6.84 ft) of the conditions that need to be checked in this rule. If the line were not designed to operate at high

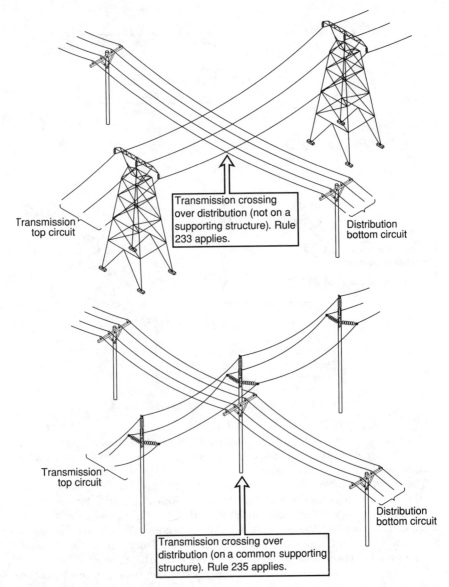

Fig. 233-1. Examples of line crossings with and without a common supporting structure (Rule 233A).

temperatures, then the 120°F, no wind, no ice, final value (5.61 ft) would be used as it is larger than the 32°F, no wind, 0.25 in ice, final value (4.49 ft). The 60°F, no wind, no ice, initial condition produces 3.14 ft of sag. The 60°F, no wind, no ice, final condition produces 4.00 ft of sag. The 60°F, 6 lb/ft², no ice,

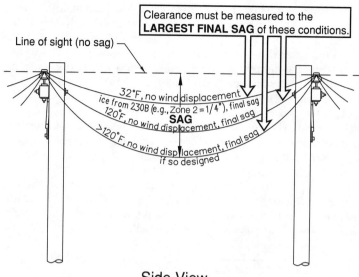

Clearance must be measured to the **LARGEST FINAL SAG** of these conditions.

Line of sight (no sag)

32°F, no wind displacement

ice from 230B (e.g., Zone 2 = 1/4"), final sag

120°F, no wind displacement, final sag

SAG

>120°F, no wind displacement, final sag
if so designed

Side View

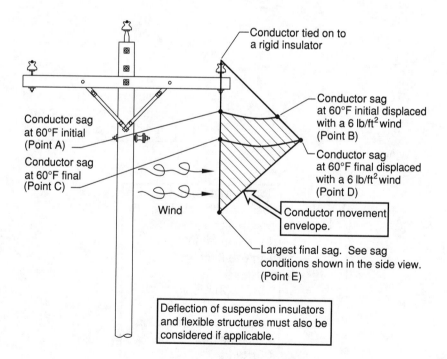

Conductor tied on to
a rigid insulator

Conductor sag
at 60°F initial
(Point A)

Conductor sag
at 60°F final
(Point C)

Wind

Conductor sag
at 60°F initial displaced
with a 6 lb/ft^2 wind
(Point B)

Conductor sag
at 60°F final displaced
with a 6 lb/ft^2 wind
(Point D)

Conductor movement
envelope.

Largest final sag. See sag
conditions shown in the side view.
(Point E)

Deflection of suspension insulators
and flexible structures must also be
considered if applicable.

Front View

Fig. 233-2. Conductor movement envelope (Rule 233A).

initial condition produces 4.03 ft. The 60°F, 6 lb/ft², no ice, final condition produces 4.66 ft of sag. The 60°F conditions with wind have more sag than without wind but the wind-blown conductor position is to the side of the span. The NESC does not provide the details of how to calculate the position of the blownout conductor at 60°F with a 6-lb/ft² wind. A transmission or distribution line design manual will typically include formulas to perform this calculation. See Rule 240 for a discussion of line design manuals. A 6-lb/ft² wind pressure on a cylindrical surface like a conductor, cable, or pole is equal to approximately a 50 mile per hour wind. The deflection of suspension insulators and flexible structures must be considered in the wind blowout calculation. See Fig. 233-3.

The footnotes to NESC Fig. 233-2 provide additional information on how to apply the conductor movement envelope. Footnote 1 requires that different wind directions be considered. The direction that produces the minimum distance between conductors must be used. Footnote 2 permits reducing the 6-lb/ft² wind force to 4 lb/ft² for sheltered areas, but trees are not considered a shelter to a line. Footnote 5 applies when one line is above the other. When the

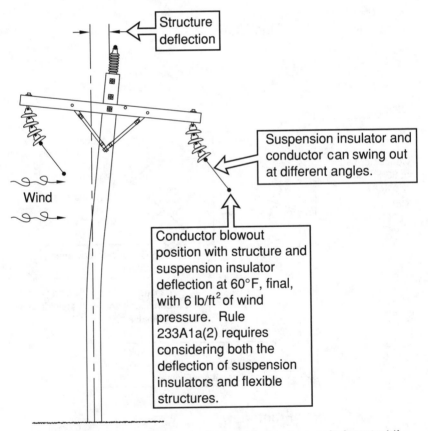

Fig. 233-3. Deflection of suspension insulators and flexible structures [Rule 233A1a(2)].

largest final sag of the upper conductor is determined, the lower conductor temperature must be equal to the ambient air temperature used to determine the largest final sag of the upper conductor. This requirement is similar to, but worded slightly different than, Rule 235C2b(1)(c). Rule 235C2b(1)(c) requires checking vertical clearances during two conditions, when the top conductor is at a maximum operating temperature (120°F, or greater if so designed) and when the top conductor is at a cold temperature with ice (32°F, with ice from Rule 230B). Footnote 3 of **NESC** Figure 233-2 only requires checking vertical clearance at one of these two conditions. Both conditions are checked to determine the greatest sag of the top conductor, but then vertical clearance is only required to be checked at the condition that produces the largest sag of the top conductor. The condition that produces the largest sag of the top conductor may not be the condition that produces the greater vertical clearance at the structure. For a thorough review, both conditions can be checked as required in Rule 235C2b(1)(c). See Rule 235C and the beginning of Sec. 23 for discussions related to conductor temperature and ambient temperature.

233B. Horizontal Clearance. Rule 233B outlines the requirements for the horizontal component of the clearance envelope. The horizontal clearance between conductors on different supporting structures is not less than 5 ft. A 0.4-in/kV clearance adder is required for voltages exceeding 22 kV between conductors involved. Unlike the vertical clearance requirements in Rule 233C, a table with various clearance values is not used for horizontal clearance. Also unlike Rule 233C and Rule 232C, Rule 233B does not require a clearance adder for elevations above 3300 ft and does not require the use of the maximum operating voltage for circuits above 50 kV. An example of horizontal clearance between wires carried on different supporting structures is shown in Fig. 233-4.

Rule 233B has an exception to the 5-ft horizontal clearance for anchor guys.

Alternate clearances may be used for voltages exceeding 98 kV AC to ground or 139 kV DC to ground. The formulas in Rule 235B3 must be used for this calculation. See Rule 232D for a discussion of the alternate clearance formulas.

233C. Vertical Clearance. Rule 233C outlines the requirements for the vertical component of the clearance envelope. **NESC** Table 233-1 is used to find the vertical clearance between conductors on different supporting structures. Rule 233C applies to crossing and adjacent wires, conductors, and cables carried on different supporting structures. Examples of vertical clearance between wires carried on different supporting structures are shown in Figs. 233-5, 233-6, and 233-7.

Rule 233C has an exception that states that no vertical clearance is required between wires, conductors, or cables that are electrically interconnected at the crossing. An example of the exception to vertical clearance between wires that are electrically connected at a crossing is shown in Fig. 233-8.

The vertical clearance adders in Rule 233C2 are similar in nature to the clearance adders used in Rule 232C. See the discussion in Rule 232C.

Alternate clearances may be used for voltages exceeding 98 kV AC to ground or 139 kV DC to ground per Rule 233C3. The formulas in Rule 233C3 must be used for this calculation. The values in **NESC** Table 233-2 can be used to check the proper application of the formulas. See Rule 232D for a discussion of the alternate clearance formulas.

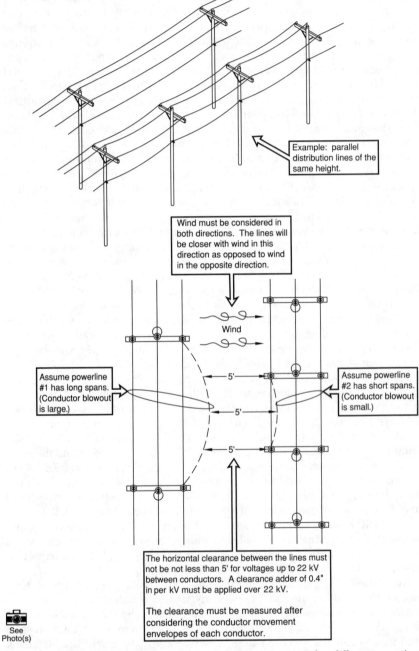

Example: parallel distribution lines of the same height.

Wind must be considered in both directions. The lines will be closer with wind in this direction as opposed to wind in the opposite direction.

Wind

Assume powerline #1 has long spans. (Conductor blowout is large.)

Assume powerline #2 has short spans. (Conductor blowout is small.)

5'

5'

5'

The horizontal clearance between the lines must not be not less than 5' for voltages up to 22 kV between conductors. A clearance adder of 0.4" in per kV must be applied over 22 kV.

The clearance must be measured after considering the conductor movement envelopes of each conductor.

See Photo(s)

Fig. 233-4. Example of horizontal clearance between wires carried on different supporting structures (Rule 233B).

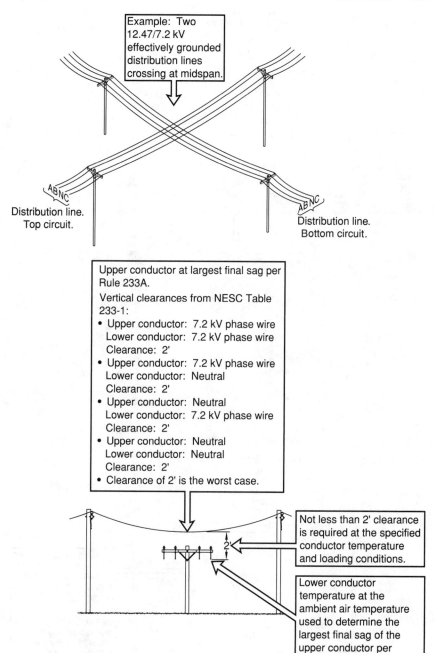

Example: Two 12.47/7.2 kV effectively grounded distribution lines crossing at midspan.

Distribution line.
Top circuit.

Distribution line.
Bottom circuit.

Upper conductor at largest final sag per Rule 233A.

Vertical clearances from NESC Table 233-1:
- Upper conductor: 7.2 kV phase wire
  Lower conductor: 7.2 kV phase wire
  Clearance: 2'
- Upper conductor: 7.2 kV phase wire
  Lower conductor: Neutral
  Clearance: 2'
- Upper conductor: Neutral
  Lower conductor: 7.2 kV phase wire
  Clearance: 2'
- Upper conductor: Neutral
  Lower conductor: Neutral
  Clearance: 2'
- Clearance of 2' is the worst case.

Not less than 2' clearance is required at the specified conductor temperature and loading conditions.

Lower conductor temperature at the ambient air temperature used to determine the largest final sag of the upper conductor per Footnote 5 to NESC Figure 233-2.

Fig. 233-5. Example of vertical clearance between wires carried on different supporting structures (Rule 233C).

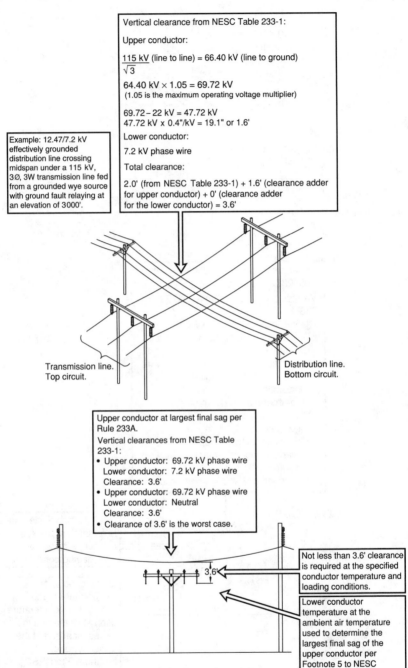

Vertical clearance from NESC Table 233-1:

Upper conductor:

$\dfrac{115\ kV}{\sqrt{3}}$ (line to line) = 66.40 kV (line to ground)

64.40 kV × 1.05 = 69.72 kV
(1.05 is the maximum operating voltage multiplier)

69.72 – 22 kV = 47.72 kV
47.72 kV x 0.4"/kV = 19.1" or 1.6'

Lower conductor:

7.2 kV phase wire

Total clearance:

2.0' (from NESC Table 233-1) + 1.6' (clearance adder for upper conductor) + 0' (clearance adder for the lower conductor) = 3.6'

Example: 12.47/7.2 kV effectively grounded distribution line crossing midspan under a 115 kV, 3Ø, 3W transmission line fed from a grounded wye source with ground fault relaying at an elevation of 3000'.

Transmission line.
Top circuit.

Distribution line.
Bottom circuit.

Upper conductor at largest final sag per Rule 233A.
Vertical clearances from NESC Table 233-1:
• Upper conductor: 69.72 kV phase wire
  Lower conductor: 7.2 kV phase wire
  Clearance: 3.6'
• Upper conductor: 69.72 kV phase wire
  Lower conductor: Neutral
  Clearance: 3.6'
• Clearance of 3.6' is the worst case.

3.6'

Not less than 3.6' clearance is required at the specified conductor temperature and loading conditions.

Lower conductor temperature at the ambient air temperature used to determine the largest final sag of the upper conductor per Footnote 5 to NESC Figure 233-2.

See
Photo(s)

Fig. 233-6. Example of vertical clearance between wires carried on different supporting structures (Rule 233C).

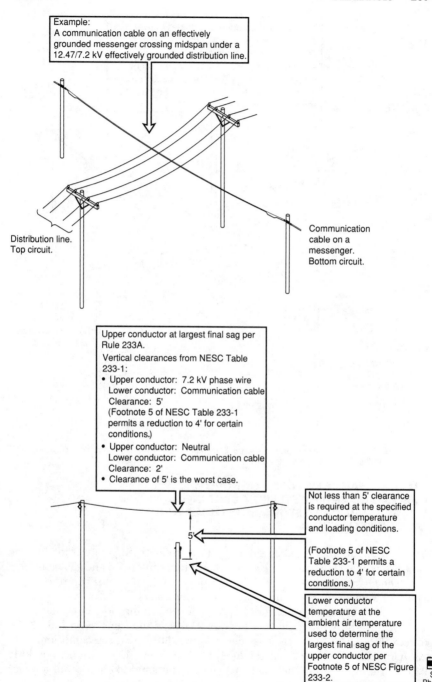

Example:
A communication cable on an effectively grounded messenger crossing midspan under a 12.47/7.2 kV effectively grounded distribution line.

Distribution line.
Top circuit.

Communication cable on a messenger.
Bottom circuit.

Upper conductor at largest final sag per Rule 233A.

Vertical clearances from NESC Table 233-1:
• Upper conductor: 7.2 kV phase wire
  Lower conductor: Communication cable
  Clearance: 5'
  (Footnote 5 of NESC Table 233-1 permits a reduction to 4' for certain conditions.)
• Upper conductor: Neutral
  Lower conductor: Communication cable
  Clearance: 2'
• Clearance of 5' is the worst case.

Not less than 5' clearance is required at the specified conductor temperature and loading conditions.

(Footnote 5 of NESC Table 233-1 permits a reduction to 4' for certain conditions.)

Lower conductor temperature at the ambient air temperature used to determine the largest final sag of the upper conductor per Footnote 5 of NESC Figure 233-2.

See Photo(s)

5'

Fig. 233-7. Example of vertical clearance between wires carried on different supporting structures (Rule 233C).

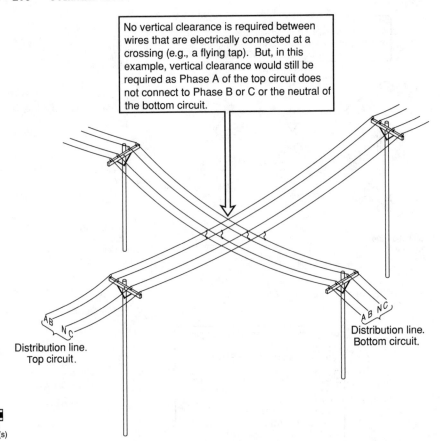

No vertical clearance is required between wires that are electrically connected at a crossing (e.g., a flying tap). But, in this example, vertical clearance would still be required as Phase A of the top circuit does not connect to Phase B or C or the neutral of the bottom circuit.

Distribution line.
Bottom circuit.

Distribution line.
Top circuit.

See Photo(s)

Fig. 233-8. Example of the exception to vertical clearance between wires that are electrically interconnected at a crossing (Rule 233C).

234. CLEARANCE OF WIRES, CONDUCTORS, CABLES, AND EQUIPMENT FROM BUILDINGS, BRIDGES, RAIL CARS, SWIMMING POOLS, AND OTHER INSTALLATIONS

234A. Application.   Rule 234A defines the conductor temperature and loading conditions that must be checked before applying the rules and tables in Sec. 234. The largest final sag conditions in Rule 234A are the same conditions provided in Rule 232A. See Rule 232A for a discussion.

Rule 234A2 provides a wind condition for checking horizontal clearance with wind. The wind condition in Rule 234A2 is the same as in Rule 233A except Rule 234A2 only requires checking the 60°F final sag condition. Rule 233A requires checking both the 60°F final and initial conditions. See Rule 233A for a discussion. The conductor temperature and loading conditions for measuring vertical and horizontal clearance of conductors to buildings and other installations are shown in Fig. 234-1.

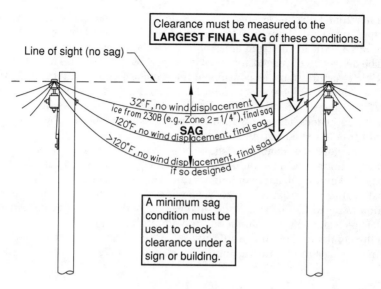

Side View

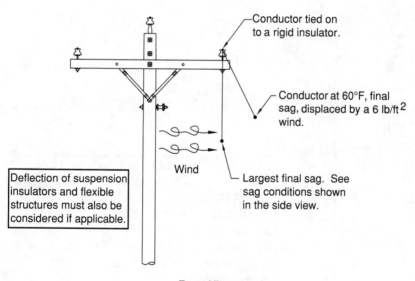

Front View

Fig. 234-1. Conductor temperature and loading conditions for measuring vertical and horizontal clearance to buildings and other installations (Rule 234A).

The horizontal clearances specified in Rule 234 must be checked at rest and after applying a 6-lb/ft² wind at 60°F, final sag. The wind pressure may be reduced to 4 lb/ft² in sheltered areas but trees are not considered a shelter to a line. The deflection of suspension insulators must be considered. The deflection of flexible structures must be considered if the highest wire, conductor, or cable attachment is 60 ft or more above grade. Rule 233A1a(2) contains similar requirements for suspension insulators and flexible structures, but unlike Rule 234A2, Rule 233A1a(2) does not specify the 60-ft height requirement. See Fig. 234-2.

In addition to the largest final sag conditions, Rule 234A1d requires a minimum conductor sag be considered at the minimum conductor temperature with no wind at initial sag. The minimum conductor temperature will be based on the expected minimum ambient temperature. Using the sample sag and tension chart at the beginning of Sec. 23, 20°F, no wind, no ice, initial tension produces a minimum sag of 1.61 ft. This is the smallest sag on the sample sag and tension chart. The clearance under and alongside a building or sign must be checked when the conductor is in this minimum sag position. An example of checking clearance using a minimum sag condition is shown in Fig. 234-3.

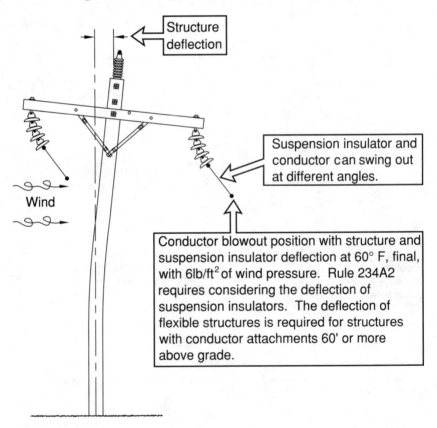

Fig. 234-2. Deflection of suspension insulators and flexible structures (Rule 234A2).

Rule 234A3 describes the method that is used to transition (T) between horizontal (H) and vertical (V) clearances. The T, H, and V values are shown in **NESC** Figs. 234-1(a), (b), and (c).

234B. Clearances of Wires, Conductors, and Cables from Other Supporting Structures.   Rule 234B is used to check clearance of wires, conductors, and cables to street lighting poles, traffic signal poles, or a pole of another power line or communication line. In these cases, the line in question is not attached to the pole in question. An example of a 12.47/7.2-kV, 3-phase, 4-wire distribution line adjacent to a street light pole is shown in Fig. 234-4.

There are times when a power line and street lighting compete for the same right of way and the only solution to meeting Rule 234B is bending the street light poles. See Fig. 234-5.

Another application of Rule 234B is a transmission line with an underbuild that has "skip span" construction. See Fig. 234-6.

The clearance adders for voltages above 22 kV are provided in Rule 234G. Alternate clearances for voltages exceeding 98 kV AC to ground or 139 kV DC to ground are provided in Rule 234H. The horizontal clearance in Rule 234B1 requires

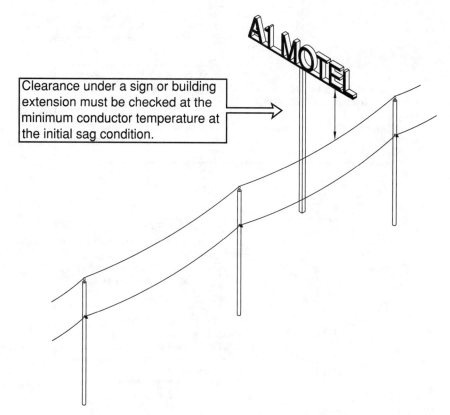

Clearance under a sign or building extension must be checked at the minimum conductor temperature at the initial sag condition.

Fig. 234-3. Example of checking clearance using a minimum sag condition (Rule 234A).

Per Rule 234B1b, the horizontal clearance between the phase conductor of a 12.47/7.2 kV, 3Ø, 4W distribution line and a street light pole must be not less than 4.5' with a 6lb/ft^2 wind applied to the conductor at 60°F final sag. No wind clearance is specified for an effectively grounded neutral (230E1), a secondary (230C3) cable below 750 V, or a communication cable. The clearance is measured after the line is placed in the blowout position.

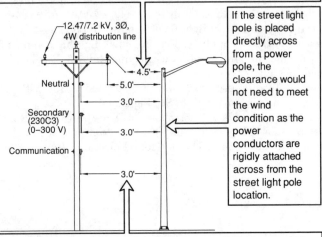

If the street light pole is placed directly across from a power pole, the clearance would not need to meet the wind condition as the power conductors are rigidly attached across from the street light pole location.

Per Rule 234B1a, the horizontal clearance between the phase conductor of a 12.47/7.2 kV, 3Ø, 4W distribution line and a street pole (without wind) must be not less than 5'.

An exception permits the neutral (230E1), secondary (230C3) up to 300 V, and communication cable to have a horizontal clearance (without wind) to be not less than 3'.

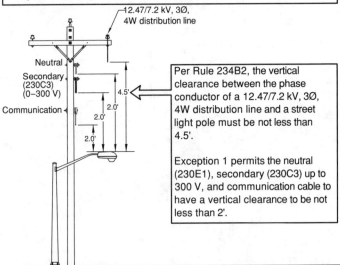

Per Rule 234B2, the vertical clearance between the phase conductor of a 12.47/7.2 kV, 3Ø, 4W distribution line and a street light pole must be not less than 4.5'.

Exception 1 permits the neutral (230E1), secondary (230C3) up to 300 V, and communication cable to have a vertical clearance to be not less than 2'.

See Photo(s)

Fig. 234-4. Example of clearance of a conductor to a street lighting pole (Rule 234B).

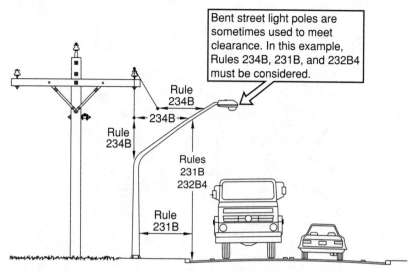

Bent street light poles are sometimes used to meet clearance. In this example, Rules 234B, 231B, and 232B4 must be considered.

Rule 234B

234B

Rule 234B

Rules 231B 232B4

Rule 231B

See Photo(s)

Fig. 234-5. Application of Rules 234B, 231B, and 232B4 to bent street lighting poles (Rule 234B).

clearance adders for voltages greater than 50 kV for conductors at rest and greater than 22 kV for conductors displaced by wind. The horizontal wind table in Rule 234B1 does not have a table number. The table does not address 230E1 (neutral conductors), 230C3 cables below 750 V (e.g., 120/240 V secondary triplex conductors), or communication cables. The vertical clearance in Rule 234B2 requires clearance adders for voltages greater than 50 kV. Separate vertical clearance values are provided for voltages less than 22 kV and voltages between 22 and 50 kV.

234C. Clearances of Wires, Conductors, Cables, and Rigid Live Parts from Buildings, Signs, Billboards, Chimneys, Radio and Television Antennas, Tanks, and Other Installations Except Bridges. Clearance from an energized conductor to a building or other installation is just as important as clearance to the surface below the line. As the title to Rule 234C states, this rule includes not only clearance to buildings, but also signs, billboards, chimneys, radio and television antennas, tanks, and other installations except bridges. This is one of the few rules in the **Code** that has graphics to help convey the **Code** requirements.

The first complete sentence of Rule 234C has four very important words that are not in the title of Rule 234C. The words are "…and any projections therefrom." This statement means that clearance must be maintained not just to the building structure, but to a gutter, awning, or any other projection from a building, sign, billboard, chimney, radio or television antenna, tank, or other installation except bridges. The "except bridges" statement is due to the fact that bridges have their own rule, Rule 234D.

The vertical and horizontal clearances for Rule 234C are determined by using **NESC** Table 234-1 with the aid of **NESC** Figs. 234-1(a), 234-1(b), and 234-1(c).

It is important to carefully read the title of **NESC** Table 234-1, "Clearance of Wires, Conductors, Cables, and Unguarded Rigid Live Parts Adjacent but Not

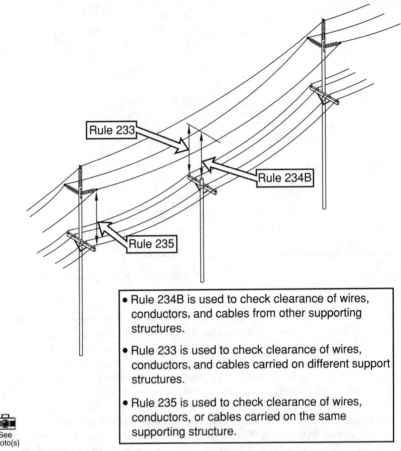

See
Photo(s)

Fig. 234-6. Application of Rules 234B, 233, and 235 to skip span construction (Rule 234B).

Attached to Buildings and Other Installations Except Bridges." The important wording is "…adjacent but not attached to…" This table is for conductors passing by a building, not a service that is attached to a building. Building services are covered in Rule 234C3.

The clearances in **NESC** Table 234-1 for conductors vertically over a building or other installation must be measured with the conductors at the largest final sag condition per Rule 234A. The vertical clearance for conductors under a sign or building projection must be measured with the conductors at the minimum sag condition as defined by Rule 234A. The clearances in **NESC** Table 234-1 for conductors to the side of (horizontal to) a building or other installation are at rest (no wind displacement) values. The horizontal clearance must also be checked using the wind condition in Rule 234A2. The horizontal clearances to conductors displaced by wind are provided in a small table in Rule 234C1. The table does not have an **NESC** table number. The table does not address 230E1 (neutral conductors), 230C3 cables below 750 V (e.g., 120/240 V secondary triplex conductors), or communication cables.

Two examples of building clearance and one example of billboard clearance are shown in Figs. 234-7, 234-8, and 234-9.

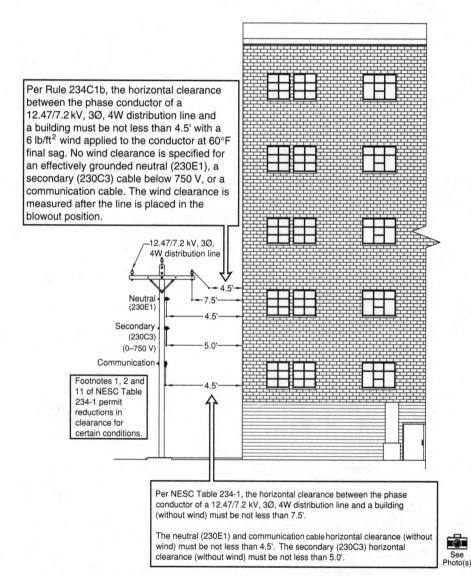

Per Rule 234C1b, the horizontal clearance between the phase conductor of a 12.47/7.2 kV, 3Ø, 4W distribution line and a building must be not less than 4.5' with a 6 lb/ft² wind applied to the conductor at 60°F final sag. No wind clearance is specified for an effectively grounded neutral (230E1), a secondary (230C3) cable below 750 V, or a communication cable. The wind clearance is measured after the line is placed in the blowout position.

12.47/7.2 kV, 3Ø, 4W distribution line

Neutral (230E1)

Secondary (230C3) (0–750 V)

Communication

Footnotes 1, 2 and 11 of NESC Table 234-1 permit reductions in clearance for certain conditions.

4.5'
7.5'
4.5'
5.0'
4.5'

Per NESC Table 234-1, the horizontal clearance between the phase conductor of a 12.47/7.2 kV, 3Ø, 4W distribution line and a building (without wind) must be not less than 7.5'.

The neutral (230E1) and communication cable horizontal clearance (without wind) must be not less than 4.5'. The secondary (230C3) horizontal clearance (without wind) must be not less than 5.0'.

See Photo(s)

Fig. 234-7. Example of horizontal clearance of conductors to a building (Rule 234C1).

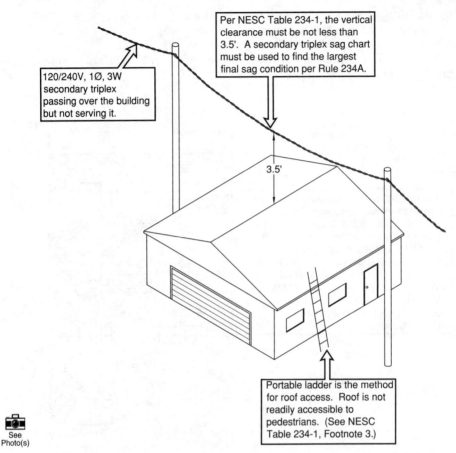

120/240V, 1Ø, 3W secondary triplex passing over the building but not serving it.

Per NESC Table 234-1, the vertical clearance must be not less than 3.5'. A secondary triplex sag chart must be used to find the largest final sag condition per Rule 234A.

3.5'

Portable ladder is the method for roof access. Roof is not readily accessible to pedestrians. (See NESC Table 234-1, Footnote 3.)

See Photo(s)

Fig. 234-8. Example of vertical clearance of a supply service drop conductor over a building but not serving it (Rule 234C1).

Rule 234C2 allows guarding as an option where **NESC** Table 234-1 clearances cannot be obtained. The note to this rule states that 230C1a cables are considered guarded. See the figure in Rule 230C for the details of a 230C1a cable.

Footnotes 1 and 2 to **NESC** Table 234-1 permit reduced clearance to buildings in certain cases. Footnote 1 has conditions related to maintenance and operations of the building or other structure. Footnote 2 permits clearance reductions for covered conductors. Covered conductor or "tree wire" is discussed in Rule 230D. Covered conductor applications along streets and in alleyways have become popular due to cramped space.

The **NESC** does not address clearance to buildings under construction. The Occupational Safety and Health Administration (OSHA) regulations may apply to building construction. Mobile cranes used in the vicinity of power lines are also addressed in the OSHA regulations. The Liquified Petroleum Gas Code (NFPA 58) specifies clearances to LP-Gas (propane) tanks that are located under power lines.

Per Rule 234C1b, the horizontal clearance between the phase conductor of a 12.47/7.2 kV, 3Ø, 4W distribution line and a billboard must be not less than 4.5' with a 6 lb/ft² wind applied to the conductor at 60°F final sag. No wind clearance is specified for an effectively grounded neutral (230E1), a secondary (230C3) cable below 750V, or a communication cable. The clearance is measured after the line is placed in the blowout position.

If the billboard is placed directly across from a power pole, the clearance would not need to meet the wind condition as the power conductors are rigidly attached across from the billboard location.

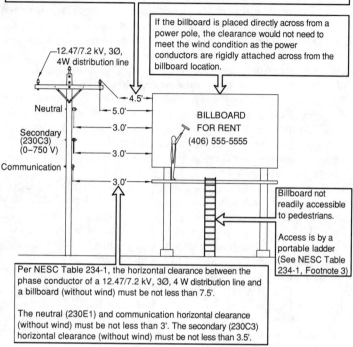

Per NESC Table 234-1, the horizontal clearance between the phase conductor of a 12.47/7.2 kV, 3Ø, 4 W distribution line and a billboard (without wind) must be not less than 7.5'.

The neutral (230E1) and communication horizontal clearance (without wind) must be not less than 3'. The secondary (230C3) horizontal clearance (without wind) must be not less than 3.5'.

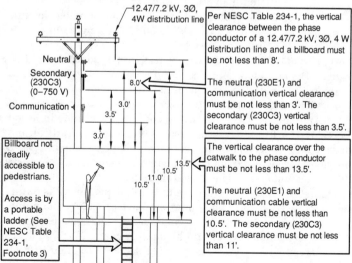

See Photo(s)

Fig. 234-9. Example of clearance of a conductor to a billboard (Rule 234C1).

The NESC does not specifically address clearance over fences or walls. Rule 012C, which requires accepted good practice, must be used. For a clearance over a wall that is wide enough to stand on, the vertical clearance over a building roof or the vertical clearance over a billboard catwalk in NESC Table 234-1 may be considered accepted good practice. For a clearance over a chain-link fence that can be climbed but is not wide enough to stand on, the vertical clearance over a sign or billboard in which personnel cannot walk may be considered accepted good practice. The choice of what accepted good practices to use depends on the given local conditions per Rule 012C.

Rule 234C3 covers supply conductors attached to buildings. The supply conductors attached to the building must connect to a service entrance into the building. The NESC and the National Electrical Code (NEC) overlap at the service entrance point. The NEC has similar rules for service entrance conductors and depending on the authority having jurisdiction (e.g., a city or state electrical inspector), the NEC rules may govern. Examples of supply conductors attached and connecting to a service entrance into a building are shown in Figs. 234-10, 234-11, and 234-12.

Rule 234C4 permits communication conductors to be attached directly to buildings without any specific clearance requirements. Rule 234C4 does not state that the rule applies only to an attachment necessary for an entrance into the building as Rule 234C3 does for supply conductors. NESC Table 234-1 does require clearances from insulated communication cables to buildings for communication cables that are not attached to the building.

Rule 234C5 requires a clear space for fire-fighting ladders for buildings exceeding three stories or 50 ft in height. A 6-ft wide zone should exist adjacent to the building or beginning not over 8 ft from the building. An exception applies if the fire department does not use ladders near supply conductors. A supply utility and the fire department should communicate with each other to discuss safety issues related to overhead lines.

The clearance adders for voltages above 22 kV are provided in Rule 234G. Alternate clearances for voltages exceeding 98 kV AC to ground or 139 kV DC to ground are provided in Rule 234H.

234D. Clearance of Wires, Conductors, Cables, and Unguarded Rigid Live Parts from Bridges. Depending on the size and design of a bridge structure, power and communication lines and equipment may be located adjacent to or within a bridge structure. NESC Table 234-2 applies to bridge structures. The clearances in Table 234-2 are for the at-rest (no wind) conditions outlined in Rule 234A.

NESC Figs. 234-1(a), (b), and (c), which focus on horizontal, vertical, and transitional clearances for signs and buildings, are also applicable to bridges. Bridges, unlike buildings and signs, do not have any required clearance to insulated communication cables and neutrals meeting Rule 230E1.

The horizontal clearance to bridges with wind displacement is outlined in a small table in Rule 234D. The table does not have an NESC table number. The table does not address 230E1 (neutral conductors), 230C3 cables below 750 V (e.g., 120/240-V secondary triplex conductors), or communication cables.

The clearance adders for voltages above 22 kV are provided in Rule 234G. Alternate clearances for voltages exceeding 98 kV AC to ground or 139 kV DC to ground are provided in Rule 234H.

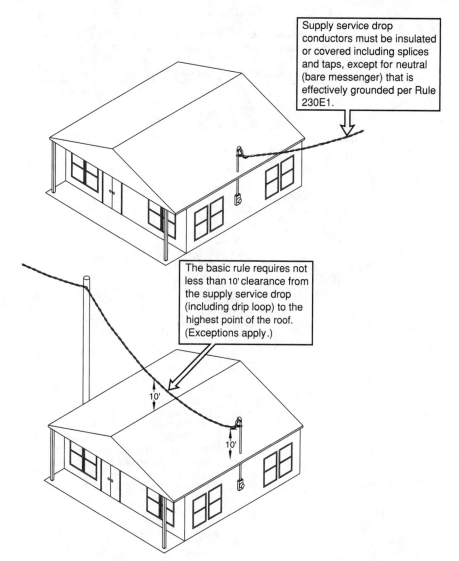

Supply service drop conductors must be insulated or covered including splices and taps, except for neutral (bare messenger) that is effectively grounded per Rule 230E1.

The basic rule requires not less than 10' clearance from the supply service drop (including drip loop) to the highest point of the roof. (Exceptions apply.)

10'

10'

Fig. 234-10. Example of clearance of supply conductors (service drops) attached to buildings (Rule 234C3).

234E. Clearance of Wires, Conductors, Cables, or Unguarded Rigid Live Parts Installed over or near Swimming Areas with No Wind Displacement. Rule 234E applies not only to swimming pools but also to swimming areas and beaches. NESC Table 234-3 and NESC Fig. 234-3 are provided in the Code to check clearance to swimming pools. The main concerns addressed in NESC Table 234-3 and NESC Fig. 234-3 are outlined in Fig. 234-13.

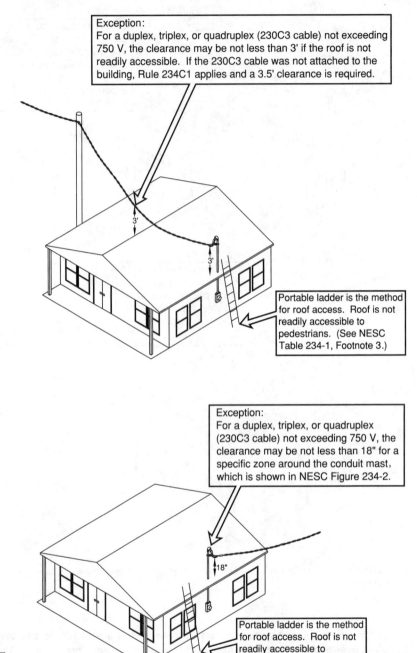

Exception:
For a duplex, triplex, or quadruplex (230C3 cable) not exceeding 750 V, the clearance may be not less than 3' if the roof is not readily accessible. If the 230C3 cable was not attached to the building, Rule 234C1 applies and a 3.5' clearance is required.

Portable ladder is the method for roof access. Roof is not readily accessible to pedestrians. (See NESC Table 234-1, Footnote 3.)

Exception:
For a duplex, triplex, or quadruplex (230C3 cable) not exceeding 750 V, the clearance may be not less than 18" for a specific zone around the conduit mast, which is shown in NESC Figure 234-2.

Portable ladder is the method for roof access. Roof is not readily accessible to pedestrians. (See NESC Table 234-1, Footnote 3.)

See Photo(s)

Fig. 234-11. Example of clearance of supply conductors (service drops) attached to buildings (Rule 234C3).

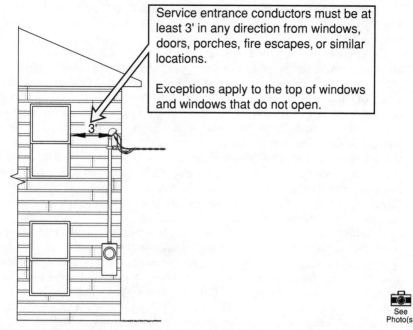

Service entrance conductors must be at least 3' in any direction from windows, doors, porches, fire escapes, or similar locations.

Exceptions apply to the top of windows and windows that do not open.

Fig. 234-12. Example of clearance of supply conductors (service drops) attached to buildings (Rule 234C3).

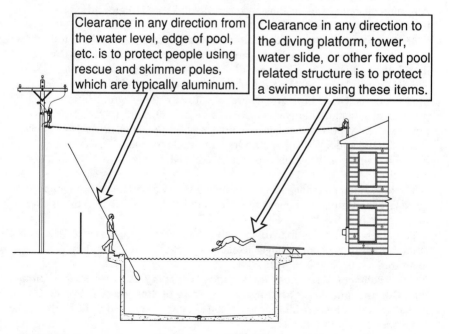

Clearance in any direction from the water level, edge of pool, etc. is to protect people using rescue and skimmer poles, which are typically aluminum.

Clearance in any direction to the diving platform, tower, water slide, or other fixed pool related structure is to protect a swimmer using these items.

Fig. 234-13. Concerns that are addressed by providing clearance to swimming pools (Rule 234E).

A fully enclosed indoor pool does not apply to Rule 234E1 per Exception 1. The National Electrical Code (NEC) contains requirements for electrical facilities around indoor pools.

Exception 2 to Rule 234E1 exempts communication conductors and cables, effectively grounded surge-protection wires, neutral conductors meeting Rule 230E1, guys and messengers, supply cables meeting Rule 230C1, and supply cables of 0 to 750 V meeting Rule 230C2 or 230C3 when these facilities are 10 ft or more horizontally from the edge of the pool, diving platform, diving tower, water slide, or other fixed pool-related structures. At that location Rule 232 (clearance above ground) clearances would apply.

The use of underground power is not required but it is certainly an acceptable design practice for routing lines near swimming pools. If underground lines are used, the underground conduit systems requirements in Sec. 32 and direct buried cable requirements in Sec. 35 both have rules for underground facilities near swimming pools.

Swimming areas that are not pools (e.g., beaches, etc.) and have lifeguards using rescue poles must also comply with Table 234-3. The clearances developed for swimming pools were based on the use of the long aluminum rescue and skimmer poles that are used near poolsides. If these poles are also used at a beach area, Table 234-3 applies. If these poles are not used, the clearances in Rule 232 (clearance above ground or water) apply. This is also true for water skiing areas. Rule 232 focuses on clearances above water as well as land and considers boating.

The clearance adders for voltages above 22 kV are provided in Rule 234G. Alternate clearances for voltages exceeding 98 kV AC to ground or 139 kV DC to ground are provided in Rule 234H.

234F. Clearance of Wires, Conductors, Cables, and Rigid Live Parts from Grain Bins. Grain bins have unique features that segregate them from Rule 234C (buildings and other installations) and place them in their own dedicated rule. The building clearance rules do apply to grain bins with some very specific modifications. Grain bins are separated into two types, grain bins loaded by permanently installed augers, conveyers, or elevators (Rule 234F1), and grain bins loaded by portable augeurs, conveyers, or elevators (Rule 234F2). NESC Figs. 234-4(a) and 234-4(b) are provided in the Code to check clearance to grain bins. The main concerns that are addressed in NESC Figs. 234-4(a) and 234-4(b) are outlined in Fig. 234-14.

The horizontal clearance to a grain bin loaded by a portable auger can become quite large for tall grain bins as the horizontal clearance is based on the height of the grain bin. See Fig. 234-15.

The clearance adders for voltages above 22 kV are provided in Rule 234G. Alternate clearances for voltages exceeding 98 kV AC to ground or 139 kV DC to ground are provided in Rule 234H.

234G. Additional Clearances for Voltages Exceeding 22 kV for Wires, Conductors, Cables, and Unguarded Rigid Live Parts of Equipment. The clearance adders in Rule 234G are similar to the clearance adders used in Rule 232C. See the discussion in Rule 232C.

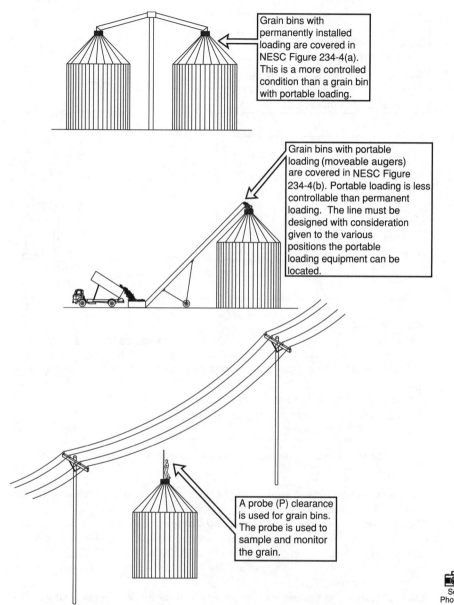

Grain bins with permanently installed loading are covered in NESC Figure 234-4(a). This is a more controlled condition than a grain bin with portable loading.

Grain bins with portable loading (moveable augers) are covered in NESC Figure 234-4(b). Portable loading is less controllable than permanent loading. The line must be designed with consideration given to the various positions the portable loading equipment can be located.

A probe (P) clearance is used for grain bins. The probe is used to sample and monitor the grain.

See Photo(s)

Fig. 234-14. Concerns that are addressed by providing clearance to grain bins (Rule 234F).

234H. Alternate Clearances for Voltages Exceeding 98 kV AC to Ground or 139 kV DC to Ground.　Alternate clearances may be used for voltages exceeding 98 kV AC to ground or 139 kV DC to ground per Rule 234H. The formulas in Rule 234H must be used for this calculation. The values in NESC Table 234-4 can be

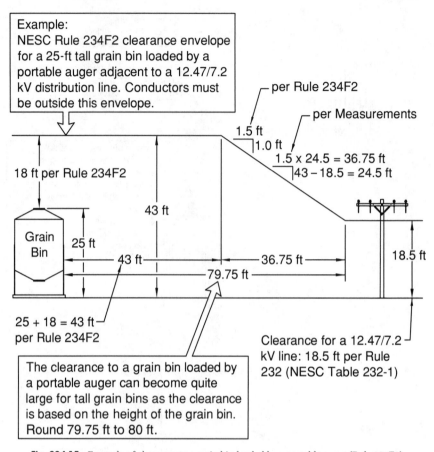

Example:
NESC Rule 234F2 clearance envelope for a 25-ft tall grain bin loaded by a portable auger adjacent to a 12.47/7.2 kV distribution line. Conductors must be outside this envelope.

per Rule 234F2

per Measurements

1.5 ft
1.0 ft
1.5 x 24.5 = 36.75 ft
43 – 18.5 = 24.5 ft

18 ft per Rule 234F2

43 ft

Grain Bin   25 ft

43 ft

36.75 ft

18.5 ft

79.75 ft

25 + 18 = 43 ft
per Rule 234F2

The clearance to a grain bin loaded by a portable auger can become quite large for tall grain bins as the clearance is based on the height of the grain bin. Round 79.75 ft to 80 ft.

Clearance for a 12.47/7.2 kV line: 18.5 ft per Rule 232 (NESC Table 232-1)

Fig. 234-15. Example of clearance to a grain bin loaded by a portable auger (Rule 234F2).

used to check the proper application of the formulas. See Rule 232D for a discussion of the alternate clearance formulas.

234I. Clearance of Wires, Conductors, and Cables to Rail Cars. NESC Fig. 234-5 is provided in the Code to aid the application of Rule 234I. Clearance to rail cars is also indirectly covered in Rule 232, NESC Table 232-1, Vertical Clearance over Track Rails of Railroads, and in Rule 231C, Clearance of Supporting Structures from Railroad Tracks.

234J. Clearance of Equipment Mounted on Supporting Structures. Unguarded rigid live parts are not subject to variations in sag. Unguarded rigid live parts are required to meet the clearances provided in Rule 234C or 234D, as applicable. Equipment cases, if effectively grounded, must meet Rule 234J2a. If an equipment case is not effectively grounded, it must meet Rule 234J2b. See Rule 215 for overhead equipment grounding requirements and see Sec. 02 for a discussion of the term "effectively grounded."

235. CLEARANCE FOR WIRES, CONDUCTORS, OR CABLES CARRIED ON THE SAME SUPPORTING STRUCTURE

235A. Application of Rule. Rule 235 is the opposite of Rule 233, Clearances between Wires, Conductors, and Cables Carried on Different Supporting Structures. Rule 235 addresses conductors on the same supporting structure (pole). Rule 235 is commonly used for checking vertical clearance on double circuit structures such as a transmission line with a distribution underbuild or a double circuit distribution line. Another common use of this rule is checking vertical clearance between the supply and communication conductors on a joint use line. Rule 235 should be used in conjunction with Rule 238, Vertical Clearance between Certain Communications and Supply Facilities Located on the Same Structure, when checking joint use (power and communication) clearances.

Rule 235A outlines the general requirements for checking clearance on the same supporting structure. Rule 235A does not immediately provide conductor temperature and loading conditions as Rules 232A, 233A, and 234A do. The conductors in Rule 235 are tied onto a structure. Conductor temperature and loading conditions will be needed to check conditions out in the span. Conductor temperature and loading conditions are provided in Rule 235B1b for horizontal clearance and Rule 235C2b(1)(c) for vertical clearance.

A multiconductor cable meeting Rule 230C or 230D is considered a single conductor for the purposes of this rule. See Fig. 235-1.

When comparing line conductors of different circuits, the voltage between the conductors is determined using the greater of the phasor difference or the phase-to-ground voltage of the higher circuit. This format is used in Rule 235 (Clearance between wires on the same supporting structure) but not in Rule 233 (Clearance between wires on different supporting structures). See Fig. 235-2.

235B. Horizontal Clearance between Line Conductors. Rule 235B applies to the horizontal clearance between line conductors attached to fixed supports (rigid insulators) on the same supporting structure. The horizontal clearance requirements apply between conductors of the same circuit or different circuits on the same supporting structure. For horizontal clearance between circuits of different voltage classifications on the same support arm (crossarm), see the additional requirements in Rule 235F. For horizontal clearance between conductors of different circuits on different supporting structures, Rule 233 applies instead of Rule 235.

The exception to Rule 235B1b states that the NESC does not specify a horizontal clearance between conductors of the same circuit when rated above 50 kV. A similar statement can be found in Rule 235C1 for vertical clearance. Typically, conductor clearance out in the span for transmission lines above 50 kV is checked by performing a galloping analysis. See Fig. 235-3.

The horizontal clearance must be checked at the support (crossarm) using NESC Table 235-1 and out in the span based on sag using either NESC Table 235-2 or

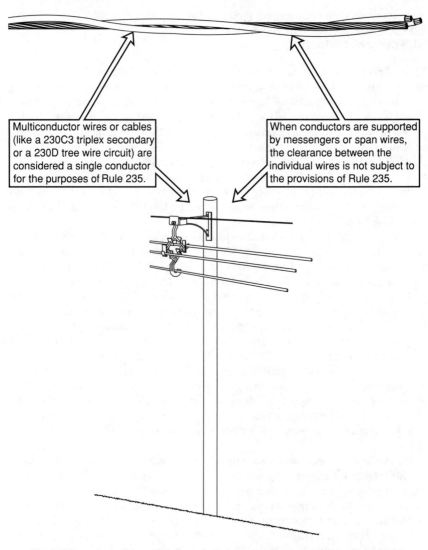

Multiconductor wires or cables (like a 230C3 triplex secondary or a 230D tree wire circuit) are considered a single conductor for the purposes of Rule 235.

When conductors are supported by messengers or span wires, the clearance between the individual wires is not subject to the provisions of Rule 235.

Fig. 235-1. Multiconductor cables and conductors supported by messengers (Rule 235A).

NESC Table 235-3 depending on the conductor size. NESC Tables 235-2 and 235-3 have a formula at the bottom of each table. The table or the formula can be used. The formulas are also presented in Rule 235B1b. The horizontal clearance at the structure will typically be controlled by the clearance required out in the span based on sag. See Fig. 235-4.

The values found in NESC Tables 235-1, 235-2, and 235-3 are clearances between conductors at supports, not spacing between supports. See Rule 230B

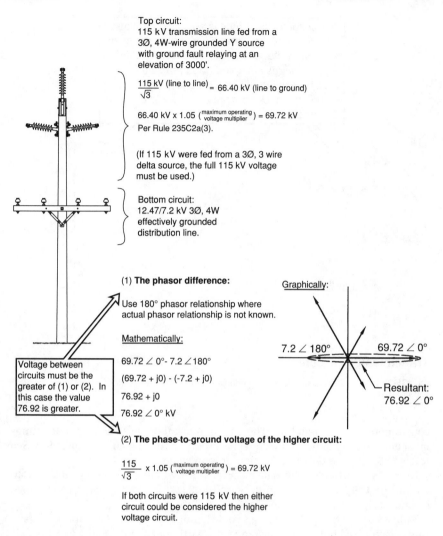

Top circuit:
115 kV transmission line fed from a
3Ø, 4W-wire grounded Y source
with ground fault relaying at an
elevation of 3000'.

$$\frac{115 \text{ kV (line to line)}}{\sqrt{3}} = 66.40 \text{ kV (line to ground)}$$

$$66.40 \text{ kV} \times 1.05 \left(\substack{\text{maximum operating} \\ \text{voltage multiplier}} \right) = 69.72 \text{ kV}$$
Per Rule 235C2a(3).

(If 115 kV were fed from a 3Ø, 3 wire
delta source, the full 115 kV voltage
must be used.)

Bottom circuit:
12.47/7.2 kV 3Ø, 4W
effectively grounded
distribution line.

(1) The phasor difference:

Graphically:

Use 180° phasor relationship where
actual phasor relationship is not known.

Mathematically:

$7.2 \angle 180°$ $69.72 \angle 0°$

$69.72 \angle 0° - 7.2 \angle 180°$

$(69.72 + j0) - (-7.2 + j0)$

$76.92 + j0$

Resultant:
$76.92 \angle 0°$

$76.92 \angle 0°$ kV

Voltage between
circuits must be the
greater of (1) or (2). In
this case the value
76.92 is greater.

(2) The phase-to-ground voltage of the higher circuit:

$$\frac{115}{\sqrt{3}} \times 1.05 \left(\substack{\text{maximum operating} \\ \text{voltage multiplier}} \right) = 69.72 \text{ kV}$$

If both circuits were 115 kV then either
circuit could be considered the higher
voltage circuit.

Fig. 235-2. Example of the voltage between line conductors of different circuits (Rule 235A).

for a discussion of spacing and clearance. Live metallic hardware electrically connected to the conductor, like a conductor tie wire on an insulator, must be addressed for the clearance at the structure but not for the clearance at the structure based on the sag out in the span. See Fig. 235-5.

Rule 235B2 has requirements for suspension insulators because **NESC** Table 235-1 is for fixed supports. The rules for suspension insulators require applying a 6-lb/ft^2 wind to the conductor that the suspension insulator is holding. The wind pressure may be reduced to 4 lb/ft^2 in sheltered areas but trees are not

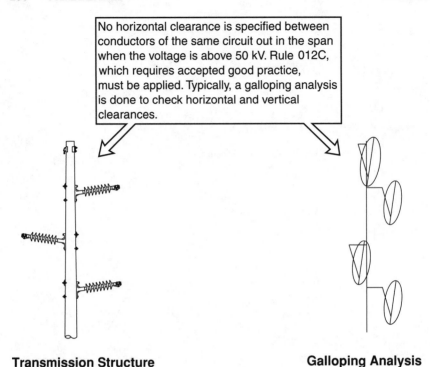

No horizontal clearance is specified between conductors of the same circuit out in the span when the voltage is above 50 kV. Rule 012C, which requires accepted good practice, must be applied. Typically, a galloping analysis is done to check horizontal and vertical clearances.

Transmission Structure

Galloping Analysis

Fig. 235-3. Horizontal clearance between conductors of the same circuit rated above 50 kV (Rule 235B1b).

considered a shelter to a line. The deflection of flexible structures and fittings must also be considered if the deflection reduces the clearance in question. See Rules 233A1a(2) and 234A2 for similar requirements.

Rule 235B3 provides alternate clearances for voltages exceeding 98 kV AC to ground or 139 kV DC to ground. See Rule 232D for a discussion of alternate clearances.

Rule 235B can be applied using two different approaches. One approach is to find the required horizontal clearance of a particular span. Another approach is to determine a maximum span based on the horizontal clearance for a particular structure type, conductor size, conductor sag, and circuit voltage.

235C. Vertical Clearance between Line Conductors.   Rule 235C applies to the vertical clearance between line conductors on the same supporting structure (pole). The vertical clearance requirements apply to conductors of the same and different circuits on the same supporting structure. For vertical clearance between conductors of different circuits on different supporting structures, Rule 233 applies instead of Rule 235. Rule 235C1a lists three items that the rule does not cover and therefore **NESC** Table 235-5 does not apply to. First, the **NESC** does not specify a vertical clearance between conductors of the same circuit when rated above 50 kV. A similar statement can be found in an exception

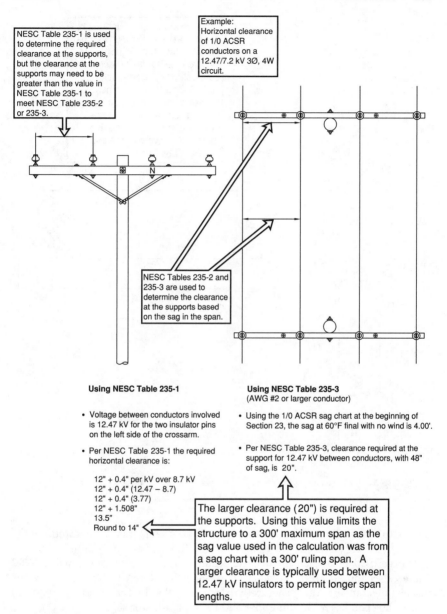

NESC Table 235-1 is used to determine the required clearance at the supports, but the clearance at the supports may need to be greater than the value in NESC Table 235-1 to meet NESC Table 235-2 or 235-3.

Example: Horizontal clearance of 1/0 ACSR conductors on a 12.47/7.2 kV 3Ø, 4W circuit.

NESC Tables 235-2 and 235-3 are used to determine the clearance at the supports based on the sag in the span.

Using NESC Table 235-1

- Voltage between conductors involved is 12.47 kV for the two insulator pins on the left side of the crossarm.

- Per NESC Table 235-1 the required horizontal clearance is:

  12" + 0.4" per kV over 8.7 kV
  12" + 0.4" (12.47 − 8.7)
  12" + 0.4" (3.77)
  12" + 1.508"
  13.5"
  Round to 14"

Using NESC Table 235-3
(AWG #2 or larger conductor)

- Using the 1/0 ACSR sag chart at the beginning of Section 23, the sag at 60°F final with no wind is 4.00'.

- Per NESC Table 235-3, clearance required at the support for 12.47 kV between conductors, with 48" of sag, is 20".

The larger clearance (20") is required at the supports. Using this value limits the structure to a 300' maximum span as the sag value used in the calculation was from a sag chart with a 300' ruling span. A larger clearance is typically used between 12.47 kV insulators to permit longer span lengths.

Fig. 235-4. Example of horizontal clearance between line conductors (Rule 235B).

to Rule 235B1b for horizontal clearance out in the span. Second, Rule 235C1a does not specify a vertical clearance between a secondary duplex, triplex, or quadruaplex (230C3) cable and a supply neutral (230E1) of the same utility. Third, Rule 235C1a does not specify a vertical clearance between ungrounded

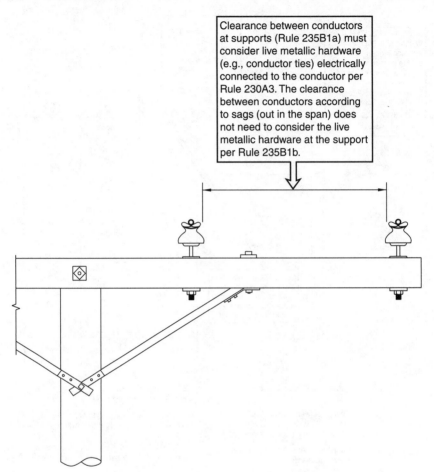

Clearance between conductors at supports (Rule 235B1a) must consider live metallic hardware (e.g., conductor ties) electrically connected to the conductor per Rule 230A3. The clearance between conductors according to sags (out in the span) does not need to consider the live metallic hardware at the support per Rule 235B1b.

Fig. 235-5. Horizontal clearance measurements between conductors at supports (Rule 235B).

open supply conductors (bare phase wires) of the same phase and circuit of the same utility (e.g., a tap off the main line of 0-50 kV). See Fig. 235-6.

The vertical clearance must be checked at the support (pole) using **NESC** Table 235-5 and out in the span based on sag using Rule 235C2b. Typically, the vertical clearance at the structure will be controlled by the clearance required out in the span based on the upper and lower conductor sags.

Per Rule 235C2b(1)(c), the upper conductor must be checked at the maximum operating temperature at final sag while the lower conductor is at the same ambient temperature as the upper conductor without electrical loading at final sag. The upper conductor must also be checked at 32°F with ice from the zone in Rule 230B at final sag, while the lower conductor is at the same ambient temperature as the upper conductor without electrical loading or ice at final sag.

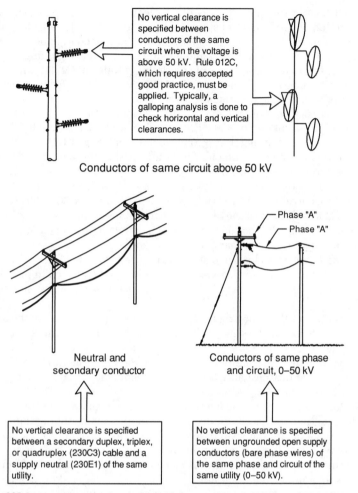

Conductors of same circuit above 50 kV

No vertical clearance is specified between conductors of the same circuit when the voltage is above 50 kV. Rule 012C, which requires accepted good practice, must be applied. Typically, a galloping analysis is done to check horizontal and vertical clearances.

Neutral and secondary conductor

Conductors of same phase and circuit, 0–50 kV

No vertical clearance is specified between a secondary duplex, triplex, or quadruplex (230C3) cable and a supply neutral (230E1) of the same utility.

No vertical clearance is specified between ungrounded open supply conductors (bare phase wires) of the same phase and circuit of the same utility (0–50 kV).

Fig. 235-6. Items not covered under vertical clearance between conductors (Rule 235C1a).

Using the 1/0 ACSR sample sag and tension chart at the beginning of Sec. 23, assume the following for an example of applying Rule 235C2b(1)(c)i:

- 1/0 ACSR phase conductor of a 12.47/7.2 kV, 3Ø, 4W, effectively grounded line is loaded to approximately 75% capacity (175 A) at both winter and summer peak periods, medium loading, clearance Zone 2, 300-ft ruling span.
- Upper phase conductor at 167°F, final. Sag is 6.25 ft.
- Lower neutral conductor at 104°F, final. Sag is 5.34 ft.

If the top and bottom conductors were attached at the same height, the top conductor would be sagging 0.91 ft (6.25 ft − 5.34 ft = 0.91 ft) below the bottom conductor. The top conductor attachment must be raised 0.91 ft (11 in) to make the sags level plus 12 in (16 in × 0.75 = 12 in per **NESC** Table 235-5 and

Rule 235C2b(1)(a)) to meet the required clearance in the span for a 7.2-kV phase conductor over a neutral conductor. The top conductor attachment must be 23 in (11 in + 12 in = 23 in) higher than the bottom conductor.

Using the 1/0 ACSR sample sag and tension chart at the beginning of Sec. 23, assume the following for an example of applying Rule 235C2b(1)(c)ii:

- 1/0 ACSR phase conductor of a 12.47/7.2 kV, 3Ø, 4W, effectively grounded line is loaded to approximately 75% capacity (175 A) at both winter and summer peak periods, medium loading, clearance Zone 2, 300-ft ruling span.
- Upper phase conductor at 32°F, 0.25 in ice, final. Sag is 4.49 ft.
- Lower neutral conductor at −20°F, no ice, final. Sag is 1.78 ft.

If the top and bottom conductors were attached at the same height, the top conductor would be sagging 2.71 ft (4.49 ft − 1.78 ft = 2.71 ft) below the bottom conductor. The top conductor attachment must be raised 2.71 ft (32.5 in) to make the sags level plus 12 in (16 in × 0.75 = 12 in per **NESC** Table 235-5 and Rule 235C2b(1)(a)) to meet the required clearance in the span for a 7.2-kV phase conductor over a neutral conductor. The top conductor attachment must be 32.5 in + 12 in = 44.5 in higher than the bottom conductor.

The greater vertical clearance at the structure between 23 in and 44.5 in is 44.5 in. A vertical clearance larger than 44.5 in will be needed to permit span lengths longer than the 300-ft ruling span in the sample sag and tension chart.

An exception to Rule 235C2b(1)(c) permits ignoring the rule when the conductors involved belong to the same utility, the conductors are the same size and type, and they are sagged the same. However, the exception is revoked if experience shows that different ice conditions do occur on the upper and lower conductor.

Examples of vertical clearance between line conductors are shown in Figs. 235-7 and Fig. 235-8.

Rule 235C can be applied using two different approaches. One approach is to find the required vertical clearance of a particular span. Another approach is to determine a maximum span based on the vertical clearance for a particular structure type, conductor size, conductor sag, and circuit voltage.

In addition to powerline-to-powerline clearance, **NESC** Table 235-5 applies to power-to-communication (joint use) clearance. Rule 235C and **NESC** Table 235-5 specify the conductor-to-conductor clearance. In addition to conductor-to-conductor clearance, a joint use **Code** review requires hardware-to-hardware clearance, which is specified in Rule 238. The vertical clearance requirements in Rule 238B and **NESC** Table 235-5 revolve around 40 in. Less than 40 in (e.g., 30 in) is acceptable for certain grounded supply facilities. Greater than 40 in is required for supply voltages above 8.7 kV to ground. The vertical clearance must be checked at the support (pole) using **NESC** Table 235-5 and out in the span based on sag using Rule 235C2b. Typically, the vertical clearance at the structure will be controlled by the clearance required out in the span based on the upper and lower conductor sags. Depending on what type of conductors are involved, a power conductor sag and tension chart and a communications cable sag and tension chart will be needed to check supply-to-communication clearance at midspan. Examples of joint use structures are shown in Figs. 235-9 through 235-13.

Example:
Vertical clearance between 1/0 ACSR phase and
neutral conductors on a 12.47/7.2 kV, 3Ø, 4W circuit.

NESC Table 235-5 is used to
determine the required
clearance at the supports, but
the clearance at the supports
may need to be greater than
the value in NESC Table 235-5
to meet Rule 235C2b(1)(a).

The vertical clearance between
the outer phase conductor and
the neutral does not need to be
checked if adequate horizontal
clearance exists per Rule 235B.

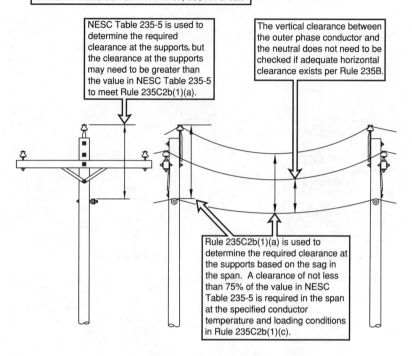

Rule 235C2b(1)(a) is used to
determine the required clearance at
the supports based on the sag in
the span. A clearance of not less
than 75% of the value in NESC
Table 235-5 is required in the span
at the specified conductor
temperature and loading conditions
in Rule 235C2b(1)(c).

Using NESC Table 235-5

- Conductor at upper level is 7.2 kV phase
  to ground.

- Conductor at lower level is a 230E1 neutral.

- Per NESC Table 235-5, the required vertical
  clearance at the support is 16".

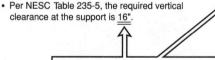

Using Rule 235C2b (1)(a)

- Per Rule 235C2b(1)(a), the required vertical clearance
  in the span is 75% of that required at the support.
  .75 x 16" = 12"

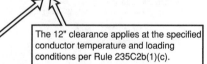

The 12" clearance applies at the specified
conductor temperature and loading
conditions per Rule 235C2b(1)(c).

The 16" vertical clearance is at the support and
the 12" vertical clearance is out in the span.
The conductor attachment locations at the
support must be determined to maintain 12" of
vertical clearance out in the span. The greater
of the two values (16" of vertical clearance at
the support or the conductor attachment
locations to maintain 12" of vertical clearance
out in the span) must be used. Typically,
more than 16" of clearance will be needed
at the structure to maintain 12" of clearance
out in the span.

Fig. 235-7. Example of vertical clearance between line conductors (Rule 235C).

Example:
Vertical clearance between a 115 kV transmission line with ground fault relaying and a 12.47/7.2 kV effectively grounded distribution line. The lines are owned by the same utility and the line is built at an elevation of 3000'. Assume the circuits are 180° out of phase.

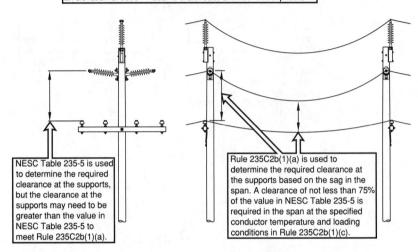

NESC Table 235-5 is used to determine the required clearance at the supports, but the clearance at the supports may need to be greater than the value in NESC Table 235-5 to meet Rule 235C2b(1)(a).

Rule 235C2b(1)(a) is used to determine the required clearance at the supports based on the sag in the span. A clearance of not less than 75% of the value in NESC Table 235-5 is required in the span at the specified conductor temperature and loading conditions in Rule 235C2b(1)(c).

Using Table 235-5

- NESC Table 235-5 only goes to 50 kV. Rule 235C2a provides clearance adders for voltages above 50 kV.

- $\dfrac{115 \text{ kV} \text{ (line to line)}}{\sqrt{3}}$ = 66.40 kV (line to ground)

  66.40 kV x 1.05 = 69.72 kV

  (1.05 is the maximum operating voltage multiplier)

- Conductor at lower level is 7.2 kV phase to ground.

- The greater of the phasor difference and the phase to ground voltage of the higher circuit is 76.92 kV. (See Fig. 235-2)

- Per NESC Table 235-5, vertical clearance at the support is:

  16" + 0.4" per kV over 8.7 kV
  16" + 0.4"/kV (50 – 8.7)
  16" + 16.52"
  32.52"

Using Rule 235C2a

32.52" + .4 (76.92 – 50)
32.52" + 10.77"
43.29"
Round to 44"

Using Rule 235C2b

- Per Rule 235C2b(1)(a), the required vertical clearance in the span is 75% of that required at the support up to 50 kV. Per Rule 235C2b(1)(b),100% (not 75%) of the required vertical clearance is required in the span for greater than 50 kV.
  (0.75 x 32.52) + 10.77" = 35.16"
  Round to 36"

The 36" clearance applies at the specified conductor temperature and loading conditions per Rule 235C2b(1)(c).

The 44" vertical clearance is at the support and the 36" vertical clearance is out in the span. The conductor attachment locations at the support must be determined to maintain 36" of vertical clearance out in the span. The greater of the two values (44" of vertical clearance at the support or the conductor attachment locations to maintain 36" of vertical clearance out in the span) must be used.

Typically, more than 44" of clearance will be needed at the structure to maintain 36" of clearance out in the span.

See Photo(s)

Fig. 235-8. Example of vertical clearance between line conductors (Rule 235C).

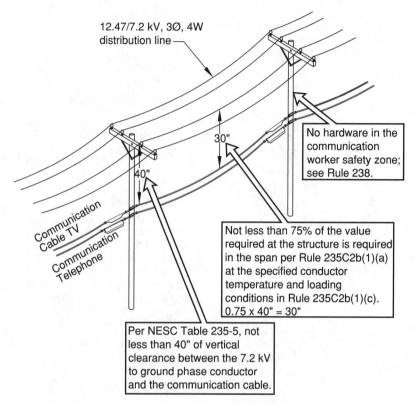

12.47/7.2 kV, 3Ø, 4W
distribution line

30"

40"

Communication
Cable TV

Communication
Telephone

No hardware in the
communication
worker safety zone;
see Rule 238.

Not less than 75% of the value
required at the structure is required
in the span per Rule 235C2b(1)(a)
at the specified conductor
temperature and loading
conditions in Rule 235C2b(1)(c).
0.75 x 40" = 30"

Per NESC Table 235-5, not
less than 40" of vertical
clearance between the 7.2 kV
to ground phase conductor
and the communication cable.

Fig. 235-9. Example of vertical clearance between joint use (supply and communication) conductors (Rule 235C).

Exceptions 2 and 3 of Rule 235C1 apply to joint use construction. See Fig. 235-14.

In addition to the sag-related clearance checks in Rule 235C2b(1), Rule 235C2b(2) requires making sag adjustments when necessary to maintain clearance. When sag is reduced, tension is increased. If sag reductions are made to maintain clearance, the tension limits in Rule 261H1 must not be exceeded. Rule 235C2b(3) requires a special clearance check for joint use structures. See Fig. 235-15.

Alternate clearances may be used for voltages exceeding 98 kV AC to ground or 139 kV DC to ground. The formulas in Rule 233C3 must be used for this calculation. See Rule 232D for a discussion of the alternate clearance formulas.

Rule 235C4 labels the area between the supply space and the communication space both at the structure and out in the span. The name for this area is the "communication worker safety zone." The work rules in Part 4 of the NESC provide the qualifications of a supply employee and a communication employee. Since the communication employee is not trained to work on supply lines, a safety zone exists for the communication employee's protection. If a communication line

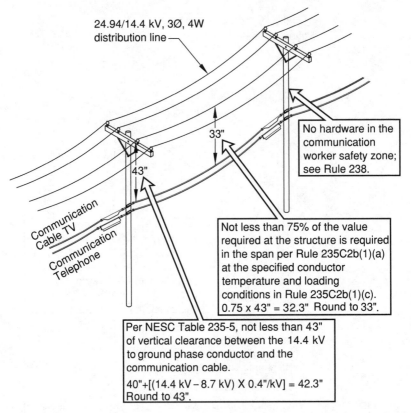

24.94/14.4 kV, 3Ø, 4W distribution line

33"

43"

No hardware in the communication worker safety zone; see Rule 238.

Not less than 75% of the value required at the structure is required in the span per Rule 235C2b(1)(a) at the specified conductor temperature and loading conditions in Rule 235C2b(1)(c). 0.75 x 43" = 32.3" Round to 33".

Communication Cable TV

Communication Telephone

Per NESC Table 235-5, not less than 43" of vertical clearance between the 14.4 kV to ground phase conductor and the communication cable.
40"+[(14.4 kV – 8.7 kV) X 0.4"/kV] = 42.3" Round to 43".

Fig. 235-10. Example of vertical clearance between joint use (supply and communication) conductors (Rule 235C).

is positioned below a supply line on an overhead structure, but the required communication worker safety zone clearance does not exist, the communication employee can correct the violation if the communication employee does not violate the approach distances and other requirements in Sec. 43. If the communication employee cannot maintain the approach distances in Sec. 43, the communication employee must contact a supply employee to correct the violation. See Sec. 43 for additional information. The communication worker safety zone is defined by both Rule 238 and Rule 235C. Rules 238C, 238D, and 239 permit only a few select items in the communications worker safety zone. See the communication worker safety zone figure in Rule 238E.

235D. Diagonal Clearance between Line Wires, Conductors, and Cables Located at Different Levels in the Same Supporting Structure.   Diagonal clearances are not specified in Rule 235, only horizontal and vertical clearances. NESC Fig. 235-1 is provided in the Code for determining diagonal clearances. If both a horizontal and vertical clearance apply to a conductor, a diagonal

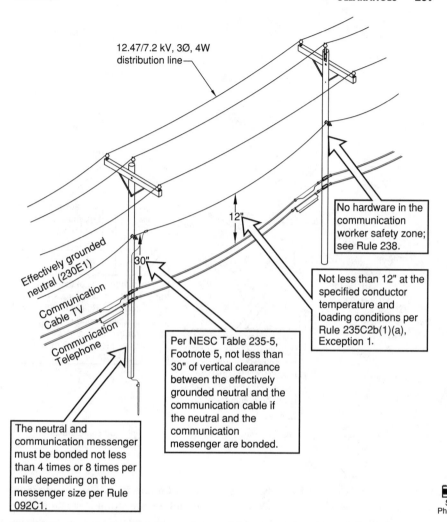

12.47/7.2 kV, 3Ø, 4W
distribution line

No hardware in the
communication
worker safety zone;
see Rule 238.

12"

Not less than 12" at the
specified conductor
temperature and
loading conditions per
Rule 235C2b(1)(a),
Exception 1.

Effectively grounded
neutral (230E1)

Communication
Cable TV

Communication
Telephone

30"

Per NESC Table 235-5,
Footnote 5, not less than
30" of vertical clearance
between the effectively
grounded neutral and the
communication cable if
the neutral and the
communication
messenger are bonded.

The neutral and
communication messenger
must be bonded not less
than 4 times or 8 times per
mile depending on the
messenger size per Rule
092C1.

See
Photo(s)

Fig. 235-11. Example of vertical clearance between joint use (supply and communication)
conductors (Rule 235C).

clearance can be determined by drawing a box with the required horizontal
clearance from Rule 235B and vertical clearance from Rule 235C.

**235E. Clearances in Any Direction from Line Conductors to Supports, and to Ver-
tical or Lateral Conductors, Span, or Guy Wires Attached to the Same Support.**
NESC Table 235-6 is used to find clearance in any direction (not just horizontal or
vertical) from line conductors to supports and to vertical or lateral conductors,
span, or guy wires attached to the support. The term line conductors refers
to the conductors spanning from pole to pole. Vertical and lateral conductors are
discussed in Rule 239. See Rule 239A for a figure showing vertical and lateral

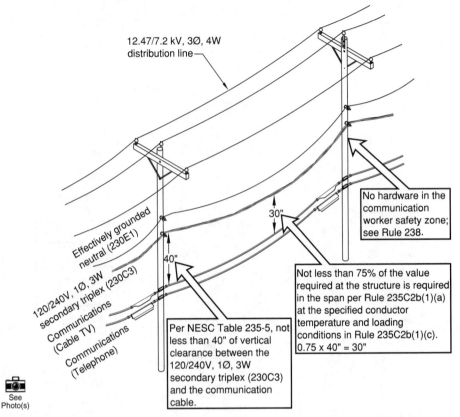

12.47/7.2 kV, 3Ø, 4W
distribution line

No hardware in the communication worker safety zone; see Rule 238.

30"

Effectively grounded neutral (230E1)

120/240V, 1Ø, 3W secondary triplex (230C3)

Communications (Cable TV)

Communications (Telephone)

40"

Not less than 75% of the value required at the structure is required in the span per Rule 235C2b(1)(a) at the specified conductor temperature and loading conditions in Rule 235C2b(1)(c).
0.75 x 40" = 30"

Per NESC Table 235-5, not less than 40" of vertical clearance between the 120/240V, 1Ø, 3W secondary triplex (230C3) and the communication cable.

See Photo(s)

Fig. 235-12. Example of vertical clearance between joint use (supply and communication) conductors (Rule 235C).

conductors. **NESC** Table 235-6 is commonly used for checking clearance to guys on complicated guying structures. Another common application is clearance of a conductor to the supporting structure (pole). See Fig. 235-16.

The notes in Rule 235E1 direct the reader to Rule 235I for clearance to communication antennas in the supply space and to Rule 236D1 for clearance to communication antennas in the communication space. Rule 238B is also a good reference as it is applicable to communication antennas in the communication space under certain conditions. See Rules 235I and 238B for additional information.

Rule 235E2 has requirements for suspension insulators because **NESC** Table 235-6 is for fixed supports. The rules for suspension insulators require applying a 6-lb/ft^2 wind to the conductor that the suspension insulator is holding. The wind pressure may be reduced to 4 lb/ft^2 in sheltered areas, but trees are not considered a shelter to a line. The deflection of flexible structures and fittings must also be considered if the deflection reduces the clearance in question. See Rules 233A1(a)(2) and 234A2 for similar requirements.

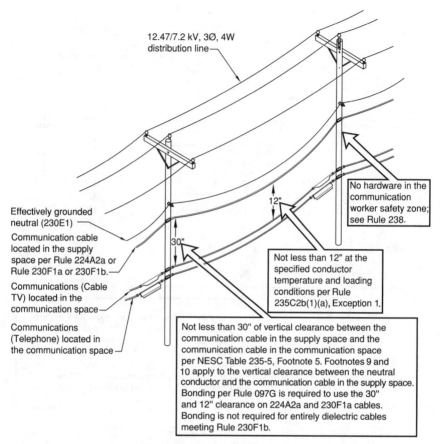

12.47/7.2 kV, 3Ø, 4W
distribution line

No hardware in the
communication
worker safety zone;
see Rule 238.

Effectively grounded
neutral (230E1)

Communication cable
located in the supply
space per Rule 224A2a or
Rule 230F1a or 230F1b.

Communications (Cable
TV) located in the
communication space

Communications
(Telephone) located in
the communication space

12"

Not less than 12" at the
specified conductor
temperature and loading
conditions per Rule
235C2b(1)(a), Exception 1.

30"

Not less than 30" of vertical clearance between the
communication cable in the supply space and the
communication cable in the communication space
per NESC Table 235-5, Footnote 5. Footnotes 9 and
10 apply to the vertical clearance between the neutral
conductor and the communication cable in the supply space.
Bonding per Rule 097G is required to use the 30"
and 12" clearance on 224A2a and 230F1a cables.
Bonding is not required for entirely dielectric cables
meeting Rule 230F1b.

Fig. 235-13. Example of vertical clearance between joint use (supply and communication)
conductors (Rule 235C).

Rule 235E3 provides alternate clearances for voltages exceeding 98 kV AC to
ground or 139 kV DC to ground. The values in NESC Table 235-7 can be used to
check the proper application of the formulas. See Rule 232D for a discussion of
alternate clearances.

**235F. Clearances between Circuits of Different Voltage Classifications Located in
the Supply Space on the Same Support Arm.**   Rule 235F recognizes that circuits of different voltage levels are not always separated vertically on a structure. Rule 220C assumes circuits of different voltage levels will be separated
vertically at different heights and does not comment on horizontal separation.
Rule 235F must be used in conjunction with Rule 235B, Horizontal Clearance
between Line Conductors, and in some cases Rule 236, Climbing Space.

The NESC uses the term support arm for a crossarm. The terms bridge arm and
sidearm are also used in this rule. A bridge arm is an arm that spans between
two poles. A sidearm is an arm to one side of the pole. A sidearm is commonly

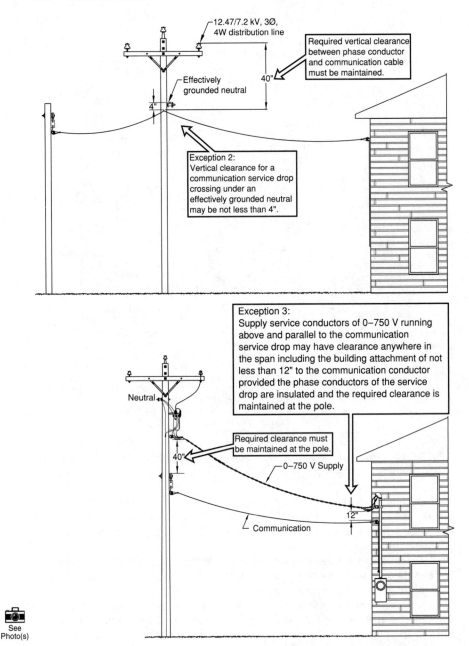

Fig. 235-14. Exceptions to vertical clearance requirements for joint use construction (Rule 235C1).

See Photo(s)

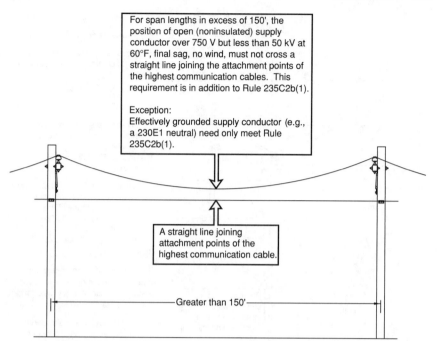

For span lengths in excess of 150', the position of open (noninsulated) supply conductor over 750 V but less than 50 kV at 60°F, final sag, no wind, must not cross a straight line joining the attachment points of the highest communication cables. This requirement is in addition to Rule 235C2b(1).

Exception:
Effectively grounded supply conductor (e.g., a 230E1 neutral) need only meet Rule 235C2b(1).

A straight line joining attachment points of the highest communication cable.

————————————Greater than 150'————————————

Fig. 235-15. Additional vertical clearance requirement for joint use lines [Rule 235C2b(3)].

called an alley arm. Examples of clearance between supply circuits of different voltages on the same support arm are outlined in Fig. 235-17.

235G. Conductor Spacing: Vertical Racks or Separate Brackets.   Reduced vertical clearance between conductors is permitted when vertical racks are used. The rules for conductor spacing on vertical racks are outlined in Fig. 235-18.

235H. Clearance and Spacing between Communication Conductors, Cables, and Equipment.   Rule 235H1 addresses the clearance and spacing between communication cables. The rule does not state if the clearance and spacing values are vertical, horizontal, or radial. Typically, communication cables are attached directly to a pole and separated vertically on the pole, but they can be separated horizontally by placing the cables on opposite sides of the pole or on a crossarm. A not less than 12-in spacing is required between messengers supporting communication cables. The 12-in value is a spacing dimension, not a clearance dimension. See Rule 230B for a discussion of clearance versus spacing.

Communication cables supported on messengers are installed by first attaching the messenger to the pole and then lashing the communication cable to the messenger. The lashing machine travels down the messenger and needs the 12-in spacing for proper operating room. The 12-in spacing between messengers may be reduced by agreement between the parties involved. It is not specified if the parties involved include only the various communication cable attachees or if the pole owner is considered an involved party. The pole owner should determine if they want to be involved.

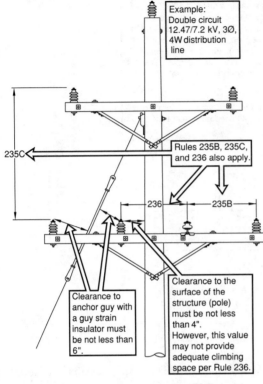

Example:
Double circuit
12.47/7.2 kV, 3Ø,
4W distribution
line

235C

Rules 235B, 235C,
and 236 also apply.

236   235B

Clearance to
anchor guy with
a guy strain
insulator must
be not less than
6".

Clearance to the
surface of the
structure (pole)
must be not less
than 4".
However, this value
may not provide
adequate climbing
space per Rule 236.

See
Photo(s)

Using Table 235-6

- Table voltages are phase to phase.

- Using over 8.7 to 50 kV column and
  row 2b (anchor guys), the clearance
  required to the anchor guy is
  6" + 0.25" per kV over 8.7 kV.
  6" + 0.25" (12.47 − 8.7) = 6.94"
  Round to 7"

- If a guy strain insulator is used,
  Footnote 11 applies and a 25%
  reduction to 6.94" is applicable.
  7" − [7" x 0.25] = 5.25"
  Round to 6"

- Using over 8.7 to 50 kV column and
  row 4b (all other), the clearance
  required to the surface of the
  structure (pole) is 3" + 0.2" per kV
  over 8.7 kV.
  3" + 0.2" (12.47 − 8.7) = 3.75"
  Round to 4"

- Greater than 4" is required to the
  surface of the structure for joint use
  poles. Less than 4" is permitted when
  footnote 10 is applicable.

- Rule 235B must be used to check
  the horizontal clearance between
  conductors. Rule 235C must be
  used to check the vertical clearance
  between conductors. Rule 236 must
  be checked for climbing space.

Fig. 235-16. Example of clearance of a conductor to an anchor guy and a conductor to a support (Rule 235E).

Rule 235H2 requires a clearance (not spacing) between cables and equipment of different communication utilities anywhere in the span to be not less than 4 in. This requirement keeps a communication circuit owned by one utility from sagging into or below the a communication circuit under it owned by a different utility. No temperature and loading conditions are specified for checking the 4-in clearance. Rule 235 does not provide conductor temperature and loading conditions in Rule 235A for use with the entire rule. Rules 232, 233, and 234 use this format but Rule 235 specifies the conductor temperature and loading conditions in each paragraph. For example, Rules 235B and 235C have within the rule the conductor temperature and loading conditions that apply. Rule 235H does not provide any conductor temperature and loading conditions to check the 4-in clearance. The 4-in clearance may be reduced (or eliminated) by agreement between the parties involved. It is not specified if the parties involved include only the various communication cable attachees or if the pole owner is considered an involved party. The pole owner should determine if they want to be involved.

The **Code** rules related to clearance and spacing between communication lines are outlined in Fig. 235-19.

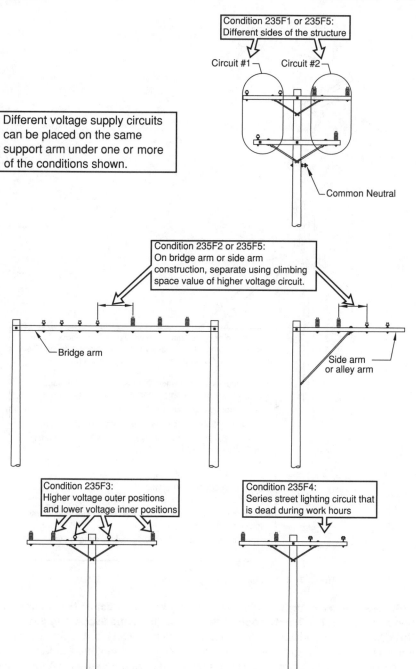

Different voltage supply circuits can be placed on the same support arm under one or more of the conditions shown.

Condition 235F1 or 235F5:
Different sides of the structure

Circuit #1   Circuit #2

Common Neutral

Condition 235F2 or 235F5:
On bridge arm or side arm construction, separate using climbing space value of higher voltage circuit.

Bridge arm

Side arm or alley arm

Condition 235F3:
Higher voltage outer positions and lower voltage inner positions

Condition 235F4:
Series street lighting circuit that is dead during work hours

See Photo(s)

Fig. 235-17. Examples of clearance between supply circuits of different voltages on the same support arm (Rule 235F).

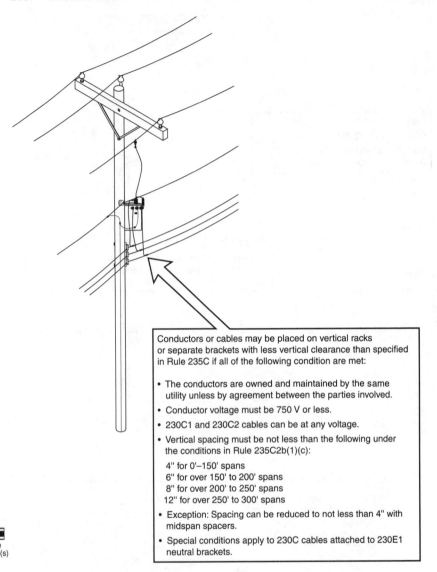

Conductors or cables may be placed on vertical racks or separate brackets with less vertical clearance than specified in Rule 235C if all of the following condition are met:

- The conductors are owned and maintained by the same utility unless by agreement between the parties involved.
- Conductor voltage must be 750 V or less.
- 230C1 and 230C2 cables can be at any voltage.
- Vertical spacing must be not less than the following under the conditions in Rule 235C2b(1)(c):

  4" for 0'–150' spans
  6" for over 150' to 200' spans
  8" for over 200' to 250' spans
  12" for over 250' to 300' spans

- Exception: Spacing can be reduced to not less than 4" with midspan spacers.
- Special conditions apply to 230C cables attached to 230E1 neutral brackets.

See Photo(s)

Fig. 235-18. Conductor spacing on vertical racks (Rule 235G).

235I. Clearances in Any Direction from Supply Line Conductors to Communication Antennas in the Supply Space Attached to the Same Supporting Structure.
Rule 235I provides clearance between supply conductors and communication antennas and communication antenna equipment cases mounted in the supply space. See Rule 238B for communication antennas mounted in the communication space. Typically, the communication antenna would be mounted at the top of the pole above the supply conductors in the supply space. Rule 235I references

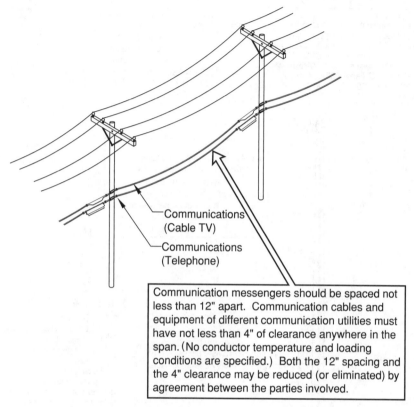

Communications
(Cable TV)

Communications
(Telephone)

Communication messengers should be spaced not less than 12" apart. Communication cables and equipment of different communication utilities must have not less than 4" of clearance anywhere in the span. (No conductor temperature and loading conditions are specified.) Both the 12" spacing and the 4" clearance may be reduced (or eliminated) by agreement between the parties involved.

See
Photo(s)

Fig. 235-19. Clearance and spacing between communication lines (Rule 235H).

specific rows in **NESC** Table 235-6 to obtain the required clearance values. Normally, a communication vertical riser would be located on the pole that has a communication antenna. Requirements for vertical communication conductors passing through the supply space on jointly used structures (poles) are found in Rule 239H. When a communication antenna is located in the supply space, the worker installing or maintaining the communication antenna must be qualified to work in the supply space per the work rules (Secs. 42 and 44) of the **NESC**. Examples of clearance between supply lines and communication antennas in the supply space are shown in Fig. 235-20.

236. CLIMBING SPACE

Rule 236, Climbing Space, and Rule 237, Working Space, may appear complicated but they revolve around one simple idea. The supply and communication worker must have adequate space to climb and work on a pole and its conductors, supports, and equipment.

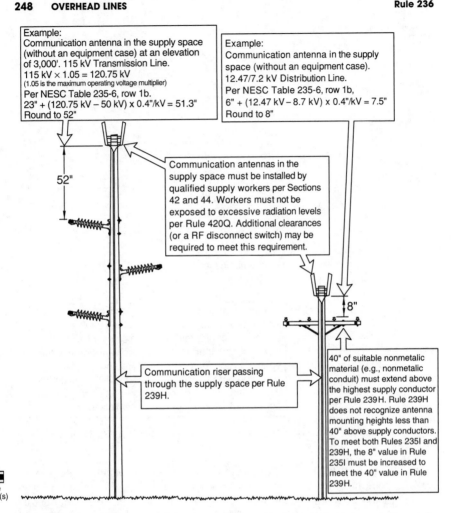

Example:
Communication antenna in the supply space (without an equipment case) at an elevation of 3,000'. 115 kV Transmission Line.
115 kV × 1.05 = 120.75 kV
(1.05 is the maximum operating voltage multiplier)
Per NESC Table 235-6, row 1b.
23" + (120.75 kV − 50 kV) x 0.4"/kV = 51.3"
Round to 52"

Example:
Communication antenna in the supply space (without an equipment case).
12.47/7.2 kV Distribution Line.
Per NESC Table 235-6, row 1b,
6" + (12.47 kV − 8.7 kV) x 0.4"/kV = 7.5"
Round to 8"

52"

Communication antennas in the supply space must be installed by qualified supply workers per Sections 42 and 44. Workers must not be exposed to excessive radiation levels per Rule 420Q. Additional clearances (or a RF disconnect switch) may be required to meet this requirement.

8"

Communication riser passing through the supply space per Rule 239H.

40" of suitable nonmetalic material (e.g., nonmetalic conduit) must extend above the highest supply conductor per Rule 239H. Rule 239H does not recognize antenna mounting heights less than 40" above supply conductors. To meet both Rules 235I and 239H, the 8" value in Rule 235I must be increased to meet the 40" value in Rule 239H.

See Photo(s)

Fig. 235-20. Examples of clearance between supply lines and communication antennas in the supply space (Rule 235I).

The first sentence of Rule 236 is critical to applying (or not applying) the entire rule. Rule 236 only applies if workers climb the pole. If the pole is worked on from a bucket truck, the climbing space rules do not apply. See Fig. 236-1.

Climbing space may be provided by temporarily moving the line conductors using live line tools per Exception 3 of Rule 236E. The NESC work rules in Part 4 must be applied when climbing structures and moving live conductors.

Rule 236 provides requirements for measuring the climbing space in Rule 236B, positioning arms in the climbing space in Rule 236C, and locating equipment in the climbing space in Rule 236D. Rule 236E provides the basic

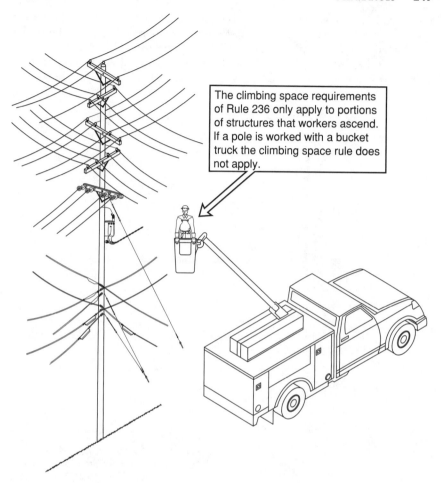

The climbing space requirements of Rule 236 only apply to portions of structures that workers ascend. If a pole is worked with a bucket truck the climbing space rule does not apply.

Fig. 236-1. Climbing space application (Rule 236).

horizontal clearances between conductors bounding the climbing space. **NESC** Table 236-1 provides horizontal clearances between conductors. The horizontal clearances in **NESC** Table 236-1 are intended to provide 24 in of clear climbing space while the conductors bounding the climbing space are covered with temporary insulation (e.g., line guards and insulator hoods). For example, a 30-in clearance on **NESC** Table 236-1 assumes the temporary insulation on both sides of the climbing space is 6 in, leaving 24 in of clear climbing space. As the voltages on the table increase, the clearances increase as the temporary insulation is assumed to be thicker and bulkier. Examples of climbing space between conductors are shown in Figs. 236-2 and 236-3.

Methods to provide climbing space on buckarm construction are provided in Rule 236F. Rule 236G provides rules for climbing space past longitudinal runs not on support arms. The location, size, and quantity of the longitudinal runs not on support arms (e.g., communication cables attached to the pole) must be considered when determining if a qualified worker can climb past them. Secondary service drops are also addressed in this rule, and **NESC** Fig. 236-1 is provided as a guide.

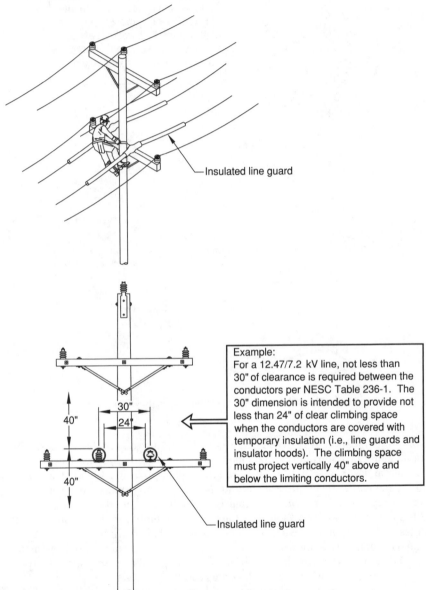

Insulated line guard

Example:
For a 12.47/7.2 kV line, not less than 30" of clearance is required between the conductors per NESC Table 236-1. The 30" dimension is intended to provide not less than 24" of clear climbing space when the conductors are covered with temporary insulation (i.e., line guards and insulator hoods). The climbing space must project vertically 40" above and below the limiting conductors.

Insulated line guard

See Photo(s)

Fig. 236-2. Examples of climbing space between conductors (Rule 236E).

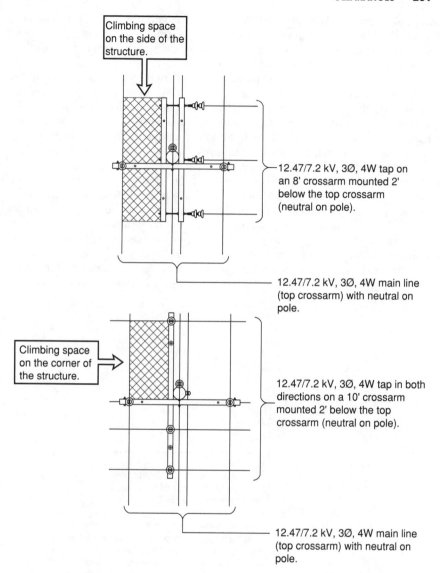

Climbing space on the side of the structure.

12.47/7.2 kV, 3Ø, 4W tap on an 8' crossarm mounted 2' below the top crossarm (neutral on pole).

12.47/7.2 kV, 3Ø, 4W main line (top crossarm) with neutral on pole.

Climbing space on the corner of the structure.

12.47/7.2 kV, 3Ø, 4W tap in both directions on a 10' crossarm mounted 2' below the top crossarm (neutral on pole).

12.47/7.2 kV, 3Ø, 4W main line (top crossarm) with neutral on pole.

Fig. 236-3. Examples of climbing space between conductors (Rule 236E).

Per Rule 236H, vertical conductors in conduit (risers) that are securely attached to the pole without spacers (offset brackets) are not considered an obstruction to the climbing space. When using the clearances in **NESC** Table 236-1, a conduit can be located within the clearances provided. Rule 362B, in Sec. 36, Risers, provides an additional requirement stating that the number, size, and location of the risers must be limited to allow adequate access for climbing. A conduit mounted on offset brackets would be considered an obstruction to the climbing space.

Conduit offset brackets are used to organize the conduit risers on a pole and make climbing easier, especially when multiple conduits or large conduits are present. The climbing space is then located on a side or corner of the pole that does not contain the riser conduits on offset brackets.

Rule 236I requires climbing space to the top conductor position where the center phase is on the pole top and the two outer phases are on a crossarm mounted below the pole top. One item that the NESC does not specifically address, but indirectly implies, is the location of obstructions at the base of the pole where the climbing space starts. Obstructions at the base of the pole should be avoided where possible. For example, telephone pedestals should be located a sufficient distance away from the base of the pole such that if a climber were to fall, the climber would not fall onto the telephone pedestal. The location of a telephone pedestal relative to a pole location can be specified in the joint use agreement between the supply and communication utilities. Other obstructions like fences can sometimes not be avoided, as many times a pole is located along a property line where a fence exists. If an obstruction like a telephone pedestal or fence exists at the base of the climbing space, the climbing space rules can be met if the climber climbs on the opposite side of the pole or rotates position on the way up and down the pole to avoid the obstruction at the base of the pole. See Fig. 236-4.

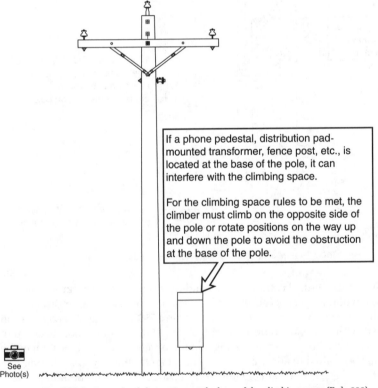

If a phone pedestal, distribution pad-mounted transformer, fence post, etc., is located at the base of the pole, it can interfere with the climbing space.

For the climbing space rules to be met, the climber must climb on the opposite side of the pole or rotate positions on the way up and down the pole to avoid the obstruction at the base of the pole.

See Photo(s)

Fig. 236-4. Example of obstructions at the base of the climbing space (Rule 236).

237. WORKING SPACE

Rule 237, Working Space, and Rule 236, Climbing Space, may appear complicated but they revolve around one simple idea. The supply and communication
worker must have adequate space to climb and work on a pole and its conductors, supports, and equipment.

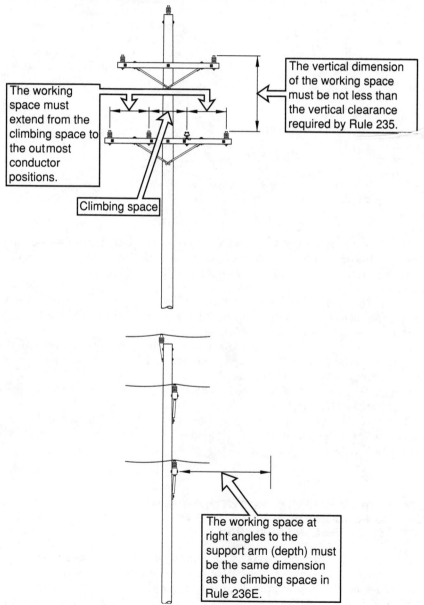

The working
space must
extend from the
climbing space to
the outmost
conductor
positions.

The vertical dimension
of the working space
must be not less than
the vertical clearance
required by Rule 235.

Climbing space

The working space at
right angles to the
support arm (depth) must
be the same dimension
as the climbing space in
Rule 236E.

Fig. 237-1. Working space dimensions (Rule 237B).

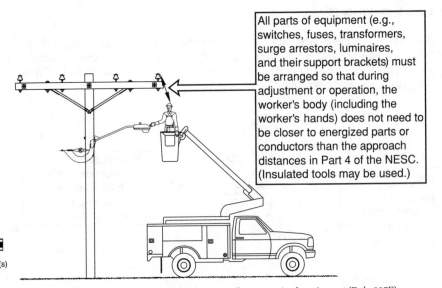

All parts of equipment (e.g., switches, fuses, transformers, surge arrestors, luminaires, and their support brackets) must be arranged so that during adjustment or operation, the worker's body (including the worker's hands) does not need to be closer to energized parts or conductors than the approach distances in Part 4 of the NESC. (Insulated tools may be used.)

See Photo(s)

Fig. 237-2. Example of working clearances from energized equipment (Rule 237F).

The working space rules are tied into the climbing space rules in Rule 236A. The working space dimensions required in Rule 237B are obtained from the climbing space requirements in Rule 236E and **NESC** Table 236-1. The working space dimension rules are outlined in Fig. 237-1.

Rule 237C provides requirements for working space relative to vertical and lateral conductors. Conductor jumpers and vertical risers that are not in conduit must be located outside the working space.

Rule 237D provides rules for working space relative to buckarm construction. This rule is to be used with the corresponding climbing space requirements in Rule 236F. **NESC** Fig. 237-1 is provided as a guide.

Energized equipment must be guarded to avoid contact if all the conditions listed in Rule 237E apply.

Rule 237F ties the working space requirements of Rule 237 to the work rules of Part 4. An example of working clearances from energized equipment is shown in Fig. 237-2.

238. VERTICAL CLEARANCE BETWEEN CERTAIN COMMUNICATION AND SUPPLY FACILITIES LOCATED ON THE SAME STRUCTURE

Rule 238 is a small rule in terms of the amount of text, but it has a big impact on clearance between supply (power) and communication facilities.

238A. Equipment.   A definition is provided in Rule 238A for equipment as it applies in Rule 238. Since the definition is specific to this rule, it is provided in Rule 238 instead of Sec. 02, "Definitions of Special Terms." The definition of equipment on a typical joint use pole is outlined in Fig. 238-1.

A wood crossarm and crossarm brace are not considered equipment per the definition of equipment in this rule. A metal support brace for a capacitor bank would be considered equipment. See Fig. 238-2.

238B. Clearances in General.   The vertical clearance between the following joint use facilities must be considered:

- Supply conductors and communication equipment
- Communication conductors and supply equipment
- Supply and communication equipment

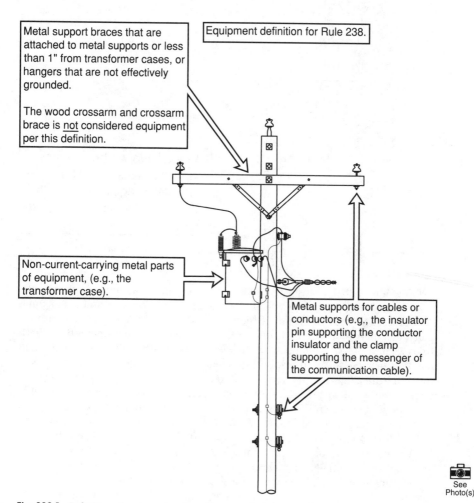

Metal support braces that are attached to metal supports or less than 1" from transformer cases, or hangers that are not effectively grounded.

The wood crossarm and crossarm brace is not considered equipment per this definition.

Equipment definition for Rule 238.

Non-current-carrying metal parts of equipment, (e.g., the transformer case).

Metal supports for cables or conductors (e.g., the insulator pin supporting the conductor insulator and the clamp supporting the messenger of the communication cable).

See Photo(s)

Fig. 238-1. Definition of equipment as it applies to vertical clearance between communication and supply facilities on the same structure (Rule 238A).

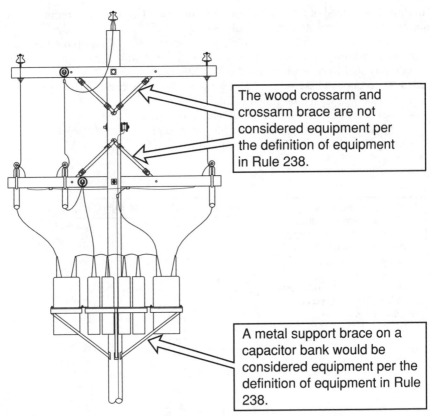

The wood crossarm and crossarm brace are not considered equipment per the definition of equipment in Rule 238.

A metal support brace on a capacitor bank would be considered equipment per the definition of equipment in Rule 238.

See Photo(s)

Fig. 238-2. Example of equipment metal supporting braces (Rule 238A).

NESC Table 238-1 is used to establish the vertical clearance requirements. NESC Table 238-1 provides vertical clearances only, not horizontal or diagonal clearances. The vertical clearance required in Rule 238 is in addition to the vertical clearance required in Rule 235C. The clearance requirements in both Rules 235C and 238 must be met. The clearances required in Rule 235 and 238 may be the same (e.g., 40 in), but the clearance is measured at different locations. In Rule 238, a supply equipment to communication equipment measurement is required. In Rule 235C, a supply conductor to communication cable measurement is required, both at the structure and out in the span. The vertical clearance requirements in Rule 238B and NESC Table 238-1 revolve around 40 in. Less than 40 in (e.g., 30 in) is acceptable for certain grounded supply facilities. Greater than 40 in is required for supply voltages above 8.7 kV to ground. Examples of vertical clearance between supply and communication equipment on the same structure are shown in Figs. 238-3 through 238-10.

Although they are not specifically referenced in Rule 238, communication antennas are a type of communication equipment. If a communication cable

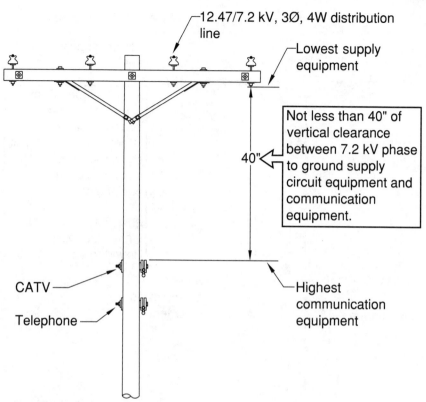

Fig. 238-3. Example of vertical clearance between supply and communication equipment on the same structure (Rule 238B).

exists on a joint use pole, then the clearance of Rule 238B is measured to the communication cable metal support hardware to establish the communication worker safety zone. The communication worker safety zone in Rule 238E does not permit a communication antenna to be located in the communication worker safety zone. If a communication antenna is mounted on a supply (power) pole without any joint use communication cables, then the communication worker safety zone is not clearly defined. In this case, accepted good practice must be used (Rule 012C). Applying accepted good practice, the communication worker safety zone can be defined by the top of the antenna instead of a metal support for communication cables. See Fig. 238-11.

238C. Clearances for Span Wires or Brackets.   The space between the supply and communication facilities is called the "Communication Worker Safety Zone." Rule 238C permits span wires or brackets carrying luminaires, traffic signals, or trolley conductors to be located in the communication worker safety zone. The **Code** recognizes that roadway lighting fixtures and traffic signals provide their own safety function and commonly the appropriate mounting height

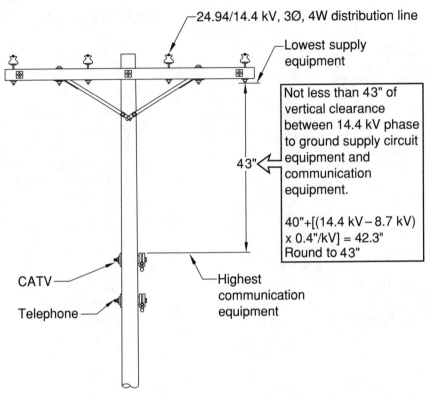

Fig. 238-4. Example of vertical clearance between supply and communications equipment on the same structure (Rule 238B).

for these item is in the space between the supply and communication facilities. Rule 238C references **NESC** Table 238-2, which provides clearances for luminaires and traffic signals.

238D. Clearance of Drip Loops of Luminaire or Traffic Signal Brackets. The space between the supply and communication facilities is called the "Communication Worker Safety Zone." Rule 238D recognizes that roadway lighting and traffic signal brackets are commonly fed from a drip loop positioned below the roadway lighting or traffic signal bracket. Special clearances for these drip loops permit them to be located in the communication worker safety zone. Examples of roadway lighting fixtures mounted above, in, and below the communication worker safety zone, along with the fixture drip loops, are shown in Figs. 238-12, 238-13, and 238-14.

An exception applies to Rule 238D that permits the clearance of a drip loop for a luminaire or traffic signal to be not less than 3 in above the communication cable or through-bolt if certain conditions apply. See Fig. 238-15.

It can be easy to lose sight of the fact that the reduced clearance to secondary drip loops only applies to drip loops feeding luminaire or traffic signal brackets.

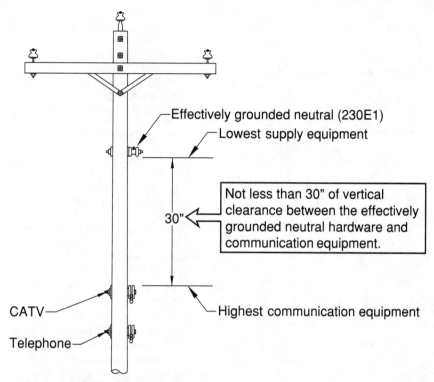

Effectively grounded neutral (230E1)

Lowest supply equipment

Not less than 30" of vertical clearance between the effectively grounded neutral hardware and communication equipment.

30"

CATV

Telephone

Highest communication equipment

See Photo(s)

Fig. 238-5. Example of vertical clearance between supply and communication equipment on the same structure (Rule 238B).

Even though a 120/240 V circuit can be used to feed a house or a street lighting luminaire, the 12-in clearance only applies to the luminaire (street lighting) drip loop entering the luminaire or traffic signal bracket. Luminaires and traffic signals serve their own safety functions so they merit special **Code** consideration. See Fig. 238-16.

238E. Communication Worker Safety Zone.   Rule 238E labels the area between the supply space and the communication space both at the structure and out in the span. The name for this area is the "Communication Worker Safety Zone." The work rules in Part 4 of the **NESC** provide the qualifications of a supply employee and a communication employee. Since the communication employee is not trained to work on supply lines, a safety zone exists for the communication employee's protection. If a communication line is positioned below a supply line on an overhead structure, but the required communication worker safety zone clearance does not exist, the communication employee can correct the violation if the communication employee does not violate the approach distances and other requuirements in Sec. 43. If the communication employee cannot maintain the approach distances and other requirements in Sec. 43, the communication employee must contact a supply employee to correct the violation. See Sec. 43

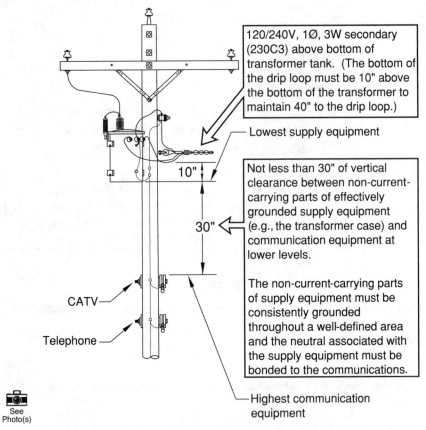

120/240V, 1Ø, 3W secondary (230C3) above bottom of transformer tank. (The bottom of the drip loop must be 10" above the bottom of the transformer to maintain 40" to the drip loop.)

— Lowest supply equipment

Not less than 30" of vertical clearance between non-current-carrying parts of effectively grounded supply equipment (e.g., the transformer case) and communication equipment at lower levels.

The non-current-carrying parts of supply equipment must be consistently grounded throughout a well-defined area and the neutral associated with the supply equipment must be bonded to the communications.

10"

30"

CATV—

Telephone —

—Highest communication equipment

See Photo(s)

Fig. 238-6. Example of vertical clearance between supply and communication equipment on the same structure (Rule 238B).

for additional information. The communication worker safety zone is defined by both Rule 238 and Rule 235C. Rules 238C, 238D, and 239 permit only a few select items in the communication worker safety zone. See Fig. 238-17.

239. CLEARANCE OF VERTICAL AND LATERAL FACILITIES FROM OTHER FACILITIES AND SURFACES ON THE SAME SUPPORTING STRUCTURE

239A. General.   Rule 239A provides a list of vertical and lateral conductors that may be placed directly on the supporting structure (pole). A conduit may also be placed directly on the pole. Rule 239D requires guarding of vertical

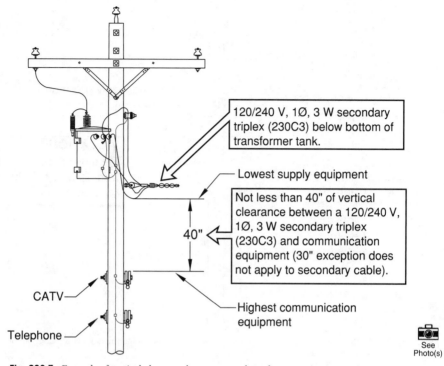

Fig. 238-7. Example of vertical clearance between supply and communication equipment on the same structure (Rule 238B).

conductors within 8 ft of the ground with certain exceptions. See Rule 239D for a discussion. Examples of vertical and lateral conductors are shown in Fig. 239-1.

Rule 239A also address various cables that can be installed in the same vertical riser duct. The rules for supply and communication cables in vertical risers are outlined in Fig. 239-2.

239B. Location of Vertical or Lateral Conductors Relative to Climbing Spaces, Working Spaces, and Pole Steps.   Climbing spaces, working spaces, and pole steps must not be obstructed with vertical and lateral conductors. Rule 239B applies to vertical conductors not in conduit. A note to this rule reminds the reader that vertical runs in conduit securely attached to the pole surface are not considered to obstruct the climbing space per Rule 236H.

239C. Conductors Not in Conduit.   Vertical and lateral conductors not in conduit must have the same clearances from conduits (enclosing other conductors) as from other structure surfaces.

239D. Guarding and Protection Near Ground.   Vertical conductors (risers) within 8 ft of the ground must be guarded. Conduit or U-guard are the most common types of guards used to protect the vertical conductors. An exception

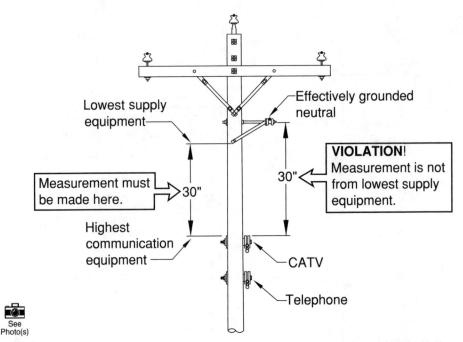

Lowest supply equipment

Effectively grounded neutral

VIOLATION! Measurement is not from lowest supply equipment.

Measurement must be made here. → 30"

30"

Highest communication equipment

CATV

Telephone

See Photo(s)

Fig. 238-8. Example of vertical clearance between supply and communication equipment on the same structure (Rule 238B).

to Rule 239D1 provides a list of conductors and cables that do not require guarding. See Fig. 239-3.

Rule 239D is very similar to Rule 093D, Guarding and Protection, for grounding conductors. Rule 239D also overlaps with Rule 362, Pole Risers—Additional Requirements. See Rules 093D and 362 for additional information and related figures.

239E. Requirements for Vertical and Lateral Supply Conductors on Supply Line Structures or within Supply Space on Jointly Used Structures. The locations and conductors that apply to Rule 239E are graphically shown in Fig. 239-4.

The general clearances for open (noninsulated) vertical and lateral supply conductors from the surface of the support are provided in **NESC** Table 239-1. **NESC** Table 239-1 also provides clearance to span, guy, and messenger wires. Clearances in Rule 235E (**NESC** Table 235-6) must also be met. **NESC** Table 235-6 provides clearances from line conductors (i.e., the conductors spanning from pole to pole) to vertical and lateral conductors attached to the same support (pole).

NESC Table 239-2 must be applied to special cases. If open line conductors (i.e., the noninsulated conductors spanning from pole to pole) are within 4 ft of the pole, vertical conductors must meet two special conditions. The special conditions in Rules 239E2a(1) and 239E2a(2) only apply if open line conductors

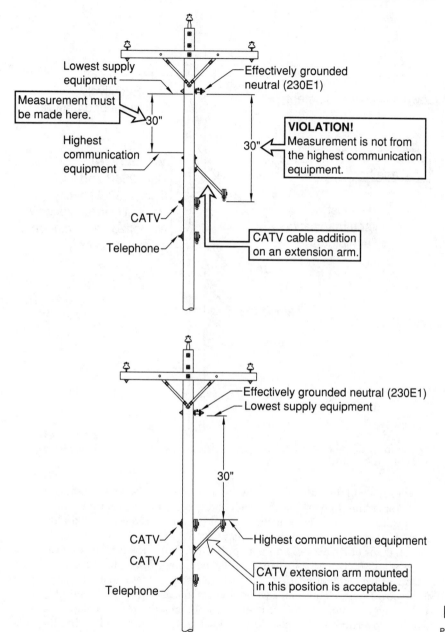

Lowest supply equipment

Effectively grounded neutral (230E1)

Measurement must be made here.

30"

VIOLATION!
Measurement is not from the highest communication equipment.

Highest communication equipment

30"

CATV

Telephone

CATV cable addition on an extension arm.

Effectively grounded neutral (230E1)
Lowest supply equipment

30"

CATV
CATV

Highest communication equipment

Telephone

CATV extension arm mounted in this position is acceptable.

See Photo(s)

Fig. 238-9. Example of vertical clearance between supply and communication equipment on the same structure (Rule 238B).

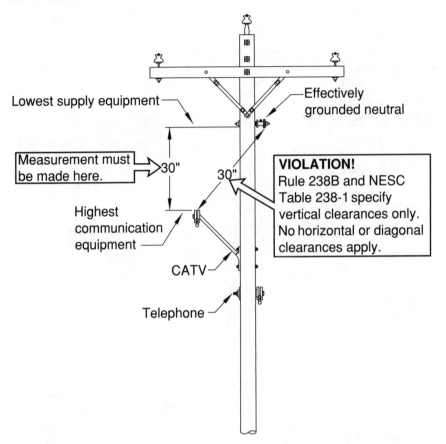

Lowest supply equipment

Effectively grounded neutral

Measurement must be made here.

30"

30"

Highest communication equipment

CATV

Telephone

VIOLATION!
Rule 238B and NESC Table 238-1 specify vertical clearances only. No horizontal or diagonal clearances apply.

Fig. 238-10. Example of vertical clearance between supply and communication equipment on the same structure (Rule 238B).

are within 4 ft of the pole, which is very common, even on an 8-ft crossarm, and if workers climb to work on the line conductors while the vertical conductors in questions are energized. If the vertical conductors in question are de-energized while climbing, or if the pole is worked on from a bucket truck, then the special cases of Rule 239E2 do not apply. NESC Table 239-2 requires clearances larger than NESC Table 239-1 to provide room for workers to climb near energized vertical conductors. NESC Table 239-2 also provides a distance from above and below the line conductor where the clearances apply. If a conduit is used within the distance specified, it must be nonmetallic or protected by a nonmetallic covering to provide additional safety for a worker climbing the pole and working near an energized vertical conductor.

239F. Requirements for Vertical and Lateral Communication Conductors on Communication Line Structures or within the Communication Space on Jointly Used Structures.   The locations and conductors that Rule 239F apply to are graphically shown in Fig. 239-5.

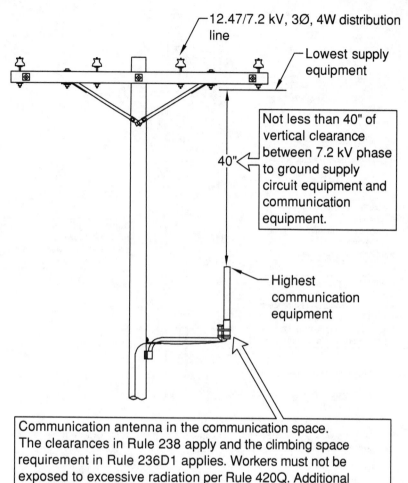

12.47/7.2 kV, 3Ø, 4W distribution line

Lowest supply equipment

Not less than 40" of vertical clearance between 7.2 kV phase to ground supply circuit equipment and communication equipment.

40"

Highest communication equipment

Communication antenna in the communication space. The clearances in Rule 238 apply and the climbing space requirement in Rule 236D1 applies. Workers must not be exposed to excessive radiation per Rule 420Q. Additional clearance or climbing space (or a RF disconnect switch) may be needed to meet this requirement.

Fig. 238-11. Example of vertical clearance between supply and communication equipment on the same structure (Rule 238B).

Uninsulated (open wire) vertical and lateral communication conductors are not as common as they once were. Open wire communication has been replaced with insulated communication cables. When an insulated communication cable riser exists on a joint use pole, the specified clearance to supply facilities in Rule 239F is essentially the same as required by Rules 235 and 238.

239G. Requirements for Vertical Supply Conductors and Cables Passing through Communication Space on Jointly Used Line Structures. The locations and conductors to which Rule 239G applies are graphically shown in Fig. 239-6.

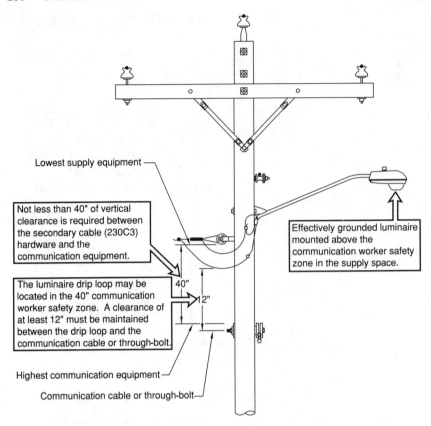

Lowest supply equipment —

Not less than 40" of vertical clearance is required between the secondary cable (230C3) hardware and the communication equipment.

Effectively grounded luminaire mounted above the communication worker safety zone in the supply space.

The luminaire drip loop may be located in the 40" communication worker safety zone. A clearance of at least 12" must be maintained between the drip loop and the communication cable or through-bolt.

40"

12"

Highest communication equipment —

Communication cable or through-bolt —

See Photo(s)

Fig. 238-12. Example of vertical clearance between a drip loop feeding roadway lighting and communication equipment (Rule 238D).

Rule 239G applies to a number of common joint use conditions. Providing a conduit over supply conductors 40 in above the highest communication attachment is a condition that relates to the communication worker safety zone discussed in Rule 238E. Examples of vertical supply conductors on joint use poles are shown in Figs. 239-7 and 239-8.

Exception 2 to Rule 239G permits omitting guarding from a supply grounding conductor (pole ground) on a joint use pole if certain conditions are met. See Fig. 239-9.

239H. Requirements for Vertical Communication Conductors Passing through Supply Space on Jointly Used Structures. The locations and conductors to which Rule 239H applies are graphically shown in Fig. 239-10.

Communication conductors passing through the supply space can be seen on poles with an overhead ground wire (static) with an embedded fiber-optic communication cable and on supply poles with communication antennas on top. See Rule 235I for additional information.

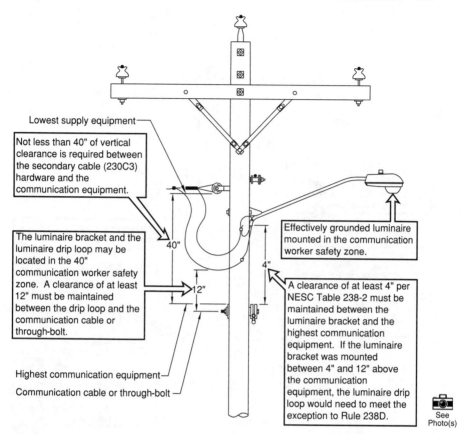

Fig. 238-13. Example of vertical clearance between a drip loop feeding roadway lighting and communication equipment (Rules 238C and 238D).

239I. Operating Rods. Operating rods for switches like a 3-phase gang operated distribution switch can pass through the communication space but they must be located outside the climbing space. See Rule 236 for additional climbing space information.

239J. Additional Rules for Standoff Brackets. Standoff brackets may be used to support conduits. Both metallic and nonmetallic conduits may be supported. The conductors inside the conduits must have sufficient insulation. Tree wire described in Rule 230D cannot be installed in a conduit riser as it is not fully insulated. Standoff brackets may be used to support cables not in conduit if the cable is a communication cable, 230C1a supply cable, or a supply cable less than 750 V with a single outer jacket or sheath. Traditional duplex, triplex, and quadruplex supply cables do not meet this criteria, nor do single insulated underground secondary conductors. The rules for positioning standoff brackets on a structure to inhibit climbing by unqualified persons are provided in Rule 217A2.

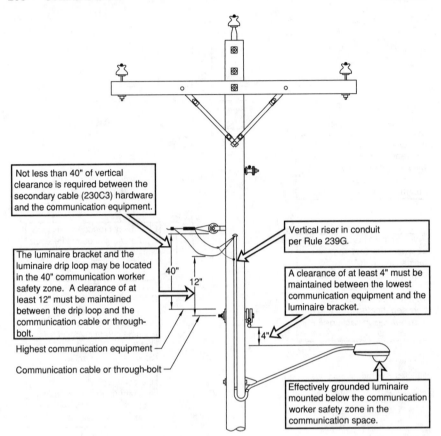

Not less than 40" of vertical clearance is required between the secondary cable (230C3) hardware and the communication equipment.

The luminaire bracket and the luminaire drip loop may be located in the 40" communication worker safety zone. A clearance of at least 12" must be maintained between the drip loop and the communication cable or through-bolt.

Highest communication equipment

Communication cable or through-bolt

Vertical riser in conduit per Rule 239G.

A clearance of at least 4" must be maintained between the lowest communication equipment and the luminaire bracket.

Effectively grounded luminaire mounted below the communication worker safety zone in the communication space.

40"

12"

4"

Fig. 238-14. Example of vertical clearance between a drip loop feeding roadway lighting and communication equipment (Rules 238C, 238D, and 239G).

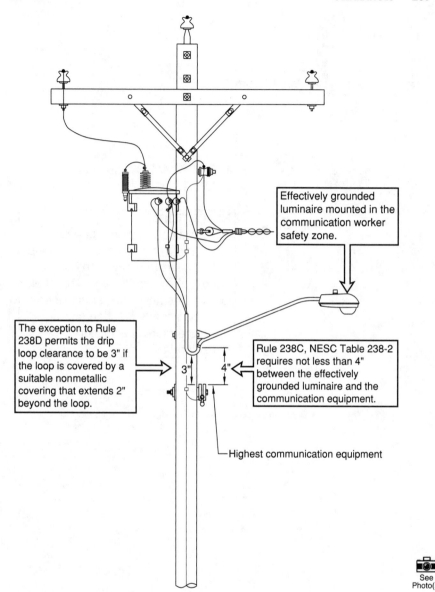

The exception to Rule 238D permits the drip loop clearance to be 3" if the loop is covered by a suitable nonmetallic covering that extends 2" beyond the loop.

Effectively grounded luminaire mounted in the communication worker safety zone.

Rule 238C, NESC Table 238-2 requires not less than 4" between the effectively grounded luminaire and the communication equipment.

3"   4"

Highest communication equipment

See Photo(s)

Fig. 238-15. Exception to vertical clearance between a drip loop feeding roadway lighting and communication equipment (Rules 238C and 238D).

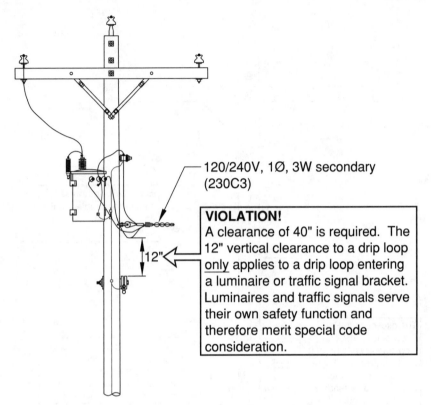

120/240V, 1Ø, 3W secondary
(230C3)

12"

VIOLATION!
A clearance of 40" is required. The 12" vertical clearance to a drip loop <u>only</u> applies to a drip loop entering a luminaire or traffic signal bracket. Luminaires and traffic signals serve their own safety function and therefore merit special code consideration.

Fig. 238-16. Example of a common joint use violation (Rule 238D).

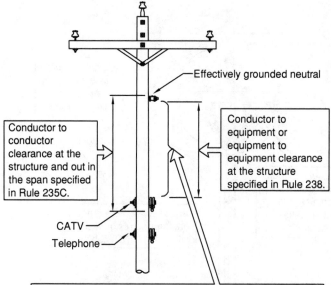

Effectively grounded neutral

Conductor to conductor clearance at the structure and out in the span specified in Rule 235C.

Conductor to equipment or equipment to equipment clearance at the structure specified in Rule 238.

CATV

Telephone

The "communication worker safety zone" is a space on the pole where nothing is located (see exceptions). Both Rules 238 and 235C are used to define the space. Both rules must be met, not just one or the other.

Exceptions that are permitted in the communication worker safety zone are as follows:
- Street lights and associated brackets or span wires (Rule 238C)
- Traffic signals and associated bracket or span wires (Rule 238C)
- Trolley conductor brackets or span wires (Rule 238C)
- Drip loops feeding street lights or traffic signals (Rule 238D)
- Vertical risers (Rule 239)

See Photo(s)

Fig. 238-17. Communication worker safety zone (Rule 238E).

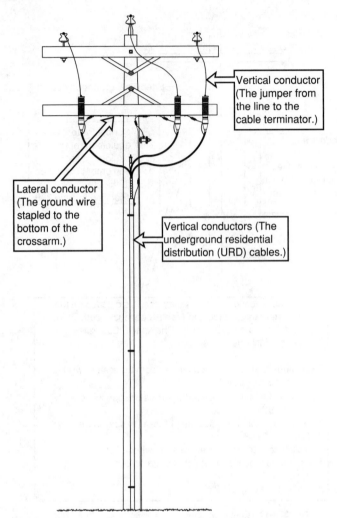

Fig. 239-1. Examples of vertical and lateral conductors (Rule 239A).

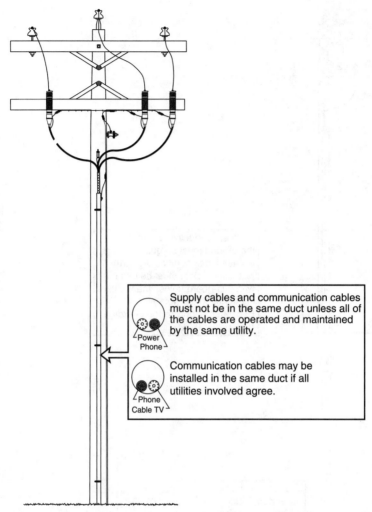

Fig. 239-2. Supply and communication cables in vertical risers (Rule 239A2d and 239A2e).

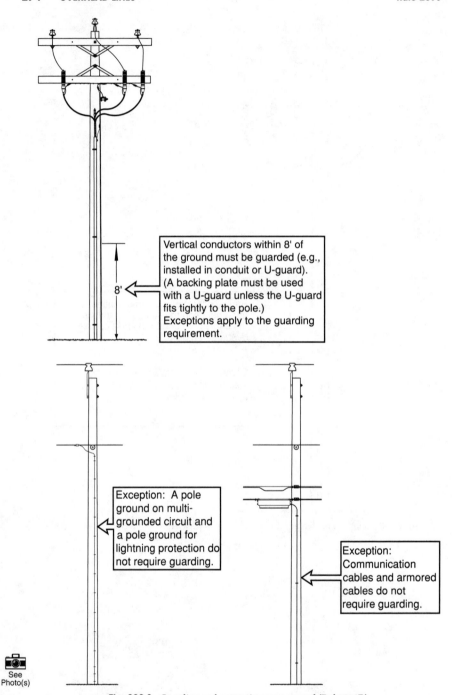

Vertical conductors within 8' of the ground must be guarded (e.g., installed in conduit or U-guard). (A backing plate must be used with a U-guard unless the U-guard fits tightly to the pole.) Exceptions apply to the guarding requirement.

8'

Exception: A pole ground on multi-grounded circuit and a pole ground for lightning protection do not require guarding.

Exception: Communication cables and armored cables do not require guarding.

See Photo(s)

Fig. 239-3. Guarding and protection near ground (Rule 239D).

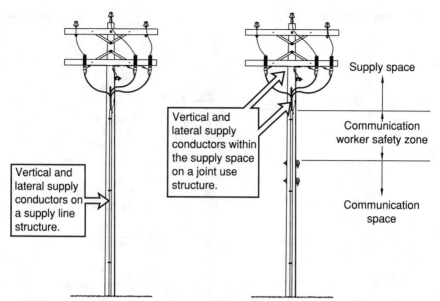

Vertical and lateral supply conductors on a supply line structure.

Vertical and lateral supply conductors within the supply space on a joint use structure.

Supply space

Communication worker safety zone

Communication space

See Photo(s)

Fig. 239-4. Location of vertical and lateral supply conductors on supply line structures or within supply space on jointly used structures (Rule 239E).

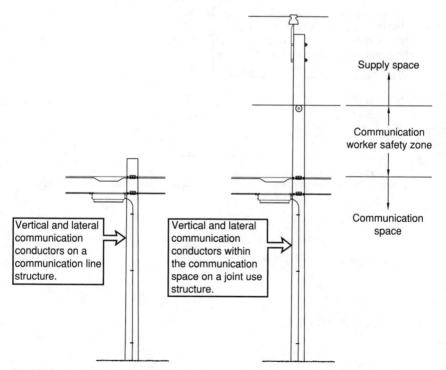

Vertical and lateral communication conductors on a communication line structure.

Vertical and lateral communication conductors within the communication space on a joint use structure.

Supply space

Communication worker safety zone

Communication space

See Photo(s)

Fig. 239-5. Location of vertical and lateral communication conductors on communication line structures or within the communication space on jointly used structures (Rule 239F).

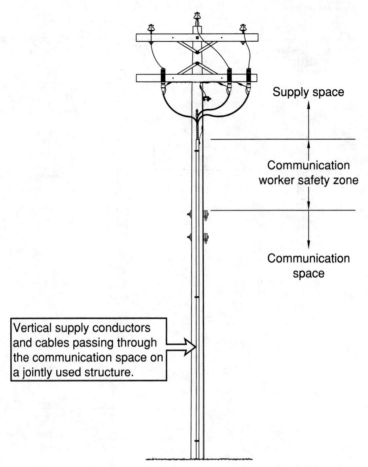

Supply space

Communication worker safety zone

Communication space

Vertical supply conductors and cables passing through the communication space on a jointly used structure.

See Photo(s)

Fig. 239-6. Location of vertical supply conductors and cables passing through the communication space on jointly used line structures (Rule 239G).

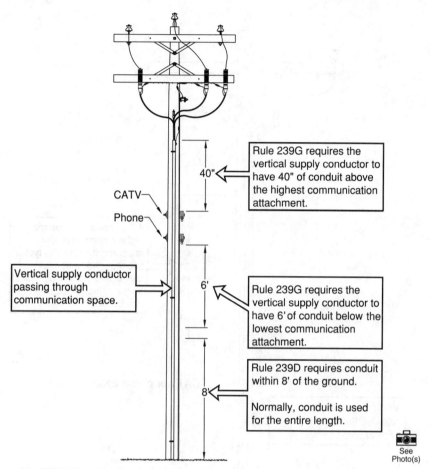

Fig. 239-7. Example of vertical supply conductors on a joint use pole (Rule 239G1).

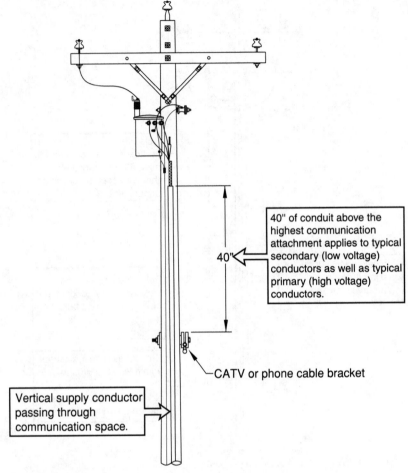

40" of conduit above the highest communication attachment applies to typical secondary (low voltage) conductors as well as typical primary (high voltage) conductors.

40"

CATV or phone cable bracket

Vertical supply conductor passing through communication space.

See Photo(s)

Fig. 239-8. Example of vertical supply conductors on a joint use pole (Rule 239G1).

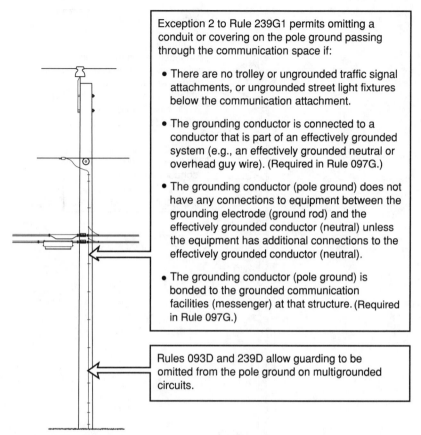

Exception 2 to Rule 239G1 permits omitting a conduit or covering on the pole ground passing through the communication space if:

• There are no trolley or ungrounded traffic signal attachments, or ungrounded street light fixtures below the communication attachment.

• The grounding conductor is connected to a conductor that is part of an effectively grounded system (e.g., an effectively grounded neutral or overhead guy wire). (Required in Rule 097G.)

• The grounding conductor (pole ground) does not have any connections to equipment between the grounding electrode (ground rod) and the effectively grounded conductor (neutral) unless the equipment has additional connections to the effectively grounded conductor (neutral).

• The grounding conductor (pole ground) is bonded to the grounded communication facilities (messenger) at that structure. (Required in Rule 097G.)

Rules 093D and 239D allow guarding to be omitted from the pole ground on multigrounded circuits.

Fig. 239-9. Exception for a vertical supply grounding conductor on a joint use pole (Rule 239G1).

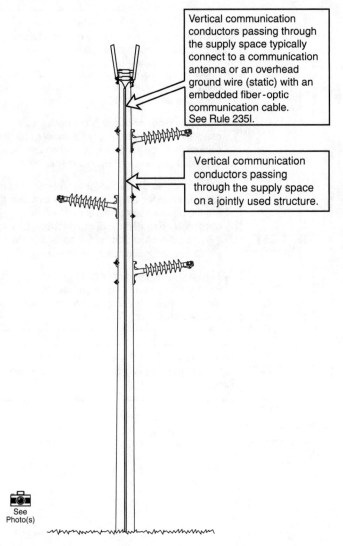

Vertical communication conductors passing through the supply space typically connect to a communication antenna or an overhead ground wire (static) with an embedded fiber-optic communication cable.
See Rule 235I.

Vertical communication conductors passing through the supply space on a jointly used structure.

See Photo(s)

Fig. 239-10. Location of vertical communication conductors passing through the supply space on jointly used structures (Rule 239H).

Section 24

Grades of Construction

240. GENERAL

Sections 24, 25, and 26 are all interrelated. Section 24, "Grades of Construction," defines the required strength of overhead line construction for safety purposes. Section 25, "Loadings for Grades B and C," defines the physical loads (i.e., ice, wind, and temperature conditions) that overhead line construction must be able to withstand and the load factors that must be applied to the physical loads. Section 26, "Strength Requirements," defines the required strength of materials used in constructing overhead lines and the strength factors that must be applied to the materials. Line insulators are not addressed in Secs. 24, 25, or 26. Insulators are covered in their own section, Sec. 27.

The details of applying load factors and strength factors are covered in Secs. 25 and 26. The purpose of Sec. 24 is to define the grade of construction that is required for different situations.

This Handbook addresses the **Code** requirements in Secs. 24, 25, and 26. Line design calculations are not presented in this Handbook. Line design calculations can be found in transmission and distribution line design manuals. Large utilities normally develop their own line design manuals. Small rural utilities typically use the transmission and distribution line design manuals published by the Rural Utilities Service (RUS), which in the past was referred to as the Rural Electrification Administration (REA). This Handbook will aid those individuals writing and using line design manuals and provide a deeper understanding of how the **Code** applies to line design calculations.

Typical strength and loading line design calculations include, but are not limited to, the following:

- Maximum wind span based on wind with ice on conductors
- Maximum weight span based on weight of conductors and ice
- Moment due to wind on pole
- Allowable resisting moment of pole
- Transverse, vertical, and total components of conductor loading
- Total ground line moment on pole
- Ruling span
- Diameter of pole at any point
- Deadend guying strength
- Bisector guying strength
- Weak-link of guy attachment, guy wire, and anchor assembly
- Crossarm strength (vertical)
- Crossarm strength (longitudinal)
- Pole buckling
- Maximum line angle based on insulator strength
- Material deflection
- Equipment loading

Single pole structures are typically easier to analyze than multiple pole and lattice structures. Line design calculations also involve clearance calculations in addition to strength and loading calculations (e.g., clearance above ground, clearance to buildings and other structures, horizontal clearance between conductors, vertical clearance between conductors, etc.). The clearance calculations will affect the line design and therefore the selection of pole heights and pole classes. The sample sag and tension chart shown at the beginning of Sec. 23 provides sag for clearance calculations and tension for strength and loading calculations. The "design tension" used for strength and loading calculations is the largest of the tensions calculated after applying Rule 250B (heavy, medium, or light loads), Rule 250C (extreme wind loads) if applicable, and Rule 250D (extreme ice with concurrent wind loads) if applicable. The 250B, 250C and 250D loads are applied to conductors in accordance with Rule 251. See Rule 251 for additional information.

If more than one condition applies to the construction of an overhead line, the condition requiring the higher grade of construction is used. The order of grades is discussed in Rule 241B.

Section 24 applies to both alternating current (AC) and direct current (DC) circuits. A DC circuit is considered equivalent to the root mean square (rms) values of an AC circuit when applying Sec. 24. This is different from Sec. 23, which requires DC circuits to be equivalent to the AC crest value. See Rule 230G for a discussion of AC rms and crest values.

241. APPLICATION OF GRADES OF CONSTRUCTION TO DIFFERENT SITUATIONS

241A. Supply Cables.   For the purposes of determining the grade of construction of a supply cable, supply (not communication) cables are broken into two types. Type 1 supply cables are cables meeting Rules 230C1, 230C2, and 230C3. See Rule 230C for a discussion of these types of supply cables. The 230C1, 230C2, and 230C3 cables are typically supported on messengers. The strength requirements for supply cable messengers are covered in Rule 261I.

Type 2 supply cables are all other supply cables that do not meet the requirements for 230C1, 230C2, and 230C3 cables. Type 2 supply cables include open wire (bare) conductors and covered (tree wire) conductors.

241B. Order of Grades.   Rule 241B outlines the order of grades. A general description of each grade of construction is outlined below:

- *Grade B.* The highest grade. The line will be "extra stout." Load and strength factors will be applied to make the strength of the line higher than any other grade. The required pole class will be larger than C and N grades (e.g., Class 3 instead of Class 5) for the same span length or the required span length will be shorter than grades C and N for the same pole class. Crossarms will be stronger than grades C and N, etc.

- *Grade C.* The medium grade. The line will be more "stout" than Grade N but not as "stout" as Grade B. Load and strength factors will be applied but they will not increase the strength of the line as much as Grade B. Grade C at crossings will be more "stout" than Grade C not at crossings.

- *Grade N.* The lowest grade. The line must be designed to support the anticipated loads but no load or strength factors need to be applied.

Figure 241-1 provides a graphical representation of the grades of construction.

241C. At Crossings.   One overhead power line is considered to be at crossings when it crosses another line, a railroad track, a limited access highway, or a navigable waterway. Joint use or collinear construction in itself is not considered at crossings. See Figs. 241-2 through 241-4.

When two lines cross, the grade of construction for the conductors and supporting structures of the lower line can be determined without considering the upper line. In other words, the grade of construction of the lower line is based on the land use. This is because the lower line cannot fall into the upper line. The upper line, however, can fall into the lower line and cause problems. The conductors and supporting structures of the upper line at the crossing require careful application of Rule 241C3 (Multiple Crossings), Rule 242 (Grades of Construction for Conductors, NESC Tables 242-1 and 242-2), and Rule 243 (Grades of Construction for Line Supports).

An example of lines at multiple crossings is shown in Fig. 241-5.

An example of lines at multiple crossings involving communications, supply, and railroad tracks is shown in Fig. 241-6.

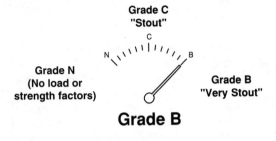

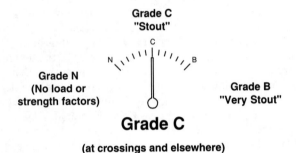

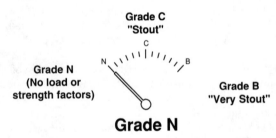

Fig. 241-1. Order of grades of construction (Rule 241B).

241D. Conflicts.   Structure conflict exists when one line can fall down and conflict with another line. See Rules 220C and 221 for additional information and figures related to structure conflict. For the purposes of determining grades of construction, conflicting lines are treated the same as line crossings since one line can fall on the other.

242. GRADES OF CONSTRUCTION FOR CONDUCTORS

Rule 242 references **NESC** Tables 242-1 and 242-2 for determining grades of construction for conductors. It is important to stress the term conductors. Rule 242

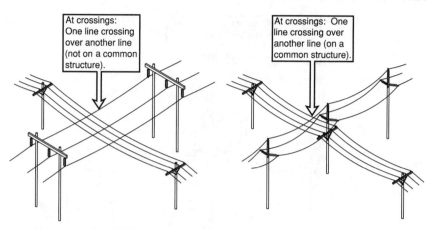

At crossings:
One line crossing
over another line
(not on a common
structure).

At crossings: One
line crossing over
another line (on a
common structure).

Fig. 241-2. Lines considered to be at crossings (Rule 241C).

and **NESC** Tables 242-1 and 242-2 are used to determine grades of construction for conductors. Rule 243 is used to determine grades of construction for line supports. The grade of construction of a line support is based on the grade of construction of the conductors attached to the line support. Rules 242A through 242F provide specific information for special types of conductors.

There are several important items of discussion related to **NESC** Tables 242-1 and 242-2.

- **NESC** Table 242-1 is for supply conductors alone, at crossing, or on the same structures with other conductors.
- **NESC** Table 242-2 is for communication conductors alone, or in the upper position of a crossing, or on joint poles.
- Supply or communication lines crossing over railroad tracks, limited access highways, or navigable waterways requiring waterway crossing permits must be built to Grade B construction.
- **NESC** Tables 242-1 and 242-2 distinguish between open and cable conductors. Cable conductors are 230C1, 230C2, and 230C3 cables. Open conductors are bare (noninsulated) conductors. See Rule 241A for a discussion.
- A 12.47/7.2-kV, 3-phase, 4-wire distribution line (open conductors) on a private right-of-way can be built to Grade N. This is not a common practice. Typically utilities build to Grade C construction.
- An effectively grounded 115-kV transmission line (with ground fault relaying) on a public right-of-way can be built to Grade C. This is not a common practice. Typically utilities build to Grade B construction.

Per **NESC** Table 242-1, a 12.47/7.2-kV, 3-phase, 4-wire distribution line (open conductors) built joint use with a communication cable must be built to Grade B unless Footnote 6 or 7 can be met. For a 12.47/7.2-kV line, Footnote 6 cannot be applied. Footnote 7 (Parts a and b) must then be met to construct a Grade C joint use (power and communication) line. Typically utilities build to Grade C construction for this application. The protection of communication equipment

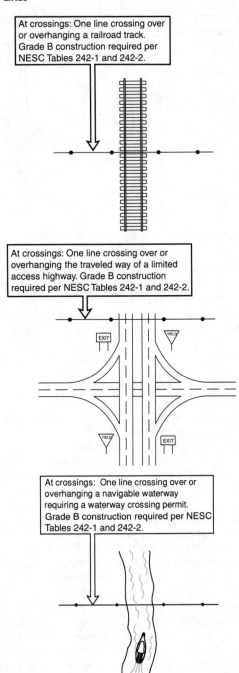

At crossings: One line crossing over or overhanging a railroad track. Grade B construction required per NESC Tables 242-1 and 242-2.

At crossings: One line crossing over or overhanging the traveled way of a limited access highway. Grade B construction required per NESC Tables 242-1 and 242-2.

At crossings: One line crossing over or overhanging a navigable waterway requiring a waterway crossing permit. Grade B construction required per NESC Tables 242-1 and 242-2.

Fig. 241-3. Lines considered to be at crossings (Rule 241C).

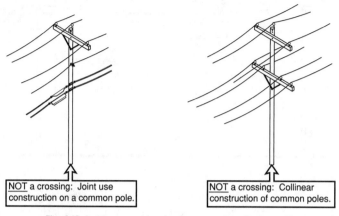

NOT a crossing: Joint use
construction on a common pole.

NOT a crossing: Collinear
construction of common poles.

Fig. 241-4. Lines considered to be at crossings (Rule 241C).

required in Footnote 7 is also required in Rule 223. Bonding of the supply neutral to the communication messenger is one of the first steps normally taken to satisfy Footnote 7 and Rule 223 (see Rule 223 for additional information).

A 120/240-V, single-phase, 3-wire triplex cable meeting Rule 230C3 falls under "cable" in **NESC** Table 242-1. Per Footnote 1 of the table, "cable" is a Type 1 cable per Rule 241A. A 120/240-V, single-phase, 3-wire triplex cable built joint use with a communication cable can be built to Grade N per **NESC** Table 242-1.

243. GRADES OF CONSTRUCTION
FOR LINE SUPPORTS

Rule 242, Grades of Construction for Conductors, applies to grades of construction for conductors and this rule, Rule 243, applies to grades of construction for line (conductor) supports. Line supports in this rule include structures, crossarms, insulators, pins, etc., basically the components that support the conductors.

243A. Structures.   The structure (e.g., pole) grade of construction must be the same as the highest grade of conductor supported. In other words, if a pole has Grade B transmission conductors, Grade C distribution conductors, and Grade N secondary conductors, all attached to the same pole, then the pole itself must be designed to Grade B construction. There are four modifications to the general rule.

243B. Crossarms and Support Arms.   The crossarm (or another type of support arm) grade of construction must be the same as the highest grade of conductor that the crossarm supports. If a crossarm carries Grade B conductors, the crossarm itself must be designed to Grade B. There are three modifications to the general rule.

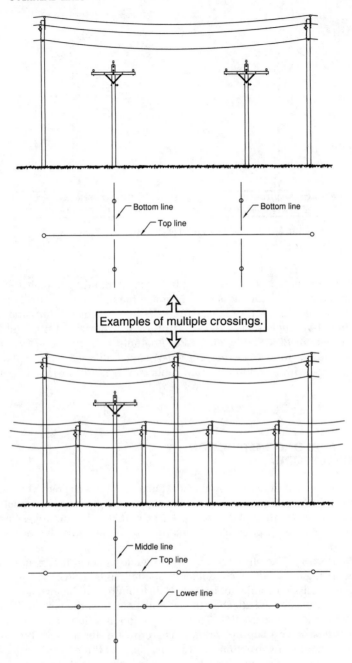

Fig. 241-5. Example of lines at multiple crossings (Rule 241C3a).

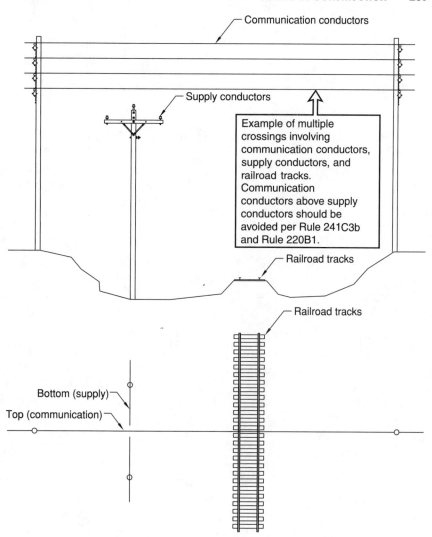

Fig. 241-6. Example of lines at multiple crossings (Rule 241C3b).

243C. Pins, Armless Construction Brackets, Insulators, and Conductor Fastenings.

The pins, armless construction brackets, insulators, and conductor fastenings grade of construction must be the same as the grade of their associated conductor. There are five modifications to the general rule. The most notable is Modification 5 (Rule 243C5). This modification states that insulators used on open conductor supply lines are covered in their own section, Sec. 27.

Section 25

Loadings for Grades B and C

250. GENERAL LOADING REQUIREMENTS AND MAPS

Sections 24, 25, and 26 are all interrelated. Section 24, "Grades of Construction," defines the required strength of overhead line construction for safety purposes. Section 25, "Loadings for Grades B and C," defines the physical loads (i.e., ice, wind, and temperature conditions) that overhead line construction must be able to withstand and the load factors that must be applied to the physical loads. Section 26, "Strength Requirements," defines the required strength of materials used in constructing overhead lines and the strength factors that must be applied to the materials. Line insulators are not addressed in Secs. 24, 25, or 26. Insulators are covered in their own section, Sec. 27.

The purpose of Sec. 24 is to define the grade of construction that is required for different situations. The details of applying load factors and strength factors are covered in Secs. 25 and 26.

This Handbook addresses the **Code** requirements in Secs. 24, 25, and 26. Line design calculations are not presented in this Handbook. Line design calculations can be found in transmission and distribution line design manuals. Large utilities normally develop their own line design manuals. Small rural utilities typically use the transmission and distribution line design manuals published by the Rural Utilities Service (RUS), which in the past was referred to as the Rural Electrification Administration (REA). This Handbook will aid those individuals writing and using line design manuals and provide a deeper understanding of how the **Code** applies to line design calculations.

Typical strength and loading line design calculations include, but are not limited to, the following:

- Maximum wind span based on wind with ice on conductors
- Maximum weight span based on weight of conductors and ice
- Moment due to wind on pole
- Allowable resisting moment of pole
- Transverse, vertical, and total components of conductor loading
- Total ground line moment on pole
- Ruling span
- Diameter of pole at any point
- Deadend guying strength
- Bisector guying strength
- Weak-link of guy attachment, guy wire, and anchor assembly
- Crossarm strength (vertical)
- Crossarm strength (longitudinal)
- Pole buckling
- Maximum line angle based on insulator strength
- Material deflection
- Equipment loading

Single pole structures are typically easier to analyze than multiple pole and lattice structures. Line design calculations also involve clearance calculations in addition to strength and loading calculations (e.g., clearance above ground, clearance to buildings and other structures, horizontal clearance between conductors, vertical clearance between conductors, etc.). The clearance calculations will affect the line design and therefore the selection of pole heights and pole classes. The sample sag and tension chart shown at the beginning of Sec. 23 provides sag for clearance calculations and tension for strength and loading calculations. The "design tension" used for strength and loading calculations is the largest of the tensions calculated after applying Rule 250B (heavy, medium, or light loads), Rule 250C (extreme wind loads) if applicable, and Rule 250D (extreme ice with concurrent wind loads) if applicable. The 250B, 250C and 250D loads are applied to conductors in accordance with Rule 251. See Rule 251 for additional information.

250A. General. If the grade of construction required in Sec. 24, "Grades of Construction," is B or C, Sec. 25 will define the physical loadings in the form of ice, wind, and load factors that increase the physical loads associated with overhead line construction. The terms "load specified in Rule 250," "Rule 250B loads," "Rule 250C loads," and "Rule 250D loads" will be used throughout Secs. 25 and 26.

When the loads in Rule 250B (heavy, medium, or light loading), Rule 250C (extreme wind loading), and Rule 250D (extreme ice with concurrent wind) must all be considered, the worst case of the three must be used. A loading map for overhead lines in the United States is provided in NESC Fig. 250-1. The United States is divided into heavy, medium, and light loading areas. Since only three types are defined, the NESC recognizes that heavier loads than specified in the map may be typical of a specific region. Using a higher loading than the map shows is certainly acceptable. Using a lower loading without approval of an administrative authority (e.g., the State Public Service Committee) is not permitted.

The weather loadings specified in Rules 250B, 250C, and 250D may not be sufficient for the forces imposed during construction and maintenance. Additional loads may need to be considered to adjust for this condition.

Rule 250A4 states that if lines are designed to meet Sec. 25 (loadings for Grades B and C) and Sec. 26 (strength requirements), the lines will have sufficient capability to resist earthquake ground motions.

250B. Combined Ice and Wind District Loading. NESC Fig. 250-1 and NESC Table 250-1 are the two basic references to determine ice and wind loading for heavy, medium, and light loading. The statement "the loads of Rule 250B" will be used throughout Secs. 25 and 26.

The requirements for heavy, medium, and light loading are outlined in Fig. 250-1.

250C. Extreme Wind Loading. NESC Fig. 250-2 (a through e) and NESC Tables 250-2 and 250-3 are the basic references for determining extreme wind loading. The statement "the loads of Rule 250C" will be used throughout Secs. 25 and 26.

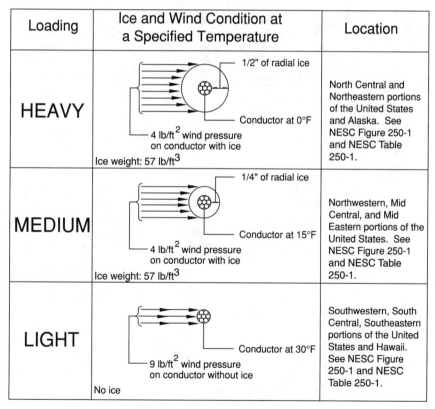

Loading	Ice and Wind Condition at a Specified Temperature	Location
HEAVY	1/2" of radial ice — Conductor at 0°F — 4 lb/ft^2 wind pressure on conductor with ice — Ice weight: 57 lb/ft^3	North Central and Northeastern portions of the United States and Alaska. See NESC Figure 250-1 and NESC Table 250-1.
MEDIUM	1/4" of radial ice — Conductor at 15°F — 4 lb/ft^2 wind pressure on conductor with ice — Ice weight: 57 lb/ft^3	Northwestern, Mid Central, and Mid Eastern portions of the United States. See NESC Figure 250-1 and NESC Table 250-1.
LIGHT	Conductor at 30°F — 9 lb/ft^2 wind pressure on conductor without ice — No ice	Southwestern, South Central, Southeastern portions of the United States and Hawaii. See NESC Figure 250-1 and NESC Table 250-1.

Fig. 250-1. Heavy, medium, and light ice and wind loading (Rule 250B).

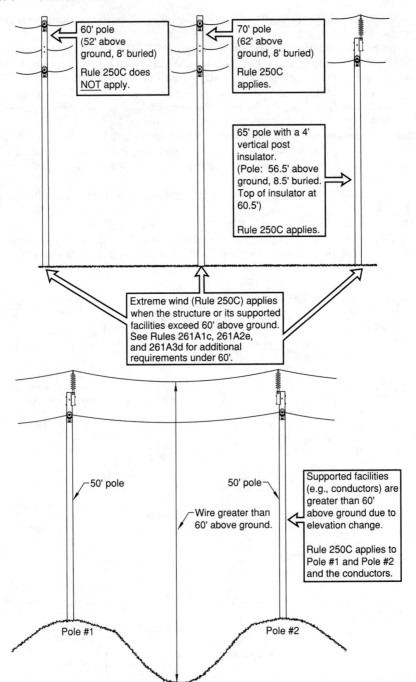

Fig. 250-2. Examples of facilities requiring extreme wind loading (Rule 250C).

Rule 250C only applies to structures (e.g., poles) and supported facilities (e.g., conductors, static wires, messengers, cables, etc.) more than 60 ft above ground or water level. See Fig. 250-2.

Rules 261A1c (metal, prestressed, and reinforced concrete poles), 261A2e (wood poles), and 261A3d (fiber-reinforced polymer poles) require the application of extreme wind to the pole, without conductors, for any height pole. Typically, applying the heavy, medium, or light loading conditions to a pole and its conductors under 60 ft above ground will result in a worse case design condition than a pole under 60 ft above ground with extreme wind applied to the pole without conductors. The under 60 ft requirement exists for special cases. If a pole is pre-engineered (e.g., a steel pole or a laminated wood pole), the design strength at the ground line could have a higher resisting moment in the transverse direction than in the longitudinal direction. For a line under 60 ft above ground, the transverse loading requires application of Rule 250B (heavy, medium, or light) loads. The transverse and longitudinal directions require application of Rule 261A1c, 261A2e, or 261A3d for extreme wind on the pole, without conductors.

Typically the heavy, medium, or light loading requirements applied to conductors in the transverse direction will require more pole strength than the extreme wind loading requirements applied to the pole (without conductors) in the longitudinal direction. If a traditional round wood pole is used, the same strength is available in both the transverse and longitudinal directions. Assuming a traditional round wood pole is sized based on the heavy, medium, or light loading with conductors in the transverse direction, adequate strength should exist for the extreme wind loading on the pole (without conductors) in any direction. Equal pole strength in all directions may not be accurate for a pre-engineered pole utilizing different strengths in different directions.

If it is determined that Rule 250C applies (i.e., the structures or the supported facilities are over 60 ft), the extreme wind must be applied to the entire structure and supported facilities, not just the portions of the structure above 60 ft. Extreme wind is applied without ice on the conductors and without ice on the structure at a 60°F temperature condition. NESC Table 250-1 specifies the temperature at which extreme wind is checked. A formula is provided to calculate the wind load on an object based on the basic wind speed, velocity pressure exposure coefficient, gust response factor, and shape factor.

The basic wind speed is found in NESC Fig. 250-2 (a through e). The velocity pressure exposure coefficient (k_z) increases with increased height due to wind friction near the earth's surface. Typical k_z values are provided in NESC Table 250-2. The formulas to calculate k_z are provided under the table. The gust response factor (G_{RF}) decreases with increased height and longer span lengths because wind velocity becomes more constant with increased height and gusts average out over longer span lengths. Typical G_{RF} values are provided in NESC Table 250-3. The formulas to calculate G_{RF} are provided under the table. Both k_z and G_{RF} have separate values for the structure height and wire height. Shape factor is defined in Rule 252B; see Rule 252B for a discussion. A separate factor may be needed for the structure and the wire if the structure is not a round pole. An example of an extreme wind calculation is shown in Fig. 250-3.

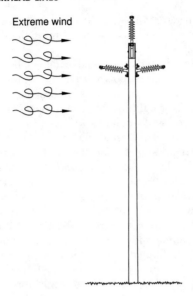

Extreme wind

Example:
- Extreme wind in central Washington State.
  V=85 miles/hour per NESC Figure 250-2(a).
- Pole height above ground line is 70'.
- Wire heights above ground line are 73' and 67'.
- Wind span length is 300'.
- The conductor and the pole have round shapes.

Wind pressure = $0.00256 \cdot (V)^2 \cdot k_z \cdot G_{RF} \cdot I \cdot C_f$	
Extreme wind pressure on the pole:	**Extreme wind pressure on the wires:**
• V = 85 miles/hour per NESC Figure 250-2a	• V = 85 miles/hour per NESC Figure 250-2a
• k_z structure = 1.10 Per NESC Table 250-2	• k_z wire = 1.20 Per NESC Table 250-2
• G_{RF} structure = 0.93 Per NESC Table 250-3	• G_{RF} wire = 0.80 Per NESC Table 250-3
• I = 1.0 for utility structures Per NESC Rule 250C	• I = 1.0 for utility structures Per NESC Rule 250C
• C_f = 1.0 for cylindrical structures per NESC Rule 252B2	• C_f = 1.0 for cylindrical shapes per NESC Rule 251A2
Wind pressure = 18.92 lb/ft^2	Wind pressure = 17.76 lb/ft^2

Fig. 250-3. Example of an extreme wind calculation (Rule 250C).

The value of 17.76 lb/ft^2 for the wire (calculated in Fig. 250-3) is entered at 60°F on the sample sag and tension chart located at the beginning of Sec. 23. This design condition is only required if the structure or the conductor is greater than 60 ft above ground or water level. A line with 45-ft poles on level ground would not require this condition to be on the conductor sag and tension chart. Additional extreme wind examples can be found in **NESC** Appendix C.

250D. Extreme Ice with Concurrent Wind Loading. NESC Fig. 250-3 (a through f) and **NESC** Table 250-4 are the basic references for determining extreme ice with concurrent wind loading. The statement "the loads of 250D" will be used throughout Secs. 25 and 26. Rule 250D only applies to structures (e.g., poles) and supported facilities (e.g., conductors, static wires, messengers, cables, etc.) more than 60 ft above ground or water level. This same requirement can be found in Rule 250C.

If it is determined that Rule 250D applies (i.e., the structure or the supported facilities are over 60 ft) the extreme ice with concurrent wind must be applied to the entire structure and supported facilities, not just the portions of the structure above 60 ft. Extreme ice with concurrent wind is applied to the conductor at a 15°F temperature condition. Only the wind portion of the extreme ice with concurrent wind needs to be applied to the structure (i.e., pole) as only the conductors are required to be covered with ice, not the pole.

NESC Table 250-1 specifies the temperature at which extreme ice with concurrent wind is checked. **NESC** Table 250-4 provides horizontal wind pressures in lb/ft^2 based on wind speed in mph. **NESC** Table 250-4 is only for use with Rule 250D, not Rule 250C, and appears to be for use with cylindrical surfaces. The appropriate shape factors for flat or latticed structures in Rule 252B2 should be considered. The ice portion for extreme ice with concurrent wind is determined from **NESC** Fig. 250-3 (a through f) and applied to the conductors in accordance with Rule 251. See Rule 251 for a discussion and figures for applying ice and wind to conductors. Rule 250D2 allows the ice (not wind) to be derated to 0.80 for Grade C construction. An extreme ice with concurrent wind value of 15°F, 0.25 in of ice, and 2.30 lb/ft^2 of wind is entered on the sample sag and tension chart located at the beginning of Sec. 23. This value was chosen from west-central Washington State. This design condition on the conductor is only required if the structure or the conductor is greater than 60 ft above ground or water level. A line with 45-ft poles on level ground would not require this condition to be on the conductor sag and tension chart.

251. CONDUCTOR LOADING

251A. General. The ice and wind loads in Rule 250, specifically Rules 250B, 250C, and 250D, must be applied to conductors. Rule 251 specifies how to apply the ice and wind loads to conductors.

If the conductor consists of a cable on a messenger, the ice and wind loads are applied to both the cable and the messenger. Examples of wind loads on a stranded conductor and on communication cables supported on a messenger are shown in Fig. 251-1.

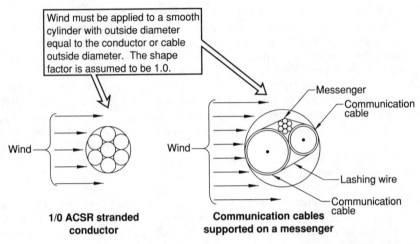

Fig. 251-1. Examples of wind loads on various conductor configurations (Rules 251A1 and 251A2).

Rule 251A3 specifies a method for determining ice loads on various conductor and cable configurations. Rule 251A4 requires that testing or a qualified engineering study be performed if the ice loading method in Rule 251A3 is reduced. Examples of ice and wind loads on a stranded conductor and on communication cables supported on a messenger are shown in Fig. 251-2.

251B. Load Components.   Rule 251B specifies the individual loads to consider when ice and wind is applied to a conductor. The load components are broken into three parts: vertical load, horizontal load, and total load.

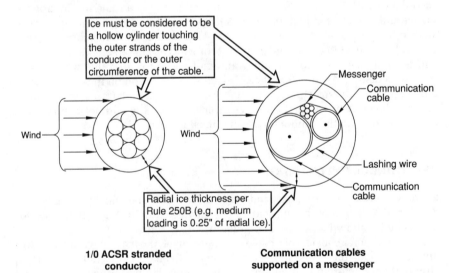

Fig. 251-2. Examples of ice and wind loads on various conductor configurations (Rule 251A3).

The vertical load must consider the weight of the conductor plus conductor spacers or any equipment that the conductor supports (e.g., Federal Aviation Administration marker balls). These items must be covered with ice to calculate weight for the medium and heavy loading districts per Rule 250B or the extreme ice loading per Rule 250D (if applicable).

The horizontal load must consider wind on the conductors plus spacers and equipment that the conductor supports (e.g., Federal Aviation Administration marker balls). These items must be covered with ice to calculate the wind force for medium and heavy loading districts per Rule 250B or the extreme ice loading per Rule 250D (if applicable). The wind pressures in Rule 250B must be considered and, if applicable, the wind pressures in Rules 250C and 250D must be considered.

The total load is the resultant of the horizontal and vertical loads plus a constant given in NESC Table 251-1. The constant varies by loading district (i.e., 0.30 lb/ft for heavy, 0.20 lb/ft for medium, 0.05 lb/ft for light). No constant is added for the extreme wind loading or extreme ice with concurrent wind loading. The conductor or cable messenger tension must be computed from the total load.

An example of the load components for the medium loading district is shown in Fig. 251-3.

The total loads for the medium loading design condition (250B), the extreme wind design condition (250C), and the extreme ice with concurrent wind design condition (250D) are shown in the sample sag and tension chart at the beginning of Sec. 23. The sample sag and tension chart indicates an initial tension of 1366 lb for 15°F, 0.25 in ice, 4 lb/ft^2 wind, and a K factor of 0.20.

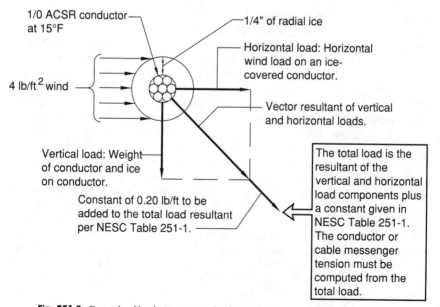

Fig. 251-3. Example of load components for the medium loading district (Rule 251B).

This tension is higher than the extreme wind (250C) initial tension of 1154 lb for 60°F and 17.76 lb/ft^2 of wind and the extreme ice with concurrent wind (250D) initial tension of 1070 lb for 15°F, 0.25 in of ice, and 2.30 lb/ft^2 of wind. The 1366-lb tension is commonly called the "design tension." It is possible for the 250C (extreme wind) or the 250D (extreme ice with concurrent wind) loads to produce a higher tension than the 250B (heavy, medium, or light) loads. If the 250C or 250D loads are applicable (i.e., the structure or conductors are more than 60 ft above ground or water), then the "design tension" must be chosen from the largest tension based on the 250B, 250C, and 250D loads.

Rule 261H1 requires the open (noninsulated) supply conductor tension at the Rule 251 condition to be not more than 60 percent of the conductor rated breaking strength. The sample sag and tension chart at the beginning of Sec. 23 indicates a percent rated tensile strength of 31.2 percent for the 1366-lb design tension. This percentage meets the not more than 60 percent requirement. Rule 261H1 also specifies an initial and final tension limit at 60°F without external load (i.e., without ice or wind) and hardware strength limits for use with extreme wind (250C) and extreme ice with concurrent wind (250D) tensions. Rules 261I (supply cable messengers) and 261K2 (communication cable messengers) specify messenger tension limits for the 250B (heavy, medium, or light loading), 250C (extreme wind loading), and 250D (extreme ice with concurrent wind loading) conditions applied in accordance with Rule 251.

252. LOADS ON LINE SUPPORTS

252A. Assumed Vertical Loads.   Ice and wind loads are first defined in Rule 250, specifically Rules 250B, 250C, and 250D. How to apply ice and wind loading to conductors is defined in Rule 251. How to apply ice and wind loading on line supports (i.e., poles, towers, foundations, crossarms, pins, insulators, and conductor fastenings) is specified in Rule 252. Rule 250A2 should be reviewed for proper consideration of construction and maintenance loads.

Rule 252A discusses how to apply vertical loads to line supports. Rule 252A does not specify any load factors. The load factors are provided in Rule 253. An example of vertical loads on line supports is shown in Fig. 252-1.

A difference in elevation of supports must be considered as it will affect the length of the vertical (weight) span. An example of vertical loads on line supports with different elevations is shown in Fig. 252-2.

252B. Assumed Transverse Loads.   Rule 252B discusses how to apply transverse (horizontal) loads to line supports. Rule 252B does not specify any load factors. The load factors are provided in Rule 253. An example of transverse loads on line supports is shown in Fig. 252-3.

Rule 252B2 specifies shape factors to be used for various surface types. Transverse wind loads on structures are calculated without ice covering. Examples of wind load on various structure types are shown in Fig. 252-4.

Rule 252B3 requires angle structure to consider both the transverse wind load and the wire tension load. The wind direction must be applied to give the

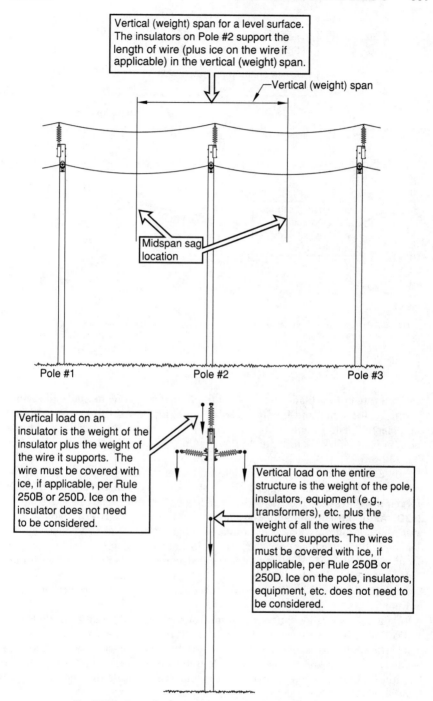

Vertical (weight) span for a level surface. The insulators on Pole #2 support the length of wire (plus ice on the wire if applicable) in the vertical (weight) span.

Vertical (weight) span

Midspan sag location

Pole #1   Pole #2   Pole #3

Vertical load on an insulator is the weight of the insulator plus the weight of the wire it supports. The wire must be covered with ice, if applicable, per Rule 250B or 250D. Ice on the insulator does not need to be considered.

Vertical load on the entire structure is the weight of the pole, insulators, equipment (e.g., transformers), etc. plus the weight of all the wires the structure supports. The wires must be covered with ice, if applicable, per Rule 250B or 250D. Ice on the pole, insulators, equipment, etc. does not need to be considered.

Fig. 252-1. Example of vertical loads on line supports (Rule 252A).

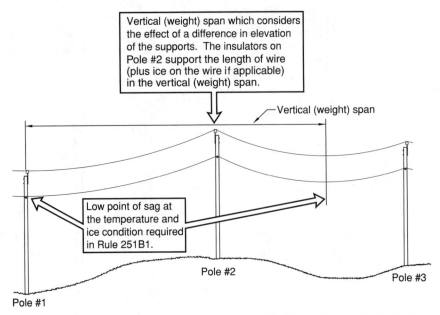

Fig. 252-2. Example of vertical loads on line supports with different elevations (Rule 252A).

maximum resultant load. The angle of the line affects how to apply the wind. Normally, for small and medium angles, an oblique wind in the direction of the bisector of the angle will provide the maximum resultant load. Proper reductions are permissible due to the angularity of the wind on the wire. See Fig. 252-5.

Rule 252B4 states that the calculated transverse load must be based on the average of the two spans adjacent to the structure concerned. Rule 252B4 does not require elevation to be considered as Rule 252A does for a vertical span.

252C. Assumed Longitudinal Loading.   Longitudinal loads are loads in line with the conductors. The most obvious longitudinal load is at a conductor deadended (single sided deadend) on a pole; however, longitudinal loading can occur on double deadends and on tangent structures under certain conditions. Rule 252C breaks longitudinal loading into seven parts (252C1 through 252C7). Rule 252C does not specify any load factors. The load factors are provided in Rule 253.

Rule 252C1 specifies longitudinal tensions to be used when a Grade B line section is located within a line of a lower grade. A common example is a Grade C distribution line crossing a limited access highway or railroad tracks. The Grade C line must be increased to Grade B for the crossing. The **Code** rules for longitudinal loads where a change in the grade of construction occurs are outlined in Fig. 252-6.

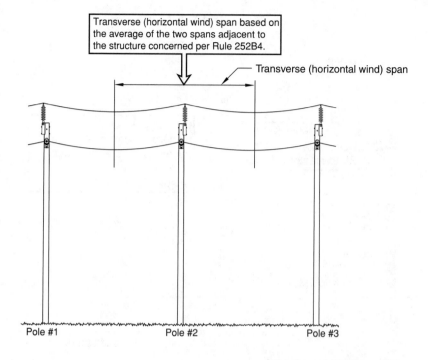

Transverse (horizontal wind) span based on the average of the two spans adjacent to the structure concerned per Rule 252B4.

Transverse (horizontal wind) span

Pole #1   Pole #2   Pole #3

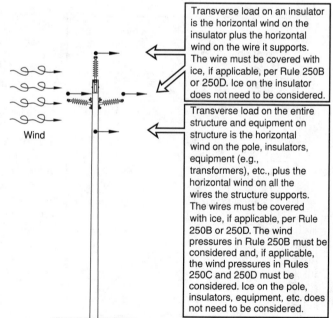

Wind

Transverse load on an insulator is the horizontal wind on the insulator plus the horizontal wind on the wire it supports. The wire must be covered with ice, if applicable, per Rule 250B or 250D. Ice on the insulator does not need to be considered.

Transverse load on the entire structure and equipment on structure is the horizontal wind on the pole, insulators, equipment (e.g., transformers), etc., plus the horizontal wind on all the wires the structure supports. The wires must be covered with ice, if applicable, per Rule 250B or 250D. The wind pressures in Rule 250B must be considered and, if applicable, the wind pressures in Rules 250C and 250D must be considered. Ice on the pole, insulators, equipment, etc. does not need to be considered.

Fig. 252-3. Example of transverse loads on line supports (Rule 252B).

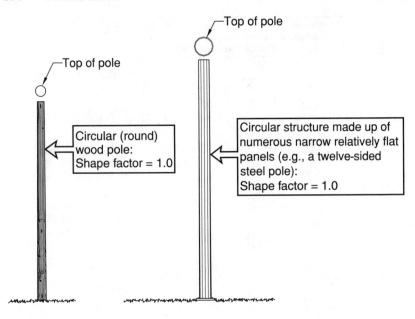

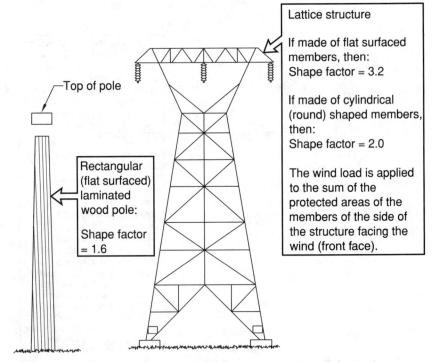

Fig. 252-4. Examples of wind load on various structure types (Rule 252B2).

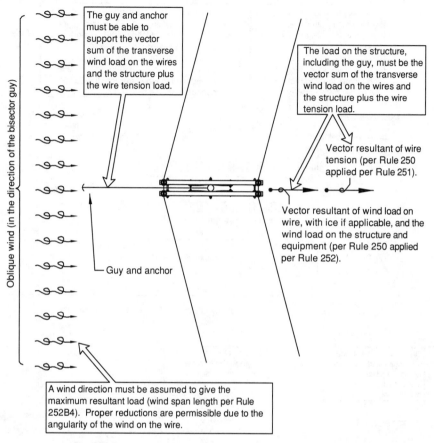

Oblique wind (in the direction of the bisector guy)

The guy and anchor must be able to support the vector sum of the transverse wind load on the wires and the structure plus the wire tension load.

The load on the structure, including the guy, must be the vector sum of the transverse wind load on the wires and the structure plus the wire tension load.

Vector resultant of wire tension (per Rule 250 applied per Rule 251).

Vector resultant of wind load on wire, with ice if applicable, and the wind load on the structure and equipment (per Rule 250 applied per Rule 252).

Guy and anchor

A wind direction must be assumed to give the maximum resultant load (wind span length per Rule 252B4). Proper reductions are permissible due to the angularity of the wind on the wire.

Fig. 252-5. Example of transverse loads at line angles (Rule 252B3).

Rule 252C2 addresses the older open-wire communication circuits with multiple conductors on a crossarm. This construction has frequently been replaced with multiple pair insulated cables and fiber-optic cables.

Rule 252C3 provides rules for longitudinal loading on deadend structures. See Fig. 252-7.

Rule 252C4 requires that a structure be capable of supporting unbalanced longitudinal loads created by unequal vertical loads or unequal spans. Extreme cases will require a double deadend structure and guying.

Rule 252C5 requires that consideration be given to longitudinal loads during wire stringing operations. Temporary deadend structures and temporary guying are typically used to meet this criterion if permanent facilities do not exist at the wire stringing locations. Rule 252C5 is similar in nature to Rule 250A2, which requires considering construction and maintenance loads.

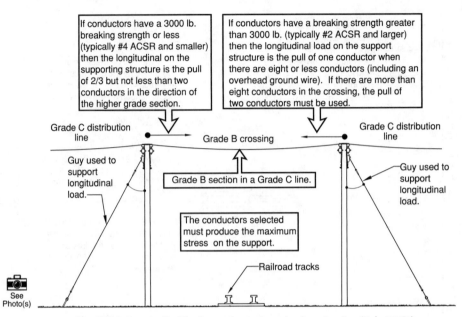

If conductors have a 3000 lb. breaking strength or less (typically #4 ACSR and smaller) then the longitudinal load on the supporting structure is the pull of 2/3 but not less than two conductors in the direction of the higher grade section.

If conductors have a breaking strength greater than 3000 lb. (typically #2 ACSR and larger) then the longitudinal load on the support structure is the pull of one conductor when there are eight or less conductors (including an overhead ground wire). If there are more than eight conductors in the crossing, the pull of two conductors must be used.

Grade C distribution line

Grade B crossing

Grade C distribution line

Guy used to support longitudinal load.

Grade B section in a Grade C line.

Guy used to support longitudinal load.

The conductors selected must produce the maximum stress on the support.

Railroad tracks

See Photo(s)

Fig. 252-6. Longitudinal loads at a change in grade of construction (Rule 252C1).

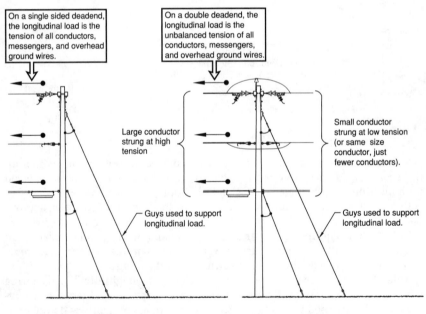

On a single sided deadend, the longitudinal load is the tension of all conductors, messengers, and overhead ground wires.

On a double deadend, the longitudinal load is the unbalanced tension of all conductors, messengers, and overhead ground wires.

Large conductor strung at high tension

Small conductor strung at low tension (or same size conductor, just fewer conductors).

Guys used to support longitudinal load.

Guys used to support longitudinal load.

See Photo(s)

Single Sided Deadend   **Double Deadend**

Fig. 252-7. Longitudinal loads at deadends (Rule 252C34).

Rule 252C6 recommends that structures with longitudinal strength capacity be provided at reasonable intervals along a line. This rule is very general in nature. The intent of this rule is to avoid a "domino effect," which could occur in long stretches of straight lines if the wires in one span were to break. Some degree of longitudinal structure strength exists in a tangent wood pole designed for transverse wind loads. If the conductors break during a nonloaded condition (i.e., no ice or wind), the actual longitudinal loads will be less than if the conductors break during ice and wind conditions. A certain amount of longitudinal strength will also be provided by application of Rules 261A1c, 261A2e, and 261A3d. Typically the transverse loading and the application of Rules 261A1c, 261A2e, and 261A3d will not be as severe as a longitudinal deadend conductor load. A line with angles, especially 90° turns, has longitudinal strength built in. A line that continues straight, say for 5 miles, may require a double deadend structure in the middle of the 5-mile stretch with enough longitudinal strength to act as a single deadend if the conductors were to break on either side of the pole. When to insert a double deadend pole in a straight line section requires accepted good practice as the **Code** does not specify a specific distance. See Fig. 252-8.

Rule 252C7 specifies longitudinal tensions for open-wire communication conductors at railroad and limited access highway crossings. This construction has frequently been replaced with multiple pair insulated cables and fiber-optic cables.

252D. Simultaneous Application of Loads.   A structure may experience a combination of vertical (weight), transverse (horizontal wind and wire tension), and longitudinal (in-line) loads. The structure must be designed to withstand the simultaneous application of these loads when they occur simultaneously. An example of a structure experiencing simultaneous loads is shown in Fig. 252-9.

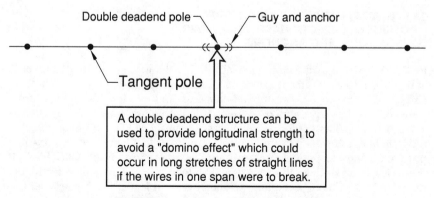

Double deadend pole — Guy and anchor

— Tangent pole

A double deadend structure can be used to provide longitudinal strength to avoid a "domino effect" which could occur in long stretches of straight lines if the wires in one span were to break.

Fig. 252-8. Longitudinal capability (Rule 252C6).

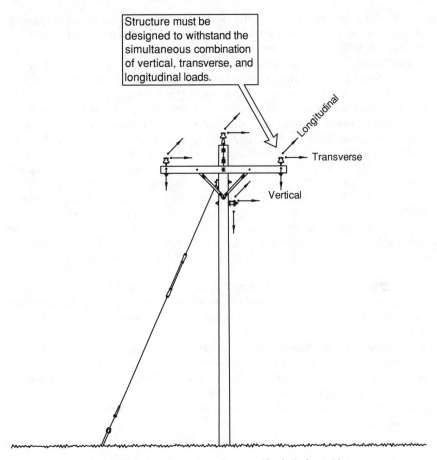

Structure must be designed to withstand the simultaneous combination of vertical, transverse, and longitudinal loads.

Longitudinal

Transverse

Vertical

Fig. 252-9. Simultaneous application of loads (Rule 252D).

253. LOAD FACTORS FOR STRUCTURES, CROSSARMS, SUPPORT HARDWARE, GUYS, FOUNDATIONS, AND ANCHORS

Rule 253 requires load factors to be applied to the loads of Rules 250B, 250C, and 250D. This requirement is also stated in a similar fashion in Rules 261A1a and 261A1b for metal, prestressed-concrete, and reinforced-concrete structures, in Rules 262A2a and 262A2b for wood structures, and in Rules 261A3a and 261A3b for fiber-reinforced polymer structures. Load factors are provided in **NESC** Table 253-1. A load factor is a load multiplier that is greater than or equal to 1.0 with one exception, which is Grade C extreme wind (Rule 250C) loads. It is applied to a base load to increase it. The load factors for Grade B are higher than Grade C with one exception, which is vertical loads. The higher load factors for Grade B produce a line that is stronger than a Grade C line because the Grade B line is required to withstand higher loads.

The load factors in **NESC** Table 253-1 in Sec. 25 must be used in combination with the strength factors in **NESC** Table 261-1A in Sec. 26. A strength factor is a material strength multiplier that is less than or equal to 1.0. It is applied to a material strength to reduce it. The strength factors for Grade B are less than or equal to Grade C. The lower strength factors for Grade B produce a line that is stronger than a Grade C line because less of the materials' full strength is permitted to be used for Grade B construction.

An alternate method is provided for wood and reinforced (not prestressed) concrete structures. The alternate method requires the use of **NESC** Table 253-2 in conjunction with **NESC** Table 261-1B. The alternate method primarily uses load factors to increase safety, as the strength factors are equal to 1.0. The alternate method does not apply to steel or fiber-reinforced polymer structures and the alternate method must not be used after July 31, 2010, which is near the middle of the 2007–2012 (5 year) **Code** cycle.

The combined effect of the load factor in **NESC** Table 253-1 and the strength factor in **NESC** Table 261-1A can be determined by dividing the load factor by the strength factor. For example, per **NESC** Table 253-1, the transverse wind load factor of a Grade B line is 2.50. Per **NESC** Table 261-1A, the wood pole strength factor of a Grade B line is 0.65. The combined effect of these factors is 2.50 ÷ 0.65 = 3.85. This value is comparable to the alternate load factor in **NESC** Table 253-2, which is 4.00.

An outline of load factors and strength factors is provided in Fig. 253-1.

Examples of applying load and strength factors are provided in Fig. 253-2.

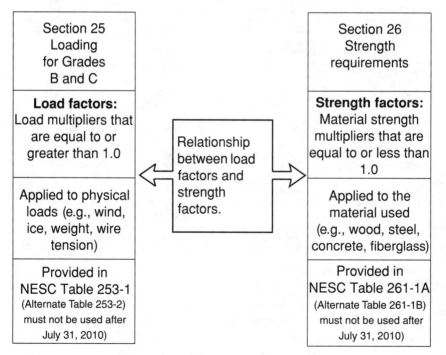

Fig. 253-1. Load and strength factors (Rules 253 and 261).

Example #1	Example #2	Example #3	Example #4
Conditions:	**Conditions:**	**Conditions:**	**Conditions:**
- Transverse wind load on 1/0 ACSR wire - Medium loading district - Wood poles - Grade B line	- Transverse wind load on 1/0 ACSR wire - Medium loading district - Wood poles - Grade C line	- Transverse wind load on 1/0 ACSR wire - Medium loading district - Steel poles - Grade B line	- Transverse wind load on 1/0 ACSR wire - Medium loading district - Steel poles - Grade C line
Load and Strength Factor:	**Load and Strength Factor:**	**Load and Strength Factor:**	**Load and Strength Factor:**
- Load factor from NESC Table 253-1: 2.50	- Load factor from NESC Table 253-1: 2.20 (at crossings) 1.75 (elsewhere)	- Load factor from NESC Table 253-1: 2.50	- Load factor from NESC Table 253-1: 2.20 (at crossings) 1.75 (elsewhere)
- Strength factor from NESC Table 261-1A: 0.65	- Strength factor from NESC Table 261-1A: 0.85	- Strength factor from NESC Table 261-1A: 1.0	- Strength factor from NESC Table 261-1A: 1.0
- Alternate load factor from NESC Table 253-2: 4.0	- Alternate load factor from NESC Table 253-2: 2.67 (at crossings) 2.0 (elsewhere)	- Alternate NESC Table 253-2 is not applicable to steel poles.	- Alternate NESC Table 253-2 is not applicable to steel poles.
- Alternate strength factor from NESC Table 261-1B: 1.0	- Alternate strength factor from NESC Table 261-1B: 1.0	- Alternate NESC Table 261-1B is not applicable to steel poles.	- Alternate NESC Table 261-1B is not applicable to steel poles.
Note: Alternate factors must not be used after July 31, 2010	Note: Alternate factors must not be used after July 31, 2010		

Example #5	Example #6	Example #7	Example #8
Conditions:	**Conditions:**	**Conditions:**	**Conditions:**
- Longitudinal wire tension load at a 1/0 ACSR deadend - Medium loading district - Wood poles - Grade B line	- Longitudinal wire tension load at a 1/0 ACSR deadend - Medium loading district - Wood poles - Grade C line	- Longitudinal wire tension load at a 1/0 ACSR deadend - Medium loading district - Steel poles - Grade B line	- Longitudinal wire tension load at a 1/0 ACSR deadend - Medium loading district - Steel poles - Grade C line
Load and Strength Factor:	**Load and Strength Factor:**	**Load and Strength Factor:**	**Load and Strength Factor:**
- Load factor from NESC Table 253-1: 1.65	- Load factor from NESC Table 253-1: 1.30	- Load factor from NESC Table 253-1: 1.65	- Load factor from NESC Table 253-1: 1.30
- Strength factor from NESC Table 261-1A: 0.65	- Strength factor from NESC Table 261-1A: 0.85	- Strength factor from NESC Table 261-1A: 1.0	- Strength factor from NESC Table 261-1A: 1.0
- Alternate load factor from NESC Table 253-2: 2.0	- Alternate load factor from NESC Table 253-2: 1.33	- Alternate NESC Table 253-2 is not applicable to steel poles.	- Alternate NESC Table 253-2 is not applicable to steel poles.
- Alternate strength factor from NESC Table 261-1B: 1.0	- Alternate strength factor from NESC Table 261-1B: 1.0	- Alternate NESC Table 261-1B is not applicable to steel poles.	- Alternate NESC Table 261-1B is not applicable to steel poles.
Note: Alternate factors must not be used after July 31, 2010	Note: Alternate factors must not be used after July 31, 2010		

Fig. 253-2. Examples of applying load and strength factors (Rules 253 and 261).

Section 26

Strength Requirements

260. GENERAL (SEE ALSO SECTION 20)

Sections 24, 25, and 26 are all interrelated. Section 24, "Grades of Construction," defines the required strength of overhead line construction for safety purposes. Section 25, "Loadings for Grades B and C," defines the physical loads (i.e., ice, wind, and temperature conditions) that overhead line construction must be able to withstand and the load factors that must be applied to the physical loads. Section 26, "Strength Requirements," defines the required strength of materials used in constructing overhead lines and the strength factors that must be applied to the materials. Line insulators are not addressed in Secs. 24, 25, or 26. Insulators are covered in their own section, Sec. 27.

The purpose of Sec. 24 is to define the grade of construction that is required for different situations. The details of applying load factors and strength factors are covered in Secs. 25 and 26.

This Handbook addresses the Code requirements in Secs. 24 to 26. Line design calculations are not presented in this Handbook. Line design calculations can be found in transmission and distribution line design manuals. Large utilities normally develop their own line design manuals. Small rural utilities typically use the transmission and distribution line design manuals published by the Rural Utilities Service (RUS), which in the past was referred to as the Rural Electrification Administration (REA). This Handbook will aid those individuals writing and using line design manuals and provide a deeper understanding of how the Code applies to line design calculations.

Typical strength and loading line design calculations include, but are not limited to, the following:

- Maximum wind span based on wind with ice on conductors
- Maximum weight span based on weight of conductors and ice
- Moment due to wind on pole
- Allowable resisting moment of pole
- Transverse, vertical, and total components of conductor loading
- Total ground line moment on pole
- Ruling span
- Diameter of pole at any point
- Deadend guying strength
- Bisector guying strength
- Weak-link of guy attachment, guy wire, and anchor assembly
- Crossarm strength (vertical)
- Crossarm strength (longitudinal)
- Pole buckling
- Maximum line angle based on insulator strength
- Material deflection
- Equipment loading

Single pole structures are typically easier to analyze than multiple pole and lattice structures. Line design calculations also involve clearance calculations in addition to strength and loading calculations (e.g., clearance above ground, clearance to buildings and other structures, horizontal clearance between conductors, vertical clearance between conductors, etc.). The clearance calculations will affect the line design and therefore the selection of pole heights and pole classes. The sample sag and tension chart shown at the beginning of Sec. 23 provides sag for clearance calculations and tension for strength and loading calculations. The "design tension" used for strength and loading calculations is the largest of the tensions calculated after applying Rule 250B (heavy, medium, or light loads), Rule 250C (extreme wind loads) if applicable, and Rule 250D (extreme ice with concurrent wind loads) if applicable. The 250B, 250C and 250D loads are applied to conductors in accordance with Rule 251. See Rule 251 for additional information.

260A. Preliminary Assumptions.   Rule 260A explains how to account for deformation, deflection, and displacement of parts of a structure when performing strength calculations. The NESC does not require every structure to have a deflection analysis. The NESC does say that when calculating stresses, allowance may be made for deformation, deflection, and displacement when the effects can be evaluated. The NESC provides a two-part approach to determine deformation, deflection, and displacement. First, the ice and wind loads of Rule 250 are applied to the structure without the load factors of Rule 253. These are the loads that are used to determine deformation, deflection, and displacement. Second, from the deformation, deflection, and displacement position, the structure strength must be analyzed using the load factors in Rule 253. See Fig. 260-1.

If allowance is made for deformation, deflection, or displacement at crossings or conflicts, the calculations used must be subject to mutual agreement of the parties involved.

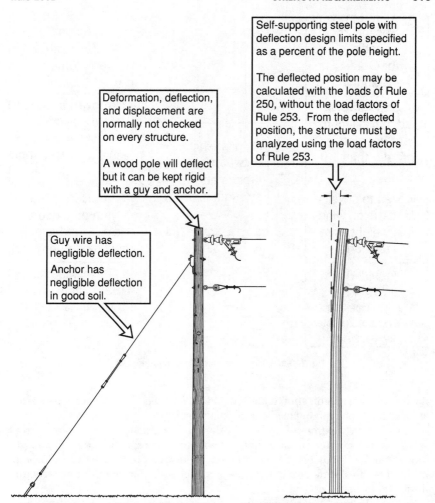

Fig. 260-1. Assumptions for deformation, deflection, or displacement of parts (Rule 260A).

The NESC recognizes that new materials may become available that are not covered in the strength requirements section. Since the NESC is on a five-year revision cycle, the Code permits trial installations of new materials. New materials must be tested and evaluated. See Rule 013 for additional requirements.

260B. Application of Strength Factors.   Rule 260B1 explains how to apply NESC load and strength factors to line design calculations. An outline of load and strength factors is provided in Fig. 260-2.

NESC Table 261-1A is the main table in Sec. 26 for determining strength factors of various materials. It has strength factors for the loads of 250B (ice and wind in heavy, medium, and light loading districts), for the loads of 250C (extreme wind), and for the loads of 250D (extreme ice with concurrent wind). NESC Table 261-1A

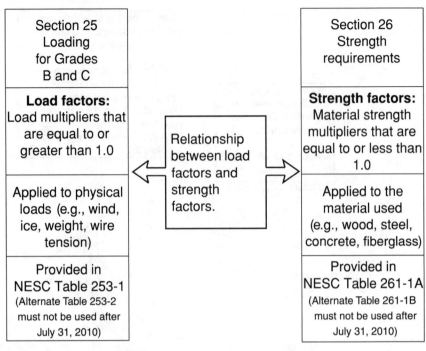

Fig. 260-2. Load and strength factors (Rule 260B).

must be used in combination with NESC Table 253-1. See Rule 253 for a discussion and examples of applying load and strength factors.

If Rule 250C (extreme wind) or Rule 250D (extreme ice with concurrent wind) applies and the strength factor is not otherwise specified (i.e., in an individual rule or Table 261-1A), a strength factor of 0.80 must be used for supported facilities. Rule 260B2 notes several standards for determining structure design capacity.

NESC Table 261-1A contains information relating to deterioration of structure strength under the title of the table and in the table footnotes. Per NESC Table 261-1A, Footnote 2, a wood pole must be replaced or rehabilitated when deterioration reduces the structure strength to $2/3$ of that required when installed. The word "required" is used to distinguish between the required strength of a structure and the actual strength of a structure. If the wood pole class was required to be Class 5, but a Class 4 pole was installed, the pole must be replaced or rehabilitated when deterioration reduces the structure strength to $2/3$ of the Class 5 rating. The most common wood pole deterioration is ground line rot. The most common rehabilitation method is pole stubbing. If a pole stub is added, the stubbed pole must have strength greater than $2/3$ of that required when installed. If the pole is replaced instead of stubbed, the full strength requirements of NESC Table 261-1A apply. Footnote 3 of NESC Table 261-1A is similar to Footnote 2. Footnote 2 specifies a strength deterioration of

$^{2}/_{3}$ for Rule 250B (heavy, medium, and light) loads. Footnote 3 specifies a deterioration of $^{3}/_{4}$ for Rule 250C (extreme wind loads) and Rule 250D (extreme ice with concurrent wind loads). The wording under the title of NESC Table 261-1A further defines how deterioration applies to existing poles that have new or changed lines or equipment added to them.

261. GRADES B AND C CONSTRUCTION

261A. Supporting Structures. The first sentence in Rule 261A states that the strength requirements of supporting structures (e.g., poles) may be provided by the structure alone or with the aid of guys or braces or both. See Fig. 261-1.

The strength requirements for guys are provided in Rules 264, 261A2c, and 261C. An example of structure strength using a structure alone or a structure with guys is shown in Fig. 261-2.

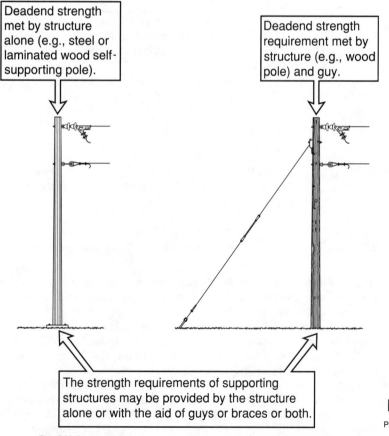

Deadend strength met by structure alone (e.g., steel or laminated wood self-supporting pole).

Deadend strength requirement met by structure (e.g., wood pole) and guy.

The strength requirements of supporting structures may be provided by the structure alone or with the aid of guys or braces or both.

See Photo(s)

Fig. 261-1. Strength requirements of supporting structure (Rule 261A).

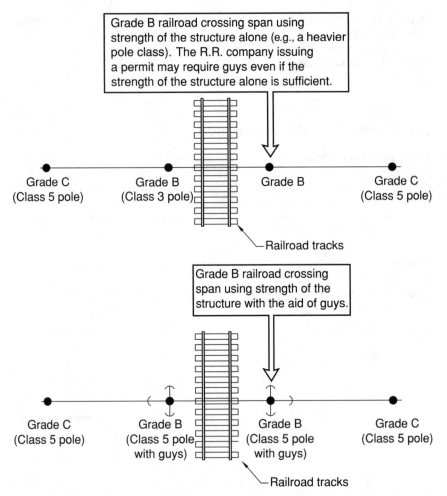

Fig. 261-2. Example of structure strength using a structure alone or a structure with guys (Rule 261A).

261A1. Metal, Prestressed-, and Reinforced-Concrete Structures. The Code rules related to strength requirements of metal, prestressed-, and reinforced-concrete structures are outlined in Fig. 261-3.

261A2. Wood Structures. The permitted stress level for natural wood poles can vary due to splices, wood grain, knots in the wood, moisture, etc. An ANSI standard, ANSI O5.1, must be used to determine the fiber stress to which strength factors in **NESC** Table 261-1A are applied.

The permitted stress level of sawed or laminated wood structural members, crossarms, and braces is determined by using the ultimate fiber stress of the material in ANSI O5.2 multiplied by the strength factors in **NESC** Table 261-1A.

The Code rules related to strength requirements of wood structures are outlined in Figs. 261-4 through 261-6.

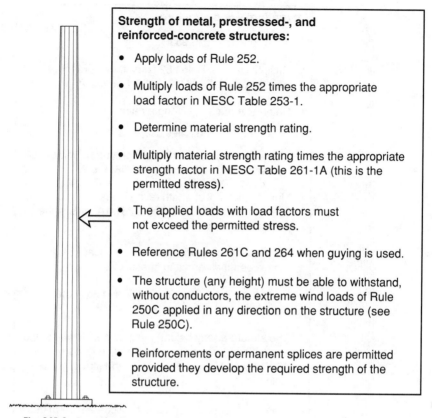

Strength of metal, prestressed-, and reinforced-concrete structures:

- Apply loads of Rule 252.

- Multiply loads of Rule 252 times the appropriate load factor in NESC Table 253-1.

- Determine material strength rating.

- Multiply material strength rating times the appropriate strength factor in NESC Table 261-1A (this is the permitted stress).

- The applied loads with load factors must not exceed the permitted stress.

- Reference Rules 261C and 264 when guying is used.

- The structure (any height) must be able to withstand, without conductors, the extreme wind loads of Rule 250C applied in any direction on the structure (see Rule 250C).

- Reinforcements or permanent splices are permitted provided they develop the required strength of the structure.

See Photo(s)

Fig. 261-3. Strength requirements of metal, prestressed-, and reinforced-concrete structures (Rule 261A1).

261A3. Fiber-Reinforced Polymer Structures. The Code rules related the strength requirements of fiber-reinforced polymer structures are outlined in Fig. 261-7.

261A4. Transverse Strength Requirements for Structures Where Side Guying Is Required, But Can Be Installed Only at a Distance. Rule 261A4 can be applied to various types (e.g., wood, steel, fiber-reinforced polymer, etc.) of supporting structures. The Code rules related to transverse structure strength where side guys are required, but can only be installed at a distance, are outlined in Fig. 261-8.

261A5. Longitudinal Strength Requirements for Sections of Higher Grade in Lines of a Lower Grade Construction. Rule 261A5 can be applied to various types (e.g., wood, steel, fiber-reinforced polymer, etc.) of supporting structures. The Code rules related to longitudinal strength requirements for sections of higher grade in lines of a lower grade construction are outlined in Fig. 261-9.

Rule 252C1 provides longitudinal loading requirements for sections of Grade B construction when located in lines of lower than Grade B construction.

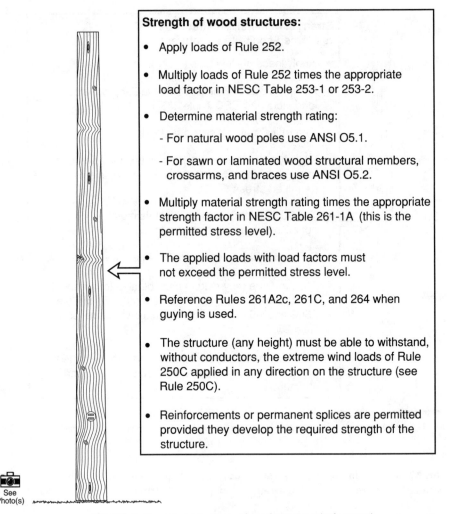

Strength of wood structures:

- Apply loads of Rule 252.

- Multiply loads of Rule 252 times the appropriate load factor in NESC Table 253-1 or 253-2.

- Determine material strength rating:

  - For natural wood poles use ANSI O5.1.

  - For sawn or laminated wood structural members, crossarms, and braces use ANSI O5.2.

- Multiply material strength rating times the appropriate strength factor in NESC Table 261-1A (this is the permitted stress level).

- The applied loads with load factors must not exceed the permitted stress level.

- Reference Rules 261A2c, 261C, and 264 when guying is used.

- The structure (any height) must be able to withstand, without conductors, the extreme wind loads of Rule 250C applied in any direction on the structure (see Rule 250C).

- Reinforcements or permanent splices are permitted provided they develop the required strength of the structure.

See Photo(s)

Fig. 261-4. Strength requirements of wood structures (Rule 261A2).

Rule 261A5b requires guying or increased clearance to overcome the increased sag that occurs when a structure is deflected.

261B. Strength of Foundations, Settings, and Guy Anchors.   The Code rules for strength of foundations, settings, and guy anchors are outlined in Fig. 261-10.

Rule 261B states that design or experience may be used to determine the strength of foundations, settings, and guy anchors. The use of experience is commonly applied to local soil conditions. A foundation, setting, or guy anchor is only as strong as the soil it is set in. If soil types are unknown, soil borings should be done to determine soil conditions. The NOTE in Rule 261B provides

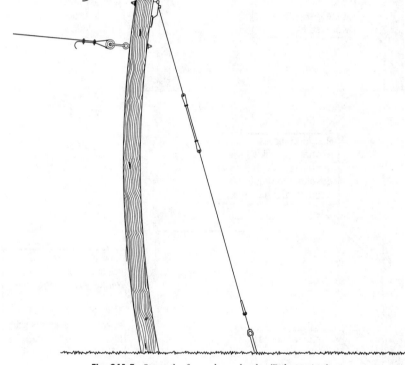

Guyed wood poles must be designed as columns, resisting the vertical component of tension in the guy and any other vertical loads. (Proper design will prevent pole buckling.)

Fig. 261-5. Strength of guyed wood poles (Rule 261A2c).

a reminder that several factors may reduce clearance and structure strength. See Fig. 261-11.

261C. Strength of Guys and Guy Insulators.   Guy strength is covered in Rule 264. Guy insulator strength is covered in Rule 279A1c. Rule 261C provides requirements relative to the integration of the guy with the structure to which the guy is attached. Rule 261A2c provides similar requirements for guys attached to wood poles only. See Fig. 261-12.

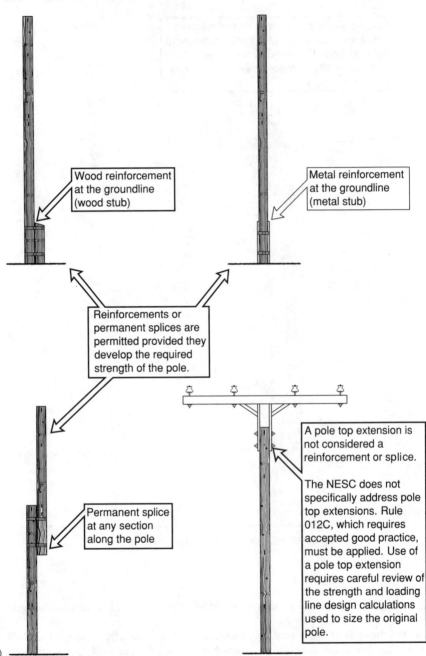

Wood reinforcement at the groundline (wood stub)

Metal reinforcement at the groundline (metal stub)

Reinforcements or permanent splices are permitted provided they develop the required strength of the pole.

A pole top extension is not considered a reinforcement or splice.

The NESC does not specifically address pole top extensions. Rule 012C, which requires accepted good practice, must be applied. Use of a pole top extension requires careful review of the strength and loading line design calculations used to size the original pole.

Permanent splice at any section along the pole

See Photo(s)

Fig. 261-6. Examples of spliced and reinforced wood poles (Rule 261A2d).

Strength of fiber-reinforced polymer structures:

- Apply loads of Rule 252.

- Multiply loads of Rule 252 times the appropriate load factor in NESC Table 253-1.

- Determine material strength rating (using 5th percentile strength or less).

- Multiply material strength rating times the appropriate strength factor in NESC Table 261-1A (this is the permitted load).

- The applied loads with load factors must not exceed the permitted load.

- Reference Rules 261C and 264 when guying is used.

- The structure (any height) must be able to withstand, without conductors, the extreme wind loads of Rule 250C applied in any direction on the structure (see Rule 250C).

- Reinforcements or permanent splices are permitted provided they develop the required strength of the structure.

See Photo(s)

Fig. 261-7. Strength requirements of fiber-reinforced polymer structures (Rule 261A3).

The NOTE in Rule 261C is similar to the NOTE in Rule 261B. Excessive movement of a guy (or movement of the anchor attached to the guy) may reduce clearance or structure capacity.

261D. Crossarms and Braces Rule 261D covers crossarms and braces made of various materials.

261D1. Concrete and Metal Crossarms and Braces. The Code rules for concrete and metal crossarms and braces are outlined in Fig. 261-13.

261D2. Wood Crossarms and Braces. The Code rules for wood crossarms and braces are outlined in Fig. 261-14.

In addition to the requirement to use NESC Table 253-1 with NESC Table 261-1A, the Code also specifies minimum dimensions for select Southern Pine and Douglas Fir wood crossarms in NESC Table 261-2. Crossarms of other wood species may be used if they provide equal strength. See Fig. 261-15.

261D3. Fiber-Reinforced Polymer Crossarms and Braces. The Code rules for fiber-reinforced polymer crossarms and braces are outlined in Fig. 261-16.

Transverse structure strength where side guys are required but can only be installed at a distance:

- The distance between the side guyed structures must not be over 800'.

- The line between the side guyed structures must be substantially straight and the average of the spans must not exceed 150'.

- The side guyed poles must be designed to support the transverse load of the entire section. Intermediate poles are assumed not to carry any transverse load.

- The line between the side guyed structures must be designed to Grade B except for the transverse strength of the poles between the side guyed structures.

- Rule 261A4 applies to Grade B only; it does not apply to Grade C.

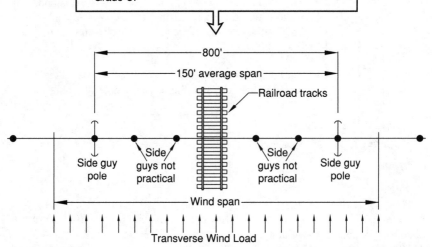

Fig. 261-8. Transverse structure strength where side guys are required but can only be installed at a distance (Rule 261A4).

261D4. Crossarms and Braces of Other Materials.   Crossarms of other materials not specified in Rule 261D (i.e., other than concrete, metal, wood, and fiber-reinforced polymer) must meet the strength requirements of wood crossarms and braces.

261D5. Additional Requirements.   Rule 261D5 provides additional rules for longitudinal strength of crossarms. Tension values and construction methods are specified. Using double wood crossarms is one method of meeting the longitudinal strength requirements of a Grade B line. A support assembly of equivalent strength (i.e., a pre-engineered crossarm assembly) is also acceptable.

Longitudinal strength requirements for sections of higher grade in lines of a lower grade construction:

- The distance between structures with the required longitudinal strength must not be over 800'.

- The distance between the higher grade section and the structure with the required longitudinal strength must not exceed 500'.

- The line between the structures with the required longitudinal strength must be designed to Grade B transverse strength and stringing requirements.

- The line between the structures with the required longitudinal strength must be approximately straight or guyed.

- The structures with the required longitudinal strength may be guyed structures (or self-supporting structures).

- Rule 261A5 applies to Grade B only; it does not apply to Grade C.

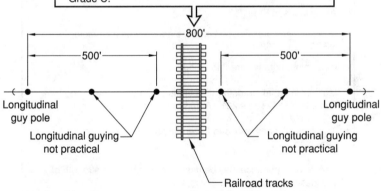

Fig. 261-9. Longitudinal strength requirements for sections of higher grade in lines of a lower grade construction (Rule 261A5).

261E. Insulators. Insulators are covered in Sec. 27. See Rule 277 for strengths of line insulators and Rule 279 for strengths of guy and span insulators.

261F. Strength of Pin-Type or Similar Construction and Conductor Fastenings. Per Rule 261F1a, the longitudinal strength of insulator pins and conductor ties can be determined by applying the loads in Rule 252 multiplied by the load factors in Rule 253, or by applying 700 lb to the pin, whichever is greater. A tangent structure can have unbalanced longitudinal loading if the spans on each side of the structure are not equal or if the span on one side is loaded with ice

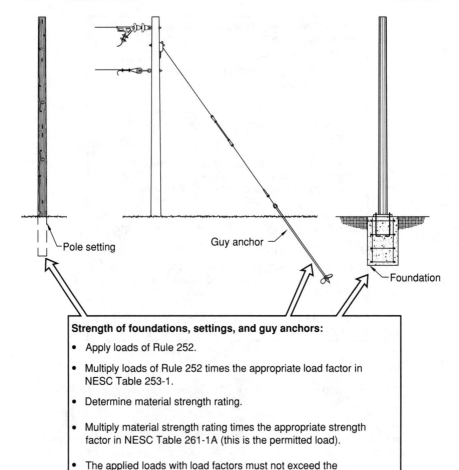

Strength of foundations, settings, and guy anchors:

- Apply loads of Rule 252.

- Multiply loads of Rule 252 times the appropriate load factor in NESC Table 253-1.

- Determine material strength rating.

- Multiply material strength rating times the appropriate strength factor in NESC Table 261-1A (this is the permitted load).

- The applied loads with load factors must not exceed the permitted load.

- Design or experience may be used to determine the strength of foundations, settings, and guy anchors.

Fig. 261-10. Strength of foundations, settings, and guy anchors (Rule 261B).

and the span on the other side has dropped its ice. Rule 261F1b specifies a construction method for meeting Rule 261F1a. Using double wood pins and ties is a construction method that is acceptable for meeting the longitudinal strength requirements of Grade B construction. The application of double wood pins in Rule 261F1b corresponds with the application of double wood arms in Rule 261D5a(2). Both rules are for conductors with tensions limited to 2000 lb. Wood insulator pins were once commonly used and can still be found on older lines.

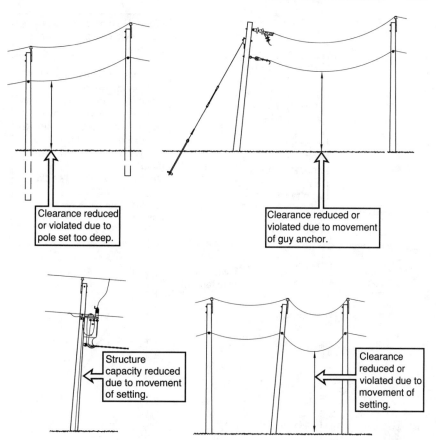

Fig. 261-11. Examples of excessive movement of foundations, settings, and guy anchors (Rule 261B).

Steel insulator pins are used in new construction. The NESC does not specify a construction application using steel pins. It is common to see steel pins used in place of wood pins on double crossarm construction. Rule 261F1c relates to Rule 261A5. Rule 261F1d relates to Rule 261A4. Rule 261F2 relates to Rule 261D5c. Rule 261F3 states that single conductor supports used instead of double wood pins must have the equivalent strength of double wood pins.

261G. Armless Construction.   The Code rules for armless construction are outlined in Fig. 261-17.

261H. Open Supply Conductors and Overhead Shield Wires.   Rule 261H1 specifies tension limits for open (bare) supply conductors and overhead shield wires. The tension limits of Rule 261H1 are shown graphically in Fig. 261-18.

The tension limits in Rule 261H1 should be entered into a conductor sag and tension program. Rule 261H1 does not specify a tension limit for supply conductors when subjected to the loads of Rule 250C (extreme wind) or 250D

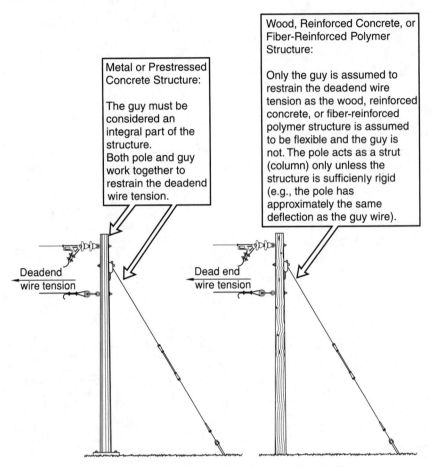

Metal or Prestressed Concrete Structure:

The guy must be considered an integral part of the structure.
Both pole and guy work together to restrain the deadend wire tension.

Wood, Reinforced Concrete, or Fiber-Reinforced Polymer Structure:

Only the guy is assumed to restrain the deadend wire tension as the wood, reinforced concrete, or fiber-reinforced polymer structure is assumed to be flexible and the guy is not. The pole acts as a strut (column) only unless the structure is sufficienly rigid (e.g., the pole has approximately the same deflection as the guy wire).

Deadend wire tension

Dead end wire tension

Fig. 261-12. Strength of guys and guy insulators (Rule 261C).

(extreme ice with concurrent wind), therefore Rule 260B2, which specifies a strength factor of 0.80, applies. The conductor manufacturer may also specify tension limits. See the sample sag and tension chart at the beginning of Sec. 23 for a discussion of NESC tension limits and user-defined design conditions.

Rule 261H2 specifies strength requirements for splices, taps, deadend fittings, and associated hardware. Rules 261F and 261M provide similar requirements. Rule 261H2 provides special requirements for crossing spans and deadends.

261I. Supply Cable Messengers.   The Code rules for strength of supply cable messengers are outlined in Fig. 261-19.

261J. Open-Wire Communication Conductors.   Open-wire communications circuits have typically been replaced with insulated communication cables or fiber-optic cables. The rules for open-wire communication circuits continue to remain in the Code. Wire tension limits for open-wire communications in

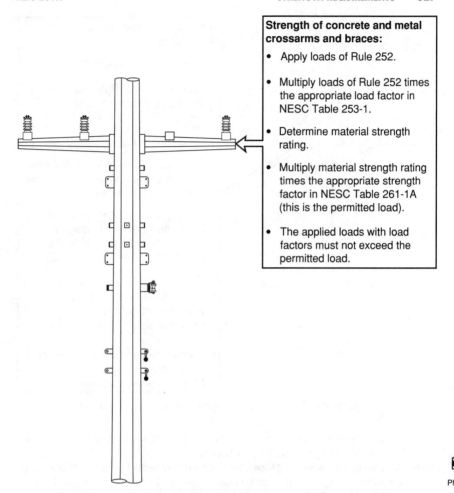

Strength of concrete and metal crossarms and braces:

- Apply loads of Rule 252.

- Multiply loads of Rule 252 times the appropriate load factor in NESC Table 253-1.

- Determine material strength rating.

- Multiply material strength rating times the appropriate strength factor in NESC Table 261-1A (this is the permitted load).

- The applied loads with load factors must not exceed the permitted load.

See
Photo(s)

Fig. 261-13. Strength of concrete and metal crossarms and braces (Rule 261D1).

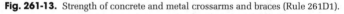

Grade B and C construction are the same as the wire tension limits for supply conductors in Rule 261H1. An example of open-wire communication conductors is shown in Fig. 261-20.

261K. Communication Cables.   The Code rules for strength of communication cables and messengers are outlined in Fig. 261-21.

261L. Paired Communication Conductors.   Rule 261L provides requirements for paired communication conductors supported on messengers and paired communication conductors not supported on messengers. Paired conductors are typically used by telephone utilities.

261M. Support and Attachment Hardware.   Rule 261M applies to all the line hardware (e.g., nuts, bolts, etc.) not covered in Rule 261F (pin-type or similar construction and conductor fastenings) or Rule 261H2 (splices, taps, deadend

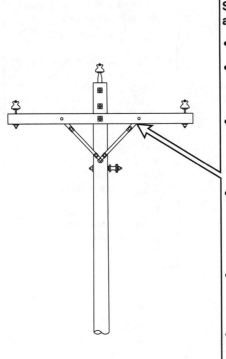

Strength of wood crossarms and braces:

• Apply loads of Rule 252.

• Multiply loads of Rule 252 times the appropriate load factor in NESC Table 253-1 or 253-2.

• Determine material strength rating:

 - For solid sawn or laminated wood, use the appropriate ultimate fiber stress of the material.

• Multiply material strength rating times the appropriate strength factor in NESC Table 261-1A or 261-1B (this is the permitted stress level).

• The applied loads with load factors must not exceed the permitted stress level.

• Crossarms of select Southern Pine or Douglas Fir must meet minimum NESC dimension values.

See Photo(s)

Fig. 261-14. Strength of wood crossarms and braces (Rule 261D2).

fittings, and associated attachment hardware). Appropriate loads, load factors, and strength factors apply to support hardware.

261N. Climbing and Working Steps and Their Attachments to the Structure. Rule 261N provides strength requirements for climbing devices (e.g., steps, ladders, platforms, and attachments). The load required is 300 lb for the weight of a lineworker and the weight of the items the lineworker carries on a structure unless the owner determines another value. Instead of derating the material strength of the steps, ladders, platforms, etc., a load factor of 2.0 is provided which increases the 300 lb load to 600 lb. The steps, ladders, platforms, etc. must not permanently deform under this load.

262. NUMBER 262 NOT USED IN THIS EDITION

263. GRADE N CONSTRUCTION

Grade N is the lowest grade of construction. Grade N supply and communication lines and structures have to withstand expected loads, including line personnel working on the structure. No load or strength factors are required for

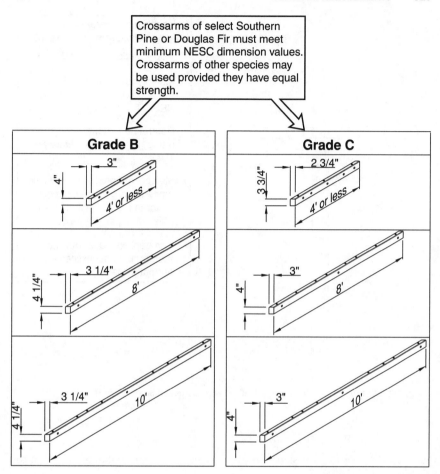

Fig. 261-15. Minimum wood crossarm dimensions for Southern Pine and Douglas Fir (Rule 261D2b).

Grade N. Said another way, the line or structure is designed to withstand the expected loads and the load factors and strength factors are equal to 1.0. Rules 263A through 263I contain minimum conductor size requirements and requirements for typical construction practices.

264. GUYING AND BRACING

264A. Where Used.   Guys and braces are used when a pole does not have sufficient strength alone. Guys and requirements for guyed poles are also covered in Rules 261A2c and 261C. Guy insulators are covered in Rule 279. Guys are

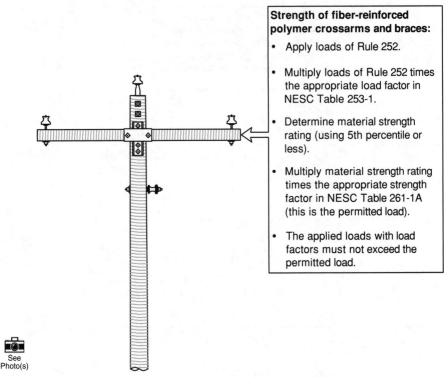

Strength of fiber-reinforced polymer crossarms and braces:

• Apply loads of Rule 252.

• Multiply loads of Rule 252 times the appropriate load factor in NESC Table 253-1.

• Determine material strength rating (using 5th percentile or less).

• Multiply material strength rating times the appropriate strength factor in NESC Table 261-1A (this is the permitted load).

• The applied loads with load factors must not exceed the permitted load.

See Photo(s)

Fig. 261-16. Strength of fiber-reinforced polymer crossarms and braces (Rule 261D3).

used to limit conductor sags and provide support for unbalanced loads at the following locations:

• Corners
• Angles
• Deadends
• Large differences in span lengths
• Changes in grades of construction

Examples of guying and bracing are shown in Fig. 264-1.

264B. Strength.   Rule 264B applies to the strength of guys. Rule 217C addresses the protection and marking requirements of guys. The Code rules for guy strength are outlined in Fig. 264-2.

264C. Point of Attachment.   The guy or brace should be attached to the structure as near as practical to the center of the conductor load. An individual guy for each conductor is typically not practical for small conductors or small angles but may be required for large conductors or large angles. Special consideration is given to insulation reduction for lines exceeding 8.7 kV. Guy insulators used exclusively for BIL insulation are discussed in Rule 279A2.

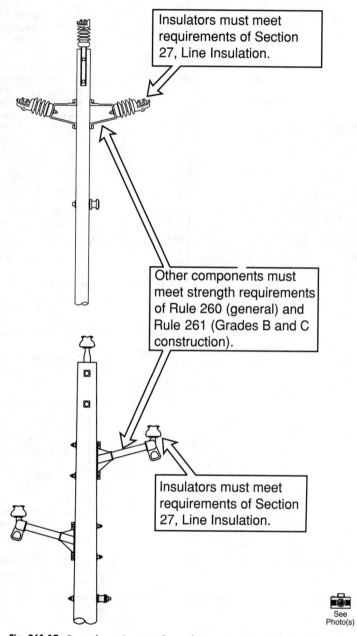

Insulators must meet requirements of Section 27, Line Insulation.

Other components must meet strength requirements of Rule 260 (general) and Rule 261 (Grades B and C construction).

Insulators must meet requirements of Section 27, Line Insulation.

See Photo(s)

Fig. 261-17. Strength requirements for armless construction (Rule 261G).

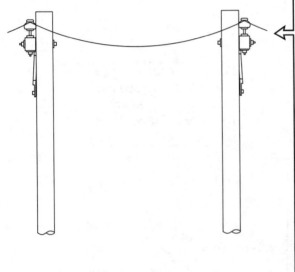

Tension limits for open supply conductors and overhead shield wires:

- The tension in the conductor must not be more than 60% of the rated breaking strength of the conductor when the loads of Rule 250B in Rule 251 are applied to the supply cable (e.g., medium loading, 15°F, 0.25" ice, 4 lb/ft^2 wind, 0.20 lb/ft K factor).

- The tension in the conductor must not be more than 35% initial, at 60°F without external load (i.e., no wind or ice).

- The tension in the conductor must not be more than 25% final, at 60°F without external load (i.e., no wind or ice).

- An exception applies to conductors of a triangular cross section.

- A note is provided stating that the above limitations may not protect the conductor from damage due to aeolian vibration.

- Rule 260B2 (0.80 strength factor) applies to extreme wind loads (Rule 250C) and extreme ice with concurrent wind loads (Rule 250D) in Rule 251 if the structure or its supported facilities exceed 60' above ground.

Fig. 261-18. Tension limits for open supply conductors and overhead shield wires (Rule 261H1).

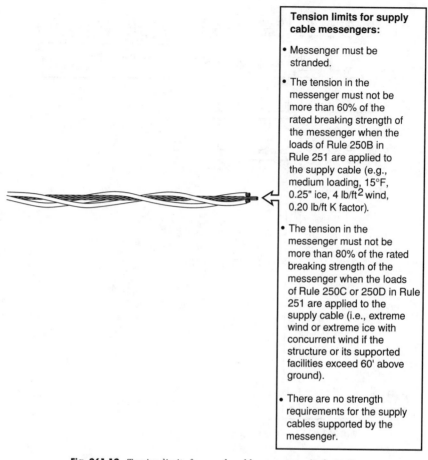

Tension limits for supply cable messengers:

• Messenger must be stranded.

• The tension in the messenger must not be more than 60% of the rated breaking strength of the messenger when the loads of Rule 250B in Rule 251 are applied to the supply cable (e.g., medium loading, 15°F, 0.25" ice, 4 lb/ft^2 wind, 0.20 lb/ft K factor).

• The tension in the messenger must not be more than 80% of the rated breaking strength of the messenger when the loads of Rule 250C or 250D in Rule 251 are applied to the supply cable (i.e., extreme wind or extreme ice with concurrent wind if the structure or its supported facilities exceed 60' above ground).

• There are no strength requirements for the supply cables supported by the messenger.

Fig. 261-19. Tension limits for supply cable messengers (Rule 261I).

The **Code** rules for the point of attachment of the guy or brace are outlined in Fig. 264-3.

264D. Guy Fastenings.   The strength rating of a guy fastening is just as important as the strength rating of the guy wire itself. Numerous guy attachment methods exist including guy plates, pole bands, wrap guys, etc. The **Code** requires the use of guy thimbles and guy shims for certain guying conditions. The entire guy and anchor system, including guy fastenings, should be reviewed to determine the weak link in the guy and anchor system. A guy is only as strong as the weakest link connected to the guy wire. An example of the weakest link in a guy and anchor system is shown in Fig. 264-4.

264E. Electrolysis.   Electrolysis (corrosion) of the anchor or anchor rod can jeopardize the strength of the structure. Corrosion normally occurs when two dissimilar metals exist in the soil (e.g., a copper ground rod and a steel anchor rod).

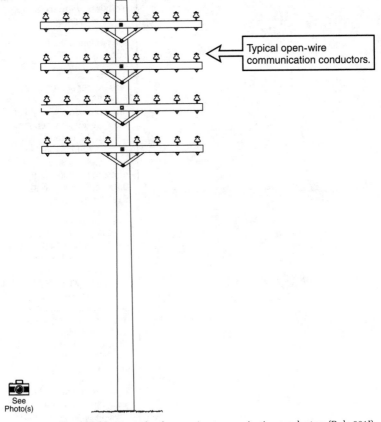

Typical open-wire communication conductors.

See Photo(s)

Fig. 261-20. Example of open-wire communication conductors (Rule 261J).

The soil conditions can have an effect on the rate of corrosion. Typical solutions to the problem are using guy insulators or using steel-coated copper ground rods. Cathodic protection in the form of a sacrificial anode is another solution. Guy insulators used exclusively for limiting corrosion are discussed in Rule 279A2.

264F. Anchor Rods.   Rule 264G1 applies to the anchor rod only. Rule 264G2 applies to the anchor and rod assembly. Rules 261B and 264B also relate to the strength of guy anchors. The Code rules for strength of anchor rods and anchors provided in Rule 264F are outlined in Fig. 264-5.

Tension limits for communication cables and messengers:

- There are no strength requirements for the communication cables supported by the messenger.

- The tension in the messenger must not be more than 60% of the rated breaking strength of the messenger when the loads of Rule 250B in Rule 251 are applied to the supply cable (e.g., medium loading, 15°F, 0.25" ice, 4 lb/ft^2 wind, 0.20 lb/ft K factor).

- The tension in the messenger must not be more than 80% of the rated breaking strength of the messenger when the loads of Rule 250C or 250D in Rule 251 are applied to the supply cable (i.e., extreme wind or extreme ice with concurrent wind if the structure or its supported facilities exceed 60' above ground).

Fig. 261-21. Tension limits for communication cables and messengers (Rule 261K).

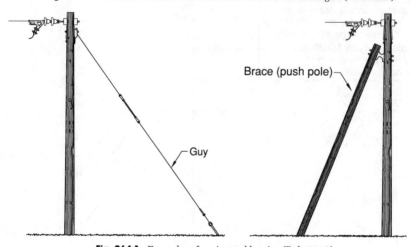

Brace (push pole)

Guy

See Photo(s)

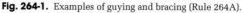

Fig. 264-1. Examples of guying and bracing (Rule 264A).

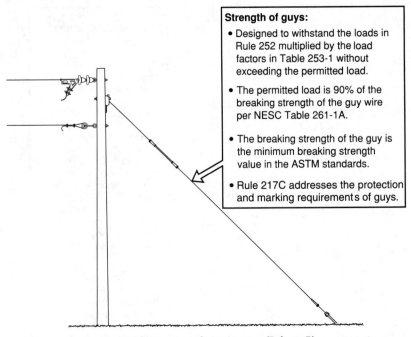

Strength of guys:

• Designed to withstand the loads in Rule 252 multiplied by the load factors in Table 253-1 without exceeding the permitted load.

• The permitted load is 90% of the breaking strength of the guy wire per NESC Table 261-1A.

• The breaking strength of the guy is the minimum breaking strength value in the ASTM standards.

• Rule 217C addresses the protection and marking requirements of guys.

Fig. 264-2. Guy strength requirements (Rule 264B).

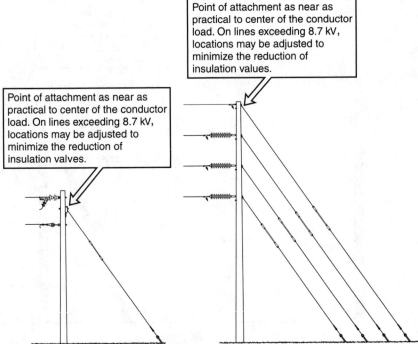

Point of attachment as near as practical to center of the conductor load. On lines exceeding 8.7 kV, locations may be adjusted to minimize the reduction of insulation values.

Point of attachment as near as practical to center of the conductor load. On lines exceeding 8.7 kV, locations may be adjusted to minimize the reduction of insulation valves.

See Photo(s)

Fig. 264-3. Point of guy attachment (Rule 264C).

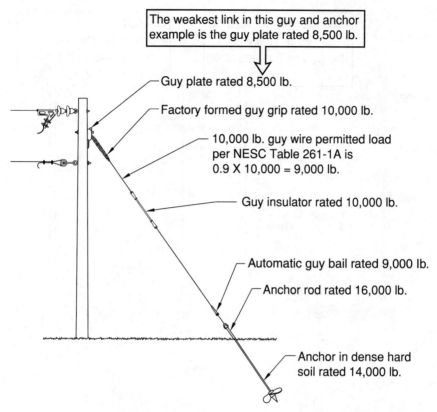

The weakest link in this guy and anchor example is the guy plate rated 8,500 lb.

Guy plate rated 8,500 lb.

Factory formed guy grip rated 10,000 lb.

10,000 lb. guy wire permitted load per NESC Table 261-1A is 0.9 X 10,000 = 9,000 lb.

Guy insulator rated 10,000 lb.

Automatic guy bail rated 9,000 lb.

Anchor rod rated 16,000 lb.

Anchor in dense hard soil rated 14,000 lb.

Fig. 264-4. Example of the weakest link in a guy and anchor system (Rule 264D).

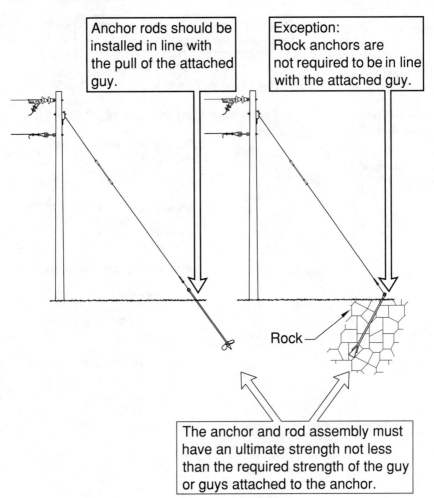

Anchor rods should be installed in line with the pull of the attached guy.

Exception:
Rock anchors are not required to be in line with the attached guy.

Rock

The anchor and rod assembly must have an ultimate strength not less than the required strength of the guy or guys attached to the anchor.

Fig. 264-5. Anchor rods (Rule 264F).

Section 27

Line Insulation

270. APPLICATION OF RULE

Line insulation in Sec. 27 applies to insulators for open conductor supply lines only, not communication lines or secondary duplex, triplex, or quadruplex (230C3) cables. Sec. 27 does not apply to substation insulators as Sec. 27 is in Part 2, Overhead Lines. Note 1 references Rule 243C5. Rule 243C5 states that the strength requirements in Sec. 27 apply to all grades of construction. There are not separate Grade B and Grade C strength factors for insulators. Note 2 references Rule 242E for insulation requirements of neutral conductors. Rule 242E states that supply neutrals, which are effectively grounded and not located above supply conductors of more than 750 V to ground, do not need to meet any insulation requirements. Even though effectively grounded neutrals do not need to meet any insulation requirements, it is common to see neutral conductors attached to insulators for support and termination purposes. Also, even though the messengers of secondary duplex, triplex, and quadruplex (230C3) cables do not need to meet any insulation requirements, it is common to see the messengers of these cables attached to insulators for support and termination purposes. The **Code** rules related to the application of line insulation are outlined in Fig. 270-1.

271. MATERIAL AND MARKING

Supply circuit insulators are required to be wet-process porcelain or other materials. Examples of insulators made of porcelain and other materials are shown in Fig. 271-1.

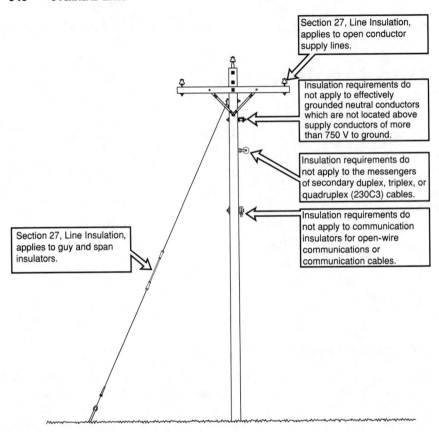

Fig. 270-1. Application of line insulation (Rule 270).

The manufacturers must mark insulators for use at or above 2.3 kV with the manufacturer name or trademark and an identification (e.g., a catalog number) that will permit determination of the electrical and mechanical properties. The Code rules for marking insulators are outlined in Fig. 271-2.

272. RATIO OF FLASHOVER TO PUNCTURE VOLTAGE

Rule 272 requires insulators to meet standards for a flashover to puncture voltage ratio. A list of applicable standards is provided. When a standard does not exist for the flashover to puncture voltage ratio, Rule 272 specifies a not-to-exceed value of 75 percent with an exception that permits not more than 80 percent in areas of high atmospheric contamination. The flashover to puncture voltage ratio is used by insulator manufacturers but it is not always published in insulator catalog data. Insulator catalog data typically does include the low-frequency dry flashover value.

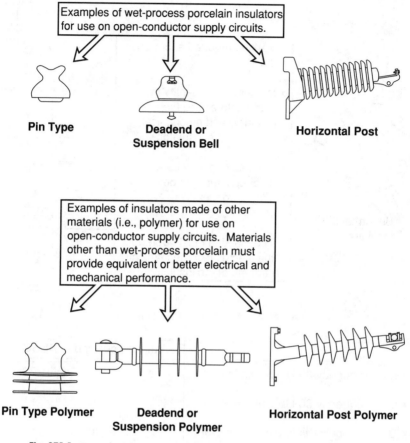

Fig. 271-1. Examples of insulators made of porcelain or other materials (Rule 271).

273. INSULATION LEVEL

Rule 273 specifies insulator dry flashover voltages in **NESC** Table 273-1. The dry flashover tests must be done in accordance with ANSI Standard C29.1. Using dry flashover voltage ratings lower than the values in **NESC** Table 273-1 requires a qualified engineering study. Values higher than shown in **NESC** Table 273-1 must be used for areas with severe lightning, high atmospheric contamination (e.g., salt water fog, industrial plant pollution, etc.) or other unfavorable circumstances. The low-frequency dry flashover rating is typically the most common value referenced when comparing insulator specifications. Additional ratings are commonly published in insulator catalog sheets and in insulator specifications. Typical insulation level ratings are listed in Fig. 273-1.

In addition to electrical ratings, the **NESC** specifies mechanical ratings for insulators in Rule 277.

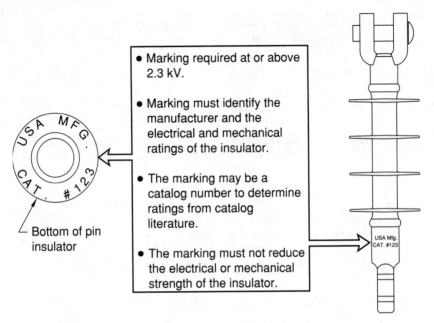

Fig. 271-2. Marking of insulators (Rule 271).

Rating	Values specified in NESC
• Low frequency 60 Hz dry flashover	Yes (see NESC Table 273-1)
• Low frequency 60 Hz wet flashover	No
• Critical impulse flashover - positive	No
• Critical impulse flashover - negative	No
• Total leakage distance	No

Fig. 273-1. Typical insulation level ratings (Rule 273).

274. FACTORY TESTS

Insulator factory tests per ANSI Standards are required for insulators at or above 2.3 kV. Accepted good practice may be used where standards do not exist.

275. SPECIAL INSULATOR APPLICATIONS

Special consideration must be given to constant-current circuits (e.g., series street lighting). The voltage rating of the insulator must be based on the full load rated voltage of the supply transformer, not just the operating voltage of the circuit.

Single-phase lines fed from three-phase lines are required to use insulators based on the three-phase line phase-to-phase voltage. See Fig. 275-1.

276. NUMBER 276 NOT USED IN THIS EDITION

277. MECHANICAL STRENGTH OF INSULATORS

Rule 277 specifies the mechanical strength requirements for insulators. This is done in the form of strength factors that are applied to the insulators. Load factors for wire tension, wire vertical (weight) loads, and wire transverse (wind) loads are not applicable. Rule 277 requires insulators to withstand the loads in Rules 250, 251, and 252. The load factors in Rule 253 are not required. Said another way, the load factor for the loads on insulators is 1.0. The strength factor percentages provided in **NESC** Table 277-1 do not distinguish between Grade B and Grade C construction. This is one reason why insulators have their own section (Sec. 27) rather than being included in Rule 261 and **NESC** Table 261-1A. The hardware for insulators is covered in Rules 261M, 261F, 261H2, and **NESC** Table 261-1A. The percent of strength rating provided in **NESC** Table 277-1 is for use with Rule 250B (heavy, medium, light) loads, not Rule 250C (extreme wind) or Rule 250D (extreme ice with concurrent wind) loads.

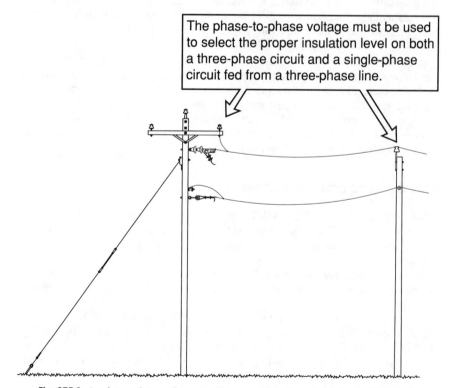

The phase-to-phase voltage must be used to select the proper insulation level on both a three-phase circuit and a single-phase circuit fed from a three-phase line.

Fig. 275-1. Insulators of singe-phase circuits connected to three-phase circuits (Rule 275).

A general statement is provided to make proper allowances for Rule 250C and 250D loads but no specifics are provided. Accepted good practice must be used.

The strength factor percentages listed in **NESC** Table 277-1 are based on manufacturers ratings when determined in accordance with specific ANSI Standards for the type of insulator (e.g., suspension or line post), the insulator material (e.g., ceramic or nonceramic), and the type of load on the insulator (e.g., combined, cantilever, tension, or compression).

The appropriate mechanical strength of an insulator that needs to be derated may not be very clear on an insulator catalog sheet. Catalog sheets may use terms like routine test load, maximum design load, average failing load, average breaking load, or maximum working load. The insulator manufacturer should be consulted or the ANSI Standards should be reviewed to verify that the percentages provided in **NESC** Table 277-1 are being applied to the proper mechanical strength rating.

An example of cantilever, compression, and tension loads on ceramic line post insulators is shown in Fig. 277-1.

278. AERIAL CABLE SYSTEMS

Aerial cable systems primarily consist of tree wire conductors described in Rule 230D. The insulators supporting aerial cables must meet Rule 273. The covered conductors of aerial cable systems are considered bare conductors for electrical insulation requirements. The insulators and spacers of an aerial cable system must meet Rule 277. Rule 278B2 requires that the insulating spacers used in aerial cable systems withstand the loads specified in Sec. 25. Section 25 includes Rule 253, which contains load factors. The 50% of ultimate rated strength requirement is for use with Rule 250B (heavy, medium, light) loads and the appropriate overload factors in Rule 253. The loads of 250C (extreme winds) and 250D (extreme ice with concurrent wind) are not addressed. Accepted good practice must be applied.

279. GUY AND SPAN INSULATORS

279A. Insulators.   Per Rule 215C2, guys must be grounded or insulated. If grounded, Rule 092C2 provides the methods for the grounding of guys. If insulated, Rule 215C5 provides requirements for the location of guy insulators. See Fig. 279-1.

The **Code** rules for the material properties of guy insulators and the electrical and mechanical strength of guy insulators per Rule 279A1 are outlined in Fig. 279-2.

The requirements for mechanical strength of a guy insulator hinge on the word "required" strength of the guy. Guy insulators do not require the application of strength factors like line insulators in Rule 277. Guy insulators indirectly have load factors applied to them via the load factors that are applied to wire tension and wind loading. See Fig. 279-3.

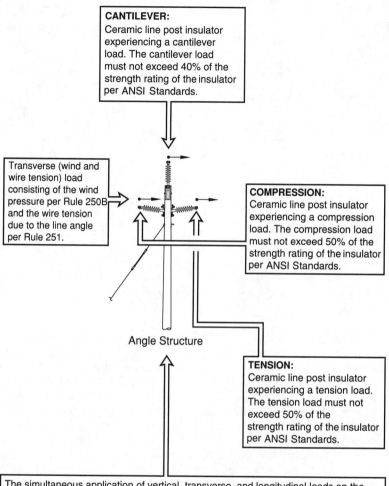

CANTILEVER:
Ceramic line post insulator experiencing a cantilever load. The cantilever load must not exceed 40% of the strength rating of the insulator per ANSI Standards.

Transverse (wind and wire tension) load consisting of the wind pressure per Rule 250B and the wire tension due to the line angle per Rule 251.

COMPRESSION:
Ceramic line post insulator experiencing a compression load. The compression load must not exceed 50% of the strength rating of the insulator per ANSI Standards.

Angle Structure

TENSION:
Ceramic line post insulator experiencing a tension load. The tension load must not exceed 50% of the strength rating of the insulator per ANSI Standards.

The simultaneous application of vertical, transverse, and longitudinal loads on the insulator should be considered. The insulators on the angle structure are experiencing vertical (weight) loads in addition to transverse loads. The vertical (weight) load of the wire and ice per Rule 250B applies a cantilever load on the two bottom insulators and a compression load on the top insulator. Unequal spans on the angle structure can also create longitudinal loads. Insulator manufacturers typically provide a method for checking simultaneous application of loads.

Fig. 277-1. Example of cantilever, compression, and tension loads on ceramic line post insulators (Rule 277).

Per Rule 279A2a, if a guy insulator is used exclusively for limiting corrosion of an anchor attached to a grounded guy on an effectively grounded system, then it is not classified as a guy insulator. Therefore, the electrical requirements in Rule 279A1b (i.e., the dry flashover and wet flashover ratings) do not apply. The guy in this case is assumed to be grounded per Rule 215C2 using the

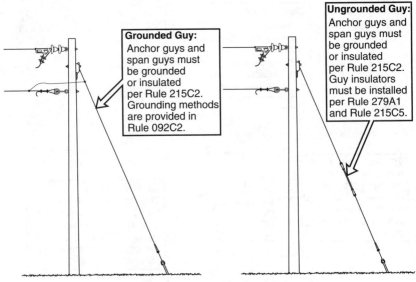

Fig. 279-1. Guy and span insulators (Rule 279).

grounding methods in Rule 092C2. The corrosion protection insulator must not reduce the mechanical strength of the guy. The **Code** is not specific in this case whether this is the required strength of the guy as stated in Rule 279A1c or the strength of the guy wire. Additional requirements for guy insulators used to limit galvanic corrosion can be found in Rule 215C7. An example of a guy insulator used exclusively for limiting galvanic corrosion is shown in Fig. 279-4.

Per Rule 279A2b, if a guy insulator is used exclusively for meeting BIL requirements for a structure on an effectively grounded system, then it is not classified as a guy insulator. However, the strength requirement of the guy insulator must meet Rule 279A1c, and either Rule 279A1 or a combination of Rule 215C5 and Rule 092C2 applies. **NESC** Fig. 279-1 is provided in the **Code** as a reference for applying this rule.

The clearance requirement between a guy (with or without a guy insulator) and an energized conductor on the same structure (pole) is provided in **NESC** Table 235-6. See Rule 235E for a discussion.

279B. Span-Wire Insulators.   The requirements of span-wire insulators are similar to guy insulators. Rule 215C is referenced for application and Rule 274 is referenced for testing. The rated ultimate strength of a span-wire insulator is based on the required strength of the span wire in which it is located.

SECTION NUMBER 28 NOT USED IN THIS EDITION

SECTION NUMBER 29 NOT USED IN THIS EDITION

Material:

- Wet-process porcelain.
- Wood.
- Fiberglass-reinforced plastic.
- Other material of suitable electrical and mechanical properties.

Electrical strength:

- The dry flashover voltage rating of the guy insulator must be at least double the voltage between conductors of the guyed circuit.

- The wet flashover voltage rating of the guy insulator must be at least as high as the voltage between conductors of the guyed circuit.

Mechanical strength:

- The rated ultimate strength of the guy insulator must be at least equal to the required strength of the guy.

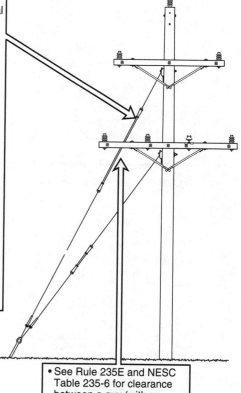

- See Rule 235E and NESC Table 235-6 for clearance between a guy (with or without a guy insulator) and an energized conductor.

- See Rules 215C2, 215C5, and 092C2 for information on grounding vs. insulating guys and the location of the guy insulator.

See Photo(s)

Fig. 279-2. Properties of guy insulators (Rule 279A1).

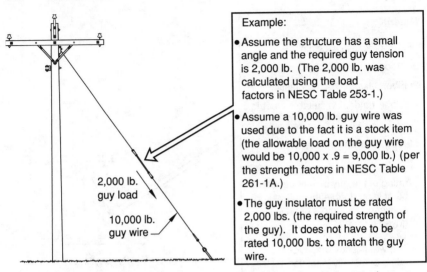

Example:

- Assume the structure has a small angle and the required guy tension is 2,000 lb. (The 2,000 lb. was calculated using the load factors in NESC Table 253-1.)

- Assume a 10,000 lb. guy wire was used due to the fact it is a stock item (the allowable load on the guy wire would be 10,000 x .9 = 9,000 lb.) (per the strength factors in NESC Table 261-1A.)

- The guy insulator must be rated 2,000 lbs. (the required strength of the guy). It does not have to be rated 10,000 lbs. to match the guy wire.

2,000 lb. guy load

10,000 lb. guy wire

Fig. 279-3. Example of the required strength of a guy insulator (Rule 279A1c).

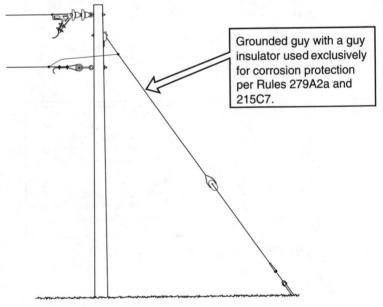

Grounded guy with a guy insulator used exclusively for corrosion protection per Rules 279A2a and 215C7.

See Photo(s)

Fig. 279-4. Example of a guy insulator used to limit galvanic corrosion (Rule 279A2a).

Part 3

Safety Rules for the Installation and Maintenance of Underground Electric Supply and Communication Lines

GENERAL SECTIONS

01 INTRODUCTION   02 DEFINITIONS

03 REFERENCES   09 GROUNDING METHODS

GENERAL SECTIONS 01, 02, 03, 09

PART 1

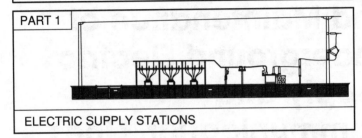

ELECTRIC SUPPLY STATIONS

PART 2

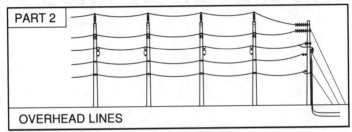

OVERHEAD LINES

PART 3

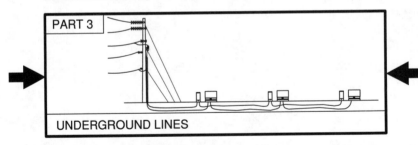

UNDERGROUND LINES

PART 4

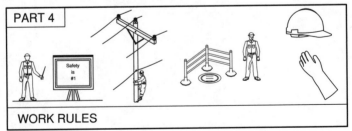

WORK RULES

Section 30

Purpose, Scope, and Application of Rules

300. PURPOSE

The purpose of Part 3, Underground Lines, is similar to the purpose of the entire NESC outlined in Rule 010, except Rule 300 is specific to underground supply and communication lines and equipment. Part 3 of the NESC focuses on the practical safeguarding of persons during the installation, operation, and maintenance of underground supply (power) and communication lines and equipment.

301. SCOPE

The scope of Part 3, Underground Lines, includes supply (power) and communications (phone, cable TV, etc.) cables and equipment in underground or buried applications. The term underground equipment covers buried equipment and pad-mounted (aboveground) equipment used with underground conductors. Separate rules apply to Electric Supply Stations (Part 1) and Overhead Lines (Part 2). Part 3 covers underground structural arrangements and extensions into buildings.

There is some overlap between Underground Lines (Part 3) and Overhead Lines (Part 2). The overlap is at the underground riser on the overhead pole. Risers are covered in Sec. 36 of Part 3 and in Sec. 23, Rule 239 of Part 2. See Fig. 301-1.

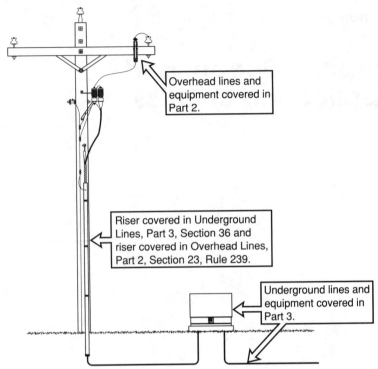

Overhead lines and equipment covered in Part 2.

Riser covered in Underground Lines, Part 3, Section 36 and riser covered in Overhead Lines, Part 2, Section 23, Rule 239.

Underground lines and equipment covered in Part 3.

See Photo(s)

Fig. 301-1. Overlap between Underground Lines (Part 3) and Overhead Lines (Part 2) (Rule 301).

There are more distinct differences between Underground Lines (Part 3) and Electric Supply Stations (Part 1). Underground lines outside the electric supply station are covered in Part 3. Conductors and conduit inside the electric supply station are covered in Sec. 16 of Part 1. See Fig. 301-2.

The second to last sentence of Rule 301 is clarifying the application of the **NESC** versus the National Electrical Code (NEC) to underground supply and communication lines. Rule 011 discusses the scope of the **NESC** and the NEC. The **NESC** covers conductors and equipment when they are serving a utility function (not an office building wiring function). The **NESC** underground lines rules cover utility functions. There are instances when utilization wiring may be associated with the utility function of underground lines. For example, utilization wiring may be needed for lighting and ventilation in a vault. The **NESC** does not provide specific rules for utilization wiring. Rule 012C, which requires accepted good practice, must be applied when specific conditions are not covered. The NEC is an excellent reference for accepted good practice in this case.

When underground supply equipment is fenced, the rules of Part 1, Electric Supply Stations, may apply. See Rule 110A for a discussion.

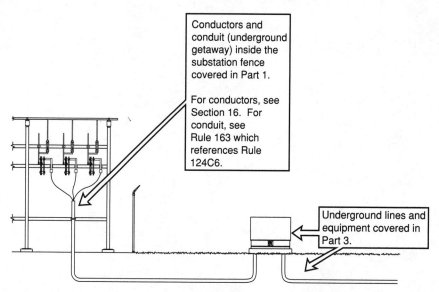

Fig. 301-2. Differences between Underground Lines (Part 3) and Electric Supply Stations (Part 1) (Rule 110).

302. APPLICATION OF RULE

Rule 302 references Rule 013 for the general application of **Code** rules. See Rule 013 for a discussion.

Section 31

General Requirements Applying to Underground Lines

310. REFERENCED SECTIONS

This rule references four sections related to Part 3, Underground Lines, so that rules do not have to be duplicated and the reader of the **Code** realizes that other sections are related to the information provided in Part 3. The related sections are:

- Introduction, Sec. 01
- Definitions, Sec. 02
- References, Sec. 03
- Grounding Methods, Sec. 09

The rules in Part 3, predominantly Rules 314, 342, 374, and 384, will provide the requirements for grounding underground lines and equipment. The grounding methods are provided in Sec. 09.

311. INSTALLATION AND MAINTENANCE

This rule requires that supply and communication utilities be able to locate their underground facilities. Locates are typically done as shown in the example in Fig. 311-1.

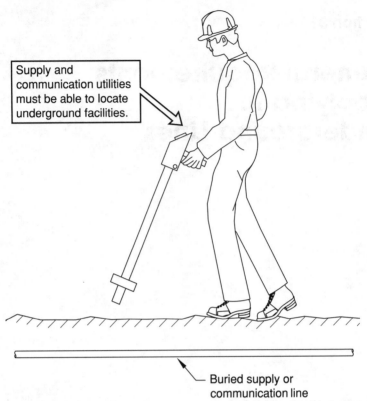

Supply and communication utilities must be able to locate underground facilities.

See Photo(s)

— Buried supply or communication line

Fig. 311-1. Example of worker locating underground lines (Rule 311).

Color-coded spray paint markings are typically used to identify power, communication, and other utilities. Most localities have a "call before you dig" system in place to aid location of underground facilities. The **NESC** does not limit locates to direct-buried cables. Any underground facility (e.g., equipment, conduit systems, duct banks, vaults, etc.) must be located. Proper mapping of facilities can aid the location of facilities and is required in Part 4, Work Rules, Rule 411A2.

The **NESC** requires advance notice (a time is not specified) to owners or operators of nearby facilities that may be adversely affected by underground digging or construction by supply or communication utilities.

Rule 311C addresses emergency installations and permits laying certain supply (power) and communication cables directly on the ground. This same permission can be found in Rule 230A2d. See the discussion and figure in Rule 230A.

312. ACCESSIBILITY

Rule 312 is the underground equivalent to overhead Rule 213.

Rule 312 generally states the requirement that underground parts needing examination or adjustment during operation must be accessible to authorized supply or communication workers. This rule requires the following, all of which are discussed in more detail throughout Part 3.

- Working spaces
- Working facilities
- Clearances

Working spaces are specifically discussed in Rule 323B. This rule dimensions the working space in manholes. Various rules throughout Part 3 discuss clearances and separations between supply and communication conductors and other facilities.

The NESC does not specify a clear working space dimension in front of a pad-mounted transformer or primary junction box. The general requirements of this rule apply, which are that adequate working space must exist. Working space is a challenge for utilities when it comes to service transformers or other equipment on private property. Some utilities apply a notification sign on the front of pad-mounted equipment that asks the landowner to keep the area in front of the equipment clear (this sign is not a Code requirement). An example of the need for adequate working space is shown in Fig. 312-1.

An example of a notification sign used on pad-mounted equipment is shown in Fig. 312-2.

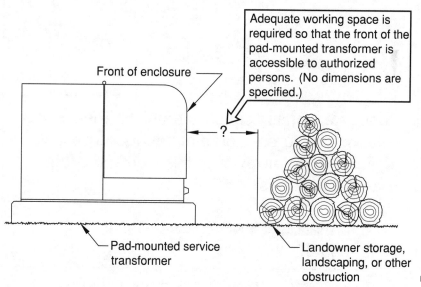

Adequate working space is required so that the front of the pad-mounted transformer is accessible to authorized persons. (No dimensions are specified.)

Front of enclosure

Pad-mounted service transformer

Landowner storage, landscaping, or other obstruction

See Photo(s)

Fig. 312-1. Example of the need for adequate working space (Rule 312).

NOTICE

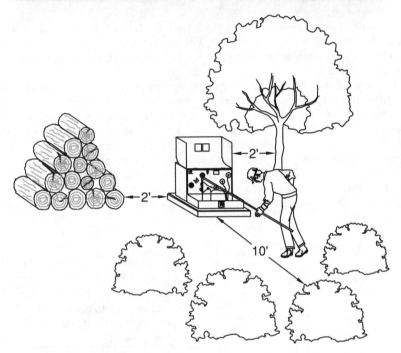

Adequate working space is required. Please keep shrubs, structures, stored materials, fences, etc. at least 10 feet away from this side and 2 feet away from other sides.

For information call USA Power Co. at 1-800-555-1212.

Fig. 312-2. Example of a notification sign used on pad-mounted equipment (Rule 312).

313. INSPECTION AND TESTS OF LINES AND EQUIPMENT

This rule is the underground equivalent of overhead Rule 214. There are slight wording differences between the overhead and underground inspection and testing rules, but the intent is the same.

Cutouts, switches, reclosers, circuit breakers, etc., are used to switch overhead lines and determine if a line is in or out of service. These same overhead devices can be used for underground switching in addition to pad-mounted switches and the termination or standoff of URD elbows.

Rule 313, like its overhead counterpart Rule 214, does not specify the inspection or testing intervals. See Rule 214 for a discussion.

314. GROUNDING OF CIRCUITS AND EQUIPMENT

Rule 314 is the underground equivalent of overhead Rule 215.

Rule 314 is broken into three main paragraphs. Rule 314A reminds us that the methods of grounding are specified in Sec. 09 and the requirements of grounding are provided in this rule for circuits and conductive parts to be grounded. Rule 314B focuses on the requirements for grounding conductive parts. Rule 314C focuses on the requirements for grounding circuits. See Sec. 02, "Definitions," for a discussion of the term "effectively grounded." Not all of the grounding requirements for Part 3 are listed in this rule. Various other grounding requirements appear throughout Part 3. One important example is the grounding and bonding requirement for aboveground apparatus (pad-mounted equipment) found in Rule 384. Other Part 3 grounding rules include Rules 342 and 374.

The conductive parts grounding requirements of Rule 314B are outlined in Fig. 314-1.

The circuit grounding requirements of Rule 314C are outlined in Fig. 314-2.

315. COMMUNICATIONS PROTECTIVE REQUIREMENTS

Rule 315 is the underground equivalent to overhead Rule 223. See Rule 223 for a discussion.

316. INDUCED VOLTAGE

Rule 316 is the underground equivalent to overhead Rule 212. See Rule 212 for a discussion.

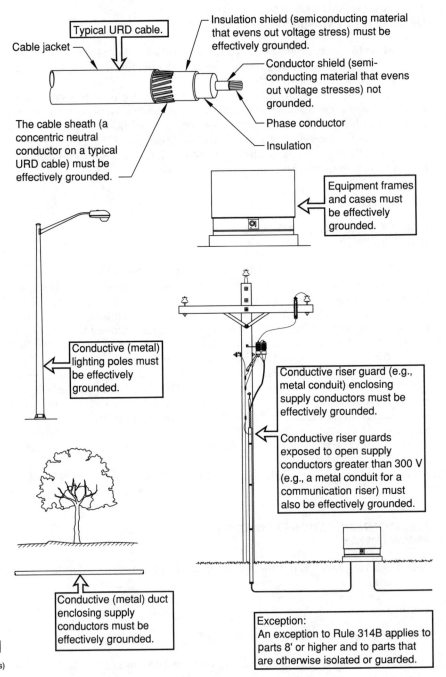

Typical URD cable.

Cable jacket

Insulation shield (semiconducting material that evens out voltage stress) must be effectively grounded.

Conductor shield (semi-conducting material that evens out voltage stresses) not grounded.

Phase conductor

Insulation

The cable sheath (a concentric neutral conductor on a typical URD cable) must be effectively grounded.

Equipment frames and cases must be effectively grounded.

Conductive (metal) lighting poles must be effectively grounded.

Conductive riser guard (e.g., metal conduit) enclosing supply conductors must be effectively grounded.

Conductive riser guards exposed to open supply conductors greater than 300 V (e.g., a metal conduit for a communication riser) must also be effectively grounded.

Conductive (metal) duct enclosing supply conductors must be effectively grounded.

Exception:
An exception to Rule 314B applies to parts 8' or higher and to parts that are otherwise isolated or guarded.

See Photo(s)

Fig. 314-1. Conductive parts to be grounded (Rule 314B).

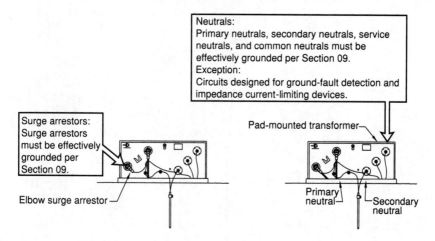

Neutrals:
Primary neutrals, secondary neutrals, service neutrals, and common neutrals must be effectively grounded per Section 09.
Exception:
Circuits designed for ground-fault detection and impedance current-limiting devices.

Surge arrestors:
Surge arrestors must be effectively grounded per Section 09.

Pad-mounted transformer—

Elbow surge arrestor—

Primary neutral—   —Secondary neutral

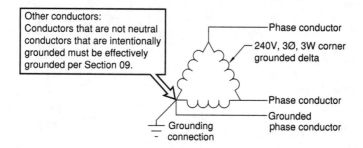

Other conductors:
Conductors that are not neutral conductors that are intentionally grounded must be effectively grounded per Section 09.

Phase conductor

240V, 3Ø, 3W corner grounded delta

Phase conductor

Grounded phase conductor

Grounding connection

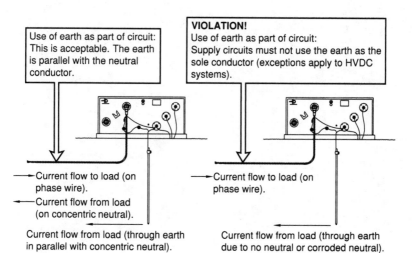

Use of earth as part of circuit:
This is acceptable. The earth is parallel with the neutral conductor.

VIOLATION!
Use of earth as part of circuit:
Supply circuits must not use the earth as the sole conductor (exceptions apply to HVDC systems).

→Current flow to load (on phase wire).
←Current flow from load (on concentric neutral).

Current flow from load (through earth in parallel with concentric neutral).

→Current flow to load (on phase wire).

Current flow from load (through earth due to no neutral or corroded neutral).

See Photo(s)

Fig. 314-2. Grounding of circuits (Rule 314C).

Section 32

Underground Conduit Systems

Note 1 under the title of Sec. 32 defines how the **NESC** uses the following terms:
- Duct
- Conduit
- Conduit system

The common trade use of these terms is slightly different from the **NESC** definitions. The use of these terms as they apply to the **NESC** is shown in Fig. 32-1.

Note 2 under the title of Sec. 32 references Rule 350G for a cable installed in a single duct that is not part of a conduit system. Rule 350G states that Sec. 35 (Direct-Buried Cable) applies to supply and communication cables that are installed in duct that is not part of a conduit system. See Fig. 32-2 and Fig. 320-1.

320. LOCATION

Section 32 applies to the underground structure only, not the cables in the structure. Section 33 applies to supply cables. Section 34 applies to the installation of cables in an underground structure. Rule 320 deals with the location of underground conduit systems, where they can be routed, and how far they must be separated from other underground installations.

320A. Routing. The general requirements for conduit system routing are outlined below:
- Subject conduit system to the least disturbance practical.
- When conduit system is parallel to another structure, do not locate directly over or under the other structure if practical.
- Align conduit without protrusions that would harm the cable.
- Provide sufficient bending radius to limit cable damage.

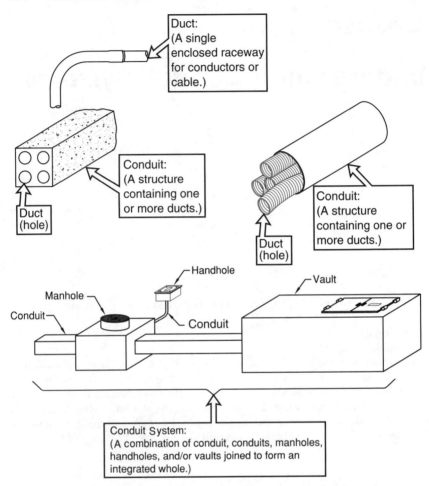

Fig. 32-1. NESC definitions of duct, conduit, and conduit system (Sec. 32).

The **NESC** does not specify burial depths for conduit systems. Direct-buried supply cables covered in Sec. 35 do have to meet the specified depths (covered in Rule 352 and Table 352-1); however, no such burial requirements exist for cables installed in conduit systems. The only specific burial depth called out in Sec. 32 is for a conduit system crossing under railroad tracks. The only other specific distance given in this section is separation between supply and communication conduits. The fact that the **NESC** does not specify burial depths for conduit systems and the relationship between several Sec. 32 and Sec. 35 rules is shown in Fig. 320-1.

To decide on the proper burial depth for conduit systems, the general rules of Sec. 32 must be applied, the strength requirements of Rules 323 and 322A3 must be considered, and Rule 012C, which requires accepted good practice, must be used. One excellent reference for accepted good practice in this case is

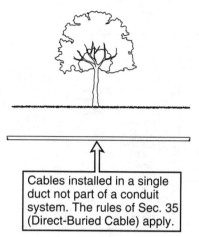

Fig. 32-2. Application of cables installed in a single duct not part of a conduit system (Secs. 32 and 35).

the National Electrical Code (NEC). The NEC specifies burial depths for various types of conduits (PVC, steel, concrete encased, etc.) containing circuits at various voltage levels. The NEC also provides methods for reducing conduit burial depth when rock is encountered.

The **NESC** is not specific on conduit bending radius requirements. Rule 320A simply states that a sufficient bending radius is needed. Rule 012C, which requires accepted good practice, must be used. The National Electrical Code (NEC) provides much more detail than the **NESC** on conduit bending radius, and the NEC can be used as a reference for accepted good practice. Typically, field bends require a larger bending radius than factory bends. A bend with too tight a radius will deform or damage the conduit material. In addition to bending radius requirements, the NEC can be referenced for conduit support, conduit fill (i.e., the number of conductors in any one conduit), and various other conduit application methods. The details of how conduit bending radius is typically measured are shown in Fig. 320-2.

The total bending radius of a conduit run can be determined by adding the angles of each bend. To determine how many bends are acceptable in a conduit run, a cable pulling calculation must be done per Rule 341. See Fig. 320-3.

Natural hazards due to unstable or corrosive soil should be avoided or proper construction methods should be used to minimize the hazard.

If a conduit can be installed outside a roadway, no interference with the road would occur. However, if the conduit is installed longitudinally under the road, using the shoulder or just one lane of traffic will make both installation and maintenance of the conduit conflict less with roadway traffic. The **NESC** does not specify a conduit system burial depth under a road for a longitudinal run or for a road crossing. The road department or highway department having jurisdiction may require specific depths or locations to obtain a permit to install utilities under the road. The rules related to conduit routing under highways and streets are outlined in Fig. 320-4.

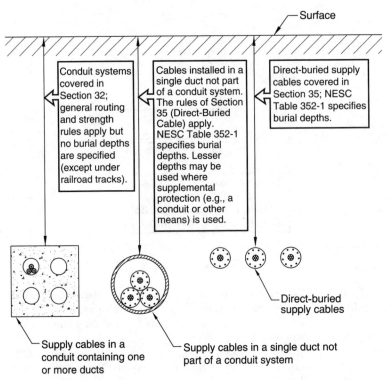

Conduit systems covered in Section 32; general routing and strength rules apply but no burial depths are specified (except under railroad tracks).

Cables installed in a single duct not part of a conduit system. The rules of Section 35 (Direct-Buried Cable) apply. NESC Table 352-1 specifies burial depths. Lesser depths may be used where supplemental protection (e.g., a conduit or other means) is used.

Direct-buried supply cables covered in Section 35; NESC Table 352-1 specifies burial depths.

Direct-buried supply cables

Supply cables in a conduit containing one or more ducts

Supply cables in a single duct not part of a conduit system

Fig. 320-1. Burial depths for conduit systems and direct-buried supply cables (Rules 320, 322A3, 323, 350G, and 352).

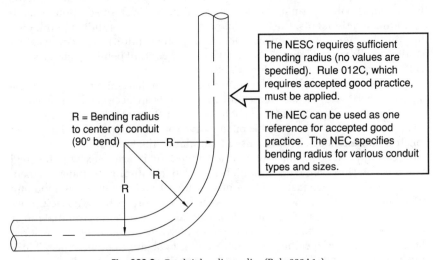

The NESC requires sufficient bending radius (no values are specified). Rule 012C, which requires accepted good practice, must be applied.

The NEC can be used as one reference for accepted good practice. The NEC specifies bending radius for various conduit types and sizes.

R = Bending radius to center of conduit (90° bend)

Fig. 320-2. Conduit bending radius (Rule 320A1c).

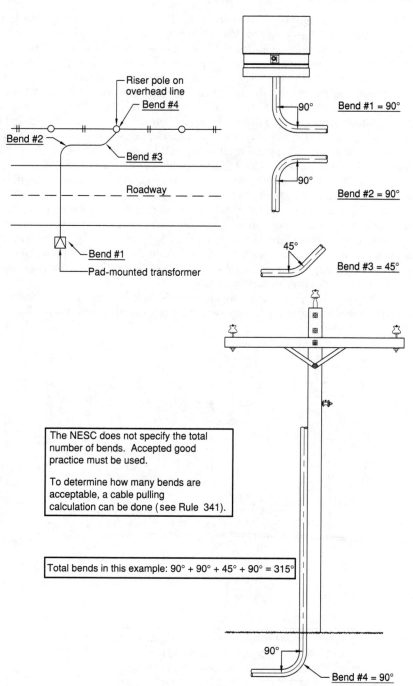

Fig. 320-3. Example of determining the total bending radius of a conduit run (Rule 320A1c).

Conduit routing in or on bridges and tunnels should not be damaged by traffic and should be located to provide safe access.

Conduits crossing under railroad tracks do have a specific burial requirement. This is the only specific burial requirement provided in Sec. 32. The burial rules for a conduit under a railroad track are outlined in Fig. 320-5.

Submarine crossings are discussed in Rule 320A6 and in Sec. 35, "Direct Buried Cable," Rule 351C5. Submarine crossings are underwater crossings usually on lake bottoms, river bottoms, or ocean bottoms. A special cable called a "submarine cable" is commonly used for underwater crossings. Conduit could be used for the entire crossing but it is typically used near the shorelines. An example of a three-phase, 15-kV submarine cable is shown in Fig. 320-6.

Higher-voltage submarine cables are also in use and can consist of oil-filled cable with a lead outer jacket.

There are no specific conduit requirements in this rule other than protecting the submarine crossing from erosion by tides or currents and locating the crossing away from where ships normally anchor. Submarine crossings frequently use conduit for some portion of the distance into the water and then the submarine cable is exposed and lies on the bottom of the body of water. An example of a submarine crossing is shown in Fig. 320-7.

Rule 96C, which requires at least four grounds in each mile, has an exception that states the rule does not apply to underwater crossings if other conditions are met.

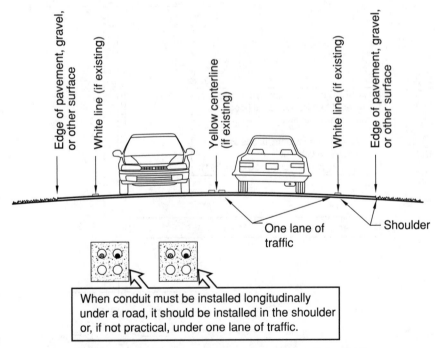

See Photo(s)

Fig. 320-4. Conduit routing under highways and streets (Rule 320A3).

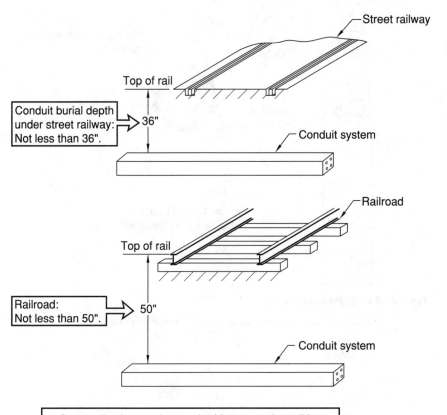

Fig. 320-5. Conduit crossings under railroad tracks (Rule 320A5).

320B. Separation from Other Underground Installations. The general requirements for separation between conduit systems and other underground structures are outlined below:
- When paralleling another structure, provide separation to permit maintenance without damaging the parallel structures.
- When crossing another structure, provide separation to limit damage to
  either structure.
- The parties involved should determine separations.
- Exception: Conduits may be supported from roofs of manholes, vaults, and
  subway tunnels with concurrence of the parties.

The term structure in this rule is used loosely to apply to many types of buried
structures including building foundations, other conduit systems, and other utility

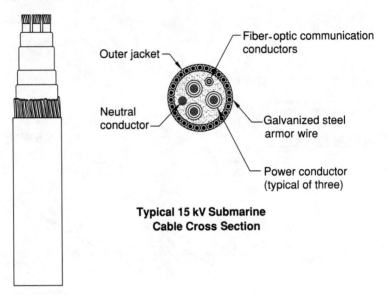

Outer jacket —

Neutral
conductor —

— Fiber-optic communication
conductors

— Galvanized steel
armor wire

— Power conductor
(typical of three)

**Typical 15 kV Submarine
Cable Cross Section**

Typical 15 kV Submarine Cable

Fig. 320-6. Example of a three-phase, 15-kV submarine cable (Rule 320A6).

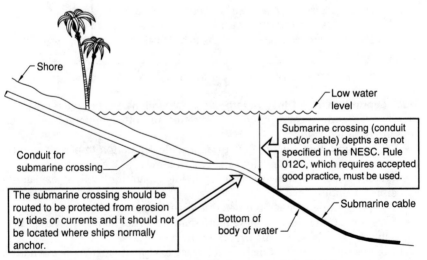

— Shore

— Low water
level

Conduit for
submarine crossing —

Submarine crossing (conduit
and/or cable) depths are not
specified in the NESC. Rule
012C, which requires accepted
good practice, must be used.

The submarine crossing should be
routed to be protected from erosion
by tides or currents and it should not
be located where ships normally
anchor.

Bottom of
body of water —

— Submarine cable

Fig. 320-7. Example of a submarine crossing (Rule 320A6).

lines such as sewer lines, water lines, gas and other lines that transport flammable material, steam lines, etc. No specific dimensions are provided. Enough separation for safe maintenance and avoiding damage are the primary concerns.

The **Code** is very specific for separations between supply and communication conduit systems as shown in Fig. 320-8.

One of the main reasons that supply and communication conduit systems need to be separated is that a fault on the supply cable in the supply conduit

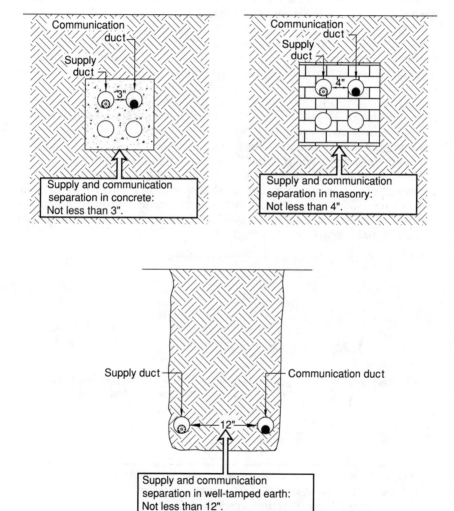

Fig. 320-8. Separation between supply and communication conduit systems (Rule 320B2).

could damage the supply conduit and the communication conduit and cable next to it. Another reason is to provide adequate space between the conduits for maintenance. These two issues must be considered in residential subdivisions. With proper fuse protection and use of concentric neutral underground residential distribution (URD) cable, the concerns related to a URD cable fault damaging a conduit can be minimized. Some supply and communication utilities apply the exception to Rule 320B2 and use no separation between the supply and communication conduits in well-tamped earth. Typical residential construction may consist of a single trench with supply and communication conduits in random separation (i.e., placed in the trench next to each other without any intentional separation). This type of installation should not be confused with random separation of direct-buried supply and communication cables. The random separation requirements for direct-buried cables are discussed in Rule 354.

The requirements for conduits crossing and paralleling sanitary and storm sewers, water lines, gas and other lines that transport flammable material, and steam lines are very general. No specific dimensions are provided. The Code is primarily concerned with providing enough separation for maintenance and avoiding damage to the lines involved. See Fig. 320-9.

The conduit system locations discussed in Rule 320A indicate that vertical placement of parallel underground lines is not preferred. The conduit system separations discussed in Rule 320B1 simply require adequate separation between paralleling lines without any reference or preference to the horizontal or vertical placement.

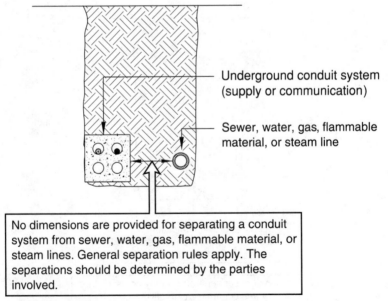

Fig. 320-9. Separation from sewer, water, gas, flammable material, and steam lines (Rules 320B3 through 320B6).

For separation between direct-buried supply and communication cables and other underground structures, see Rule 353.

321. EXCAVATION AND BACKFILL

The trench and backfill rules for conduit systems are outlined in Fig. 321-1.

For trench and backfill requirements related to direct-buried supply and communication cables, see Rule 352.

322. DUCTS AND JOINTS

322A. General.   The general rules for ducts and joints are outlined in Fig. 322-1.

322B. Installation.   The installation rules for ducts and joints are outlined in Fig. 322-2.

323. MANHOLES, HANDHOLES, AND VAULTS

323A. Strength.   Manholes, handholes, and vaults must be designed to withstand a variety of loads as outlined in Fig. 323-1.

In general, dead loads are the weight of the structure, constant weight on the structure, and permanent attachments to the structure. In general, live loads are loads that are not permanent, like a truck passing over the top of a manhole.

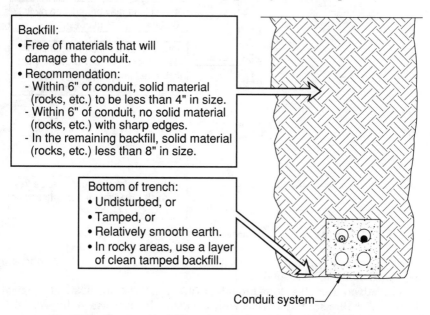

Backfill:
- Free of materials that will damage the conduit.
- Recommendation:
  - Within 6" of conduit, solid material (rocks, etc.) to be less than 4" in size.
  - Within 6" of conduit, no solid material (rocks, etc.) with sharp edges.
  - In the remaining backfill, solid material (rocks, etc.) less than 8" in size.

Bottom of trench:
- Undisturbed, or
- Tamped, or
- Relatively smooth earth.
- In rocky areas, use a layer of clean tamped backfill.

Conduit system

Fig. 321-1.   Trench and backfill requirements for conduit systems (Rule 321).

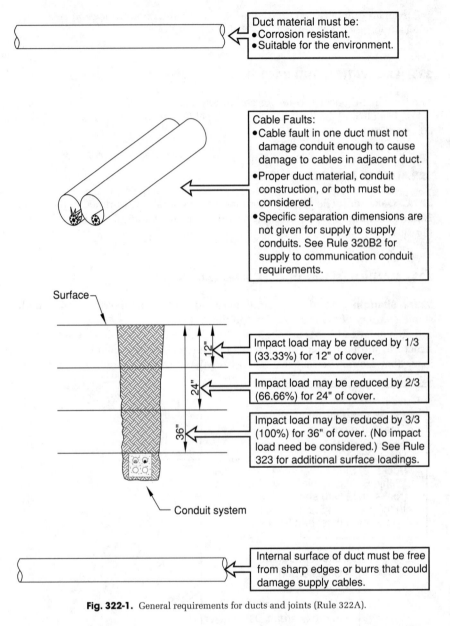

Fig. 322-1. General requirements for ducts and joints (Rule 322A).

Impact loads are generally applied in a similar fashion as live loads and are expressed as a percentage of live loads.

For roadway areas the **Code** provides **NESC** Figs. 323-1 and 323-2 of a typical 10-wheel semitractor trailer truck with associated loads to use for live-load calculations. A 10-wheel semi is used as opposed to an 18-wheeler, as a 10-wheel

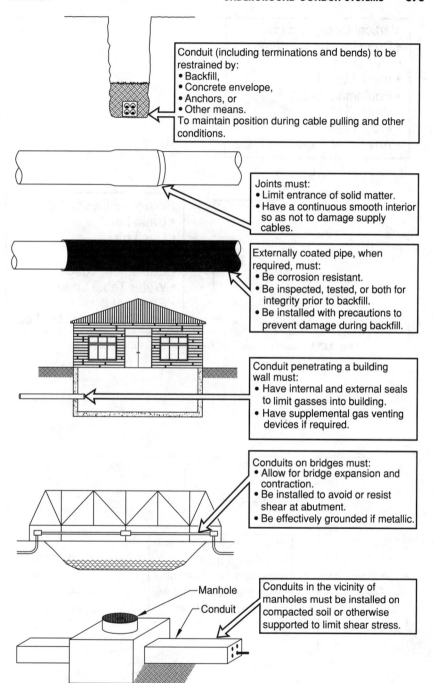

Conduit (including terminations and bends) to be restrained by:
• Backfill,
• Concrete envelope,
• Anchors, or
• Other means.
To maintain position during cable pulling and other conditions.

Joints must:
• Limit entrance of solid matter.
• Have a continuous smooth interior so as not to damage supply cables.

Externally coated pipe, when required, must:
• Be corrosion resistant.
• Be inspected, tested, or both for integrity prior to backfill.
• Be installed with precautions to prevent damage during backfill.

Conduit penetrating a building wall must:
• Have internal and external seals to limit gasses into building.
• Have supplemental gas venting devices if required.

Conduits on bridges must:
• Allow for bridge expansion and contraction.
• Be installed to avoid or resist shear at abutment.
• Be effectively grounded if metallic.

Manhole
Conduit

Conduits in the vicinity of manholes must be installed on compacted soil or otherwise supported to limit shear stress.

Fig. 322-2. Installation requirements for ducts and joints (Rule 322B).

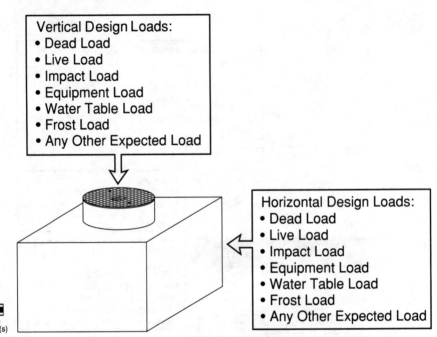

Fig. 323-1. Loads on manholes, handholes, and vaults (Rule 323A).

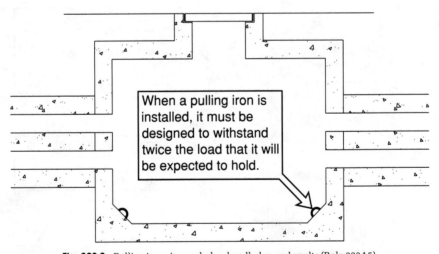

Fig. 323-2. Pulling irons in manholes, handholes, and vaults (Rule 323A5).

semi will have a larger force (weight) allocated to each wheel. A note in Rule 323A1 is also provided, reminding the designer that road construction equipment may exceed the typical truck loads.

For areas not subject to vehicle traffic, the **Code** requires that the design live load be 300 lb/ft^2. The **Code** requires that the live loads be increased 30 percent for impact. Impact loads can be reduced for specific burial depths (see Rule 322A).

The weight of the manhole, handhole, or vault must be sufficient to withstand hydraulic, frost, or other uplift forces. If the structure weight is not sufficient, some type of restraint or anchorage must be provided. The weight of equipment inside the structure cannot be considered, as the equipment may be removed or modified.

When a pulling iron is installed in a manhole, handhold, or vault, it must be rated to withstand twice the expected load. See Fig. 323-2.

323B. Dimensions.   The **Code** is very specific about working space dimensions in manholes. The **Code** does not specifically mention vaults in Rule 323B but the manhole dimensions are also appropriate for vaults. The general requirement for working space in Part 3 is found in Rule 312. Rule 323B is one of the few rules in Part 3 that actually dimensions the working space.

The rules for working space in manholes are outlined in Figs. 323-3 and 323-4.

It is important to note that the clear working space dimensions are measured after cables and equipment are placed in the manhole. They are not the dimensions to the bare manhole wall.

323C. Manhole Access.   The rules for manhole access are outlined in Figs. 323-5 through 323-8.

The access dimensions required in Rule 323C are for the manhole opening. They are not the dimensions of the lid or cover. The lid or cover could be larger than the manhole opening.

Additional rules related to manhole access require the openings to be free of protrusions. Where practical, they should be located outside of highways, intersections, and crosswalks to reduce traffic hazards to the workers.

The manhole access openings should not be located directly over cables or equipment. If the manhole opening is located over cables (not equipment) due to curb locations, a safety sign or protective barrier over the cables or a fixed ladder must be provided.

A manhole greater than 4 ft deep must be designed so it can be entered by means of a ladder or other suitable climbing device. If proper working space exists in the manhole per Rule 323B and the manhole opening is above the working space, ladder access should not be difficult. The ladder is not required to be permanent. Ladder requirements are also covered in Rule 323F. The **Code** does not consider the cables and equipment in the manhole to be suitable climbing devices.

323D. Covers.   The rules for manhole and handhole covers are outlined in Fig. 323-9.

323E. Vault and Utility Tunnel Access.   The general access rules for vaults and utility tunnels are similar to the general manhole access rules discussed in Rule 323C. No dimensions are provided for vault access openings. Accepted good practice must be applied. The manhole access dimensions are certainly one of the best references for accepted good practice in this case. Typically, manholes

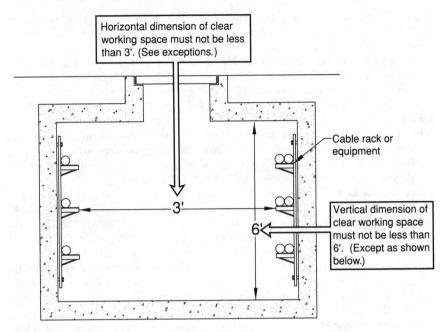

Horizontal dimension of clear working space must not be less than 3'. (See exceptions.)

Cable rack or equipment

3'

Vertical dimension of clear working space must not be less than 6'. (Except as shown below.)

6'

The 6' vertical dimension may be reduced (no value specified) if the manhole opening is within 1' horizontally of the adjacent interior side wall.

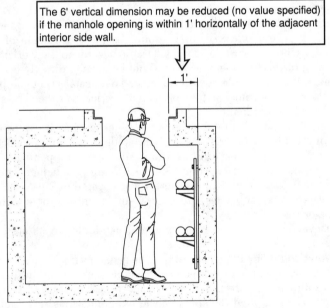

1'

Fig. 323-3. Manhole working space dimensions (Rule 323B).

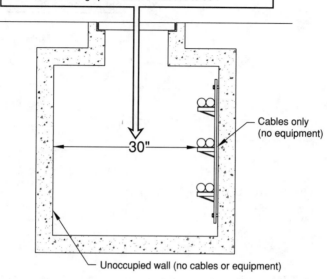

Exception 1:
If cables only (no equipment) are located on one wall and the
opposite wall is unoccupied, then the 3' horizontal dimension
of the clear working space can be reduced to 30".

30"

Cables only
(no equipment)

Unoccupied wall (no cables or equipment)

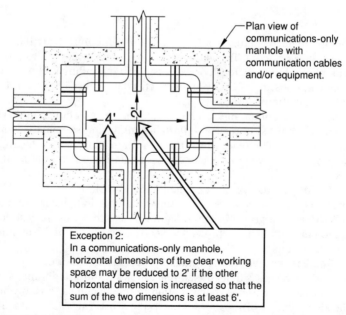

Plan view of
communications-only
manhole with
communication cables
and/or equipment.

4'

2'

Exception 2:
In a communications-only manhole,
horizontal dimensions of the clear working
space may be reduced to 2' if the other
horizontal dimension is increased so that the
sum of the two dimensions is at least 6'.

Fig. 323-4. Exceptions to manhole working space dimensions (Rule 323B).

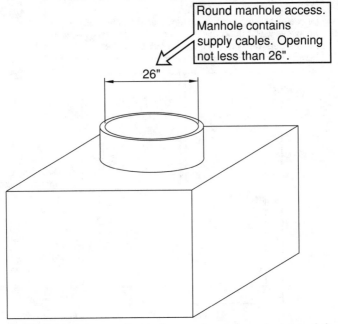

Fig. 323-5. Manhole access diameter for a round manhole opening in a manhole containing supply cables (Rule 323C1).

are thought to have openings or lids, and vaults and tunnels are thought to have openings, lids, or doors. The vault and utility tunnel access rules provide some additional requirements for doors accessible to the public including locking of the doors and posting of safety signs. Section 39, Rules 390 and 391, are dedicated to installations in tunnels.

323F. Ladder Requirements.   Fixed ladders in manholes and vaults are required to be corrosion-resistant. Portable ladders must meet the work rules in Part 4, Rule 420J. The work rules in Part 4, Rule 423B require testing for flammable gases and oxygen deficiency before entering a manhole or an unventilated vault.

323G. Drainage.   Drainage for manholes, handholes, and vaults is not required to drain into sewers. Many times, natural drainage is provided by utilizing a dry well or gravel base. When drainage into sewers is desired, a trap is a common method that can be used to keep sewer gas out of the manhole, vault, or tunnel. See Fig. 323-10.

323H. Ventilation.   Ventilation must be provided if the public can somehow be affected by the gases that collect in manholes, vaults, and tunnels. An exception applies to areas under water or other impractical locations. If ventilation is not needed (typical of most manholes), the work rules in Part 4, Rule 423B require an adequate air supply for the employee while the employee is working in the manhole or unventilated vault.

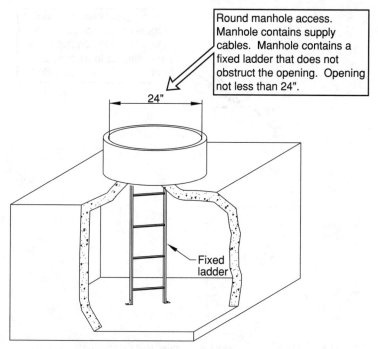

> Round manhole access.
> Manhole contains supply
> cables. Manhole contains a
> fixed ladder that does not
> obstruct the opening. Opening
> not less than 24".

Fig. 323-6. Manhole access diameter for a round manhole opening in a manhole containing supply cables (Rule 323C1).

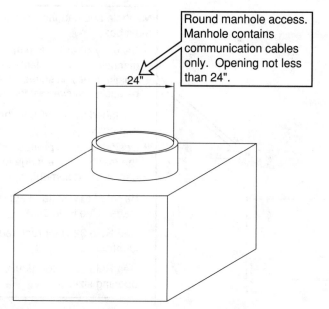

> Round manhole access.
> Manhole contains
> communication cables
> only. Opening not less
> than 24".

Fig. 323-7. Manhole access diameter for round manhole opening in a manhole containing communication cables only (Rule 323C1).

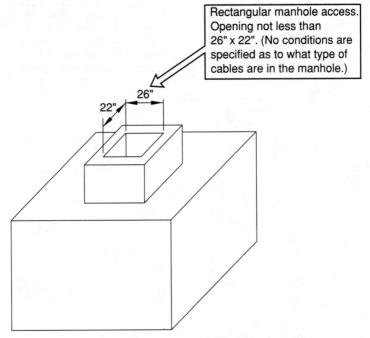

Rectangular manhole access. Opening not less than 26" x 22". (No conditions are specified as to what type of cables are in the manhole.)

26"

22"

Fig. 323-8. Manhole access size for a rectangular manhole opening (Rule 323C1).

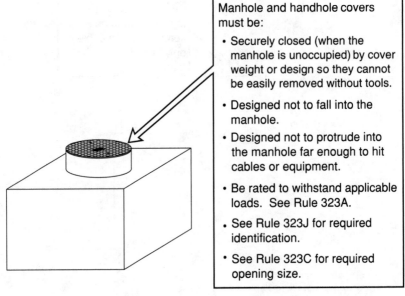

Manhole and handhole covers must be:

• Securely closed (when the manhole is unoccupied) by cover weight or design so they cannot be easily removed without tools.

• Designed not to fall into the manhole.

• Designed not to protrude into the manhole far enough to hit cables or equipment.

• Be rated to withstand applicable loads. See Rule 323A.

• See Rule 323J for required identification.

• See Rule 323C for required opening size.

Fig. 323-9. Requirements for manhole and handhole covers (Rule 323D).

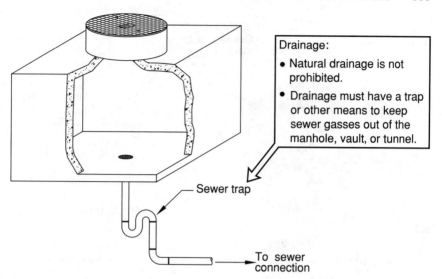

Fig. 323-10. Drainage requirements for manholes, vaults, and tunnels (Rule 323G).

323I. Mechanical Protection.   If a manhole, handhole, or vault has a grated cover that allows objects to fall on the supply cables and equipment, the supply cables and equipment must be located (positioned) or guarded (physically protected) to prevent damage. This rule does not mention communication cables or equipment. Grated covers are typically used for venting. Cables and equipment must not be located directly under any type of manhole opening for personnel access per Rule 323C.

323J. Identification.   The rules for identifying manhole and handhole covers are outlined in Fig. 323-11.

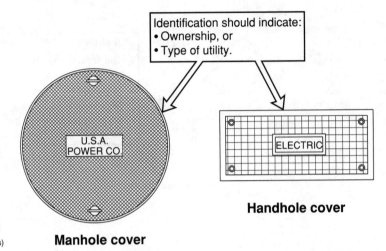

See Photo(s)

Manhole cover

Handhole cover

Fig. 323-11. Requirements for identification of manhole and handhole covers (Rule 323J).

Section 33

Supply Cable

330. GENERAL

Section 33 is only applicable to supply cable. It does not address communication cables. Rule 330 starts out with a recommendation that supply cables be tested in accordance with applicable standards. The rule also provides additional general requirements that are applicable to both the supply cable specifier and the supply cable manufacturer. The supply cable should be rated for the particular application and environment. The cable must also be able to withstand the fault current to which it will be subjected.

331. SHEATHS AND JACKETS

Cable jackets protect cables from adverse environmental conditions. Older underground residential distribution (URD) cables did not have a cable jacket, and concentric neutral corrosion became a widespread problem in some soils. Newer URD cables can be purchased with an outer jacket to protect the concentric neutral conductors. URD cables with a jacket do not have the concentric neutral in contact with the earth. The jacket may have to be stripped back to ground the concentric neutral. See Rule 096 for a grounding discussion. An example of a sheath, jacket, and shield on a typical URD supply cable is shown in Fig. 331-1.

332. SHIELDING

General shielding guidelines are provided in this rule. Organizations that specialize in developing cable standards are noted for a reference. See the figure in Rule 331 for a description of the individual components that make up a typical underground residential distribution (URD) supply cable.

333. CABLE ACCESSORIES AND JOINTS

Cable accessories and joints are the elbows, splices, and terminations used on supply cables. These accessories must be able to withstand stresses similar to the cable to which they are connected. The requirements for cable accessories in this rule focus on material specifications. Cable accessories are also discussed in Sec. 37, "Supply Cable Terminations." Section 37 focuses on the installation of the cable terminations. Examples of cable accessories and joints are shown in Fig. 333-1.

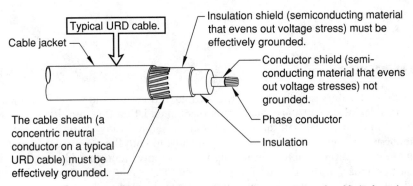

Typical URD cable.

Cable jacket —

Insulation shield (semiconducting material that evens out voltage stress) must be effectively grounded.

Conductor shield (semiconducting material that evens out voltage stresses) not grounded.

The cable sheath (a concentric neutral conductor on a typical URD cable) must be effectively grounded. —

Phase conductor

Insulation

See Photo(s)

Figure 331-1. Example of a sheath, jacket, and shield on a typical URD supply cable (Rule 331).

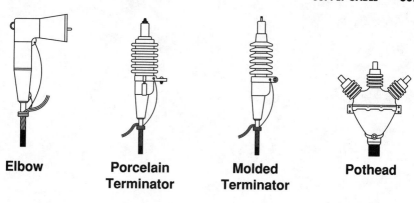

Elbow   **Porcelain**   **Molded**   **Pothead**
              Terminator   **Terminator**

Cable Splice

Figure 333-1. Examples of cable accessories and joints (Rule 333).

Section 34

Cable in Underground Structures

340. GENERAL

Section 34 provides rules for underground cables when they are installed in underground structures (e.g., conduit systems). Section 33 provides rules for the components of a supply cable. Section 32 provides rules for the conduit system. Section 34 does not apply to direct-buried cables. See Fig. 340-1.

Per Rule 340B, a supply cable over 2 kV to ground in a conduit system consisting of nonmetallic conduit should consider the need for an effectively grounded shield, sheath, or both. When a supply cable is direct-buried, Rule 350B requires an effectively grounded shield, sheath, or concentric neutral with more force (via the word *shall* instead of *should*), and the requirement is for above 600 V, not 2 kV.

341. INSTALLATION

341A. General. Supply cables are mechanically stressed when they are bent. The NESC does not specify a cable bending radius. Only a general statement requiring bending to be controlled to avoid damage is provided. The cable manufacturer's recommendations should be used. A bending radius is usually expressed as a multiple of the cable diameter. When the supply cable is installed in conduit, the cable bending radius should always be compatible with the conduit bending radius. See Fig. 341-1.

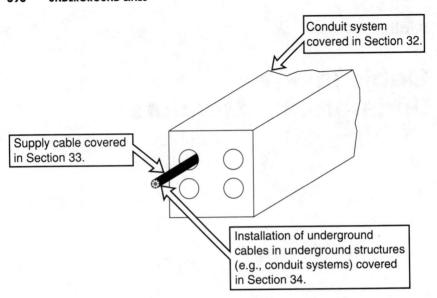

Conduit system covered in Section 32.

Supply cable covered in Section 33.

Installation of underground cables in underground structures (e.g., conduit systems) covered in Section 34.

Fig. 340-1. Application of Secs. 32, 33, and 34 (Rule 340).

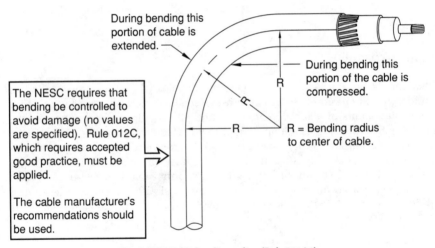

During bending this portion of cable is extended.

During bending this portion of the cable is compressed.

R = Bending radius to center of cable.

The NESC requires that bending be controlled to avoid damage (no values are specified). Rule 012C, which requires accepted good practice, must be applied.

The cable manufacturer's recommendations should be used.

See Photo(s)

Fig. 341-1. Cable bending radius (Rule 341A1).

The NESC does not specify conduit fill requirements. Conduit fill is the number of conductors or cables that fit in a specific conduit size. Conduit fill is normally limited to some percentage of the conduit cross-sectional area. The National Electrical Code (NEC) is a good reference for acceptable good practice in this case. Most utilities develop conduit fill tables for the conduits and cables they normally stock.

The **Code** requires limiting pulling tensions and sidewall pressures to avoid cable damage. A note suggests using the cable manufacturer's recommendations. Cable manufacturers can also supply engineering guidelines for making pulling calculations. Cable lubricants are used to make cable pulling easier, therefore reducing the pulling tension. Additional ways to reduce pulling tensions include pulling from an uphill point to a downhill point and starting the pull near a conduit bend instead of away from a conduit bend. Sidewall pressures are the forces that push on the side of the cable around a conduit bend. The same methods used to reduce pulling tension can be used to reduce sidewall pressure. Sidewall pressure is a function of the cable tension at the bend and the radius of the bend, not the diameter of the conduit. See Fig. 341-2.

The **Code** specifically discusses pulling tensions and sidewall pressures relative to supply cables. The **Code** does not discuss pulling tensions for communication cables, but communication cables require the same considerations.

The **Code** requires ducts to be cleaned of foreign material. This is typically referred to as "swabbing" the duct. The cable lubricants must be safe for the type of conduit and conduit materials. Too little friction can also be a problem. If a cable is installed on a downhill slope or vertical run, the cable may need to be restrained to prevent a cable elbow or termination from being stressed or pulled out.

Rule 320B requires separation between supply and communication conduit systems. Rule 341A6 reinforces that idea by saying that supply and communication cables must not be installed in the same duct unless they are operated and maintained by the same utility. Rule 341A7 permits the installation of communication

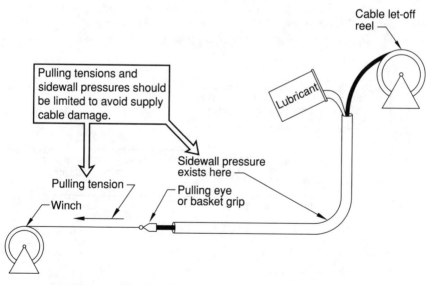

Fig. 341-2. Supply cable pulling tension and sidewall pressure (Rule 341A2).

cables of different communication utilities in the same duct provided the utilities are in agreement. See the beginning of Sec. 32 for definitions of duct, conduit, and conduit system. The requirements of Rules 341A6 and 341A7 related to supply and communication cables in duct are outlined in Fig. 341-3.

341B. Cable in Manholes and Vaults. Rule 341B covers the installation of cable in manholes and vaults. The rule is divided into three main parts, supports, clearance, and identification. Rule 341B1 covers supports for cable in manholes and vaults. The rules for supporting cable in manholes and vaults are outlined in Fig. 341-4.

Rule 341B2 covers clearance for cables in manholes and vaults. The rules for clearance of cables in manholes and vaults focus on clearance between joint use (supply and communication) facilities. Joint-use manholes and vaults are usually not preferred by either utility. If a joint-use duct bank is used, the communication conduits can branch off of the duct bank before the duct bank enters the supply manhole. The communication conduits can then enter a separate communication manhole adjacent to the supply manhole. If this system is not used and a joint-use manhole is desired, it must be done with the concurrence of all the parties involved. The work rules in Part 4, Rule 433, must be applied when a communications worker is in a joint-use manhole. The preferred location of supply cables in a joint-use manholes is below communication cables. This is the opposite position of overhead lines as supply conductors are preferred at the higher level on an overhead line per Rule 220B. The rules for clearance of cables (and equipment per Rule 341B2b) in joint-use manholes and vaults are outlined in Figs. 341-5 and 341-6.

Rule 341B3 requires the cables in manholes or other access openings of a conduit system (i.e., vaults and handholes) to be identified by tags that are corrosion-resistant and suitable for the environment in which they are installed. The quality of the tag must be good enough to read with portable lighting. Brass tags and plastic compounds are both popular solutions. An exception is

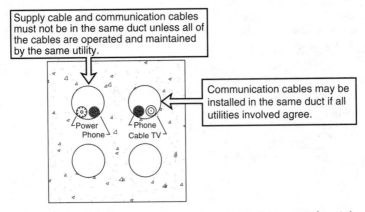

Fig. 341-3. Supply and communication cables in duct (Rules 341A6 and 341A7).

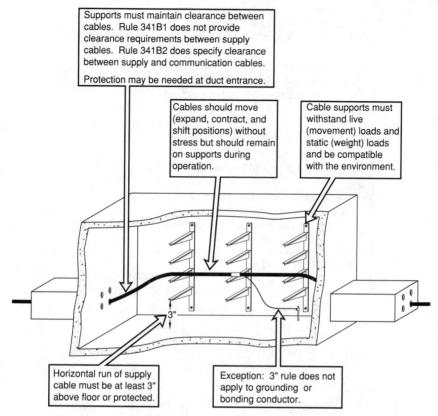

Supports must maintain clearance between cables. Rule 341B1 does not provide clearance requirements between supply cables. Rule 341B2 does specify clearance between supply and communication cables.

Protection may be needed at duct entrance.

Cables should move (expand, contract, and shift positions) without stress but should remain on supports during operation.

Cable supports must withstand live (movement) loads and static (weight) loads and be compatible with the environment.

Horizontal run of supply cable must be at least 3" above floor or protected.

Exception: 3" rule does not apply to grounding or bonding conductor.

Fig. 341-4. Supporting cable in manholes and vaults (Rule 341B1).

provided that eliminates the need for tags if maps or diagrams combined with cable position provide sufficient identification.

For joint-use (i.e., supply and communication) manholes or vaults, a cable identification tag or marking of some type must be provided denoting the utility name and the type of cable used. This requirement could also be applied to a manhole or vault with multiple utilities of the same kind. The identification requirements of Rule 341B3 are outlined in Fig. 341-7.

342. GROUNDING AND BONDING

Rule 342 provides grounding requirements in addition to the general grounding requirements in Rule 314. Cable or joints (splices) with bare metallic shields, sheaths, or concentric neutrals that are exposed to personnel contact must be

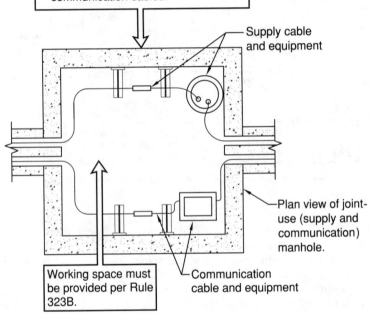

- All parties concerned must agree on joint use of manhole or vault.

- Supply and communication cables and equipment must be installed to permit access to either without moving the other.

- Supply and communication cables should be racked on separate walls and crossings should be avoided.

- Where supply and communication cables must be racked on the same wall, supply cables should be below communication cables.

Supply cable and equipment

Plan view of joint-use (supply and communication) manhole.

Working space must be provided per Rule 323B.

Communication cable and equipment

Fig. 341-5. Clearance of cables and equipment in joint-use manholes and vaults (Rule 341B2).

effectively grounded. See the figure in Rule 331 for a description of the individual components that make up a typical underground residential distribution (URD) supply cable. Cable sheaths or shields must have a common ground to keep different cables at equal ground potential, and the grounding materials must be corrosion-resistant or suitably protected.

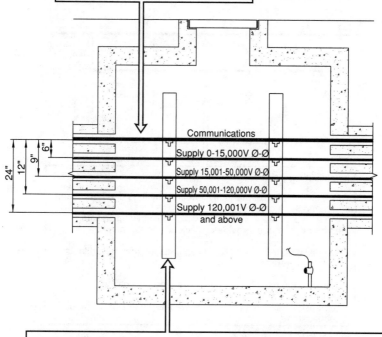

Clearance of communication cable to:
- Supply cable of 0–15,000V Ø-Ø:
  Not less than 6".
- Supply cable of 15,001–50,000V Ø-Ø:
  Not less than 9".
- Supply cable of 50,001–120,000V Ø-Ø:
  Not less than 12".
- Supply cable of 120,001 Ø-Ø and above:
  Not less than 24".

Communications

Supply 0-15,000V Ø-Ø

Supply 15,001-50,000V Ø-Ø

Supply 50,001-120,000V Ø-Ø

Supply 120,001V Ø-Ø
and above

- Supply and communication should be racked on separate walls.

- Where supply and communication cables must be racked on the same wall, supply cables should be below communication cables.

- Clearances from NESC Table 341-1 are surface to surface clearances between supply and communication cables and supply and communication equipment. For cable to cable clearance, cable diameters must be considered in addition to rack positions.

- Exception: Clearances do not apply to grounding conductors.

- Exception: Lesser clearances are acceptable with concurrence of both parties when barriers or guards are installed.

Fig. 341-6. Clearance of cables and equipment in joint-use manholes and vaults (Rule 341B2).

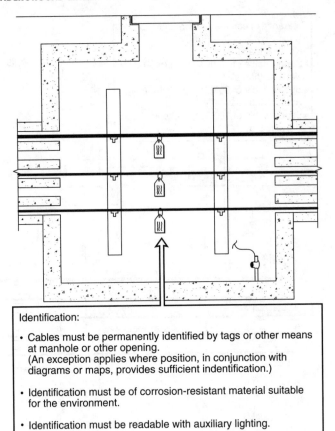

Identification:

- Cables must be permanently identified by tags or other means at manhole or other opening.
  (An exception applies where position, in conjunction with diagrams or maps, provides sufficient indentification.)

- Identification must be of corrosion-resistant material suitable for the environment.

- Identification must be readable with auxiliary lighting.

- Joint-use locations must identify the utility name and type of cable use.

Fig. 341-7. Identification of cables in manholes and vaults (Rule 341B3).

343. FIREPROOFING

Although fireproofing is not required and typically not needed due to noncombustible surroundings (e.g., concrete, dirt, etc.), this rule serves as a reminder for special cases.

344. COMMUNICATION CABLES CONTAINING SPECIAL SUPPLY CIRCUITS

Rule 344 is the underground equivalent to the overhead Rule 224B. Both this rule and Rule 224B provide special conditions for embedding a supply circuit in a communication cable. Rule 224B lists five special conditions (Rule 224B2a through 224B2e) for an overhead cable construction. Rule 344 lists six special conditions (Rules 344A1 through 344A6) for an underground cable construction. The extra condition for Rule 344 is for identification of the underground cable. Both Rule 344A and Rule 224B provide exceptions when the power within the communication cable is limited.

Section 35

Direct-Buried Cable

350. GENERAL

Section 35 provides rules for direct-buried cable (both supply and communication). If a cable is installed in a conduit system, the rules of Sec. 35 are not applicable. Sections 32, 33, and 34 apply to conduit systems. If cables are installed in a single duct that is not part of a conduit system, the rules of Sec. 35 apply. See the discussion and Figs. 32-1, 32-2, and 320-1 at the beginning of Sec. 32.

Per Rule 350B, a supply cable over 600 V to ground must have an effectively grounded shield, sheath, or concentric neutral that is continuous. See the figure in Rule 331 for a description of the individual components that make up an underground residential distribution (URD) supply cable.

The exception to Rule 350B recognizes that the standard splicing methods used on concentric neutral cable do not allow the concentric neutral to remain in a concentric pattern across the splice. See Fig. 350-1.

Rule 350C states that supply cables meeting Rule 350B of the same supply circuit (e.g., the same three-phase feeder) may be direct-buried with no deliberate separation. See Fig. 350-2.

Rule 350D states that underground direct-buried cables below 600 V to ground without an effectively grounded shield or sheath must be placed in close proximity to each other. Underground direct-buried secondary service conductors typically have an insulated (not bare) neutral. The requirement to place the conductors in close proximity with no intentional separation helps reduce the path that a fault current would need to follow if a cable fault occurred. The cables that meet Rule 350B are permitted to be buried with deliberate separation but they

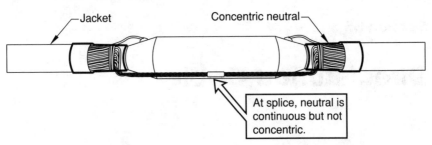

Fig. 350-1. Exception related to splices on concentric neutral cables (Rule 350B).

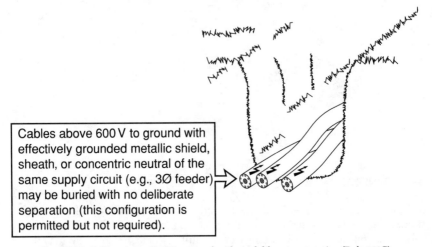

Fig. 350-2. Cables above 600 V to ground with no deliberate separation (Rule 350C).

are not required to be installed in that manner per Rule 350C. Rule 350C differs from Rule 350D as the cables in 350D are required to be installed in close proximity with no intentional separation. See Fig. 350-3.

Rule 350E references Rules 344A1 through 344A5 for rules related to communication cables containing special supply circuits. These rules also apply to direct-buried communication cables.

Rule 350F requires marking of direct-buried supply and communication cable with the "lightning bolt" and the "telephone handset." NESC Fig. 350-1 provides a diagram of the requirements. The telephone handset is used for all communications, even data and cable TV. This requirement is for direct-buried cables. Cables installed in conduit systems do not need to meet this requirement, and Sec. 32, "Underground Conduit Systems," does not have any conduit labeling or color coding requirements. Exception 1 to this rule addresses the fact that some cables are too small in diameter to mark or have some other marking constraints. Exception 2 states that unmarked cable from stock can be used to repair existing buried unmarked cables.

Rule 350G is related to Note 2 at the beginning of Sec. 32. See the discussion and Figs. 32-1, 32-2, and 320-1 at the beginning of Sec. 32.

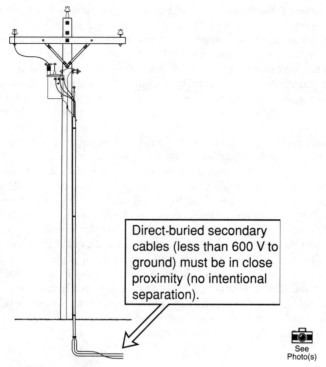

Direct-buried secondary cables (less than 600 V to ground) must be in close proximity (no intentional separation).

See Photo(s)

Fig. 350-3. Cables below 600 V to ground buried in close proximity (Rule 350D).

351. LOCATION AND ROUTING

The rules for locating and routing direct-buried cables are similar to, but slightly stricter than, the rules for locating conduit systems in Rule 320.

351A. General. The general requirements for direct-buried cable location and routing are outlined below:

- Subject cables to the least disturbance practical.
- When paralleling and directly over or under other buried facilities, the separation requirements in Rule 353 or 354 must be met.
- Install cables as straight as practical.
- Where required, cable bends must be large enough to avoid cable damage (see Rule 341 for a discussion).
- Route for safe access during construction, inspection, and maintenance.
- Plan route and structure conflicts before trenching, plowing, or boring.

351B. Natural Hazards. Natural hazards are discussed here for direct-buried cables and in Rule 320 for conduit systems. The cable must be constructed and installed to be protected from damage. If the installation method requires switching from a direct-buried cable installation to a conduit system installation, Rule 320 would apply.

351C. Other Conditions. Swimming pools get a lot of attention in the NESC. Part 2, Overhead Lines, and Part 3, Underground Lines, both have specific rules for swimming pool areas. Rule 351C1 addresses direct-buried supply (not communication) cable in the vicinity of in-ground swimming pools. This installation is very common in residential subdivisions in warmer parts of the country. For the purposes of locating and routing direct buried cables, aboveground pools are treated as buildings or other structures. The rules for locating and routing direct-buried cables in the vicinity of in-ground swimming pools are outlined in Fig. 351-1.

Buildings or other structures (including aboveground pools) should not have direct-buried cables installed under them. If they must, the structure must be properly supported so it will not damage the cables. If a conduit system is chosen to protect the cables installed under or near a building, Rule 320B1 would apply.

Direct-buried cable installed longitudinally under railroad tracks is to be avoided. Where it must be done, the direct-buried cable is to be 50 in below the top rail. An exception to Rule 351C3 states that this clearance may be reduced by agreement of the parties concerned. A permit is normally required to cross or parallel a railroad right-of-way, and to obtain the permit, the railroad agency usually requires the utilities to be located as far away from the tracks or as deep as possible. If a direct-buried cable is crossing (not paralleling) the railroad tracks, the same burial depths required for conduit apply. Many utilities install underground railroad crossings in a conduit or a conduit system, but it is not a Code requirement. See the figure in Rule 320A.

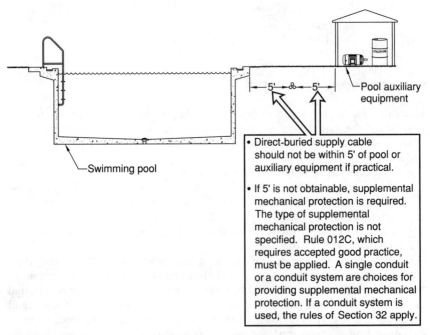

Fig. 351-1. Locating and routing direct-buried supply cables near in-ground swimming pools (Rule 351C1).

The **Code** states in Rule 351C4 that the installation of direct-buried cable longitudinally under highways and streets should be avoided. The equivalent rule in the underground conduit systems section (Rule 320A3) does not have the "should be avoided" language. Therefore, it seems appropriate to install cables under streets in conduit. Both the direct-buried and underground conduit systems sections define where to put the direct-buried cable or conduit. See the figure in Rule 320A.

Submarine crossings are addressed here in Sec. 35, "Direct-Buried Cable," and in Sec. 32, "Underground Conduit Systems." The wording is nearly identical, and neither one specifically addresses the terms "conduit" or "cable." Only the term "crossing" is used. See the submarine crossings discussion and the figures in Rule 320A.

352. INSTALLATION

Direct-buried cable installation can be done using three primary methods—trenching, plowing, and boring. Installation of underground conduit systems in Sec. 32 focused on trenching (see Rule 321). Rule 352 for the installation of direct-buried cables focuses on trenching, plowing, and boring. The rules for installation of direct-buried cable by trenching, plowing, and boring are outlined in Figs. 352-1, 352-2, and 352-3.

The terms "cable shall be adequately protected" or "protected from damage" or "supplemental mechanical protection must be provided" are used frequently in this rule and other rules in Sec. 35. The **Code** is not specific about how to accomplish these tasks. Possible solutions to these statements are installing a direct-buried cable in a conduit or conduit system. If a conduit system is selected (not just a single conduit), then Sec. 32, "Underground Conduit Systems," applies to the installation. See the discussion and Figs. 32-1, 32-2, and 320-1 at the beginning of Sec. 32.

Rule 352D, Depth of Burial, is one of the most referenced rules in Part 3. The rule starts out by defining how to measure the depth of burial of a direct-buried cable. The measurement is from the surface to the top of the cable. The general requirement is that burial depths must be sufficient to protect the cable from damage by the expected surface usage. The specific burial depth requirements are presented in **NESC** Table 352-1 for supply cables only.

There is not a table for communication conductor burial depth. Communication conductors do fall into the 0- to 600-V range, but **NESC** Table 352-1 is titled Supply Cable or Conductor Burial Depth. It does not apply to communications. Depths for communication conductors must only meet the general requirements for burial depth outlined in Rule 352D1. In addition to Rule 352D1, a communication utility must also use Rule 012C, which requires accepted good practice, to develop standards for communication cable burial depths. Communication cables laid on the ground do not meet the **Code** requirements for burial depth or overhead clearances.

Supply cables laid on the ground do not meet the **Code** requirements for burial depth or overhead clearances. Supply and communication cables may be

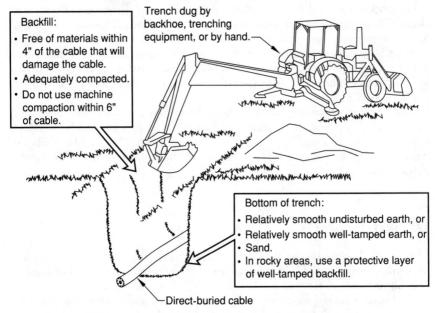

Backfill:
- Free of materials within 4" of the cable that will damage the cable.
- Adequately compacted.
- Do not use machine compaction within 6" of cable.

Trench dug by backhoe, trenching equipment, or by hand.

Bottom of trench:
- Relatively smooth undisturbed earth, or
- Relatively smooth well-tamped earth, or
- Sand.
- In rocky areas, use a protective layer of well-tamped backfill.

Direct-buried cable

See Photo(s)

Fig. 352-1. Trenching requirements for direct-buried cables (Rule 352A).

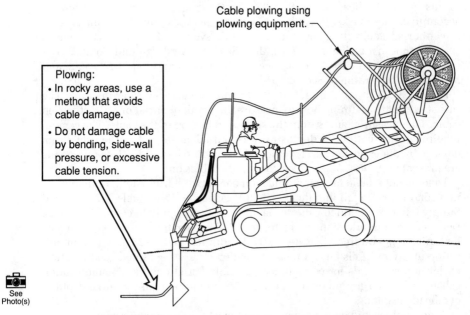

Cable plowing using plowing equipment.

Plowing:
- In rocky areas, use a method that avoids cable damage.
- Do not damage cable by bending, side-wall pressure, or excessive cable tension.

See Photo(s)

Fig. 352-2. Plowing requirements for direct-buried cables (Rule 352B).

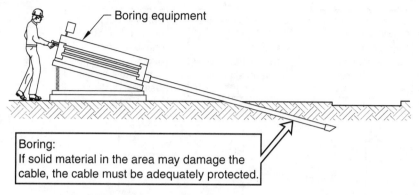

Boring equipment

Boring:
If solid material in the area may damage the
cable, the cable must be adequately protected.

See
Photo(s)

Fig. 352-3. Boring requirements for direct-buried cables (Rule 352C).

laid on the ground for emergency (not temporary) installations. See Rules 230A2d and 311C. The requirements in **NESC** Table 352-1 for supply cable or conductor burial depth are graphically outlined in Fig. 352-4.

Some utilities establish supply cable burial depth values by using the **Code** clearance plus an adder. The adder (for example 6 in) could be thought of as a design or construction tolerance adder to maintain the required **Code** burial depth over time. There are several factors that could jeopardize a supply cable burial depth. One factor could be soil erosion. Installing a 15-kV phase-to-phase direct-buried supply cable at 36 in deep can help maintain the **Code**-required 30-in depth over the life of the installation.

Rule 352D2b allows lesser burial depths if supplemental protection is used. The decision of what to use for supplemental protection is up to the designer. Rule 012C, which requires accepted good practice, must be applied. Conduit is one form of supplemental protection. Concrete slabs, wood planks, or other barriers are also forms of supplemental protection. The type of conduit (rigid steel, PVC, etc.) will affect the amount of protection. If a conduit system is used (not just a single conduit), then the rules of Sec. 32 apply. See the discussion and Figs. 32-1, 32-2, and 320-1 at the beginning of Sec. 32. One of the best references for accepted good practice for a single conduit that is not part of a conduit system is the National Electrical Code (NEC). It provides specific burial depths for various conduit types.

Rule 352E states that supply and communication cables must not be installed in the same duct unless they are operated and maintained by the same utility. Rule 352F permits the installation of communication cables of different communication utilities in the same duct provided the utilities are in agreement. Rules related to installations in ducts appear in Sec. 35, "Direct-Buried Cables," due to Rule 350G. See Rule 350G and the beginning of Sec. 32 for more information and for the definitions of duct, conduit, and conduit system. Rules 352E and 352F are similar in nature to Rules 341A6 and 341A7 (Cable in Underground Structures) and to Rules 239A2d and 239A2e (Vertical Risers). The requirements of Rules 352E and 352F are outlined in Fig. 352-5.

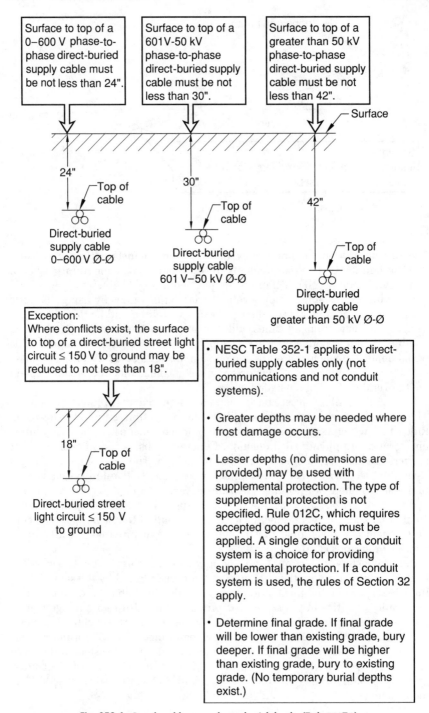

Fig. 352-4. Supply cable or conductor burial depths (Rule 352D2).

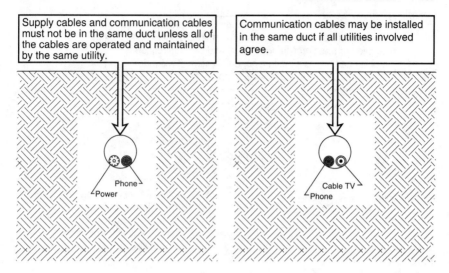

Fig. 352-5. Supply and communication cables in duct (Rules 352E and 352F).

353. DELIBERATE SEPARATIONS—EQUAL TO OR GREATER THAN 300 MM (12 IN) FROM UNDERGROUND STRUCTURES OR OTHER CABLES

Rules 353 and 354 are closely related. As their titles indicate, Rule 353 applies to 12 in or greater separation and Rule 354 applies to less than 12 in of separation. Rule 353 for direct-buried cable is similar to Rule 320B for underground conduit systems. In Rule 353 the NESC uses the term underground structures to mean sewers, water lines, gas and other lines that transport flammable material, building foundations, steam lines, etc.

The requirement for radial (any direction) separation between a direct-buried cable and another underground structure or cable in Rule 353 is 12 in or more. This value is assumed to permit maintenance on both facilities without damaging the other. If 12 in of separation is not obtainable, Rule 354 applies. Another option to use when 12 in of separation is not obtainable is to install the direct-buried cable in a conduit system. When a direct-buried cable is parallel and directly over or under another underground structure or cable, the parties involved must be in agreement to the method used. No such agreement is specified if a side-by-side configuration is used. Crossings require support or sufficient vertical separation to limit the transfer of detrimental loads from one system to the other. The rules related to the radial separation of direct-buried cables from another underground structure or cable are outlined in Fig. 353-1.

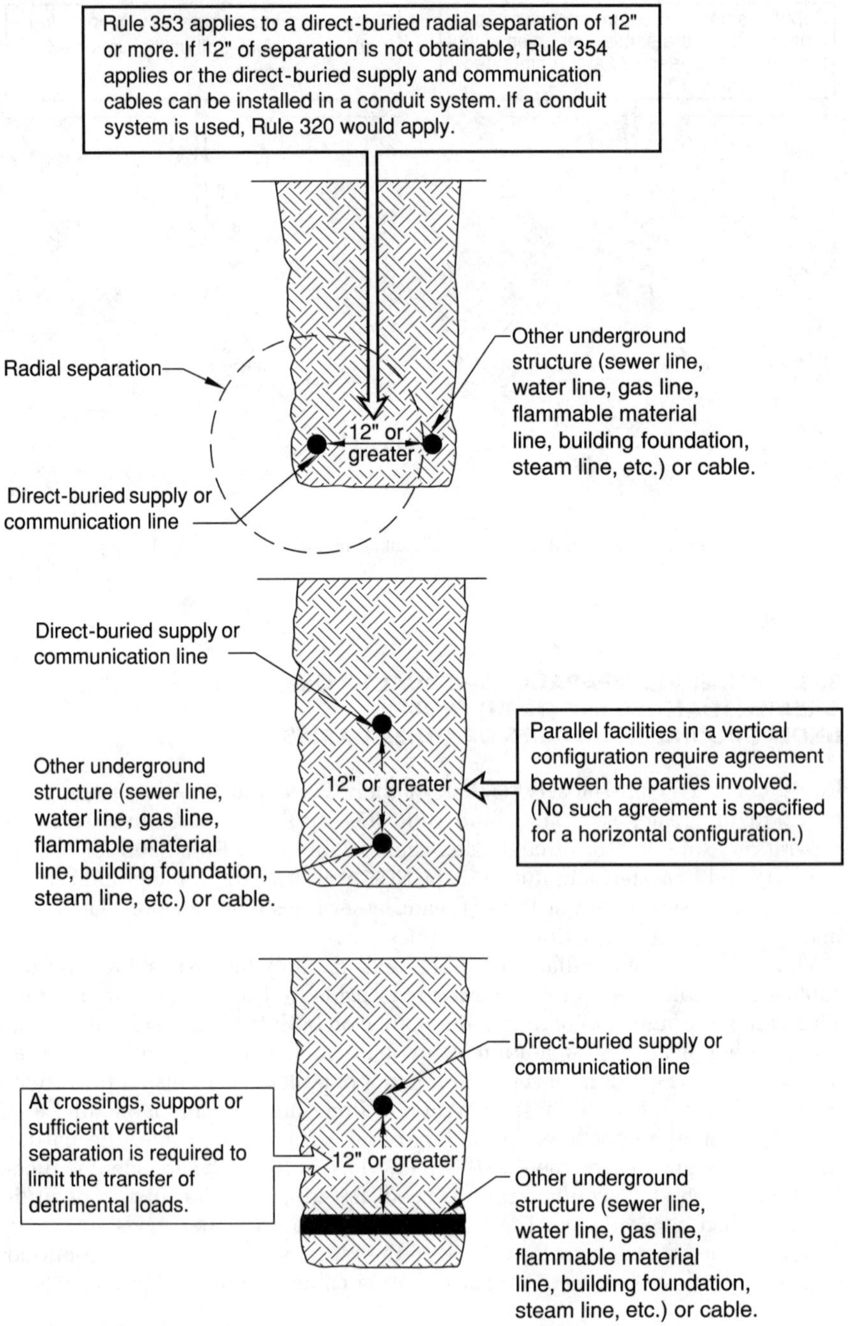

Rule 353 applies to a direct-buried radial separation of 12"
or more. If 12" of separation is not obtainable, Rule 354
applies or the direct-buried supply and communication
cables can be installed in a conduit system. If a conduit
system is used, Rule 320 would apply.

Radial separation—

Other underground
structure (sewer line,
water line, gas line,
flammable material
line, building foundation,
steam line, etc.) or cable.

12" or
greater

Direct-buried supply or
communication line

Direct-buried supply or
communication line

Other underground
structure (sewer line,
water line, gas line,
flammable material
line, building foundation,
steam line, etc.) or cable.

12" or greater

Parallel facilities in a vertical
configuration require agreement
between the parties involved.
(No such agreement is specified
for a horizontal configuration.)

Direct-buried supply or
communication line

At crossings, support or
sufficient vertical
separation is required to
limit the transfer of
detrimental loads.

12" or greater

Other underground
structure (sewer line,
water line, gas line,
flammable material
line, building foundation,
steam line, etc.) or cable.

Fig. 353-1. Deliberate separation of 12 in or more between a direct-buried cable and another underground structure or cable (Rule 353).

Vertical separation between supply and communication lines is typically used by utilities when plowing in direct-buried cables. The separations in Rule 353 are from surface to surface, not center to center. Therefore, a 12-in center-to-center spacing of the plow chutes may not provide a 12-in separation between cables depending on the cable diameters.

When paralleling or crossing a line that involves hot or cold temperatures like a steam line or a cryogenic (cold) line, consideration must be given to the heat or cold so it does not damage the crossing or paralleling line. Adequate separation is to be used. If separation is not obtainable, a thermal barrier is required.

354. RANDOM SEPARATION—SEPARATION LESS THAN 300 MM (12 IN) FROM UNDERGROUND STRUCTURES OR OTHER CABLES

Rules 354 and 353 are closely related. As their titles indicate, Rule 354 applies to less than 12 in of separation and Rule 353 applies to 12 in or greater separation. Random separation is sometimes referred to as random lay. The main requirements for this rule came from a series of tests performed on specific types of supply and communication cables; therefore, the rules for the type of cable insulation and grounding are very specific. The focus of the random separation rules is to prevent damage to direct-buried communication lines when a fault occurs on direct-buried supply lines. The intent is not to reduce or minimize conductive interference or "noise."

The general rules for random separation are outlined below:
- The rules apply to a direct-buried cable and another underground structure or cable with a radial (any direction) separation less than 12 in.
- The radial separation between a direct-buried supply or communication conductor and a steam line, gas line, or other line that transports flammable material must not be less than 12 in. (Rule 353 must be used in this case or Rule 320 can be applied if a conduit system is used.)
- Supply circuits, when faulted, need to be de-energized by a protective device (i.e., fuse, recloser, circuit breaker, etc.).
- Supply and communication cables and conductors in random separation are treated as one system when considering separation from other conductors.

Supply cables above 600 V to ground with an effectively grounded continuous metallic shield, sheath, or concentric neutral of the same supply circuit may be buried in random separation (no deliberate separation). See Rule 350C. Supply cables of the same circuit operating below 600 V to ground without an effectively grounded shield or sheath must be buried in random separation (in close proximity with no deliberate separation). See Rule 350D.

Supply cables of multiple supply circuits may be buried in random separation (no deliberate separation) if all parties involved are in agreement. See Fig. 354-1.

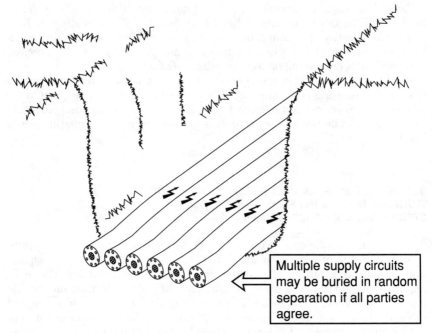

Fig. 354-1. Multiple supply circuits in random separation (Rule 354B).

Communication cables of multiple communication circuits may be buried in random separation (no deliberate separation) if all parties involved are in agreement. See Fig. 354-2.

Rule 354D focuses on supply and communication cables or conductors that are buried less than 12 in apart. If supply and communication cables are direct-buried less than 12 in apart, a variety of rules must be met. The rules include special consideration of voltage limitations, grounding and bonding requirements, protection requirements, type of cable jacket, and concentric neutral material and size. The steps needed to install direct-buried supply and communication conductors in random separation (less than 12 in apart) are outlined in Fig. 354-3.

The tests used to develop the random separation rules were done before all-dielectric fiber-optic communication cables became commonplace. An exception is provided in Rule 354D for all-dielectric fiber-optic cables.

Rule 354D4 allows a supply cable in a nonmetallic conduit and a communication cable direct-buried to have less than 12 in of separation. The supply cable used in this application must meet special conditions that are provided in Rule 354D3. If both the supply and communication cables are in conduit, Rule 320B2 applies. Rule 320B2 contains a simple exception which allows less than 12 in between supply and communication conduit systems if the parties concur.

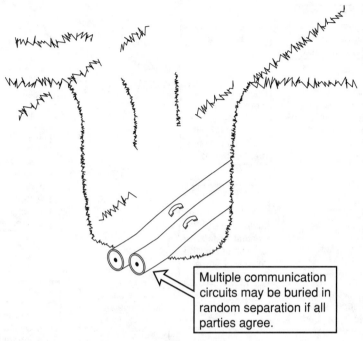

Fig. 354-2. Multiple communication circuits in random separation (Rule 354C).

Rule 354E provides rules for less than 12 in of separation between a direct-buried supply or communication cable and a nonmetallic water or sewer line. Metallic water or sewer lines must have 12 in or more separation (per Rule 353) or the direct-buried cables must be installed in a conduit system (per Rule 320). Building foundations are not specifically addressed in Rule 354.

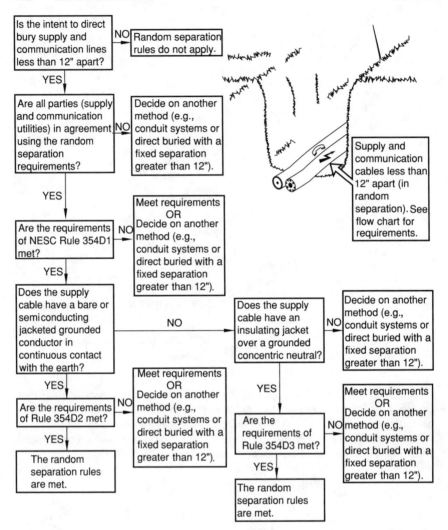

Fig. 354-3. Direct-buried supply and communication random separation requirements (Rule 354D).

Section 36

Risers

360. GENERAL

The rules for risers overlap in Part 3, Underground Lines, and Part 2, Overhead Lines. Rule 239D is referenced as it focuses on mechanical protection of a riser on an overhead pole. See Rule 239 for a discussion. Rule 360 extends the protection required in Rule 239D at least 1 ft below grade. See Fig. 360-1.

Cable bending must be considered when transitioning from a buried horizontal position to a vertical riser position. See Rule 341 for a discussion of cable bending.

Risers in metallic conduit or under metallic guards (i.e., U-Guard) need to have the metallic conduit or metallic guard grounded in accordance with Rule 314. Rule 360C does not require risers containing supply conductors to be metal, but if they are metal, they must be grounded.

361. INSTALLATION

Rule 361 provides general installation rules for risers. Rule 362 provides additional rules for pole risers and Rule 363 provides additional rules for risers entering pad-mounted equipment. The general rules for the installation of risers are outlined in Fig. 361-1.

362. POLE RISERS—ADDITIONAL REQUIREMENTS

Rules 360 and 361 apply to risers in general. Rule 362 applies specifically to pole risers. The rules for pole risers are outlined in Fig. 362-1.

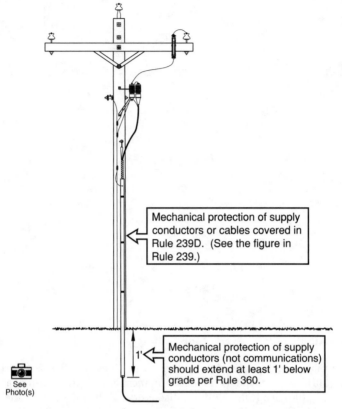

Mechanical protection of supply
conductors or cables covered in
Rule 239D. (See the figure in
Rule 239.)

Mechanical protection of supply
conductors (not communications)
should extend at least 1' below
grade per Rule 360.

1'

See
Photo(s)

Fig. 360-1. Mechanical protection of supply risers below grade (Rule 360A).

Standoff brackets are used by some utilities to make climbing easier. If standoff brackets are used on risers, Rule 217A2c must be met. See the figure in Rule 217.

363. PAD-MOUNTED INSTALLATIONS

Rules 360 and 361 apply to risers in general. Rule 363 applies specifically to underground cables rising from a horizontal position up to pad-mounted equipment. The rules for pad-mounted installations are outlined in Fig. 363-1.

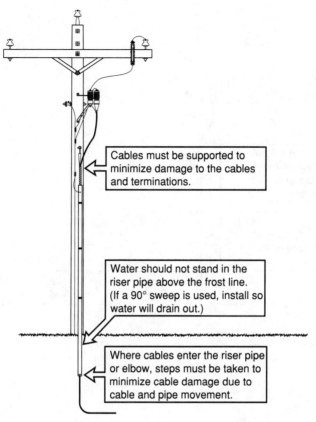

Fig. 361-1. Riser general installation requirements (Rule 361).

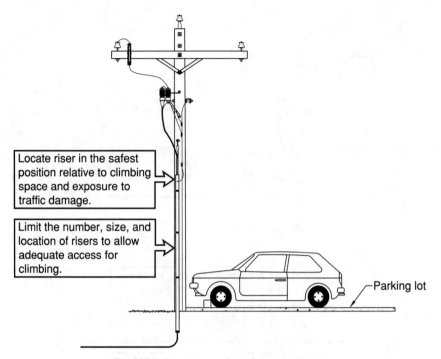

Locate riser in the safest position relative to climbing space and exposure to traffic damage.

Limit the number, size, and location of risers to allow adequate access for climbing.

—Parking lot

Fig. 362-1. Requirements for pole risers (Rule 362).

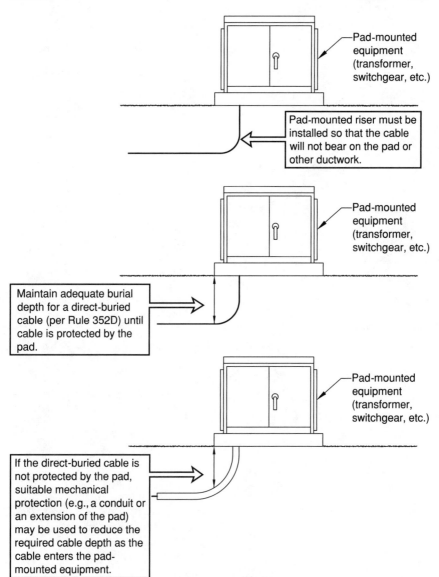

Fig. 363-1. Requirements for risers to pad-mounted equipment (Rule 363).

Section 37

Supply Cable Terminations

370. GENERAL

Supply cable terminations are the various items used to terminate underground cables such as elbows, cable terminators, potheads, etc. A reference to Rule 333 is made to tie Sec. 37 to Rule 333. Rule 333 focuses on the material specifications of the termination. Section 37 focuses on the installation of the termination. Cable terminators protect conductors and conductor insulation from mechanical damage, moisture, and electrical stress. Examples of supply cable terminations are shown in Fig. 370-1.

If a cable terminator is located inside a vault or pad-mounted equipment, the rules of Part 3, Underground Lines, apply. If the cable terminator is on a riser that terminates overhead inside a substation, the clearance rules found in Part 1, Electric Supply Stations, apply. If the cable terminator is on a pole outside of a substation, the clearance rules in Part 2, Overhead Lines, apply.

Equipment manufacturers determine clearance between cable terminators inside pad-mounted equipment by referencing ANSI and NEMA standards. The NESC provides a general statement (no specific distances) that suitable clearance must exist for the voltage and basic impulse level (BIL). Fully insulated terminations (i.e., elbows) or insulated barriers may be used to meet the necessary requirements.

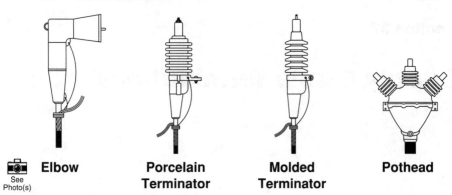

| Elbow | Porcelain Terminator | Molded Terminator | Pothead |

See Photo(s)

Fig. 370-1. Examples of supply cable terminations (Rule 370).

371. SUPPORT AT TERMINATIONS

Cable terminators must be installed to maintain their position. Underground residential distribution (URD) elbows must not pull out of the bushing wells they are installed in and riser pole cable terminators must not be moving around on the pole. The cable attached to the URD elbow or riser pole terminator may need to be supported to minimize stress on the cable terminator. Requirements for supporting cables are also stated in Rules 341A5 and 361B.

Supply utilities commonly use both porcelain cable terminators that are supported on a bracket or crossarm and lightweight polymer terminators that are not supported independent of the cable. The Code does not specify exactly how to support the cable termination, it only states that the termination must maintain its installed position.

372. IDENTIFICATION

Identification of a cable at a termination point is a simple method to maintain an organized electrical system, and it is a Code requirement for supply cables. Rule 372 does not apply to communication cables, as it is in Sec. 37, which is titled "Supply Cable Terminations." An exception is provided that eliminates the need for circuit identification if the position of the termination combined with maps or diagrams provides sufficient identification.

Additional identification rules for underground lines can be found in Rules 311 (underground locates), 323J (manhole and handhole covers), 341B3 (cables in manholes), 350F (marking of cables), and 385 (equipment operating in multiple). An example of supply cable identification at termination points is shown in Fig. 372-1.

The Code does not contain a rule requiring the use of underground warning ribbons or tapes. The Code also does not have a requirement for cable route markers. The Code certainly does not prohibit the use of these items and these items are installed by some utilities. The lack of a Code rule related to underground warning ribbons or tapes and cable route markers is shown in Fig. 372-2.

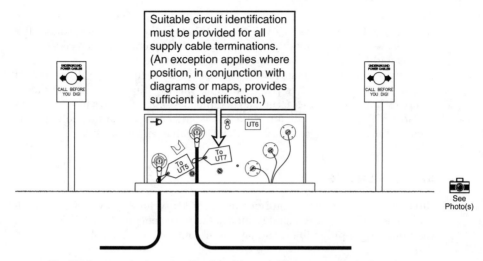

Fig. 372-1. Example of supply cable identification at termination points (Rule 372).

373. CLEARANCES IN ENCLOSURES OR VAULTS

Clearance between terminators depends on numerous factors including the type of terminator, the insulation method, and the voltage level.

This rule provides only a general statement that adequate clearance is required between supply terminations and between terminations and ground. No specific distances are provided. If terminators are in an enclosure (like a pad-mounted switch, junction box, transformer enclosure, etc.), the manufacturer will reference industry standards for the type of equipment. Adequate clearance or insulated barriers are needed for live parts in a manufactured enclosure.

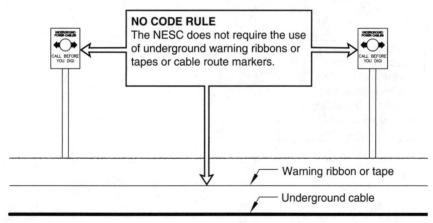

Fig. 372-2. Absence of Code rule related to underground warning ribbons or tapes and cable route markers (Rule N/A).

The burden is on the manufacturer to determine the specific clearances. The same is true for live parts in a vault, only this time the burden is on the utility designer. Guarding or isolating live parts in a vault is required. Using dead front fittings and terminations or metal-clad switchgear are methods of guarding or isolating live parts. Adequate physical clearance above the floor is also a method of guarding or isolating live parts. Since no specific dimensions are provided, Rule 012C, which requires accepted good practice, applies. The substation rules in Part 1 of the NESC, specifically Rule 124, would be a good starting point for the accepted good practice of guarding or isolating live parts in a vault.

374. GROUNDING

Rule 374 has grounding requirements specifically for supply cable terminations in addition to the general grounding requirements in Rule 314. The rules for grounding supply cable terminations are outlined in Fig. 374-1.

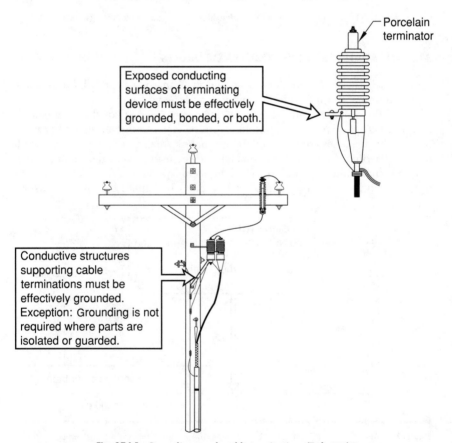

Porcelain terminator

Exposed conducting surfaces of terminating device must be effectively grounded, bonded, or both.

Conductive structures supporting cable terminations must be effectively grounded. Exception: Grounding is not required where parts are isolated or guarded.

Fig. 374-1. Grounding supply cable terminations (Rule 374).

Section 38

Equipment

380. GENERAL

Rule 380 starts out by providing examples of supply and communication equipment relevant to this section. See Fig. 380-1.

Equipment located in a joint-use (supply and communication) manhole can only be installed with the concurrence of the parties concerned. Similar wording appears in Rule 341B2b(1) for cables in joint-use manholes. If the parties do not concur, separate conduit systems and manholes can be used or a joint conduit system (duct bank) can be used where the conduits leave the joint duct bank and enter separate manholes or pedestals at every termination point.

The pads, supports, and foundations used to support equipment must be rated for the load and stress of the equipment and the equipment's operation. For example, the forces associated with installing and removing underground residential distribution (URD) elbows must be considered.

Rule 380D is similar to Rule 231A. Rule 380D specifies a distance from pad-mounted equipment to a fire hydrant. Rule 231A specifies a distance from a supporting structure (i.e., pole) to a fire hydrant. Rule 231A has an exception for reducing clearance by agreement with the local fire authority, Rule 380D does not contain this exception. The main purpose of these rules is to allow adequate working space for the fire department to connect hoses to the fire hydrant. See Fig. 380-2.

The Code does not address clearance from pad-mounted equipment to a roadway or clearance of an oil-filled pad-mounted transformer to a building. Since the Code does not provide dimensions for these installation conditions, accepted good practice (Rule 012C) must be used. The absence of a Code rule related to pad-mounted equipment location adjacent to roads and buildings is outlined in Fig. 380-3.

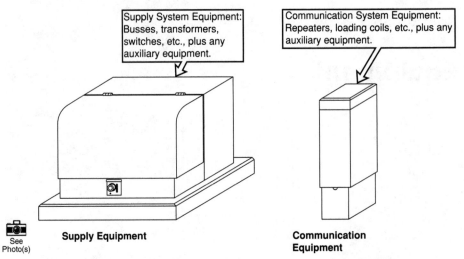

Fig. 380-1. Examples of supply and communication equipment (Rule 380).

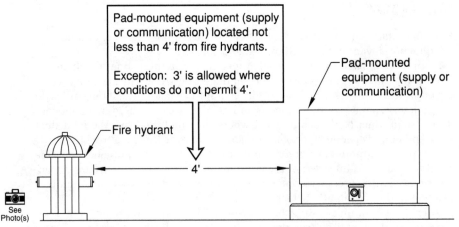

Fig. 380-2. Clearance of pad-mounted equipment to fire hydrants (Rule 380D).

381. DESIGN

Equipment design and mounting must consider the following:

- Thermal conditions
- Chemical conditions
- Mechanical conditions
- Environmental conditions

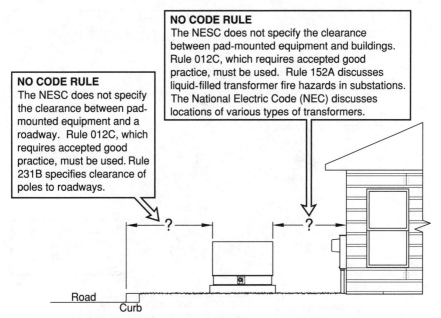

NO CODE RULE
The NESC does not specify the clearance between pad-mounted equipment and buildings. Rule 012C, which requires accepted good practice, must be used. Rule 152A discusses liquid-filled transformer fire hazards in substations. The National Electric Code (NEC) discusses locations of various types of transformers.

NO CODE RULE
The NESC does not specify the clearance between pad-mounted equipment and a roadway. Rule 012C, which requires accepted good practice, must be used. Rule 231B specifies clearance of poles to roadways.

Road
Curb

See
Photo(s)

Fig. 380-3. Absence of Code rule related to underground pad-mounted equipment location adjacent to roads and buildings (Rule N/A).

Equipment and auxiliary devices must also be rated for the expected:
• Normal conditions
• Emergency conditions
• Fault conditions

The requirement for switching underground equipment is similar in nature to switching requirements for overhead switches discussed in Rule 216. Switches for underground lines must provide a clear indication of the switch contact position, and the switch handle must be marked with operating directions. The recommendation in Rule 381C is similar to Rule 216D. Uniform switch handle positions throughout the system can help minimize errors. Rule 381 makes this statement as a recommendation, a provision that is desirable but not mandatory. Rule 216D makes this statement as part of the Code rule but permits marking as an option to uniform handle positions. The rules for underground switches are outlined in Fig. 381-1.

Remotely controlled or automatic switching devices such as pad-mounted vacuum fault interrupt switches must have provisions on the equipment to render remote or automatic controls inoperable. This feature protects workers from accidental operation or energization during maintenance.

When applying equipment containing fuses and interrupting contacts, the following must be considered during operation:
• Normal conditions
• Emergency conditions
• Fault conditions

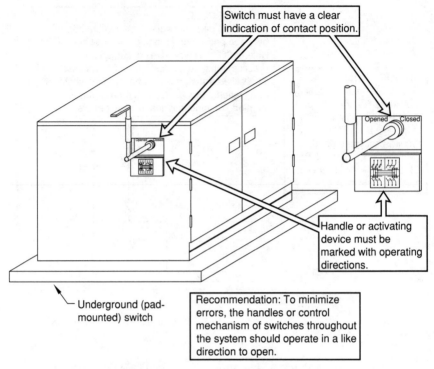

Switch must have a clear indication of contact position.

Handle or activating device must be marked with operating directions.

Underground (pad-mounted) switch

Recommendation: To minimize errors, the handles or control mechanism of switches throughout the system should operate in a like direction to open.

Fig. 381-1. Design of switches (Rule 381C).

When tools such as insulated shotgun sticks are used to handle energized devices in underground equipment, physical space or barriers must provide adequate clearance from ground or between phases.

Rule 381G provides the locking and access requirements for pad-mounted and other aboveground equipment. Pad-mounted and other aboveground equipment is typically not fenced in like substation equipment and it is not elevated like overhead equipment. Pad-mounted equipment has more exposure to the public than most other supply and communication facilities. Rule 381G1 applies to both supply and communication pad-mounted equipment. Pad-mounted and other aboveground equipment must have an enclosure that is locked or otherwise secured. The purpose of this is to keep out unauthorized persons (i.e., the public). Typical locking or securing methods used by both supply and communication utilities include keyed padlocks, disposable locks that can be cut with bolt cutters, penta head bolts, tamper-proof bolts, etc. Rule 381G2 applies to supply pad-mounted equipment over 600 V. This requirement applies to almost all supply utility transformers, switches, junction boxes, etc., except for low-voltage secondary enclosures. Rule 381G2 requires two separate conscious acts to access exposed live parts. The first act is opening the pad-mounted enclosure that was locked or secured in Rule 381G1. It does not matter how many locks or penta head bolts are handled to open the enclosure; opening the enclosure is the first conscious act. The second conscious act must be the opening of a door (not the enclosure door in the first conscious act) or the removal of a barrier.

A pad-mounted switch with fuses is a good example of a device that requires two conscious acts to access exposed live parts. After opening the enclosure door (the first conscious act), a separate steel door or a separate fiberglass barrier exists that must be opened or removed (the second conscious act) before the live parts are exposed. A slightly different example is a typical pad-mounted service transformer that uses load break elbows for terminating the high-voltage cables. This type of transformer does not have any exposed live parts over 600 V and is usually referred to as "dead front." Opening the enclosure door is the first conscious act. No exposed live parts in excess of 600 V exist when the transformer enclosure is opened. The transformer secondary lugs (i.e., 120/240 V) may be exposed but they are under 600 V. The pulling of the insulated elbow is the second conscious act. The barrier for the second conscious act is the insulated elbow itself. See Fig. 381-2.

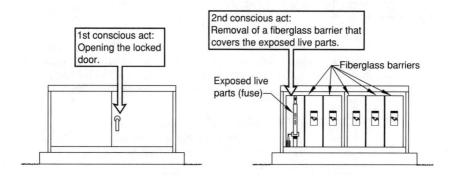

Pad Mounted-Fused Switch

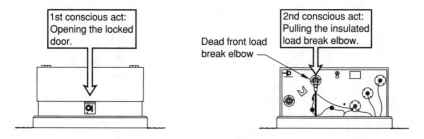

Pad-Mounted Service Transformer

Fig. 381-2. Examples of access to exposed live parts in excess of 600 V (Rule 381G).

The recommendation in Rule 381G2 states that a safety sign be visible when the first door or barrier is opened or removed. The ANSI Z535 series of signing standards are referenced in a note. Signage on pad-mounted equipment is just as critical as signage for an electric supply station. Some utilities use a warning sign on the outside of the pad-mounted enclosure and a danger sign on the inside of the pad-mounted enclosure. This approach uses the ANSI Z535 philosophy that warning is appropriate on the outer barrier (enclosure) and if that barrier is breached, a danger sign is then appropriate. Utilities should consult the ANSI documents, federal or state regulatory agencies, and the utility's insurance company for signing application recommendations. It is important to note that **NESC** Rule 381G2 does not require a safety sign on the outside of pad-mounted equipment. Safety signs are only a RECOMMENDATION (see Rule 015) for the inside of the equipment. See the discussion and the figure in Rule 110A for additional information on ANSI signing requirements. The safety sign requirements for pad-mounted equipment are outlined in Fig. 381-3.

382. LOCATION IN UNDERGROUND STRUCTURES

When equipment is located in the underground structures (i.e., manholes and vaults) discussed in Rule 320, the equipment must not obstruct the personnel access openings discussed in Rule 323C. The equipment must also not impede the egress of a person working in the manhole or vault per Rule 382A.

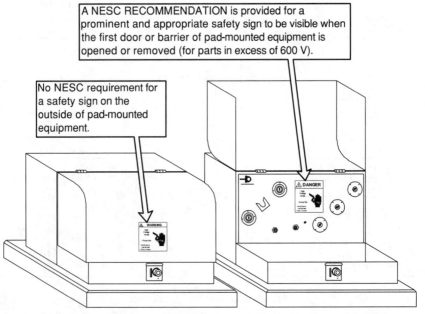

A NESC RECOMMENDATION is provided for a prominent and appropriate safety sign to be visible when the first door or barrier of pad-mounted equipment is opened or removed (for parts in excess of 600 V).

No NESC requirement for a safety sign on the outside of pad-mounted equipment.

See Photo(s)

Fig. 381-3. Safety sign requirements for pad-mounted equipment (Rule 381G2).

The only dimension provided in Rule 382 is the 8-in clearance from equipment to the back of a fixed ladder. Fixed ladders are not required except for the special conditions specified in Rules 323C1 and 323C4. Ladder requirements are also discussed in Rules 323C5 and 323F. No dimension is given from equipment to the front of the ladder. The clearance to the front of the ladder must meet the general requirement that the equipment must not interfere with the proper use of the ladder. The remaining clearance requirements in this rule are also general wording requirement without any stated dimensions. Equipment arrangement must consider installation, operation, and maintenance requirements. Switching equipment must be operable from a safe position. Equipment must not interfere with the drainage or ventilation of the underground structure.

Although Rule 382 does not reference Rule 341B2 and NESC Table 341-1, the clearances in this table apply between joint-use equipment.

383. INSTALLATION

The installation requirements for underground equipment (both pad-mounted equipment and equipment installed in manholes and vaults) are very general. No dimensions are provided for any of the requirements. Equipment installation must consider the following:

- Equipment weight (lifting, rolling, and mounting considerations)
- Guarding or isolating live parts from persons
- Easy access to operate, inspect, and test facilities
- Isolating or protecting live parts from conductive liquids or other materials
- Locking or securing operating controls of supply equipment

384. GROUNDING AND BONDING

Rules 384A and 384B have grounding requirements similar to the general grounding requirements in Rule 314.

Rule 384C requires bonding between aboveground metallic power and communications apparatus that are 6 ft or less apart. Examples of the term apparatus are provided in the rule. The intent of this rule is to not have different potentials between adjacent metallic enclosures. Bonding between fiberglass enclosures or a metallic enclosure and a fiberglass enclosure is not required. A size is not specified for the bonding jumper. Rule 012C, which requires accepted good practice, must be used. Rule 099C, which requires an AWG No. 6 copper bond, is one option of an accepted good practice in this case. The 6-ft spacing is an average reach for an adult person. Rule 384C clarifies that a pole ground is not be required to be bonded to an adjacent metallic communications pedestal. The rules for bonding aboveground metallic supply and communications apparatus are outlined in Fig. 384-1.

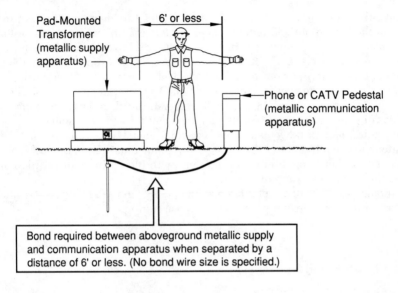

Pad-Mounted Transformer (metallic supply apparatus) —

6' or less

Phone or CATV Pedestal (metallic communication apparatus)

Bond required between aboveground metallic supply and communication apparatus when separated by a distance of 6' or less. (No bond wire size is specified.)

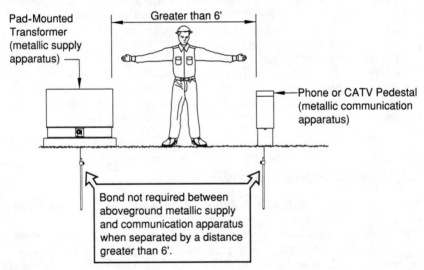

Pad-Mounted Transformer (metallic supply apparatus) —

Greater than 6'

Phone or CATV Pedestal (metallic communication apparatus)

Bond not required between aboveground metallic supply and communication apparatus when separated by a distance greater than 6'.

See Photo(s)

Fig. 384-1. Bonding aboveground metallic supply and communication apparatus (Rule 384C).

385. IDENTIFICATION

Identification of equipment that operates in multiple eliminates confusion and adds safety.

Section 39

Installation in Tunnels

390. GENERAL

Installation in tunnels must meet applicable rules found in Part 3 and the additional rules of this section. Rule 323E, Vault and Utility Tunnel Access, provides additional requirements. If unqualified persons (i.e., the general public) have access to the tunnel, the applicable requirements of Part 2 must be met. The Code is referring to a tunnel that could be a traffic tunnel on a roadway with a utility line passing through it, or the tunnel could be a dedicated utility tunnel under the surface of the ground. Utility tunnels can typically be found connecting buildings on university campuses. In some cases, the tunnels are used just for utilities accessible to qualified personnel. In other cases, the utility tunnel doubles as a public walkway. In addition to the NESC rules, the National Electrical Code (NEC) can be referenced for accepted good practice. A utility tunnel used as a public walkway between buildings will require adherence to the National Electrical Code (NEC). No matter what type of tunnel is involved, all parties concerned must agree on the design of the tunnel structure and the design of the utilities within the structure.

391. ENVIRONMENT

If the tunnel is accessible to the public or workers, the environment within the tunnel must be suitable for people. If the tunnel is not accessible to the public or workers, the construction would be similar to duct bank construction where cables are pulled in and out of ducts but access is only obtainable at pulling or splice locations. Rule 391A provides requirements for general environmental safety and working space. Rule 391B provides additional rules for joint-use (supply and communication) tunnels.

Part 4

Rules for the Operation of Electric Supply and Communications Lines and Equipment

GENERAL SECTIONS

01 INTRODUCTION   02 DEFINITIONS

03 REFERENCES   09 GROUNDING METHODS

GENERAL SECTIONS 01, 02, 03, 09

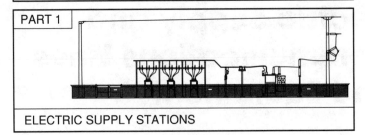

PART 1

ELECTRIC SUPPLY STATIONS

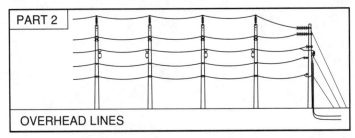

PART 2

OVERHEAD LINES

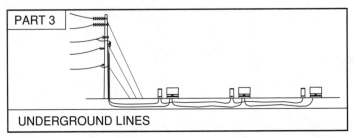

PART 3

UNDERGROUND LINES

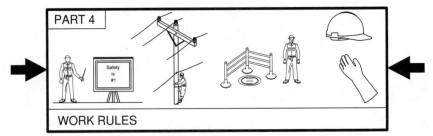

PART 4

WORK RULES

Section 40

Purpose and Scope

400. PURPOSE

The purpose of Part 4, Work Rules, is similar to the purpose of the entire NESC outlined in Rule 010, except Rule 400 is specific to the work rules for the operation of electric supply and communication lines and equipment. Part 4 of the NESC focuses on practical work rules as a means of safeguarding employees and the public. The Code states that the intent of Part 4 is not to require unreasonable steps to comply with the rules; however, reasonable steps must be taken. A more specific statement is given in Rule 410A. Rule 410A requires that the work rules be used, but if strict enforcement of the work rules seriously impedes safety, the employee in charge may temporarily modify the work rules without increasing hazards. This flexibility is needed because every conceivable situation cannot be covered in the work rules. Flexibility in applying the work rules cannot be abused in situations where the work rules apply.

401. SCOPE

The scope of Part 4, Work Rules, includes work rules to be used in the installation, operation, and maintenance of both electric supply and communications systems. Part 4 is broken down into four sections that are all interrelated. The four sections are outlined below:

- Section 41, "Supply and Communications Systems—Rules for Employ*ers.*" These rules apply to the supply and communications company.
- Section 42, "General Rules for Employ*ees.*" These rules apply to the supply and communications employee working for the supply and communications company.
- Section 43, "Additional Rules for *Communications* Employ*ees.*" These rules are additional rules for communications employees only. Note that Sec. 42 was for both supply and communications employees.
- Section 44, "Additional Rules for *Supply* Employ*ees.*" These rules are additional rules for the supply employees only. Note that Sec. 42 was for both supply and communications employees.

The titles of **NESC** Sections 41, 42, 43, and 44 are graphically represented in Fig. 401-1.

The format of this Handbook for Part 4, Work Rules, is different from the format of the other parts of this book. The format for Part 4 summarizes the **Code** text in an outline bulleted list. A graphic is used to visually represent which section, rule, and paragraph the outline applies to. A brief general discussion of each section is provided at the beginning of each section.

Occupational Safety and Health Administration (OSHA) regulations also apply to the operation, maintenance, and construction of electric supply and

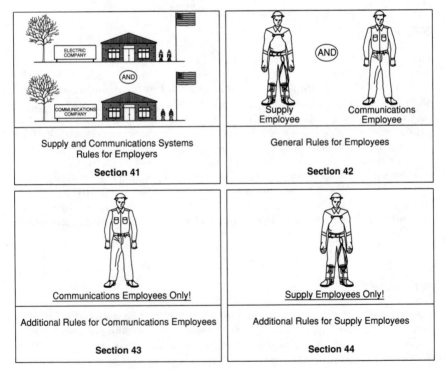

Fig. 401-1. Titles of Sections 41, 42, 43, and 44 (Rule 401).

communications systems. See Rule 402 for a discussion of applicable OSHA standards.

402. REFERENCED SECTIONS

This rule references four **NESC** sections related to Part 4, Work Rules, so that rules do not have to be duplicated and the reader of the **Code** realizes that other sections are related to the information provided in Part 4. The related sections are:

- Introduction, Sec. 01
- Definitions, Sec. 02
- References, Sec. 03
- Grounding Methods, Sec. 09

The Work Rules in Part 4 of the **NESC** have been coordinated with (but they are not exactly the same as) the Occupational Safety and Health Administration (OSHA) regulations related to electric supply and communications systems.

The scope of this **NESC** Handbook does not include specific comments on the OSHA regulations. Many utilities tend to place more emphasis on the OSHA regulations than the work rules in Part 4 of the **NESC**, as OSHA performs both routine and accident investigations. In general, utilities are required to meet the **NESC** rules and the OSHA standards; therefore, both apply, not one or the other. The work rules in **NESC** Part 4 are tied to the rules in other parts of the **NESC**.

The OSHA standards related to the **NESC** Work Rules are listed below and are reprinted in Appendix B of this Handbook as a reference.

- 1910.268 Telecommunications (Operation and Maintenance)
- 1910.269 (including Appendix A–E) Electric Power Generation, Transmission, and Distribution (Operation and Maintenance)
- 1926.950 through 1926.960 Power Transmission and Distribution (Construction)

The OSHA 1910 series is for operation and maintenance. OSHA Standards 1910.268 and 1910.269 are listed under 1910 Subpart R—Special Industries. The OSHA 1926 series is for construction. OSHA Standards 1926.950 through 1926.960 are listed under 1926 Subpart V—Power Transmission and Distribution. OSHA uses the terms electric power and telecommunications instead of the **NESC** terms electric supply and communications. OSHA separates its standards into operation and maintenance standards and construction standards, the **NESC** does not. OSHA has separate operations and maintenance standards and construction standards for the electric power industry. OSHA only has operation and maintenance standards for the telecommunications industry. When OSHA does not have a construction standard for a specific industry, as in the case of telecommunications, the general OSHA construction standards apply. Where specific rules do not exist in the general OSHA construction standards, applying the 1910.268 operation and maintenance standards to construction work would be one way of meeting OSHA's General Duty Clause.

The OSHA 1910.268 Telecommunications (Operations and Maintenance) standard became effective in 1975. The OSHA 1910.269 Electric Power Generation, Transmission, and Distribution (Operations and Maintenance) standard became effective in 1994. The OSHA 1926.950 through 1926.960 Power Transmission and Distribution (Construction) standards became effective in 1972. Amendments have been made to the standards over the years. The OSHA 1926.950 through 1926.960 Power Transmission and Distribution (Construction) standards are currently (at the time this Handbook is being published) in the process of a major revision. This revision will also affect the OSHA 1910.269 Electric Power Generation, Transmission, and Distribution (Operations and Maintenance) standard as the two standards are closely related. As of the publication date of this Handbook, the updated OSHA standards have not been finalized

There are several other OSHA standards referenced throughout 1910.268, 1910.269, and 1926.950 through 1926.960. These related standards are referenced either without amendment or referenced and supplemented with additional information. The most commonly referenced related standards are listed below:

- 1910 Subpart S—Electrical (1910.301 to 1910.399)
- 1910.151—Medical Services and First Aid (Part of 1910 Subpart K)
- 1910.147—The Control of Hazardous Energy (Lockout/Tagout) (Part of 1910 Subpart J)
- 1910 Subpart I—Personal Protective Equipment (1910.132 to 1910.138)
- 1910.137—Electrical Protective Devices (Part of 1910 Subpart I)
- 1926 Subpart M—Fall Protection (1926.500 to 1926.503)
- 1926 Subpart P—Excavations (1926.650 to 1926.652)
- 1910 Subpart D—Walking-Working Surfaces (1910.21 to 1910.30)
- 1910.67 Vehicle-Mounted Elevating and Rotating Work Platforms (Part of 1910 Subpart F)

OSHA standards are very easy to access from the Internet. The phone numbers and addresses of OSHA offices are also easily accessed from the Internet. Each standard also has a list of requested interpretations. Each interpretation lists the question asked and the OSHA response. All of this information is available at www.osha.gov.

Section 41

Supply and Communications Systems—Rules for Employers

The following general discussion applies to Secs. 41 and 42. This discussion is repeated at the beginning of Sec. 42. Both employers and employees have an obligation to safety. Section 41 provides rules for supply and communications employers. Section 42 provides general rules for supply and communications employees. Section 43 provides additional rules for communications employees. Section 44 provides additional rules for supply employees.

Sections 41 and 42 are written to put the responsibility for safety on both the employer (company) and the employee (worker). The employer must designate an employee in charge to represent the company. The employee in charge is responsible for making sure employees adhere to the work rules. The employees must also assume responsibility for following safety rules. This system builds redundancy by putting the safety requirement on both the employer and employee.

Two specific examples of the relationship between Secs. 41 and 42 (employer and employee) are listed below:

- The *employer* must inform each employee of the safety rules (Rule 410A). The *employee* must read and study the safety rules (Rule 420A).
- The *employer* must have an adequate supply of protective devices (e.g., hard hats) and equipment sufficient to enable employees to meet the requirements of the work to be undertaken, and first aid equipment and materials shall be available in readily accessible and, where practical, conspicuous places (Rule 411B).

  The *employee* must use personal protective equipment, the protective devices, and the special tools provided for their work and inspect these devices and tools before starting work to verify that they are in good condition (Rule 420H).

When the employee is required to perform a task, for example, inspecting personal protective equipment, the employer must have a designated employee in charge responsible for making sure the employees perform the inspection. The duties of the designated employee in charge are outlined in Rule 421A.

Section 41 (rules for employers) provides a list of typical protective devices and equipment for employees (who are covered in Sec. 42) to use. The **Code** does not dictate what protective devices and equipment must be used for a particular task. The choice of what protective devices and equipment need to be used is site-specific. The **NESC** cannot cover every conceivable site-specific situation.

Section 41 (rules for employers), Rule 411A2, requires that diagrams (i.e., maps) be maintained and on file for employees (who are covered in Sec. 42) to use. The diagrams aid the identification of structures required in Part 2, Overhead Lines, Rule 217A3, and the location of underground facilities required in Part 3, Underground Lines, Rule 311A. Accurate diagrams are useful for minimizing errors and accidents. Inaccurate maps can increase errors and accidents. Diagrams cannot be used as a substitute for applying proper safety procedures.

Special attention should be given to the employer clothing rules in Sec. 41, Rule 410A3 (and the corresponding rule for employees in Sec. 42, Rule 420I2), which will become effective approximately 2 years into the **Code** cycle on January 1, 2009. These rules deal with performing an arc hazard analysis and selecting appropriate flame-resistant clothing for the hazards involved.

410. GENERAL REQUIREMENTS

410A. General

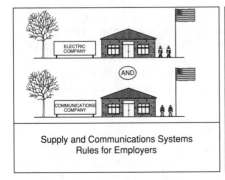

Supply and Communications Systems Rules for Employers

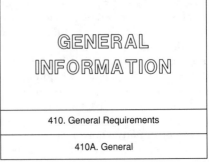

GENERAL INFORMATION

410. General Requirements

410A. General

- Employer must inform each employee of the safety rules.
- Employer must train employees.
- Employer must ensure that each employee has demonstrated proficiency in required tasks.
- Employer must retrain employees who are not following work rules.
- Effective Jan 1, 2009 the employer must ensure an assessment is performed to determine the potential exposure to electric arcs for employees who work on or near energized lines or equipment.

- Assessments that determine a potential employee exposure greater than 2 cal/cm² require the employee to wear clothing or a clothing system that has an effective arc rating at least equal to the anticipated level of arc energy.
- When exposed to an electric arc or flame, the following materials must not be worn:
  - ✓ Acetate
  - ✓ nylon
  - ✓ polyester
  - ✓ polypropylene
- Effective arc ratings of clothing or clothing systems above 1000 volts shall be determined using NESC Tables 410-1 and 410-2 or an arc hazard analysis must be performed.
- An arc hazard analysis calculation of the estimated arc energy must be based on the available fault current, the duration of the arc in cycles, and the distance of the arc to the employee.
- An exception applies if the clothing required creates a hazard greater than the possible exposure to the heat energy of the electric arc.
- An exception applies for secondary systems below 1000 volts. Work rules and engineering controls must be used to limit exposure. In lieu of performing an arc hazard analysis, clothing with an effective arc rating of 4 cal/cm² must be used to limit the likelihood of ignition.
- Note that a clothing system in multiple layers may block more heat than a single layer. The effect of a combination of multiple layers is referred to as the effective arc rating.
- Note that energy levels may be excessive in secondary systems and applicable work rules and engineering controls must be utilized.
- Employers must use procedures to assure compliance with safety rules.
- If strict enforcement of rules seriously impedes safety, the employee in charge may temporarily modify rules as long as hazards are not increased.
- If a disagreement on how to apply operating rules occurs, the employer's decision will be final; however, the decision must not be hazardous to the employee.

410B. Emergency Procedures and First Aid Rules

Supply and Communications Systems
Rules for Employers

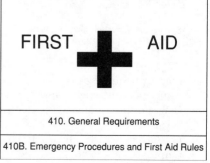

410. General Requirements

410B. Emergency Procedures and First Aid Rules

- Employer must inform employee of emergency procedures and first aid rules including CPR methods.
- Copies of emergency procedures and first aid rules must be kept in vehicles and other locations.
- Employers must regularly instruct employees on the methods of first aid and emergency procedures.

410C. Responsibility

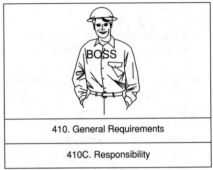

Supply and Communications Systems Rules for Employers

410. General Requirements

410C. Responsibility

- Employer must select a designated person to be in charge of operations and responsible for safety.
- A crew must have only one person in charge.
- For multiple locations, one person may be in charge at each location.

411. PROTECTIVE METHODS AND DEVICES

411A. Methods

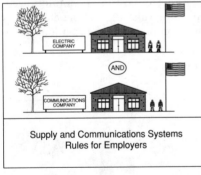
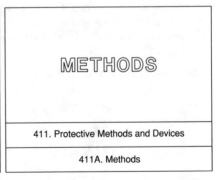

METHODS

Supply and Communications Systems Rules for Employers

411. Protective Methods and Devices

411A. Methods

- Employer must restrict employees from access to energized or rotating equipment unless the employee is authorized.
- Diagrams (i.e., maps) of the electric system must be available to authorized employees.
- Employees are to be instructed before work starts.
- Employees are to be instructed to take additional precautions for unusual hazards.

411B. Devices and Equipment

Supply and Communications Systems
Rules for Employers

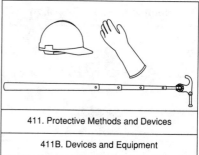

411. Protective Methods and Devices

411B. Devices and Equipment

- Employer must have an adequate supply of protective devices (e.g., hard hats, rubber gloves, insulated tools, body belts, etc.).
- Employer must have an adequate supply of first aid equipment.
- Protective devices must conform to applicable standards.

411C. Inspection and Testing of Protective Devices

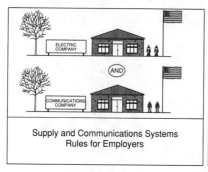

Supply and Communications Systems
Rules for Employers

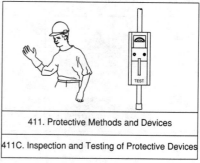

411. Protective Methods and Devices

411C. Inspection and Testing of Protective Devices

- Inspect or test protective devices and equipment.
- Inspect insulating gloves, sleeves, and blankets before use.
- Test insulated gloves and sleeves as required.
- Inspect line worker's body belts, lanyards, positioning straps, etc., to ensure safety.

411D. Signs and Tags for Employee Safety

Supply and Communications Systems
Rules for Employers

411. Protective Methods and Devices

411D. Signs and Tags for Employee Safety

- Safety signs and tags must meet ANSI Z535 Standards.

411E. Identification and Location

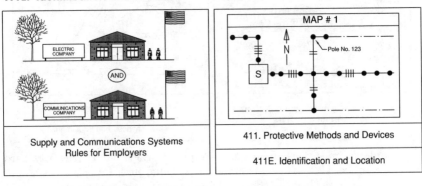

- Provide a means to identify lines before they are worked on.
- Be able to locate underground facilities.

411F. Fall Protection

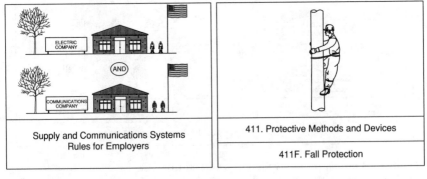

- Employer must develop, implement, and maintain an effective fall protection program.
- The fall protection program must include all of the following:
  - ✓ Training, retraining, and documentation
  - ✓ Guidance on equipment selection, inspection, care, and maintenance
  - ✓ Considerations concerning structural design and integrity, with particular reference to anchorages and their availability
  - ✓ Rescue plans and related training
  - ✓ Hazard recognition
- The employer must not permit employees to use 100 percent leather positioning straps.

Section 42

General Rules
for Employees

The following general discussion applies to Secs. 42 and 41. This discussion is repeated at the beginning of Sec. 41. Both employers and employees have an obligation to safety. Section 41 provides rules for supply and communications employers. Section 42 provides general rules for supply and communications employees. Section 43 provides additional rules for communications employees. Section 44 provides additional rules for supply employees.

Sections 41 and 42 are written to put the responsibility for safety on both the employer (company) and the employee (worker). The employer must designate an employee in charge to represent the company. The employee in charge is responsible for making sure employees adhere to the work rules. The employees must also assume responsibility for following safety rules. This system builds redundancy by putting the safety requirement on both the employer and employee.

Two specific examples of the relationship between Secs. 41 and 42 (employer and employee) are listed below:

- The *employer* must inform each employee of the safety rules (Rule 410A). The *employee* must read and study the safety rules (Rule 420A).
- The *employer* must have an adequate supply of protective devices (e.g., hard hats) and equipment, sufficient to enable employees to meet the requirements of the work to be undertaken, and first aid equipment and materials shall be available in readily accessible and, where practical, conspicuous places (Rule 411B).

The *employee* must use personal protective equipment, the protective devices, and the special tools provided for their work and inspect these devices and tools before starting work to verify that they are in good condition (Rule 420H).

When the employee is required to perform a task, for example, inspecting personal protective equipment, the employer must have a designated employee in charge responsible for making sure the employees perform the inspection. The duties of the designated employee in charge are outlined in Rule 421A.

Section 41 (rules for employers) provides a list of typical protective devices and equipment for employees (who are covered in Sec. 42) to use. The Code does not dictate what protective devices and equipment must be used for a particular task. The choice of what protective devices and equipment need to be used is site-specific. The NESC cannot cover every conceivable site-specific situation.

Section 41 (rules for employers), Rule 411A2, requires that diagrams (i.e., maps) be maintained and on file for employees (who are covered in Sec. 42) to use. The diagrams aid the identification of structures required in Part 2, Overhead Lines, Rule 217A3, and the location of underground facilities required in Part 3, Underground Lines, Rule 311A. Accurate diagrams are useful for minimizing errors and accidents. Inaccurate maps can increase errors and accidents. Diagrams cannot be used as a substitute for applying proper safety procedures.

Special attention should be given to the employer clothing rules in Sec. 41, Rule 410A3 (and the corresponding rule for employees in Sec. 42, Rule 420I2), which will become effective approximately 2 years into the Code cycle on January 1, 2009. These rules deal with performing an arc hazard analysis and selecting appropriate flame-resistant clothing for the hazards involved.

420. PERSONAL GENERAL PRECAUTIONS

420A. Rules and Emergency Methods

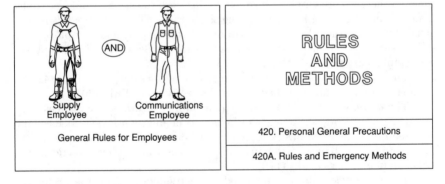

- Read and study safety rules.
- Show knowledge of safety rules.
- Be familiar with first aid, rescue techniques, and fire extinguishing.

420B. Qualification of Employees

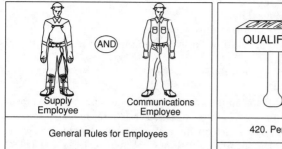

- Employees must only perform tasks for which they are trained, equipped, authorized, and directed.
- Inexperienced employees must work under experienced employees and perform only directed tasks.
- Employees operating mechanical equipment must be qualified to perform those tasks.
- If safety is in doubt, request instructions from supervisor.
- Employees who only occasionally work on electric supply lines can only do work when authorized.

420C. Safeguarding Oneself and Others

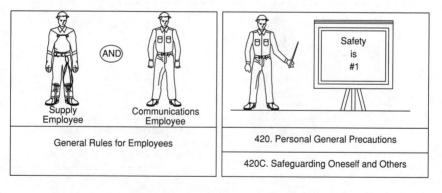

- Heed safety signs and signals.
- Warn others who are in danger or near energized lines.
- Report line or equipment defects (e.g., low clearance, broken insulators, etc.).
- Report accidentally energized items.
- Report any defect that may cause danger.
- Employees who do not work on lines and equipment must keep away from them and keep away from worksites with falling objects.

- Employees who work on energized lines must:
  - ✓ Consider the effects of their actions.
  - ✓ Account for their own safety.
  - ✓ Account for the safety of other employees on the job site.
  - ✓ Account for the safety of other employees remote from the job site but affected by the work.
  - ✓ Account for the property of others.
  - ✓ Account for the public.
- Communications employees must not approach energized parts closer than the approach distances shown in Rule 431 (**NESC** Table 431).
- Communications employees must not take conductive objects near energized parts closer than the approach distances shown in Rule 431 (**NESC** Table 431).
- Supply employees must not approach energized parts closer than the approach distances shown in Rule 441 (**NESC** Tables 441-1 through 441-4).
- Supply employees must not take conductive objects near energized parts (without an insulating handle) closer than the approach distances shown in Rule 441 (**NESC** Tables 441-1 through 441-4).
- Employees must use care when working with metal ropes, tapes, or wires in the vicinity of energized high-voltage lines due to energization and induced voltages.
- Clearance measurements from energized lines must be done with approved devices (e.g., insulated measuring sticks).

420D. Energized or Unknown Conditions

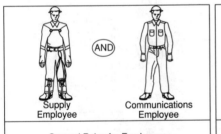

Supply Employee   Communications Employee
General Rules for Employees

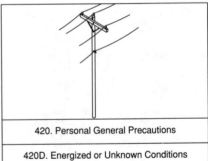

420. Personal General Precautions
420D. Energized or Unknown Conditions

- Consider equipment and lines to be energized unless they are positively known to be de-energized.
- Determine existing conditions before starting work by inspection or tests.
- Determine the operating voltage of equipment and lines before starting work.

420E. Ungrounded Metal Parts

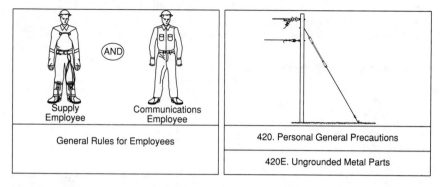

- Consider ungrounded metal parts energized at the highest voltage to which they are exposed.

420F. Arcing Conditions

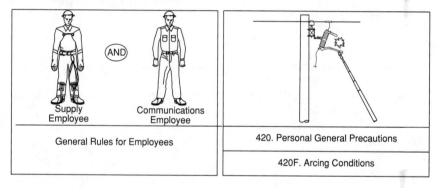

- Keep body parts far away from devices that produce arcs during operation, such as switches.

420G. Liquid-Cell Batteries

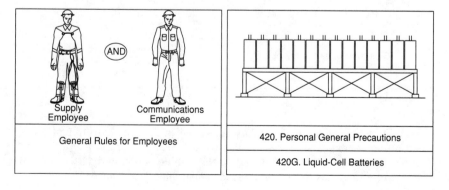

- Determine that battery areas are adequately ventilated.
- Avoid smoking, open flames, or tools that produce sparks.
- Use eye and skin protection during handling.
- Take precautions to avoid short circuits and electric shocks.

420H. Tools and Protective Equipment

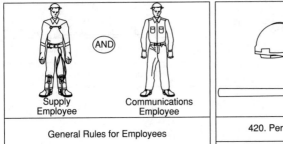

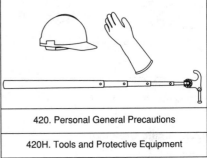

- Use the personal protective equipment, devices, and tools provided for the work.
- Before starting work, carefully inspect the personal protective equipment, devices, and tools to verify they are in good condition.

420I. Clothing

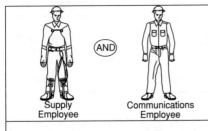

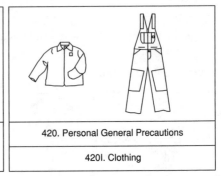

- Wear clothing suitable for the assigned task and work environment.
- Employees exposed to an electric arc must wear clothing or a clothing system in accordance with Rule 410A (**NESC** Tables 410-1 and 410-2).
- Avoid wearing exposed metal articles near energized lines.

420J. Ladders and Supports

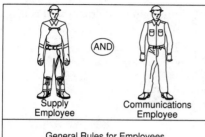

Supply Employee	Communications Employee
General Rules for Employees	

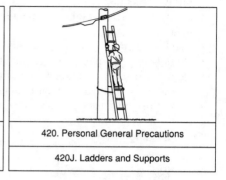

420. Personal General Precautions
420J. Ladders and Supports

- Verify ladders, aerial lifts, etc., are strong, in good condition, and secure before using them.
- Do not paint portable wood ladders except with a clear nonconductive coating.
- Do not reinforce portable wood ladders with metal.
- Do not use portable metal ladders near energized parts.
- Conductive portable ladders for specialized work must only be used for the work intended.

420K. Fall Protection

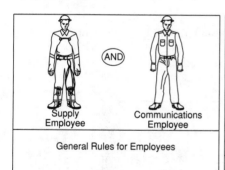

Supply Employee	Communications Employee
General Rules for Employees	

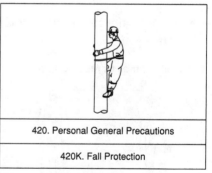

420. Personal General Precautions
420K. Fall Protection

- Climbers must use fall protection systems above 10 ft. This includes aerial bucket trucks, helicopters, and other elevated items.
- Qualified climbers can be unattached while climbing, transferring, or transitioning.
- Unqualified climbers must remain attached while climbing, transferring, or transitioning.
- Fall protection equipment must be inspected before use.
- Fall arrest equipment must be suitably anchored.
- Determine that the fall protection system is engaged and secure.

- Be aware of accidental disengagement of the snap hook from the D-ring by foreign objects.
- Be aware of accidental disengagement of the snap hook from the D-ring by rollout.
- Use locking snap hooks and compatible hardware.
- Do not connect snap hooks to each other.
- Do not use 100 percent leather positioning straps.
- Use wire rope lanyards where the lanyard could be cut.
- Do not use wire rope lanyards near energized lines.

420L. Fire Extinguishers

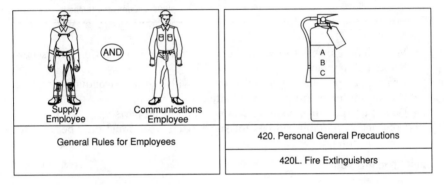

- Use fire extinguishers or materials rated for energized parts or de-energize the parts first.

420M. Machines or Moving Parts

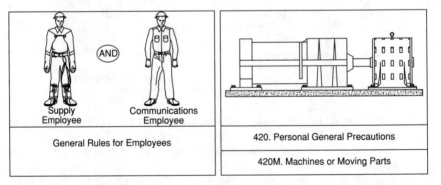

- When working on moving parts, verify accidental startup will not occur by using lockout/tagout procedures.
- When working on automatic switches, stay clear of moving parts.

420N. Fuses

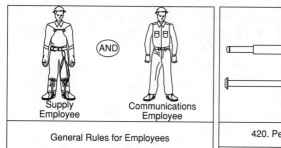
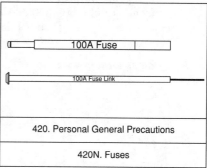

General Rules for Employees	420. Personal General Precautions
	420N. Fuses

- Use insulated gloves or tools when installing fuses on energized lines.
- Use eye protection and stand clear when installing expulsion-type fuses on energized lines.

420O. Cable Reels

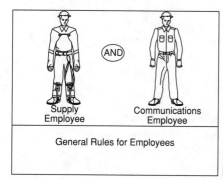
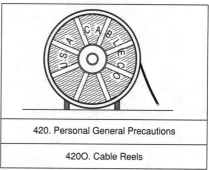

General Rules for Employees	420. Personal General Precautions
	420O. Cable Reels

- Block cable reels so they do not accidentally roll.

420P. Street and Area Lighting

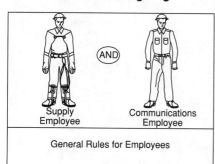
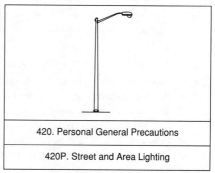

General Rules for Employees	420. Personal General Precautions
	420P. Street and Area Lighting

- Periodically examine lowering ropes or chains, supports, and fastenings.
- A device must be provided to safely disconnect each lamp on a series lighting circuit of more than 300 V before the lamp is handled.
- An exception applies when insulated devices or tools are used and the circuit is treated as a full-voltage circuit.

420Q. Communication Antennas

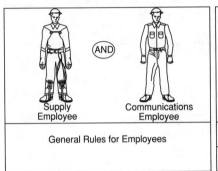

Supply Employee **AND** Communications Employee

General Rules for Employees

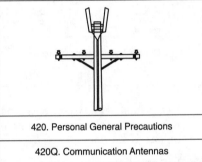

420. Personal General Precautions

420Q. Communication Antennas

- Employees working near communication antennas must not be exposed to radiation levels exceeding regulatory requirements.
- Regulatory standards in OSHA, FCC, and IEEE can be referenced for radiation level limits.

421. GENERAL OPERATING ROUTINES

421A. Duties of a First-Level Supervisor or Person in Charge

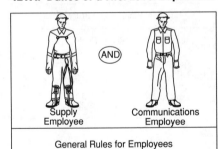

Supply Employee **AND** Communications Employee

General Rules for Employees

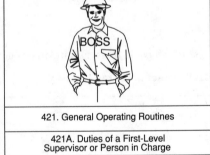

421. General Operating Routines

421A. Duties of a First-Level Supervisor or Person in Charge

- Duties of the individual in charge:
    - ✓ Adopt precautions to prevent accidents.
    - ✓ See that employees are observing safety rules and operating procedures.
    - ✓ Keep records and make reports.
    - ✓ Prevent unauthorized persons from approaching the workplace.
    - ✓ Do not allow tools or devices unsuitable for the work.
    - ✓ Do not allow tools or devices to be used without testing or inspecting first.

421B.   Area Protection

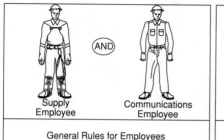

Supply Employee   AND   Communications Employee

General Rules for Employees

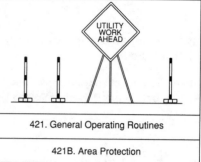

421. General Operating Routines

421B. Area Protection

- Areas accessible to vehicular and pedestrian traffic:
  - ✓ Prevent vehicles and pedestrians from approaching the work site.
  - ✓ Warn the public of openings or obstructions.
  - ✓ Openings or obstructions exposed at night must have warning lights and must be enclosed with protective barricades.
- Areas accessible to employees only:
  - ✓ If the work exposes energized or moving parts that are normally protected, safety signs must be displayed.
  - ✓ If the work exposes energized or moving parts that are normally protected, barricades must be erected.
  - ✓ Work on one section of a switchboard with multiple sections or work on one portion of a substation with several portions requires barriers to prevent contact with energized parts.
- Locations with crossed or fallen wires:
  - ✓ If an employee encounters crossed or fallen wires, the employee must remain on guard or use other means to prevent accidents.
  - ✓ The proper authority must be notified.
  - ✓ If qualified, and if safety rules can be met, the employee may correct the condition.

421C.   Escort

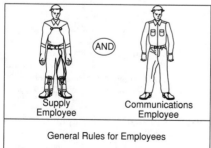

Supply Employee   AND   Communications Employee

General Rules for Employees

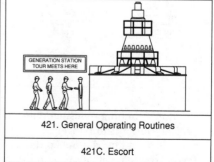

GENERATION STATION TOUR MEETS HERE

421. General Operating Routines

421C. Escort

- An employee responsible for safety must escort nonqualified employees or visitors near electrical lines or equipment.

422. OVERHEAD LINE OPERATING PROCEDURES

422A. Setting, Moving, or Removing Poles in or near Energized Electric Supply Lines

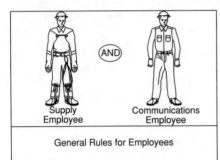

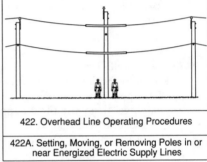

- Avoid direct contact of the pole with the energized conductors.
- Wear insulating gloves.
- Do not contact the pole with uninsulated body parts.
- Avoid touching trucks or other equipment not bonded to an effective ground unless suitable protective equipment is used.

422B. Checking Structures Before Climbing

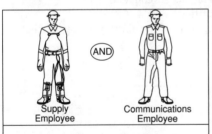

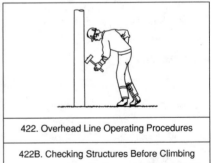

- Before climbing poles, ladders, etc., verify the structure is capable of handling the additional weight and unbalanced forces.
- Poles must not be climbed if unsafe unless guying, bracing, or other means are used to create a safe condition.

422C. Installing and Removing Wires or Cables

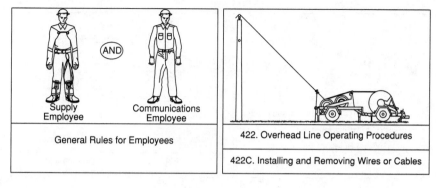

Supply Employee AND Communications Employee

General Rules for Employees

422. Overhead Line Operating Procedures

422C. Installing and Removing Wires or Cables

- Wires being installed or removed must be kept clear from energized wires.
- Wires being installed or removed that are not bonded to an effective ground must be considered energized.
- Control sag of wires being installed or removed to prevent pedestrian and vehicle traffic damage.
- Verify structures can handle the forces associated with installing or removing wires.
- Avoid contact with moving winch lines.
- Consider the effect of a higher-voltage line on a lower-voltage line. Verify that the line being worked on is free from dangerous leakage and induction voltages or verify that it is effectively grounded.

423. UNDERGROUND LINE OPERATING PROCEDURES

423A. Guarding Manhole and Street Openings

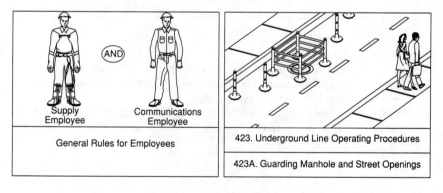

Supply Employee AND Communications Employee

General Rules for Employees

423. Underground Line Operating Procedures

423A. Guarding Manhole and Street Openings

- Open manholes, handholes, and vaults must be protected with a barrier, temporary cover, or guard.

423B. Testing for Gas in Manholes and Unventilated Vaults

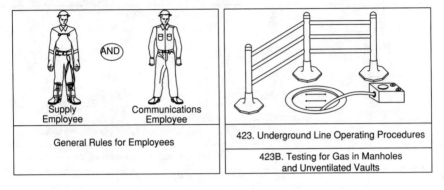

Supply Employee **AND** Communications Employee

General Rules for Employees

423. Underground Line Operating Procedures

423B. Testing for Gas in Manholes and Unventilated Vaults

- Test manhole for combustible or flammable gases before entry.
- If combustible or flammable gases exist, ventilate before entry.
- Test for oxygen deficiency.
- Make provisions for a good air supply during work.

423C. Flames

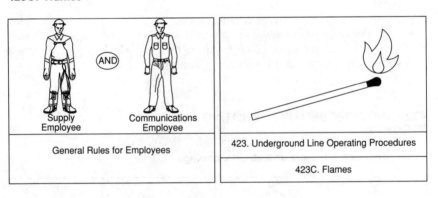

Supply Employee **AND** Communications Employee

General Rules for Employees

423. Underground Line Operating Procedures

423C. Flames

- Do not smoke in manholes.
- Use extra precaution to ensure ventilation when flames are required for work.
- Test excavation areas for combustible gases or liquids (e.g., near a gasoline service station) before using open flames.
- Provide adequate air space or a barrier to protect lines that transport flammable material when flames are required for work and the lines are exposed.

423D. Excavation

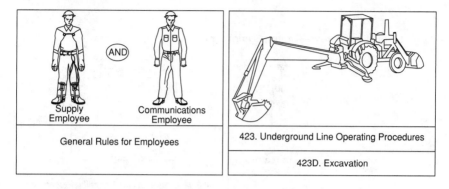

Supply Employee (AND) Communications Employee

General Rules for Employees

423. Underground Line Operating Procedures

423D. Excavation

- Locate buried utilities prior to excavation.
- Existing utilities should be exposed where the bore path of guided boring or direction drilling machines crosses existing utilities.
- Hand tools used for manual excavation near supply cables must have non-conductive handles.
- Hand digging must be used when close to cables or other utilities.
- If lines that transport flammable material are broken or damaged, the employee must:
  - ✓ Leave the excavation open.
  - ✓ Extinguish all flames.
  - ✓ Notify the proper authority.
  - ✓ Keep the public away.
- When an employee is working in a trench or excavation in excess of 5 ft deep, or when a trench or excavation presents a cave-in hazard, shoring, sloping, or shielding methods must be used for employee protection.

423E. Identification

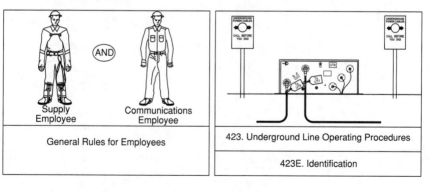

Supply Employee (AND) Communications Employee

General Rules for Employees

423. Underground Line Operating Procedures

423E. Identification

- Identify and protect exposed buried utilities.
- When working on one cable, protect other cables from damage.
- Before cutting a cable or opening a splice, verify its identity.

- Where multiple cables exist, the cable to be worked on must be positively identified.

423F. Operation of Power-Driven Equipment

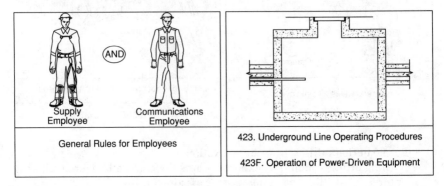

Supply Employee (AND) Communications Employee

General Rules for Employees

423. Underground Line Operating Procedures

423F. Operation of Power-Driven Equipment

- Keep out of manholes when power rodding.

Section 43

Additional Rules for Communications Employees

The following general discussion applies to Secs. 43 and 44. This discussion is repeated at the beginning of Sec. 44. Sections 43 and 44 provide additional rules for employees. Section 43 provides additional rules for communications employees only. Section 44 provides additional rules for supply employees only. Section 41 provides rules for supply and communications employers and Sec. 42 provides rules for supply and communications employees.

The additional rules in Sec. 43 for communications employees primarily focus on keeping the communications employee safe when the communication lines are on joint-use structures or in joint-use manholes with electric supply conductors. Section 43 requires that communications employees maintain minimum approach distances between the communications employee and electric supply lines and equipment. In addition to the minimum approach distances to electric supply conductors, communications employees must not position themselves above the lowest electric supply conductor exclusive of vertical runs (risers) and street lights.

The additional rules in Sec. 44 for supply employees primarily focus on minimum approach distances to energized parts, switching control procedures, work on energized lines, de-energizing lines, protective grounding, and live-line work.

If a communications line is located in the supply space in accordance with the overhead line clearance rules in Sec. 23, the worker who enters the supply space to work on the communications line must be trained as a supply employee.

If a communications line is located below a supply line on an overhead structure and the proper clearances in Sec. 23 are met, the worker who is maintaining the communications line must be trained as a communications employee.

If a communications line is positioned below a supply line on an overhead structure, but the proper clearances in Sec. 23 are not met, the communications employee can correct the violation if the communications employee does not violate the minimum approach distances and other requirements in Sec. 43. If the communications employee cannot maintain the minimum approach distances in Sec. 43, the communications employee must contact a supply employee to correct the violation.

For example, the minimum approach distance rules in Sec. 43 (NESC Table 431-1) require the communications worker to have a 2 ft–2 in minimum approach distance to a 12.47/7.2 kV line. A common supply to communications clearance value in Rules 235 and 238 is 40 in (see Rules 235 and 238 for specific applications). This value is intended to provide enough space on the pole for the communications worker's hands to be on the communications cable and the communication worker's head to be slightly above the communications cable and still maintain the 2 ft–2 in minimum approach distance between the communications worker and the 12.47/7.2 kV line. If a violation exists on the pole and the 40 in clearance requirement between supply and communications is only 38 in, the communications worker can probably lower the communications cable down 2 in (from 38 in to 40 in) without violating the 2 ft–2 in minimum approach distance between the communications worker and the 12.47/7.2 kV line. However, if the communications cable was in violation and mounted only 1 ft below the 12.47/7.2 kV line (say right under the crossarm brace) the communications worker could not correct this violation. The communications worker would have to call on a trained supply worker to fix the problem. The supply worker has the same 2 ft–2 in minimum approach distance in Sec. 44 (NESC Table 441-1) but the supply worker has additional rules for live-line work that allow the supply worker to work inside the minimum approach distance.

If a communications employee is not a qualified employee (i.e., if the worker is not properly trained in recognizing the hazards involved, is not able to identify the nominal voltage of exposed live parts, is not knowledgeable of the minimum approach distances involved with the work, etc.), then the worker is considered an unqualified employee and the worker must maintain at least a 10-ft distance from energized lines (more than 10 ft for greater than 50 kV) per OSHA Standard 1910.333. The OSHA 10-ft requirement applies to other workers in the vicinity of power lines such as painters, roofers, etc. The NESC does not provide distances for unqualified workers but the OSHA standards do. The bottom line is that communications workers need training in recognizing the hazards involved, identifying the nominal voltage of exposed live parts, knowledge of minimum approach distances involved with the work, etc. or they cannot work on a communications line that is constructed jointly with a power line.

The concept of minimum approach distance is outlined in Fig. 43-1.

Minimum Approach Distance:
- Employees must not approach or bring any conductive object within the minimum approach distance.
- The minimum approach distance contains an electrical component and an inadvertent (i.e., unintentional or unexpected) movement component.
- Communications worker minimum approach distances are found in NESC Table 431-1. (A similar table is found in OSHA 1910.268.)
- Supply worker minimum approach distances are found in NESC Tables 441-1, 441-2, 441-3, and 441-4. (Similar tables are found in OSHA 1910.269.)
- Supply workers can work inside the minimum approach distance if additional live-line rules are followed.
- Communications workers may never approach closer than the minimum approach distance.

Expected Reach of Employee:
The expected reach of the employee includes intentional and expected movements, including such things as adjusting a hard hat, maneuvering tools, reaching for items being passed to the employee, adjusting parts and work positions, etc.

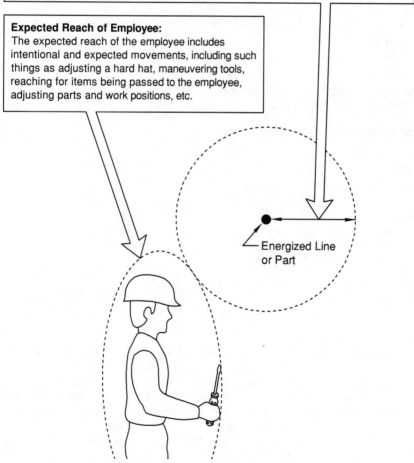

Energized Line or Part

Fig. 43-1. Minimum approach distance (Sec. 43).

430. GENERAL

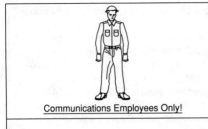

Communications Employees Only!

Additional Rules for Communications Employees

GENERAL INFORMATION

430. General

- Section 42, "General Rules for Employees" (both supply and communications) also apply.

431. APPROACH TO ENERGIZED CONDUCTORS OR PARTS

431A. No Employee Shall Approach...

Communications Employees Only!

Additional Rules for Communications Employees

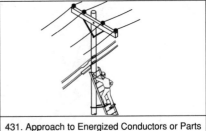

431. Approach to Energized Conductors or Parts

431A. No Employee Shall Approach...

- Communications employees must not approach energized parts closer than the approach distances shown in **NESC** Table 431-1.
- Communications employees must not take conductive objects near energized parts closer than the approach distances shown in **NESC** Table 431-1.
- Communications employees repairing storm damage to communications lines that are joint-use with electric supply lines must:
  - ✓ Treat the supply and communications lines as energized to the highest voltage to which they are exposed, or
  - ✓ Assure that the electric supply lines are de-energized and grounded per the work rules of **NESC** Sec. 44.

431B. Altitude Correction

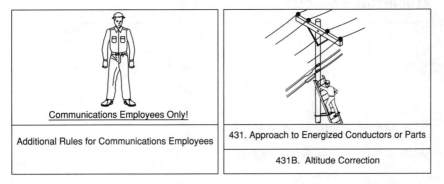

Communications Employees Only!

Additional Rules for Communications Employees

431. Approach to Energized Conductors or Parts

431B. Altitude Correction

- The approach distances in **NESC** Table 431-1 are increased for altitudes above 3000 ft (for some voltage levels).
- **NESC** Table 441-5 provides altitude correction factors, which must be applied to the electrical component of the approach distance (for some voltage levels).

432. JOINT-USE STRUCTURES

Communications Employees Only!

Additional Rules for Communications Employees

432. Joint-Use Structures

- On joint-use structures (power and communications), communications employees must not approach energized parts closer than the distances shown in **NESC** Table 431-1.
- On joint-use structures (power and communications), communications employees must not take conductive objects near energized parts closer than the distances shown in **NESC** Table 431-1.
- Communications employees must not climb above the lowest electric supply conductor (not including vertical risers and street lighting).
- An exception applies to this rule when fixed rigid barriers are installed between the supply and communications facilities if the supply voltage is 140 kV or below.

433. ATTENDANT ON SURFACE
AT JOINT-USE MANHOLES

Communications Employees Only!

Additional Rules for Communications Employees

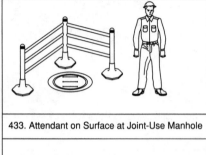

433. Attendant on Surface at Joint-Use Manhole

- Work in joint-use (power and communications) manholes requires an employee to be available on the surface to assist the worker in the manhole.

434. SHEATH CONTINUITY

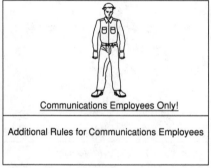

Communications Employees Only!

Additional Rules for Communications Employees

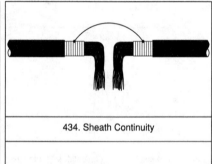

434. Sheath Continuity

- Metallic or semiconductive sheath continuity must be maintained when working on underground cables.

Section 44

Additional Rules for Supply Employees

The following general discussion applies to Secs. 44 and 43. This discussion is repeated at the beginning of Sec. 43. Sections 44 and 43 provide additional rules for employees. Section 44 provides additional rules for supply employees only. Section 43 provides additional rules for communications employees only. Section 41 provides rules for supply and communications employers and Sec. 42 provides rules for supply and communications employees.

The additional rules in Sec. 44 for supply employees primarily focus on minimum approach distances to energized parts, switching control procedures, work on energized lines, de-energizing lines, protective grounding, and live-line work.

The additional rules in Sec. 43 for communications employees primarily focus on keeping the communications employee safe when the communications lines are on joint-use structures or in joint-use manholes with electric supply conductors. Section 43 requires that communications employees maintain minimum approach distances between the communications employee and electric supply lines and equipment. In addition to the minimum approach distances to electric supply conductors, communications employees must not position themselves above the lowest electric supply conductor exclusive of vertical runs (risers) and street lights.

If a communications line is located in the supply space in accordance with the overhead line clearance rules in Sec. 23, the worker who enters the supply space to work on the communications line must be trained as a supply employee.

If a communications line is located below a supply line on an overhead structure and the proper clearances in Sec. 23 are met, the worker who is maintaining the communications line must be trained as a communications employee.

If a communications line is positioned below a supply line on an overhead structure, but the proper clearances in Sec. 23 are not met, the communications employee can correct the violation if the communications employee does not violate the minimum approach distances and other requirements in Sec. 43. If the communications employee cannot maintain the minimum approach distances in Sec. 43, the communications employee must contact a supply employee to correct the violation.

For example, the minimum approach distance rules in Sec. 43 (NESC Table 431-1) require the communications worker to have a 2 ft–2 in minimum approach distance to a 12.47/7.2 kV line. A common supply to communications clearance value in Rules 235 and 238 is 40 in (see Rules 235 and 238 for specific applications). This value is intended to provide enough space on the pole for the communications worker's hands to be on the communications cable and the communications worker's head to be slightly above the communications cable and still maintain the 2 ft–2 in minimum approach distance between the communications worker and the 12.47/7.2 kV line. If a violation exists on the pole and the 40 in clearance requirement between supply and communications is only 38 in, the communications worker can probably lower the communications cable down 2 in (from 38 in to 40 in) without violating the 2 ft–2 in minimum approach distance between the communications worker and the 12.47/7.2 kV line. However, if the communications cable was in violation and mounted only 1 ft below the 12.47/7.2 kV line (say right under the crossarm brace) the communications worker could not correct this violation. The communications worker would have to call on a trained supply worker to fix the problem. The supply worker has the same 2 ft–2 in minimum approach distance in Sec. 44 (NESC Table 441-1) but the supply worker has additional rules for live-line work that allow the supply worker to work inside the minimum approach distance.

If a communications employee is not a qualified employee (i.e., if the worker is not properly trained in recognizing the hazards involved, is not able to identify the nominal voltage of exposed live parts, is not knowledgeable of the minimum approach distances involved with the work, etc.), then the worker is considered an unqualified employee and the worker must maintain at least a 10-ft distance from energized lines (more than 10 ft for greater than 50 kV) per OSHA Standard 1910.333. The OSHA 10-ft requirement applies to other workers in the vicinity of power lines such as painters, roofers, etc. The NESC does not provide distances for unqualified workers but the OSHA standards do. The bottom line is that communications workers need training in recognizing the hazards involved, identifying the nominal voltage of exposed live parts, knowledge of minimum approach distances involved with the work, etc. or they cannot work on a communications line that is constructed jointly with a power line.

The concept of minimum approach distance is outlined in Fig. 44-1.

Minimum Approach Distance:
- Employees must not approach or bring any conductive object within the minimum approach distance.
- The minimum approach distance contains an electrical component and an inadvertent (i.e., unintentional or unexpected) movement component.
- Communications worker minimum approach distances are found in NESC Table 431-1. (A similar table is found in OSHA 1910.268.)
- Supply worker minimum approach distances are found in NESC Tables 441-1, 441-2, 441-3, and 441-4. (Similar tables are found in OSHA 1910.269.)
- Supply workers can work inside the minimum approach distance if additional live-line rules are followed.
- Communications workers may never approach closer than the minimum approach distance.

Expected Reach of Employee:
The expected reach of the employee includes intentional and expected movements, including such things as adjusting a hard hat, maneuvering tools, reaching for items being passed to the employee, adjusting parts and work positions, etc.

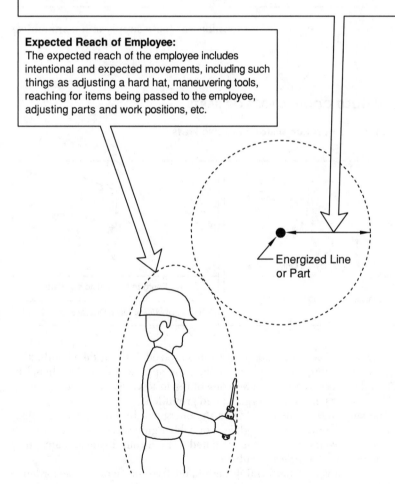

Energized Line or Part

Fig. 44-1. Minimum approach distance (Sec. 44).

440. GENERAL

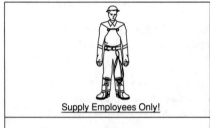

Supply Employees Only!

Additional Rules for Supply Employees

GENERAL
INFORMATION

440. General

- Section 42, "General Rules for Employees" (both supply and communications) also apply.

441. ENERGIZED CONDUCTORS OR PARTS

441A. Minimum Approach Distance to Live Parts

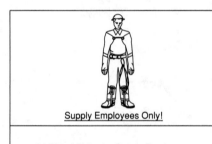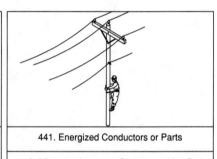

Supply Employees Only!

Additional Rules for Supply Employees

441. Energized Conductors or Parts

441A. Minimum Approach Distance to Live Parts

- Supply employees must not approach energized parts or take conductive objects near energized parts closer than the approach distances listed in NESC Table 441-1 or 441-4 unless one of the following conditions is met:
  ✓ The line or part is de-energized and grounded.
  ✓ The employee is insulated from the energized line or part using insulated tools, gloves, rubber gloves, or rubber gloves with sleeves.
  ✓ The energized line or part is insulated from the employee and any other line or part at a different voltage.
  ✓ The employee is performing bare-hand live-line work in accordance with Rule 446.

- Precaution for approaching voltages from 51 to 300 V:
  - ✓ Do not contact exposed energized parts unless one of the above conditions has been met.
- Precautions for approaching voltages from 301 V to 72.5 kV:
  - ✓ Employees must be protected from phase-to-phase and phase-to-ground differences in voltage (**NESC** Table 441-1 applies).
  - ✓ Exposed grounded lines, conductors, or parts must be guarded or insulated.
  - ✓ When the rubber glove method is used, the gloves must be insulated for the maximum use voltage in **NESC** Table 441-6 and worn whenever the employee is within reach or extended reach of the approach distances in **NESC** Table 441-1.
  - ✓ When the rubber glove method is used, it must be used with one of the two following methods:
    - ▪ Rubber insulating sleeves which are insulated for the maximum use voltage in **NESC** Table 441-6.
    - ▪ Insulating exposed energized lines or parts within the employee's maximum reach (this does not apply to the part being worked on).
  - ✓ Exceptions apply to sleeves and cover-up when the voltage is less than 750 V.
  - ✓ When the rubber glove method is used on voltages above 15 kV phase to phase, an insulated aerial device, insulated structure-mounted platform, or other supplementary insulation must be used to support the worker.
  - ✓ Insulating cover-up shall be rated for the phase to phase voltage of the circuit or the phase to ground voltage of the circuit depending on the exposure of the circuit being worked.
  - ✓ Determination of the phase to phase or phase to ground exposure must consider factors such as work rules, conductor spacing, worker position, tasks being performed, and any other relative factors.
  - ✓ When insulated cover-up is used, it must be applied as the employee first approaches the energized line and it must be removed in the reverse order.
  - ✓ Insulated cover-up must extend beyond the reach of the employee's anticipated work position or extended reach position.
- Precautions for approaching voltages above 72.5 kV:
  - ✓ **NESC** Tables 441-2, 441-3, and 441-4 must be used for approach distances.
  - ✓ If the per unit over voltage factor (T) at the work site has not been determined by engineering analysis, specified values must be applied.
  - ✓ If the per unit over voltage factor (T) at the work site has been determined by engineering analysis, specified conditions must be met.
- A temporary (transient) over voltage control device (TTOCD) which is designed and tested for installation adjacent to a work site to limit the TOV at the work site may be used to obtain a lower value of T. (Example formulas are provided in the rule, and **NESC** Appendix D provides additional information.)
- The approach distances in **NESC** Tables 441-2 through 441-4 must be increased for altitudes above 3000 ft.

- NESC Table 441-5 provides altitude correction factors, which must be applied to the electrical component of the approach distance.
- Approach distances contain an electrical component and an inadvertent movement component.
  - ✓ Approach distances calculated for 0.301 kV to 0.750 kV contain an electrical component plus a 1-ft inadvertent movement component.
  - ✓ Approach distances calculated for 0.751 kV to 72.5 kV contain an electrical component plus a 2-ft inadvertent movement component.
  - ✓ Approach distances calculated for voltages above 72.5 kV contain an electrical component plus a 1-ft inadvertent movement component.
  - ✓ Data for approach distances of voltages between 1.1 and 72.5 kV was obtained from rod gap data measurements.
  - ✓ Data for approach distances of voltages above 72.5 kV was determined from IEEE standards.
  - ✓ Voltage ranges are contained in ANSI Standards.

441B. Additional Approach Requirements

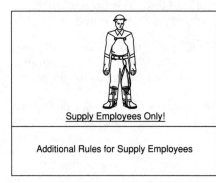

Supply Employees Only!

Additional Rules for Supply Employees

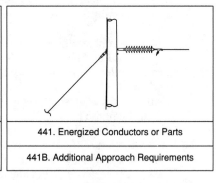

441. Energized Conductors or Parts

441B. Additional Approach Requirements

- The clear insulation distance of insulators is defined as the shortest straight-line air-gap distance from the nearest energized part to the nearest grounded part.
- When working on insulators using rubber gloves or insulated tools, the clear insulation distance must not be less than the straight-line distance in Rule 441A.
- To work on the grounded end of an open switch, all of the following conditions must be met:
  - ✓ The full air-gap distance of the switch must be maintained.
  - ✓ The air-gap distance of the switch must not be less than the electrical component of the approach distance (inadvertent movement components are not required).
  - ✓ The approach distance to the energized portion of the switch must be maintained.
- Special rules apply to working on insulator assemblies operating above 72.5 kV.

441C. Live-Line Tool Clear Insulation Length

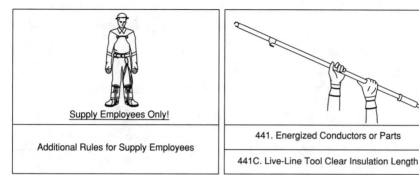

Supply Employees Only!

Additional Rules for Supply Employees

441. Energized Conductors or Parts

441C. Live-Line Tool Clear Insulation Length

- For live-line tools (i.e., hot sticks) the approach distances of **NESC** Tables 441-1 through 441-4 must be maintained or exceeded between the conductive end of the tool and the employee's hands or other body parts.
- Insulated conductor support tools may be used if the clear insulation distance is at least as long as the insulator string or the distances specified in Rule 441A.
- When installing insulated conductor support tools, employee approach distances must be maintained.
- Note that the conductive portion of an insulated tool can decrease the insulation value of the tool more than just the length of the conductive portion.

442. SWITCHING CONTROL PROCEDURES

442A. Designated Person

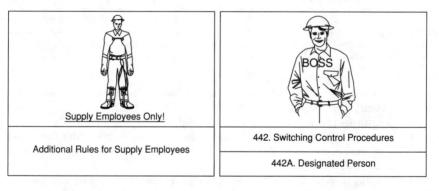

Supply Employees Only!

Additional Rules for Supply Employees

442. Switching Control Procedures

442A. Designated Person

- A designated person must authorize switching.
- The designated person must:
  - ✓ Keep informed of operating conditions to maintain safety.
  - ✓ Maintain suitable records.
  - ✓ Issue or deny authorization for switching to maintain safety.

442B. Specific Work

Supply Employees Only!

Additional Rules for Supply Employees

442. Switching Control Procedures

442B. Specific Work

- The designated person shall give authorization before beginning work.
- The designated person shall be notified when work ends.
- Exceptions to the switching control procedures exist for emergencies and catastrophic service disruptions.

442C. Operations at Stations

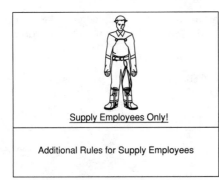

Supply Employees Only!

Additional Rules for Supply Employees

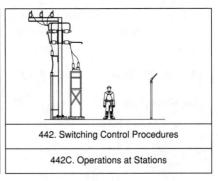

442. Switching Control Procedures

442C. Operations at Stations

- Qualified employees must obtain authorization from the designated person before switching.
- Specific operating schedules may be used.
- If specific operating schedules do not exist, authorization from the designated person must be obtained for switching or starting and stopping of equipment.
- Exceptions exist for switching sections of distribution circuits and for emergency situations.

442D. Re-energizing after Work

Supply Employees Only!

Additional Rules for Supply Employees

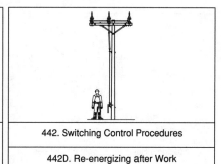

442. Switching Control Procedures

442D. Re-energizing after Work

- Instructions to re-energize after work is complete are to be given by the designated person after the employees who requested the de-energization have reported clear.
- Employees who requested de-energization of lines for other employees or crews cannot request re-energization until the other employees or crews have reported clear.
- The above procedure must be used when more than one location is involved.

442E. Tagging Electric Supply Circuits Associated with Work Activities

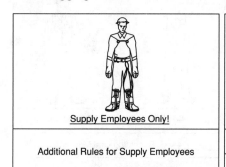

Supply Employees Only!

Additional Rules for Supply Employees

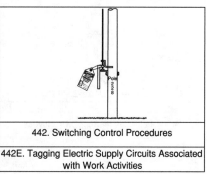

442. Switching Control Procedures

442E. Tagging Electric Supply Circuits Associated with Work Activities

- De-energized and grounded circuits must be tagged at all points where the circuit or equipment can be energized.
- When reclosers or circuit breakers are put on one-shot, a tag must be placed at the reclosing device location.
- Exceptions exist for SCADA-controlled systems at both the SCADA operating point and the reclosing device location.
- Tags shall clearly identify the equipment or circuits being worked on.

442F. Restoration of Service after Automatic Trip

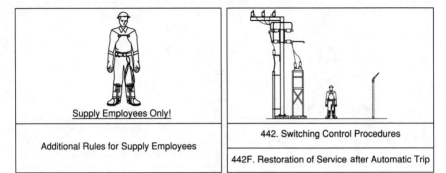

Supply Employees Only!

Additional Rules for Supply Employees

442. Switching Control Procedures

442F. Restoration of Service after Automatic Trip

- Tagged circuits that open automatically (i.e., a circuit with a recloser set to one-shot) must not be closed without authorization.
- When circuits open automatically, local operating rules determine how they may be closed back in with safety.

442G. Repeating Oral Messages

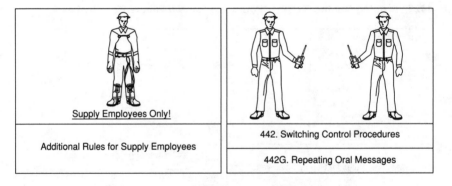

Supply Employees Only!

Additional Rules for Supply Employees

442. Switching Control Procedures

442G. Repeating Oral Messages

- Oral messages associated with line switching must be repeated back to the sender and the sender's identity must be obtained by the receiver.
- The sender of an oral message associated with line switching must require that the message be repeated back by the receiver and must obtain the receiver's identify.

443. WORK ON ENERGIZED LINES AND EQUIPMENT

443A. General Requirements

Supply Employees Only!

Additional Rules for Supply Employees

GENERAL INFORMATION

443. Work on Energized Lines and Equipment

443A. General Requirements

- When working on energized lines, one of the following must be done:
  - ✓ Insulate the employee from energized parts.
  - ✓ Isolate or insulate the employee from ground and other voltages other than the one being worked on.
- Treat covered conductors (i.e., tree wire that is covered with nonrated insulation) as bare energized conductors.
- Consider the effect of a higher-voltage line on a lower-voltage line. Verify that the line being worked on is free from dangerous leakage and induction voltages or verify that it is effectively grounded.
- Insulated supply cables that cannot be positively identified as de-energized must be pierced or severed with an appropriate tool.
- Consider the operating voltage and take appropriate precautions before cutting an insulated energized supply cable.
- When the insulating covering on an energized cable must be cut, an appropriate tool must be used.
- When the insulating coupling on an energized cable must be cut, suitable eye protection and insulating gloves with protectors must be used.
- When the insulating covering on an energized cable must be cut, extreme care must be taken to prevent short-circuiting conductors.
- Metal measuring tapes and tapes or ropes containing metal must not be used closer to energized parts than the approach distances in NESC Tables 441-1 through 441-4.
- Metal measuring tapes and tapes or ropes containing metal must be used with care near energized lines due to the effect of induced voltage.
- Metallic equipment or material that is not bonded to an effective ground and could approach energized parts closer than the approach distances must be treated as if it were energized at the voltage to which it is exposed.

443B. Requirement for Assisting Employee

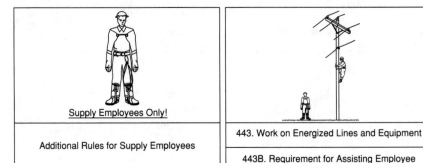

Supply Employees Only!	443. Work on Energized Lines and Equipment
Additional Rules for Supply Employees	443B. Requirement for Assisting Employee

- Employees shall not work alone in bad weather or at night on systems of more than 750 V.
- An exception applies permitting one employee to do certain tasks.

443C. Opening and Closing Switches

Supply Employees Only!	443. Work on Energized Lines and Equipment
Additional Rules for Supply Employees	443C. Opening and Closing Switches

- A smooth continuous motion must be used to close manual switches and disconnects.
- Care must be applied to opening switches to avoid serious arcing.

443D. Working Position

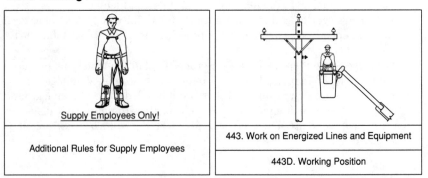

Supply Employees Only!	443. Work on Energized Lines and Equipment
Additional Rules for Supply Employees	443D. Working Position

- Avoid working positions in which a shock or slip will bring the worker's body toward energized parts at a potential different from the employee's body.
- The work position should generally be from below.

443E. Protecting Employees by Switches and Disconnectors

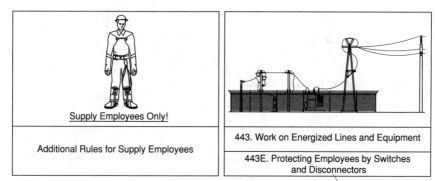

- Load break switches must be opened before non-load-break disconnects.
- Non-load-break disconnects must be closed before closing the load break switch.

443F. Making Connections

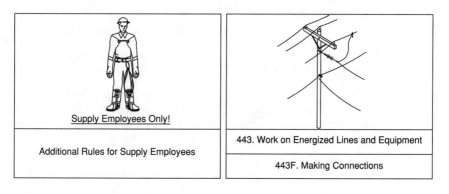

- When connecting de-energized lines or equipment to energized lines, the jumper wire should first be attached to the de-energized part.
- When disconnecting a line or equipment from an energized line, the source end of the jumper should be removed first.
- Loose conductors (i.e., jumpers) should be kept away from energized parts.

443G. Switchgear

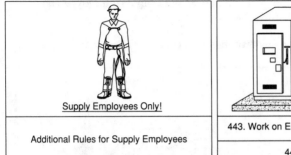

Supply Employees Only!

Additional Rules for Supply Employees

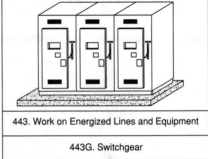

443. Work on Energized Lines and Equipment

443G. Switchgear

- Switchgear must be de-energized and grounded before doing work that involves removing protective barriers unless other safety means are used.
- The switchgear safety features must be replaced after work is completed.

443H. Current Transformer Secondaries

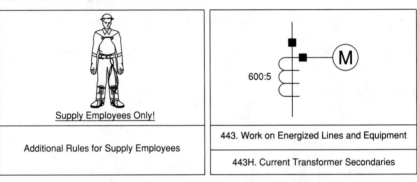

Supply Employees Only!

Additional Rules for Supply Employees

600:5

443. Work on Energized Lines and Equipment

443H. Current Transformer Secondaries

- A current transformer secondary must be de-energized before working on it.
- If the current transformer secondary cannot be properly de-energized, the secondary circuit must be bridged (shorted) so that the secondary will not be opened.

443I. Capacitors

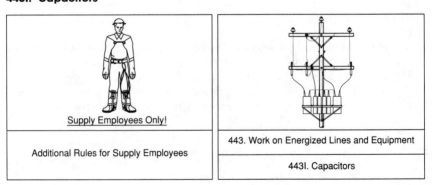

Supply Employees Only!

Additional Rules for Supply Employees

443. Work on Energized Lines and Equipment

443I. Capacitors

- Before working on capacitors, they must be disconnected from the energized source, shorted, and grounded.
- Any line that has capacitors must be short-circuited and grounded before it is considered de-energized.
- Before capacitors are handled, each unit must be shorted between all insulated terminals and the capacitor tank due to series-parallel operation.
- Where capacitors are installed on ungrounded racks, the racks must be grounded before working on the capacitors.
- The internal resistor of a capacitor must not be depended on to discharge the capacitor.

443J. Gas-Insulated Equipment

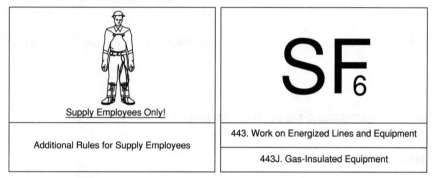

Supply Employees Only!

Additional Rules for Supply Employees

443. Work on Energized Lines and Equipment

443J. Gas-Insulated Equipment

- Employees must be instructed on the special precautions related to handling SF_6 gas.
- The by-products resulting from arcing in SF_6 gas are generally toxic and irritant.

443K. Attendant on Surface

Supply Employees Only!

Additional Rules for Supply Employees

443. Work on Energized Lines and Equipment

443K. Attendant on Surface

- When one employee is in a manhole, another employee must be on the surface to render assistance as required.
- The employee on the surface may enter the manhole to provide short-term assistance.
- An exception permits working alone to do certain tasks.

443L. Unintentional Grounds on Delta Circuits

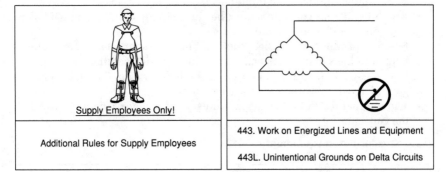

Supply Employees Only!

Additional Rules for Supply Employees

443. Work on Energized Lines and Equipment

443L. Unintentional Grounds on Delta Circuits

- Unintentional grounds on delta circuits must be removed as soon as practical.

444. DE-ENERGIZING EQUIPMENT OR LINES TO PROTECT EMPLOYEES

444A. Application of Rule

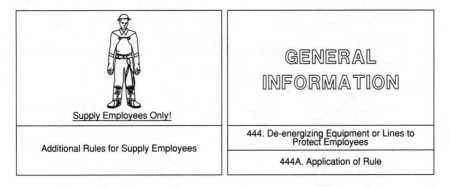

Supply Employees Only!

Additional Rules for Supply Employees

GENERAL INFORMATION

444. De-energizing Equipment or Lines to Protect Employees

444A. Application of Rule

- When employees depend on others to operate switches to de-energize circuits or when employees must secure authorization to operate a switch, Rules 444A through 444H must be followed in order.
- If an employee is in sole charge of a circuit section and is responsible for directing the disconnecting of the section, the portions of Rules 444A through 444H that deal with a designated person may be omitted.
- Records must be kept on utility interactive systems and these systems must be capable of being visibly disconnected.

444B. Employee's Request

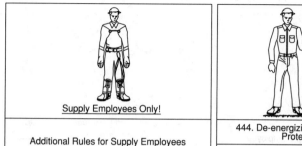
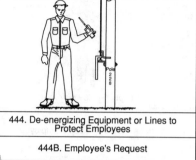

Supply Employees Only!

Additional Rules for Supply Employees

444. De-energizing Equipment or Lines to Protect Employees

444B. Employee's Request

- The employee in charge of the work must make a request to the designated switching person to have the line or equipment de-energized.
- The switching request must properly identify the switch or line section by position, letter, color, number, or other means.

444C. Operating Switches, Disconnectors, and Tagging

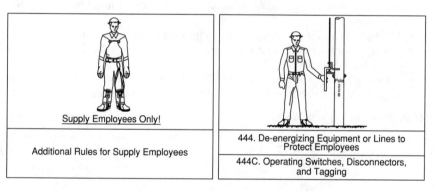

Supply Employees Only!

Additional Rules for Supply Employees

444. De-energizing Equipment or Lines to Protect Employees

444C. Operating Switches, Disconnectors, and Tagging

- The designated switching person must direct the operation of switches and disconnects to de-energize the circuit.
- The designated switching person must request that the switches and disconnect be rendered inoperable and tagged.
- If switches that are controlled automatically or remotely can be rendered inoperable, they must be tagged at the switch location.
- If switches that are controlled automatically or remotely cannot be rendered inoperable, then they must be tagged at all points of control.
- The following information must be recorded when placing a tag:
  ✓ Time of disconnection.
  ✓ Name of person making the disconnection.
  ✓ Name of person who requested the disconnection.
  ✓ Name or title or both of the designated switching person.

444D. Employee's Protective Grounds

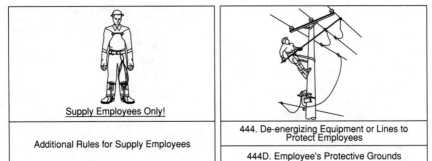

Supply Employees Only!

Additional Rules for Supply Employees

444. De-energizing Equipment or Lines to Protect Employees

444D. Employee's Protective Grounds

- After switching, rendering inoperable where practical, and tagging have been completed and the employee in charge has been given permission by the designated switching person to proceed, protective grounds must be applied.
- Approach distances must be maintained for making voltage tests.
- Approach distances must be maintained for applying grounds.
- Grounds must be placed on each side of the work location and as close as practical to the work location, or a worksite ground must be used at the work location.
- Multiple work locations must have a ground at each work location.
- The approach distances must be maintained from ungrounded conductors at the work location.
- Where making protective grounds is not possible or creates a hazardous condition, grounds may be omitted by special permission.

444E. Proceeding with Work

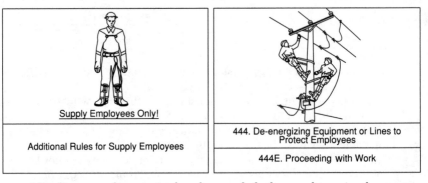

Supply Employees Only!

Additional Rules for Supply Employees

444. De-energizing Equipment or Lines to Protect Employees

444E. Proceeding with Work

- After lines are de-energized and grounded, the employee in charge may direct the work to be done.
- Lines may be re-energized for testing under the supervision of the employee in charge and authorization of the designated switching person.
- Additional employees in charge desiring the same equipment or lines to be de-energized and grounded must follow these same procedures.

444F. Reporting Clear—Transferring Responsibility

Supply Employees Only!

Additional Rules for Supply Employees

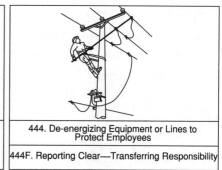

444. De-energizing Equipment or Lines to Protect Employees

444F. Reporting Clear—Transferring Responsibility

- Once work is completed, the employee in charge must verify that the crew assigned to the employee in charge is in the clear.
- Once work is completed, the employee in charge must remove protective grounds.
- Once work is completed, the employee in charge must report to the designated switching person that tags may be removed.
- The employee in charge may transfer responsibilities to another employee after receiving permission from the designated person and personally informing the affected persons of the transfer.

444G. Removal of Tags

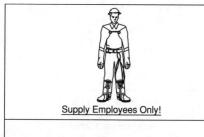

Supply Employees Only!

Additional Rules for Supply Employees

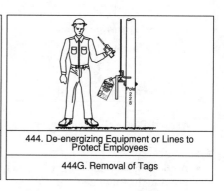

444. De-energizing Equipment or Lines to Protect Employees

444G. Removal of Tags

- The designated switching person must direct the removal of tags.
- The removal of tags must be reported back to the designated switching person by the person removing them.
- Upon tag removal, the record keeping must include the following:
  ✓ Name of the person requesting removal.
  ✓ Time of removal.
  ✓ Name of the person removing the tag.

- The name of the person requesting removal must match the name of the person who requested placement of the tag unless responsibility was properly transferred.

444H. Sequence of Re-energizing

Supply Employees Only!

Additional Rules for Supply Employees

444. De-energizing Equipment or Lines to Protect Employees

444H. Sequence of Re-energizing

- The designated switching person may direct the re-energization of the line only after the removal of protective grounds and tags.

445. PROTECTIVE GROUNDS

445A. Installing Grounds

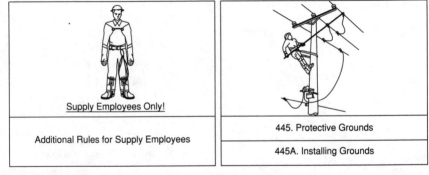

Supply Employees Only!

Additional Rules for Supply Employees

445. Protective Grounds

445A. Installing Grounds

- A specific sequence must be used when applying protective grounds to a previously energized part.
- An exception to the sequence may apply to some high-voltage towers.
- 1st, grounding conductors and devices must be sized to carry anticipated fault currents.
- 2nd, one end of the grounding device must be connected to an effective ground. Grounding switches may be used.
- 3rd, a voltage test must be done utilizing appropriate (insulated) tools and approach distances.
- 4th, if the line shows no voltage, grounding may be completed. If voltage is present, the source must be determined. To complete grounding, the ground device must be brought into contact with the previously energized part

using insulating handles and securely attached. Where bundled conductors exist, each conductor of the bundle should be grounded. Only after completion of the fourth step can the employee approach closer than the approach distances or proceed with work on the parts as grounded parts.

445B. Removing Grounds

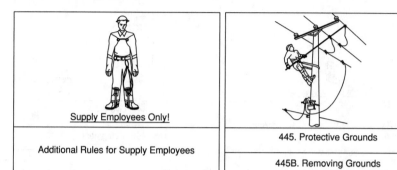

Supply Employees Only!

Additional Rules for Supply Employees

445. Protective Grounds

445B. Removing Grounds

- Grounding devices must first be removed from the de-energized parts using tools with insulated handles.
- If multiple ground cables are connected to the same grounding point, all phase connections must be removed before removing any of the ground connections.
- An exception applies to removing all of the phase connections before removing any of the ground connections if this produces a hazard such as unintentional contact of the ground with ungrounded parts. In this case the grounds may be removed individually from each phase and ground connection.
- Extreme caution must be used to assure that the proper sequence is used when installing or removing grounds.
- Do not remove the connection from the effective ground first. This may result in electric shock and injury.

446. LIVE WORK

446A. Training

Supply Employees Only!

Additional Rules for Supply Employees

Safety
is
#1

446. Live Work

446A. Training

- Employees must be trained to use rubber gloves, hot sticks, or the bare-hand method before using these techniques on energized lines.

446B. Equipment

Supply Employees Only!

Additional Rules for Supply Employees

446. Live Work

446B. Equipment

- Insulated bucket trucks, ladders, and other support equipment must be evaluated for performance at the voltages involved. Tests must be done to ensure the equipment's integrity.
- Insulated bucket trucks and other insulated aerial devices used in bare-hand work must be tested before work is started.
- IEEE Standards and ANSI Standards must be referenced for equipment operation and testing.
- Bucket trucks and other insulated aerial devices must be kept clean.
- Tools and other equipment must not be used in a manner that would reduce the insulating strength of the insulated bucket truck or other insulated aerial device.

446C. When Working on Insulators...

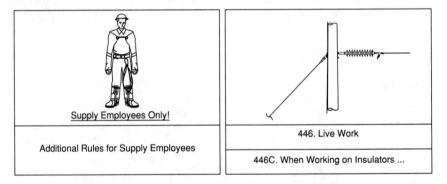

Supply Employees Only!

Additional Rules for Supply Employees

446. Live Work

446C. When Working on Insulators ...

- When insulators are worked on using live-line procedures, the clear insulation distance must not be less than the approach distances of **NESC** Tables 441-1 through 441-4.

446D. Bonding and Shielding for Bare-Hand Method

Supply Employees Only!

Additional Rules for Supply Employees

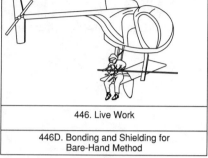

446. Live Work

446D. Bonding and Shielding for
Bare-Hand Method

- A conductive bucket liner or other suitable conducting device must be provided to bond the insulated bucket or other insulated aerial device to the energized line.
- The employee must be bonded to the insulated bucket or other insulated aerial device by using conducting shoes, leg clips, or other means.
- Protective clothing for electrostatic shielding must be used where necessary. IEEE Standards provide additional information on clothing designed for electrostatic shielding.
- The aerial device must be bonded to the energized conductor before the employee contacts the energized part.

447. PROTECTION AGAINST ARCING AND OTHER DAMAGE WHILE INSTALLING AND MAINTAINING INSULATORS AND CONDUCTORS

Supply Employees Only!

Additional Rules for Supply Employees

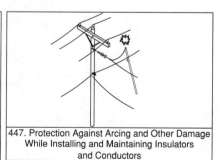

447. Protection Against Arcing and Other Damage
While Installing and Maintaining Insulators
and Conductors

- When installing and maintaining insulators and conductors, use precautions to limit damage that would cause the conductors or insulators to fall.
- Use precautions to prevent arcs from forming.
- If an arc does form, use precautions to prevent injuring or burning any parts of the supporting structure, insulators, or conductors.

Appendix A

Photographs of NESC Applications

The photos shown correspond to figures in the text with the following icon:

See
Photo(s)

Photo 011-2. Typical dividing lines between the NESC and the NEC (see Fig. 011-2).

Photo 017-1. Example of nominal values (see Fig. 017-1).

Photo 092-4. Grounded connections for nonshielded cables over 750 V (see Fig. 092-4).

Photo 092-5. Surge arrester cable—shielding interconnection (see Fig. 092-5).

Photo 092-6. Grounding points for a shielded cable without an insulating jacket (see Fig. 092-6).

Photo 092-7. Grounding points for a shielded cable with an insulating jacket (see Fig. 092-7).

Photo 092-9. Grounding of messenger wires (see Fig. 092-9).

Photo 092-12. Example of fence grounding (see Fig. 092-12).

Photo 093-1. Connection of grounding conductor to grounded conductor (see Fig. 093-1).

Photo 093-5. Example of pole ground ampacity (see Fig. 093-5).

Photo 093-6(1). Requirements for grounding conductors with or without guards (see Fig. 093-6).

Photo 093-6(2). Requirements for grounding conductors with or without guards (see Fig. 093-6).

Photo 093-8. Example of aluminum grounding conductor transitioning to copper for burial (see Fig. 093-8).

Photo 093-9. Example of common grounding conductor for neutral and equipment (see Fig. 093-9).

Photo 094-4. Made electrodes-driven ground rods (see Fig. 094-4).

Photo 095-1. Connection of grounding conductor to grounding electrode (see Fig. 095-1).

Photo 097-2. Example of a common neutral with single grounding (see Fig. 097-2).

Photo 097-4. Bonding of communication systems to electric supply systems on a joint use structure (see Fig. 097-4).

Photo 110-1. Example of how Rule 110A applies to Part 1, Electric Supply Stations, or Part 2, Overhead Lines (see Fig. 110-1).

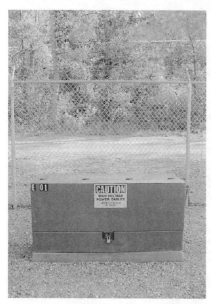

Photo 110-2. Example of how Rule 110A applies to Part 1, Electric Supply Stations, or Part 3, Underground Lines (see Fig. 110-2).

Photo 110-5. Fence height requirements (see Fig. 110-5).

Photo 110-6. Example of how to apply the safety clearance zone to fences per NESC Table 110-1 and NESC Fig. 110-1 (see Fig. 110-6).

Photo 124-1. Example of how to apply vertical clearances per NESC Table 124-1 and NESC Fig. 124-1 (see Fig. 124-1).

Photo 124-3. Clearance measurements made to a permanent supporting surface (see Fig. 124-3).

Photo 124-4. Absence of Code rule related to bus to bus clearances (see Fig. 124-4).

Photo 124-5. Example of how to apply NESC Table 124-1 and NESC Fig. 124-2 (see Fig. 124-5).

Photo 125-3. VIOLATION! Storage materials must not be in the working space (see Fig. 125-3).

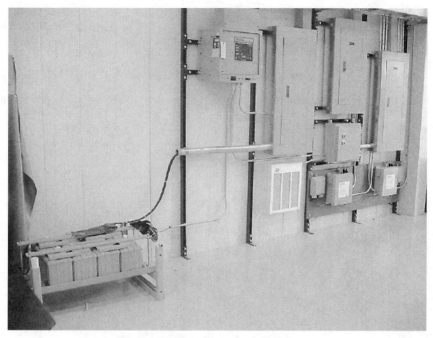

Photo 140-1. Storage batteries (see Fig. 140-1).

Photo 150-1. Example of protecting current-transformer secondary circuits (see Fig. 150-1).

Photo 153-1. Choices for short-circuit protection of power transformers (see Fig. 153-1).

Photo 162-1(1). Substation conductor mechanical protection and support (see Fig. 162-1).

Photo 162-1(2). Substation conductor mechanical protection and support (see Fig. 162-2).

Photo 163-1. Guarding a substation conductor with insulation (see Fig. 163-1).

Photo 170-1. Example of a switch position indicator (see Fig. 170-1).

Photo 180-4. Adequate clearance for reading meters on control switchboards (see Fig. 180-4).

Photo 193-1. Arrester installation (see Fig. 193-1).

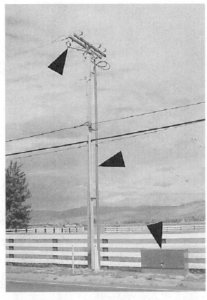

Photo 201-1. Overlap between Overhead Lines (Part 2) and Underground Lines (Part 3) (see Fig. 201-1).

Photo 201-2. Overlap between Overhead Lines (Part 2) and Electric Supply Stations (Part 1) (see Fig. 201-2).

Photo 214-1. VIOLATION! Inspection and tests of lines and equipment when in and out of service (see Fig. 214-1).

Photo 215-5. Example of a guy insulator used to limit galvanic corrosion (see Fig. 215-5).

Photo 216-1. Arrangement of overhead line switches (see Fig. 216-1).

Photo 217-2. Readily climbable supporting structures (see Fig. 217-2).

Photo 217-3. Permanently mounted steps on supporting structures (see Fig. 217-3).

Photo 217-4. Arrangement of standoff brackets (see Fig. 217-4).

Photo 217-5. Identification of supporting structures (see Fig. 217-5).

Photo 217-6(1). VIOLATION! Obstructions on supporting structures (see Fig. 217-6).

Photo 217-9(1). Protection and marking of guys (see Fig. 217-9).

Photo 217-6(2). VIOLATION! Obstructions on supporting structures (see Fig. 217-6).

Photo 217-9(2). Protection and marking of guys (see Fig. 217-9).

Photo 218-1. General vegetation management requirements (see Fig. 218-1).

Photo 218-2. VIOLATION! Vegetation management at line, railroad, and limited access highway crossings (see Fig. 218-2).

Photo 220-1. Preferred supply and communication conductor levels (see Fig. 220-1).

Photo 220-2. Relative levels of supply lines of different voltages at crossings (see Fig. 220-2).

Photo 220-3. Relative levels of supply lines of different voltages at structure conflict locations (see Fig. 220-3).

Photo 220-4. Relative levels of supply circuits of different voltages owned by one utility (see Fig. 220-4).

Photo 221-1(1). Avoiding conflict between two separate lines (see Fig. 221-1).

Photo 221-1(2). Avoiding conflict between two separate lines (see Fig. 221-1).

Photo 222-1. Joint use (power and communication) of structures (see Fig. 222-1).

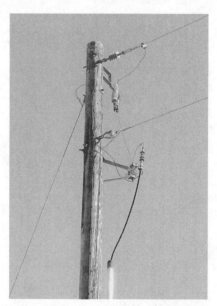

Photo 230-5. Clearance measurements for parts of conductors and parts of supporting structure (see Fig. 230-5).

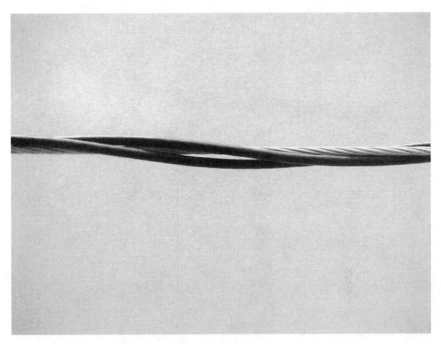

Photo 230-7. Construction of 230C3 supply cables (see Fig. 230-7).

Photo 230-8. Covered conductors (see Fig. 230-8).

Photo 230-9. Neutral conductors (see Fig. 230-9).

Photo 231-1. Clearance of a supporting structure from a fire hydrant (see Fig. 231-1).

Photo 231-2. VIOLATION! Clearance of a supporting structure from a redirectional curb (see Fig. 231-2).

Photo 231-3(1). Clearance of a supporting structure from a swale-type curb (see Fig. 231-3).

Photo 231-3(2). VIOLATION! Clearance of a supporting structure from a swale-type curb (see Fig. 231-3).

Photo 231-4. Clearance of a supporting structure from a roadway without a curb (see Fig. 231-4).

Photo 231-5. Clearance of a supporting structure from railroad tracks (see Fig. 231-5).

Photo 232-7. Common clearance values from NESC Table 232-1 (see Fig. 232-7).

Photo 232-8. Common clearance values from NESC Table 232-1 (see Fig. 232-8).

Photo 232-11. Common clearance values from NESC Table 232-1 (see Fig. 232-11).

Photo 232-17. Example of clearance to equipment cases and unguarded rigid live parts (see Fig. 232-17).

Photo 232-18. Example of clearance to street lighting (see Fig. 232-18).

Photo 232-19. Vertical clearance to land surface (see Fig. 232-19).

Photo 232-20. Federal Aviation Administration (FAA) requirements (see Fig. 232-20).

Photo 233-4. Example of horizontal clearance between wires carried on different supporting structures (see Fig. 233-4).

Photo 233-6. Example of vertical clearance between wires carried on different supporting structures (see Fig. 233-6).

Photo 233-7. Example of vertical clearance between wires carried on different supporting structures (see Fig. 233-7).

Photo 233-8. Example of the exception to vertical clearance between wires that are electrically interconnected at a crossing (see Fig. 233-8).

Photo 234-4. Example of clearance of a conductor to a street lighting pole (see Fig. 234-4).

Photo 234-5(1). Application of Rules 234B, 231B, and 232B4 to bent street lighting poles (see Fig. 234-5).

Photo 234-5(2). Application of Rules 234B, 231B, and 232B4 to bent street lighting poles (see Fig. 234-5).

Photo 234-6. Application of Rules 234B, 233, and 235 to skip span construction (see Fig. 234-6).

Photo 234-7. Example of horizontal clearance of conductors to a building (see Fig. 234-7).

Photo 234-8. Example of vertical clearance of a supply service drop conductor over a building but not serving it (see Fig. 234-8).

Photo 234-9. Example of clearance of a conductor to a billboard (see Fig. 234-9).

Photo 234-11(1). Example of clearance of supply conductors (service drops) attached to buildings (see Fig. 234-11).

Photo 234-11(2). Example of clearance of supply conductors (service drops) attached to buildings (see Fig. 234-11).

Photo 234-12. VIOLATION! Example of clearance of supply conductors (service drops) attached to buildings (see Fig. 234-12).

Photo 234-14(1). Example of providing clearance to grain bins (see Fig. 234-14).

Photo 234-14(2). Example of providing clearance to grain bins (see Fig. 234-14).

Photo 235-8. Example of vertical clearance between line conductors (see Fig. 235-8).

Photo 235-11. Example of vertical clearance between joint use (power and communication) conductors (see Fig. 235-11).

Photo 235-12. Example of vertical clearance between joint-use (power and communication) conductors (see Fig. 235-12).

Photo 235-14. Exceptions to vertical clearance requirements for joint-use construction (see Fig. 235-14).

Photo 235-16. Example of clearance of a conductor to an anchor guy and a conductor to a support (see Fig. 235-16).

Photo 235-17. Examples of clearance between supply circuits of different voltages on the same support arm (see Fig. 235-17).

Photo 235-18. Conductor spacing on vertical racks (see Fig. 235-18).

Photo 235-19. Clearance and spacing between communication lines (see Fig. 235-19).

Photo 235-20. Example of clearance between supply lines and communication antennas in the supply space (see Fig. 235-20).

Photo 236-2. Example of climbing space between conductors (see Fig. 236-2).

Photo 236-4. Example of obstructions at the base of the climbing space (see Fig. 236-4).

Photo 237-2. Example of working clearances from energized equipment (see Fig. 237-2).

Photo 238-1. Definition of equipment as it applies to vertical clearance between communication and supply facilities on the same structure (see Fig. 238-1).

Photo 238-2. Example of equipment metal supporting braces (see Fig. 238-2).

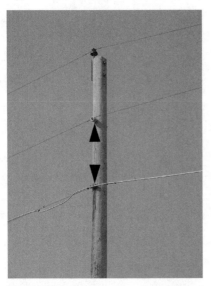

Photo 238-5. Example of vertical clearance between supply and communication equipment on the same structure (see Fig. 238-5).

Photo 238-6. Example of vertical clearance between supply and communication equipment on the same structure (see Fig. 238-6).

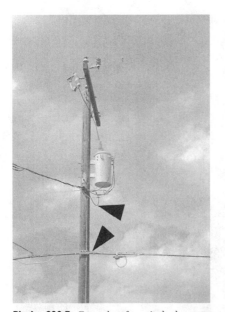

Photo 238-7. Example of vertical clearance between supply and communication equipment on the same structure (see Fig. 238-7).

Photo 238-8. Example of vertical clearance between supply and communication equipment on the same structure (see Fig. 238-8).

Photo 238-9(1). Example of vertical clearance between supply and communication equipment on the same structure (see Fig. 238-9).

Photo 238-9(2). Example of vertical clearance between supply and communication equipment on the same structure (see Fig. 238-9).

Photo 238-12. Example of vertical clearance between a drip loop feeding roadway lighting and communication equipment (see Fig. 238-12).

Photo 238-13. Example of vertical clearance between a drip loop feeding roadway lighting and communication equipment (see Fig. 238-13).

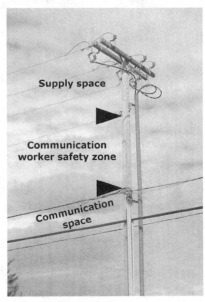

Photo 238-15. Exception to vertical clearance between a drip loop feeding roadway lighting and communication equipment (see Fig. 238-15).

Photo 238-17. Communication worker safety zone (see Fig. 238-17).

Photo 239-1. Examples of vertical and lateral conductors (see Fig. 239-1).

Photo 239-3(1). Guarding and protection near ground (see Fig. 239-3).

Photo 239-3(2). Guarding and protection near ground (see Fig. 239-3).

Photo 239-3(3). Guarding and protection near ground (see Fig. 239-3).

Photo 239-4. Location of vertical and lateral supply conductors on supply-line structures or within supply space on jointly used structures (see Fig. 239-4).

Photo 239-5. Location of vertical and lateral communication conductors on communication line structures (see Fig. 239-5).

Photo 239-6. Location of vertical supply conductors passing through the communication space (see Fig. 239-6).

Photo 239-7. Example of vertical supply conductors on a joint-use pole (see Fig. 239-7).

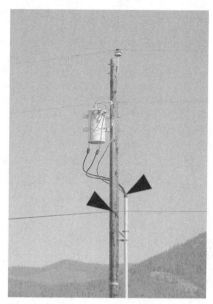

Photo 239-8. VIOLATION! Example of vertical supply conductors on a joint-use pole (see Fig. 239-8).

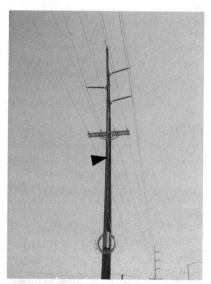

Photo 239-10. Location of vertical communication conductors passing through the supply space (see Fig. 239-10).

Photo 252-6. Longitudinal loads at a change in grade of construction (see Fig. 252-6).

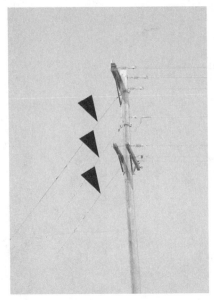

Photo 252-7(1). Longitudinal loads at deadends (see Fig. 252-7).

Photo 252-7(2). Longitudinal loads at dead-ends (see Fig. 252-7).

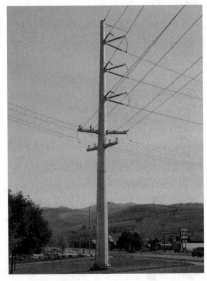

Photo 261-1(1). Strength requirements of supporting structure (see Fig. 261-1).

Photo 261-1(2). Strength requirements of supporting structure (see Fig. 261-1).

Photo 261-4. Strength requirements of wood structures (see Fig. 261-4).

Photo 261-3. Strength requirements of metal, prestressed, and reinforced concrete structures (see Fig. 261-3).

Photo 261-6(1). Spliced and reinforced wood poles (see Fig. 261-6).

Photo 261-6(2). Spliced and reinforced wood poles (see Fig. 261-6).

Photo 261-6(3). Spliced and reinforced wood poles (see Fig. 261-6).

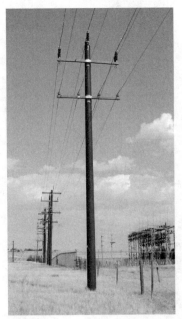

Photo 261-7. Strength requirements of fiber-reinforced polymer structures (see Fig. 261-7).

Photo 261-13. Strength of concrete and metal crossarms and braces (see Fig. 261-13).

Photo 261-14. Strength of wood crossarms and braces (see Fig. 261-14).

Photo 261-16. Strength of fiber-reinforced polymer crossarms and braces (see Fig. 261-16).

Photo 261-17. Strength requirements for armless construction (see Fig. 261-17).

Photo 261-20. Example of open-wire communication conductors (see Fig. 261-20).

Photo 264-1(1). Examples of guying and bracing (see Fig. 264-1).

Photo 264-1(2). Examples of guying and bracing (see Fig. 264-1).

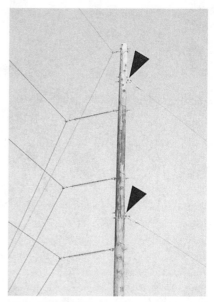

Photo 264-3(1). Point of guy attachment (see Fig. 264-3).

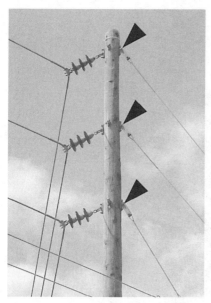

Photo 264-3(2). Point of guy attachment (see Fig. 264-3).

Photo 279-2. Properties of guy insulators (see Fig. 279-2).

Photo 279-4. Example of a guy insulator used to limit galvanic corrosion (see Fig. 279-4).

Photo 301-1. Overlap between Underground Lines (Part 3) and Overhead Lines (Part 2) (see Fig. 301-1).

Photo 311-1. Example of worker locating underground lines (see Fig. 311-1).

Photo 312-1. Example of the need for adaquate working space (see Fig. 312-1).

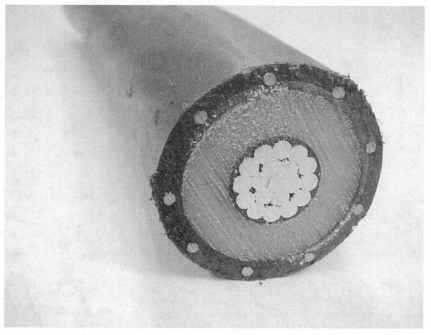

Photo 314-1. Conductive parts to be grounded (see Fig. 314-1).

Photo 314-2. Grounding of circuits (see Fig. 314-2).

Photo 320-4. Conduit routing under highways and streets (see Fig. 320-4).

Photo 323-1. Loads on manholes, handholes, and vaults (see Fig. 323-1).

Photo 323-11(1). Requirements for identification of manhole and handhole covers (see Fig. 323-11).

Photo 323-11(2). Requirements for identification of manhole and handhole covers (see Fig. 323-11).

Photo 331-1. Example of a sheath, jacket, and shield on a typical URD supply cable (see Fig. 331-1).

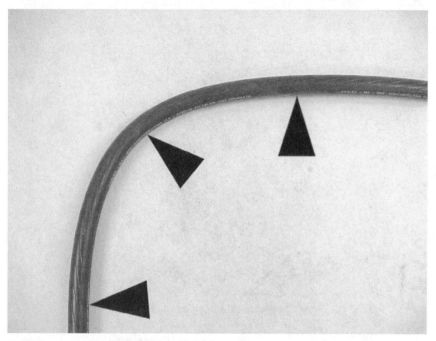

Photo 341-1. Cable bending radius (see Fig. 341-1).

Photo 350-3. Cables below 600 V to ground buried in close proximity (see Fig. 350-3).

Photo 352-1. Trenching requirements for direct-buried cables (see Fig. 352-1).

Photo 352-2. Plowing requirement for direct-buried cables (see Fig. 352-2).

Photo 352-3. Boring requirement for direct-buried cables (see Fig. 352-3).

Photo 360-1. Mechanical protection of supply risers below grade (see Fig. 360-1).

Photo 370-1(1). Examples of supply cable terminations (see Fig. 370-1).

Photo 370-1(2). Examples of supply cable terminations (see Fig. 370-1).

Photo 370-1(3). Examples of supply cable terminations (see Fig. 370-1).

Photo 372-1. Example of supply cable identification at termination points (see Fig. 372-1).

Photo 372-2. Absence of Code rule related to cable route markers (see Fig. 372-2).

Photo 380-1. Examples of supply and communications equipment (see Fig. 380-1).

Photo 380-2. Clearance of pad-mounted equipment to fire hydrants (see Fig. 380-2).

Photo 380-3(1). Absence of Code rule related to underground pad-mounted equipment location adjacent to roads and buildings (see Fig. 380-3).

Photo 380-3(2). Absence of Code rule related to underground pad-mounted equipment location adjacent to roads and buildings (see Fig. 380-3).

Photo 381-3(1). Safety sign requirements for pad-mounted equipment (see Fig. 381-3).

Photo 381-3(2). Safety sign requirements for pad-mounted equipment (see Fig. 381-3).

Photo 384-1. Bonding above ground metallic supply and communications apparatus (see Fig. 384-1).

Appendix B

OSHA Standards Related to the NESC Work Rules

1910.268
Telecommunications (Operation and Maintenance)
1910 Subpart R—Special Industries

1910.269 (Includes Appendix A–E)
Electric Power Generation, Transmission, and Distribution (Operation and Maintenance)
1910 Subpart R—Special Industries

1926.950 through 1926.960
Power Transmission and Distribution (Construction)
1926 Subpart V—Power Transmission and Distribution

Appendix B

OSHA Standards Related to the NESC Work Rules

1910.268

Telecommunications (Operation and Maintenance)
1910 Subpart R—Special Industries

OSHA 1910.268 Paragraph Titles:

(a) Application
(b) General
(c) Training
(d) Employee protection in public work areas
(e) Tools and personal protective equipment
(f) Rubber insulating equipment
(g) Personal climbing equipment
(h) Ladders
(i) Other tools and personal protective equipment
(j) Vehicle-mounted material handling devices and other mechanical equipment
(k) Materials handling and storage
(l) Cable fault locating and testing
(m) Grounding for employee protection—pole lines
(n) Overhead lines
(o) Underground lines
(p) Microwave transmission
(q) Tree trimming—electrical hazards
(r) Buried facilities—Communications lines and power lines in the same trench
(s) Definitions

REGULATIONS (STANDARDS - 29 CFR)

TELECOMMUNICATIONS. -1910.268

- **Part Number:** 1910
- **Part Title:** Occupational Safety and Health Standards
- **Subpart:** R
- **Subpart Title:** Special Industries
- **Standard Number:** 1910.268
- **Title:** Telecommunications.

1910.268(a)

Application.

1910.268(a)(1)

This section sets forth safety and health standards that apply to the work conditions, practices, means, methods, operations, installations and processes performed at telecommunications centers and at telecommunications field installations, which are located outdoors or in building spaces used for such field installations. *Center* work includes the installation, operation, maintenance, rearrangement, and removal of communications equipment and other associated equipment in telecommunications switching centers. *Field* work includes the installation, operation, maintenance, rearrangement, and removal of conductors and other equipment used for signal or communication service, and of their supporting or containing structures, overhead or underground, on public or private rights of way, including buildings or other structures.

..*1910.268(a)(2)*

1910.268(a)(2)

These standards do not apply:

1910.268(a)(2)(i)

To construction work, as defined in § 1910.12, nor

1910.268(a)(2)(ii)

to installations under the exclusive control of electric utilities used for the purpose of communications or metering, or for generation, control, transformation, transmission, and distribution of electric energy, which are located in buildings used exclusively by the electric utilities for such purposes, or located outdoors on property owned or leased by the electric utilities or on public highways, streets, roads, etc., or outdoors by established rights on private property.

1910.268(a)(3)

Operations or conditions not specifically covered by this section are subject to all the applicable standards contained in this Part 1910. See § 1910.5(c). Operations which involve construction work, as defined in 1910.12 are subject to all the applicable standards contained in Part 1926 of this chapter.

1910.268(b)

General—

1910.268(b)(1)

Buildings containing telecommunications centers—

1910.268(b)(1)(i)

Illumination. Lighting in telecommunication centers shall be provided in an adequate amount such that continuing work operations, routine observations, and the passage of employees can be carried out in a safe and healthful manner. Certain specific tasks in centers, such as splicing cable and the maintenance and repair of equipment frame lineups, may require a higher level of illumination. In such cases, the employer shall install permanent lighting or portable supplemental lighting to attain a higher level of illumination shall be provided as needed to permit safe performance of the required task.

..1910.268(b)(1)(ii)

1910.268(b)(1)(ii)

Working surfaces. Guard rails and toe boards may be omitted on distribution frame mezzanine platforms to permit access to equipment. This exemption applies only on the side or sides of the platform facing the frames and only on those portions of the platform adjacent to equipped frames.

1910.268(b)(1)(iii)

Working spaces. Maintenance aisles, or wiring aisles, between equipment frame lineups are working spaces and are not an exit route for purposes of 29 CFR 1910.34.

1910.268(b)(1)(iv)

Special doors. When blastproof or power actuated doors are installed in specially designed hardsite security buildings and spaces, they shall be designed and installed so that they can be used as a means of egress in emergencies.

1910.268(b)(1)(v)

Equipment, machinery and machine guarding. When power plant machinery in telecommunications centers is operated with commutators and couplings uncovered, the adjacent housing shall be clearly marked to alert personnel to the rotating machinery.

1910.268(b)(2)

Battery handling.

1910.268(b)(2)(i)

Eye protection devices which provide side as well as frontal eye protection for employees shall be provided when measuring storage battery specific gravity or handling electrolyte, and the employer shall ensure that such devices are used by the employees. The employer shall also ensure that acid resistant gloves and aprons shall be worn for protection against spattering. Facilities for quick drenching or flushing of the eyes and body shall be provided unless the storage batteries are of the enclosed type and equipped with explosion proof vents, in which case sealed water rinse or neutralizing packs may be substituted for the quick drenching or flushing facilities. Employees assigned to work with storage batteries shall be instructed in emergency procedures such as dealing with accidental acid spills.

1910.268(b)(2)(ii)

Electrolyte (acid or base, and distilled water) for battery cells shall be mixed in a well ventilated room. Acid or base shall be poured gradually, while stirring, into the water. Water shall never be poured into concentrated (greater than 75 percent) acid solutions. Electrolyte shall never be placed in metal containers nor stirred with metal objects.

1910.268(b)(2)(iii)

When taking specific gravity readings, the open end of the hydrometer shall be covered with an acid resistant material while moving it from cell to cell to avoid splashing or throwing the electrolyte.

1910.268(b)(3)

Employers must provide employees with readily accessible, adequate, and appropriate first aid supplies. A non-mandatory example of appropriate supplies is listed in Appendix A to 29 CFR 1910.151.

..1910.268(b)(4)
1910.268(b)(4)

Hazardous materials. Highway mobile vehicles and trailers stored in garages in accordance with § 1910.110 may be equipped to carry more than one LP-gas container, but the total capacity of LP-gas containers per work vehicle stored in garages shall not exceed 100 pounds of LP-gas. All container valves shall be closed when not in use.

1910.268(b)(5)

Compressed gas. When using or transporting nitrogen cylinders in a horizontal position, special compartments, racks, or adequate blocking shall be provided to prevent cylinder movement. Regulators shall be removed or guarded before a cylinder is transported.

1910.268(b)(6)

Support structures. No employee, or any material or equipment, may be supported or permitted to be supported on any portion of a pole structure, platform, ladder, walkway or other elevated structure or aerial device unless the employer ensures that the support structure is first inspected by a competent person and it is determined to be adequately strong, in good working condition and properly secured in place.

1910.268(b)(7)

Approach distances to exposed energized overhead power lines and parts. The employer shall ensure that no employee approaches or takes any conductive object closer to any electrically energized overhead power lines and parts than prescribed in Table R-2, unless:

1910.268(b)(7)(i)

The employee is insulated or guarded from the energized parts (insulating gloves rated for the voltage involved shall be considered adequate insulation), or

1910.268(b)(7)(ii)

The energized parts are insulated or guarded from the employee and any other conductive object at a different potential, or

1910.268(b)(7)(iii)

The power conductors and equipment are deenergized and grounded.

Table R-2   Approach Distances to Exposed Energized Overhead Power Lines and Parts

Voltage range (phase to phase, RMS)	Approach distance (inches)
300 V and less	([1])
Over 300V, not over 750V	12
Over 750V not over 2 kV	18
Over 2 kV, not over 15 kV	24
Over 15 kV, not over 37 kV	36
Over 37 kV, not over 87.5 kV	42
Over 87.5 kV, not over 121 kV	48
Over 121 kV, not over 140 kV	54

[1]Avoid contact.

1910.268(b)(8)

Illumination of field work. Whenever natural light is insufficient to adequately illuminate the worksite, artificial illumination shall be provided to enable the employee to perform the work safely.

..1910.268(c)

1910.268(c)

Training. Employers shall provide training in the various precautions and safe practices described in this section and shall insure that employees do not engage in the activities to which this section applies until such employees have received proper training in the various precautions and safe practices required by this section. However, where the employer can demonstrate that an employee is already trained in the precautions and safe practices required by this section prior to his employment, training need not be provided to that employee in accordance with this section. Where training is required, it shall consist of on-the-job training or classroom-type training or a combination of both. The employer shall certify that employees have been trained by preparing a certification record which includes the identity of the person trained, the signature of the employer or the person who conducted the training, and the date the training was completed. The certification record shall be prepared at the completion of training and shall be maintained on file for the duration of the employee's employment. The certification record shall be made available upon request to the Assistant Secretary for Occupational Safety and Health. Such training shall, where appropriate, include the following subjects:

1910.268(c)(1)

Recognition and avoidance of dangers relating to encounters with harmful substances and animal, insect, or plant life;

1910.268(c)(2)

Procedures to be followed in emergency situations; and,

1910.268(c)(3)

First aid training, including instruction in artificial respiration.

1910.268(d)

Employee protection in public work areas.

1910.268(d)(1)

Before work is begun in the vicinity of vehicular or pedestrian traffic which may endanger employees, warning signs and/or flags or other traffic control devices shall be placed conspicuously to alert and channel approaching traffic. Where further protection is needed, barriers shall be utilized. At night, warning lights shall be prominently displayed, and excavated areas shall be enclosed with protective barricades.

1910.268(d)(2)

If work exposes energized or moving parts that are normally protected, danger signs shall be displayed and barricades erected, as necessary, to warn other personnel in the area.

1910.268(d)(3)

The employer shall insure that an employee finding any crossed or fallen wires which create or may create a hazardous situation at the work area:

1910.268(d)(3)(i)

Remains on guard or adopts other adequate means to warn other employees of the danger and

1910.268(d)(3)(ii)

has the proper authority notified at the earliest practical moment.

1910.268(e)

Tools and personal protective equipment—Generally. Personal protective equipment, protective devices and special tools needed for the work of employees shall be provided and the employer shall ensure that they are used by employees. Before each day's use the employer shall ensure that these personal protective devices, tools, and equipment are carefully inspected by a competent person to ascertain that they are in good condition.

..1910.268(f)

1910.268(f)

Rubber insulating equipment.

1910.268(f)(1)

Rubber insulating equipment designed for the voltage levels to be encountered shall be provided and the employer shall ensure that they are used by employees as required by this section. The requirements of § 1910.137, Electrical Protective Equipment, shall be followed except for Table I-6.

1910.268(f)(2)

The employer is responsible for the periodic retesting of all insulating gloves, blankets, and other rubber insulating equipment. This retesting shall be electrical, visual and mechanical. The following maximum retesting intervals shall apply:

Gloves, blankets, and other insulating equipment	Natural rubber	Synthetic rubber
	Months	
New	12	18
Re-issued	9	15

1910.268(f)(3)

Gloves and blankets shall be marked to indicate compliance with the retest schedule, and shall be marked with the date the next test is due. Gloves found to be defective in the field or by the tests set forth in paragraph (f)(2) of this section shall be destroyed by cutting them open from the finger to the gauntlet.

1910.268(g)

Personal climbing equipment—

1910.268(g)(1)

General. Safety belts and straps shall be provided and the employer shall ensure their use when work is performed at positions more than 4 feet above ground, on poles, and on towers, except as provided in paragraphs (n)(7) and (n)(8) of this section. No safety belts, safety straps or lanyards acquired after July 1, 1975 may be used unless they meet the tests set forth in paragraph (g)(2) of this section. The employer shall ensure that all safety belts and straps are inspected by a competent person prior to each day's use to determine that they are in safe working condition.

1910.268(g)(2)

Telecommunication lineman's body belts, safety straps, and lanyards—

..1910.268(g)(2)(i)

1910.268(g)(2)(i)

General requirements.

1910.268(g)(2)(i)(A)

Hardware for lineman's body belts, safety straps, and lanyards shall be drop forged or pressed steel and shall have a corrosion resistant finish tested to meet the requirements of the American Society for Testing and Materials B117-64, which is incorporated by reference as specified in § 1910.6 (50-hour test). Surfaces shall be smooth and free of sharp edges. Production samples of lineman's safety straps, body belts and lanyards shall be approved by a nationally recognized testing laboratory, as having been tested in accordance with and as meeting the requirements of this paragraph.

1910.268(g)(2)(i)(B)

All buckles shall withstand a 2,000-pound tensile test with a maximum permanent deformation no greater than one sixty-forth inch.

1910.268(g)(2)(i)(C)

D rings shall withstand a 5,000 pound tensile test without cracking or breaking.

1910.268(g)(2)(i)(D)

Snaphooks shall withstand a 5,000 pound tensile test, or shall withstand a 3,000-pound tensile test and a 180° bend test. Tensile failure is indicated by distortion of the snaphook sufficient to release the keeper; bend test failure is indicated by cracking of the snaphook.

1910.268(g)(2)(ii)

Specific requirements.

1910.268(g)(2)(ii)(A)(1)

All fabric used for safety straps shall be capable of withstanding an A.C. dielectric test of not less than 25,000 volts per foot "dry" for 3 minutes, without visible deterioration.

1910.268(g)(2)(ii)(A)(2)

All fabric and leather used shall be tested for leakage current. Fabric or leather may not be used if the leakage current exceeds 1 milliampere when a potential of 3,000 volts is applied to the electrodes positioned 12 inches apart.

1910.268(g)(2)(ii)(A)(3)

In lieu of alternating current tests, equivalent direct current tests may be performed.

..1910.268(g)(2)(ii)(B)

1910.268(g)(2)(ii)(B)

The cushion part of the body belt shall:

1910.268(g)(2)(ii)(B)(1)

Contain no exposed rivets on the inside. This provision does not apply to belts used by craftsmen not engaged in line work.

1910.268(g)(2)(ii)(B)(2)

Be at least three inches in width;

1910.268(g)(2)(ii)(B)(3)

Be at least five thirty-seconds (5/32) inch thick, if made of leather; and

1910.268(g)(2)(ii)(C)

[Reserved]

1910.268(g)(2)(ii)(D)

Suitable copper, steel, or equivalent liners shall be used around the bars of D rings to prevent wear between these members and the leather or fabric enclosing them.

1910.268(g)(2)(ii)(E)

All stitching shall be done with a minimum 42 pound weight nylon or equivalent thread and shall be lock stitched. Stitching parallel to an edge may not be less than three-sixteenths (3/16) inch from the edge of the narrowest member caught by the thread. The use of cross stitching on leather is prohibited.

..1910.268(g)(2)(ii)(F)

1910.268(g)(2)(ii)(F)

The keepers of snaphooks shall have a spring tension that will not allow the keeper to begin to open when a weight of 2 1/2 pounds or less is applied, but the keepers shall begin to open when a weight of four pounds is applied. In making this determination, the weight shall be supported on the keeper against the end of the nose.

1910.268(g)(2)(ii)(G)

Safety straps, lanyards, and body belts shall be tested in accordance with the following procedure:

1910.268(g)(2)(ii)(G)(1)

Attach one end of the safety strap or lanyard to a rigid support, and the other end to a 250 pound canvas bag of sand;

1910.268(g)(2)(ii)(G)(2)

Allow the 250 pound canvas bag of sand to free fall 4 feet when testing safety straps and 6 feet when testing lanyards. In each case, the strap or lanyard shall stop the fall of the 250 pound bag;

1910.268(g)(2)(ii)(G)(3)

Failure of the strap or lanyard shall be indicated by any breakage or slippage sufficient to permit the bag to fall free from the strap or lanyard.

1910.268(g)(2)(ii)(G)(4)

The entire "body belt assembly" shall be tested using on D ring. A safety strap or lanyard shall be used that is capable of passing the "impact loading test" described in paragraph (g)(2)(ii)(G)(*2*) of this section and attached as required in paragraph (g)(2)(ii)(G)(*1*) of this section. The body belt shall be secured to the 250 pound bag of sand at a point which simulates the waist of a man and shall be dropped as stated in paragraph (g)(2)(ii)(G)(*2*) of this section. Failure of the body belt shall be indicated by any breakage or slippage sufficient to permit the bag to fall free from the body belt.

1910.268(g)(3)

Pole climbers.

1910.268(g)(3)(i)

Pole climbers may not be used if the gaffs are less than 1 1/4 inches in length as measured on the underside of the gaff. The gaffs of pole climbers shall be covered with safety caps when not being used for their intended use.

1910.268(g)(3)(ii)

The employer shall ensure that pole climbers are inspected by a competent person for the following conditions: Fractured or cracked gaffs or leg irons, loose or dull gaffs, broken straps or buckles. If any of these conditions exist, the defect shall be corrected before the climbers are used.

1910.268(g)(3)(iii)

Pole climbers shall be inspected as required in this paragraph (g)(3) before each day's use and a gaff cut-out test performed at least weekly when in use.

..1910.268(g)(3)(iv)

1910.268(g)(3)(iv)

Pole climbers may not be worn when:

1910.268(g)(3)(iv)(A)

Working in trees (specifically designed tree climbers shall be used for tree climbing),

1910.268(g)(3)(iv)(B)

Working on ladders,

1910.268(g)(3)(iv)(C)

Working in an aerial lift,

1910.268(g)(3)(iv)(D)

Driving a vehicle, nor

1910.268(g)(3)(iv)(E)

Walking on rocky, hard, frozen, brushy or hilly terrain.

1910.268(h)

Ladders.

1910.268(h)(1)

The employer shall ensure that no employee nor any material or equipment may be supported or permitted to be supported on any portion of a ladder unless it is first determined, by inspections and checks conducted by a competent person that such ladder is adequately strong, in good condition, and properly secured in place, as required in Subpart D of this part and as required in this section.

1910.268(h)(2)

The spacing between steps or rungs permanently installed on poles and towers shall be no more than 18 inches (36 inches on any one side). This requirement also applies to fixed ladders on towers, when towers are so equipped. Spacing between steps shall be uniform above the initial unstepped section, except where working, standing, or access steps are required. Fixed ladder rungs and step rungs for poles and towers shall have a minimum diameter of 5/8". Fixed ladder rungs shall have a minimum clear width of 12 inches. Steps for poles and towers shall have a minimum clear width of 4 1/2 inches. The spacing between detachable steps may not exceed 30 inches on any one side, and these steps shall be properly secured when in use.

1910.268(h)(3)

Portable wood ladders intended for general use may not be painted but may be coated with a translucent nonconductive coating. Portable wood ladders may not be longitudinally reinforced with metal.

1910.268(h)(4)

Portable wood ladders that are not being carried on vehicles and are not in active use shall be stored where they will not be exposed to the elements and where there is good ventilation.

1910.268(h)(5)

The provisions of § 1910.25(c)(5) shall apply to rolling ladders used in telecommunications centers, except that such ladders shall have a minimum inside width, between the side rails, of at least eight inches.

..1910.268(h)(6)
1910.268(h)(6)

Climbing ladders or stairways on scaffolds used for access and egress shall be affixed or built into the scaffold by proper design and engineering, and shall be so located that their use will not disturb the stability of the scaffold. The rungs of the climbing device shall be equally spaced, but may not be less than

12 inches nominal nor more than 16 inches nominal apart. Horizontal end rungs used for platform support may also be utilized as a climbing device if such rungs meet the spacing requirement of this paragraph (h)(6), and if there is sufficient clearance between the rung and the edge of the platform to afford an adequate handhold. If a portable ladder is affixed to the scaffold, it shall be securely attached and shall have rungs meeting the spacing requirements of this paragraph (h)(6). Clearance shall be provided in the back of the ladder of not less than 6 inches from center of rung to the nearest scaffold structural member.

1910.268(h)(7)

When a ladder is supported by an aerial strand, and ladder hooks or other supports are not being used, the ladder shall be extended at least 2 feet above the strand and shall be secured to it (e.g. lashed or held by a safety strap around the strand and ladder side rail). When a ladder is supported by a pole, it shall be securely lashed to the pole unless the ladder is specifically designed to prevent movement when used in this application.

1910.268(h)(8)

The following requirements apply to metal manhole ladders.

1910.268(h)(8)(i)

Metal manhole ladders shall be free of structural defects and free of accident hazards such as sharp edges and burrs. The metal shall be protected against corrosion unless inherently corrosion-resistant.

1910.268(h)(8)(ii)

These ladders may be designed with parallel side rails, or with side rails varying uniformly in separation along the length (tapered), or with side rails flaring at the base to increase stability.

1910.268(h)(8)(iii)

The spacing of rungs or steps shall be on 12-inch centers.

1910.268(h)(8)(iv)

Connections between rungs or steps and siderails shall be constructed to insure rigidity as well as strength.

1910.268(h)(8)(v)

Rungs and steps shall be corrugated, knurled, dimpled, coated with skid-resistant material, or otherwise treated to minimize the possibility of slipping.

1910.268(h)(8)(vi)

Ladder hardware shall meet the strength requirements of the ladder's component parts and shall be of a material that is protected against corrosion unless inherently corrosion-resistant. Metals shall be so selected as to avoid excessive galvanic action.

1910.268(i)

Other tools and personal protective equipment—

..1910.268(i)(1)

1910.268(i)(1)

Head protection. Head protection meeting the requirements of ANSI Z89.2-1971, "Safety Requirements for Industrial Protective Helmets for Electrical Workers, Class B" shall be provided whenever there is exposure to possible high voltage electrical contact, and the employer shall ensure that the head protection is used by employees. ANSI Z89.2-1971 is incorporated by reference as specified in § 1910.6.

1910.268(i)(2)

Eye protection. Eye protection meeting the requirements of §1910.133 (a)(2) thru (a)(6) shall be provided and the employer shall ensure its use by employees where foreign objects may enter the eyes due to work operations such as but not limited to:

1910.268(i)(2)(i)

Drilling or chipping stone, brick or masonry, breaking concrete or pavement, etc. by hand tools (sledgehammer, etc.) or power tools such as pneumatic drills or hammers;

1910.268(i)(2)(ii)

Working on or around high speed emery or other grinding wheels unprotected by guards;

1910.268(i)(2)(iii)

Cutting or chipping terra cotta ducts, tile, etc.;

1910.268(i)(2)(iv)

Working under motor vehicles requiring hammering;

1910.268(i)(2)(v)

Cleaning operations using compressed air, steam, or sand blast;

..1910.268(i)(2)(vi)

1910.268(i)(2)(vi)

Acetylene welding or similar operations where sparks are thrown off;

1910.268(i)(2)(vii)

Using powder actuated stud drivers;

1910.268(i)(2)(viii)

Tree pruning or cutting underbrush;

1910.268(i)(2)(ix)

Handling battery cells and solutions, such as taking battery readings with a hydrometer and thermometer;

1910.268(i)(2)(x)

Removing or rearranging strand or open wire; and

1910.268(i)(2)(xi)

Performing lead sleeve wiping and while soldering.

1910.268(i)(3)

Tent heaters. Flame-type heaters may not be used within ground tents or on platforms within aerial tents unless:

..1910.268(i)(3)(i)
1910.268(i)(3)(i)

The tent covers are constructed of fire resistant materials, and

1910.268(i)(3)(ii)

Adequate ventilation is provided to maintain safe oxygen levels and avoid harmful buildup of combustion products and combustible gases.

1910.268(i)(4)

Torches. Torches may be used on aerial splicing platforms or in buckets enclosed by tents provided the tent material is constructed of fire resistant material and the torch is turned off when not in actual use. Aerial tents shall be adequately ventilated while the torch is in operation.

1910.268(i)(5)

Portable power equipment. Nominal 120V, or less, portable generators used for providing power at work locations do not require grounding if the output circuit is completely isolated from the frame of the unit.

1910.268(i)(6)

Vehicle-mounted utility generators. Vehicle-mounted utility generators used for providing nominal 240V AC or less for powering portable tools and equipment need not be grounded to earth if all of the following conditions are met:

1910.268(i)(6)(i)

One side of the voltage source is solidly strapped to the metallic structure of the vehicle;

1910.268(i)(6)(ii)

Grounding-type outlets are used, with a "grounding" conductor between the outlet grounding terminal and the side of the voltage source that is strapped to the vehicle;

1910.268(i)(6)(iii)

All metallic encased tools and equipment that are powered from this system are equipped with three-wire cords and grounding-type attachment plugs, except as designated in paragraph (i)(7) of this section.

1910.268(i)(7)

Portable lights, tools, and appliances. Portable lights, tools, and appliances having noncurrent-carrying external metal housing may be used with power equipment described in paragraph (i)(5) of this section without an equipment grounding conductor. When operated from commercial power such metal parts of these devices shall be grounded, unless these tools or appliances are protected by a system of double insulation, or its equivalent. Where such a system is employed, the equipment shall be distinctively marked to indicate double insulation.

1910.268(i)(8)

Soldering devices. Grounding shall be omitted when using soldering irons, guns or wire-wrap tools on telecommunications circuits.

..1910.268(i)(9)

1910.268(i)(9)

Lead work. The wiping of lead joints using melted solder, gas fueled torches, soldering irons or other appropriate heating devices, and the soldering of wires or other electrical connections do not constitute the welding, cutting and brazing described in Subpart Q of this part. When operated from commercial power the metal housing of electric solder pots shall be grounded. Electric solder pots may be used with the power equipment described in paragraph (i)(5) of this section without a grounding conductor. The employer shall ensure that wiping gloves or cloths and eye protection are used in lead wiping operations. A drip pan to catch hot lead drippings shall also be provided and used.

1910.268(j)

Vehicle-mounted material handling devices and other mechanical equipment—

1910.268(j)(1)

General.

1910.268(j)(1)(i)

The employer shall ensure that visual inspections are made of the equipment by a competent person each day the equipment is to be used to ascertain that it is in good condition.

1910.268(j)(1)(ii)

The employer shall ensure that tests shall be made at the beginning of each shift by a competent person to insure the vehicle brakes and operating systems are in proper working condition.

1910.268(j)(2)

Scrapers, loaders, dozers, graders and tractors.

1910.268(j)(2)(i)

All rubber-tired, self-propelled scrapers, rubber-tired front end loaders, rubber-tired dozers, agricultural and industrial tractors, crawler tractors, crawler-type loaders, and motor graders, with or without attachments, that are used in telecommunications work shall have rollover protective structures that meet the requirements of Subpart W of Part 1926 of this Title.

1910.268(j)(2)(ii)

Eye protection shall be provided and the employer shall ensure that it is used by employees when working in areas where flying material is generated.

1910.268(j)(3)

Vehicle-mounted elevating and rotating work platforms. These devices shall not be operated with any conductive part of the equipment closer to exposed energized power lines than the clearances set forth in Table R-2 of this section.

1910.268(j)(4)

Derrick trucks and similar equipment.

1910.268(j)(4)(i)

This equipment shall not be operated with any conductive part of the equipment closer to exposed energized power lines than the clearances set forth in Table R-2 of this section.

1910.268(j)(4)(ii)

When derricks are used to handle poles near energized power conductors, these operations shall comply with the requirements contained in paragraphs (b)(7) and (n)(11) of this section.

1910.268(j)(4)(iii)

Moving parts of equipment and machinery carried on or mounted on telecommunications line trucks shall be guarded. This may be done with barricades as specified in paragraph (d)(2) of this section.

..1910.268(j)(4)(iv)

1910.268(j)(4)(iv)

Derricks and the operation of derricks shall comply with the following requirements:

1910.268(j)(4)(iv)(A)

Manufacturer's specifications, load ratings and instructions for derrick operation shall be strictly observed.

1910.268(j)(4)(iv)(B)

Rated load capacities and instructions related to derrick operation shall be conspicuously posted on a permanent weather-resistant plate or decal in a location on the derrick that is plainly visible to the derrick operator.

1910.268(j)(4)(iv)(C)

Prior to derrick operation the parking brake must be set and the stabilizers extended if the vehicle is so equipped. When the vehicle is situated on a grade, at least two wheels must be chocked on the downgrade side.

1910.268(j)(4)(iv)(D)

Only persons trained in the operation of the derrick shall be permitted to operate the derrick.

1910.268(j)(4)(iv)(E)

Hand signals to derrick operators shall be those prescribed by ANSI B30.6-1969, "Safety Code for Derricks", which is incorporated by reference as specified in § 1910.6.

1910.268(j)(4)(iv)(F)

The employer shall ensure that the derrick and its associated equipment are inspected by a competent person at intervals set by the manufacturer but in no case less than once per year. Records shall be maintained including the dates of inspections, and necessary repairs made, if corrective action was required.

..1910.268(j)(4)(iv)(G)

1910.268(j)(4)(iv)(G)

Modifications or additions to the derrick and its associated equipment that alter its capacity or affect its safe operation shall be made only with written certification from the manufacturer, or other equivalent entity, such as a nationally recognized testing laboratory, that the modification results in the equipment being safe for its intended use. Such changes shall require the changing and posting of revised capacity and instruction decals or plates. These new ratings or limitations shall be as provided by the manufacturer or other equivalent entity.

1910.268(j)(4)(iv)(H)

Wire rope used with derricks shall be of improved plow steel or equivalent. Wire rope safety factors shall be in accordance with American National Standards Institute B30.6-1969.

1910.268(j)(4)(iv)(I)

Wire rope shall be taken out of service, or the defective portion removed, when any of the following conditions exist:

1910.268(j)(4)(iv)(I)(1)

The rope strength has been significantly reduced due to corrosion, pitting, or excessive heat, or

1910.268(j)(4)(iv)(I)(2)

The thickness of the outer wires of the rope has been reduced to two-thirds or less of the original thickness, or

1910.268(j)(4)(iv)(I)(3)

There are more than six broken wires in any one rope lay, or

1910.268(j)(4)(iv)(I)(4)

There is excessive permanent distortion caused by kinking, crushing, or severe twisting of the rope.

..1910.268(k)

1910.268(k)

Materials handling and storage—

1910.268(k)(1)

Poles. When working with poles in piles or stacks, work shall be performed from the ends of the poles as much as possible, and precautions shall be taken for the safety of employees at the other end of the pole. During pole hauling operations, all loads shall be secured to prevent displacement. Lights, reflectors and/or flags shall be displayed on the end and sides of the load as necessary. The requirements for installation, removal, or other handling of poles in pole lines are prescribed in paragraph (n) of this section which pertains to overhead lines. In the case of hoisting machinery equipped with a positive stop load-holding device, it shall be permissible for the operator to leave his position at the controls (while a load is suspended) for the sole purpose of assisting in positioning the load prior to landing it. Prior to unloading steel, poles, crossarms, and similar material, the load shall be thoroughly examined to ascertain that the load has not shifted, that binders or stakes have not broken, and that the load is not otherwise hazardous to employees.

1910.268(k)(2)

Cable reels. Cable reels in storage shall be checked or otherwise restrained when there is a possibility that they might accidentally roll from position.

1910.268(l)

Cable fault locating and testing.

1910.268(l)(1)

Employees involved in using high voltages to locate trouble or test cables shall be instructed in the precautions necessary for their own safety and the safety of other employees.

1910.268(l)(2)

Before the voltage is applied, cable conductors shall be isolated to the extent practicable. Employees shall be warned, by such techniques as briefing and tagging at all affected locations, to stay clear while the voltage is applied.

1910.268(m)

Grounding for employee protection—pole lines—

1910.268(m)(1)

Power conductors.   Electric power conductors and equipment shall be considered as energized unless the employee can visually determine that they are bonded to one of the grounds listed in paragraph (m)(4) of this section.

1910.268(m)(2)

Nonworking open wire.   Nonworking open wire communications lines shall be bonded to one of the grounds listed in paragraph (m)(4) of this section.

1910.268(m)(3)

Vertical power conduit, power ground wires and street light fixtures.

1910.268(m)(3)(i)

Metal power conduit on joint use poles, exposed vertical power ground wires, and street light fixtures which are below communications attachments or less than 20 inches above these attachments, shall be considered energized and shall be tested for voltage unless the employee can visually determine that they are bonded to the communications suspension strand or cable sheath.

1910.268(m)(3)(ii)

If no hazardous voltage is shown by the voltage test, a temporary bond shall be placed between such street light fixture, exposed vertical power grounding conductor, or metallic power conduit and the communications cable strand. Temporary bonds used for this purpose shall have sufficient conductivity to carry at least 500 amperes for a period of one second without fusing.

1910.268(m)(4)

Suitable protective grounding. Acceptable grounds for protective grounding are as follows:

1910.268(m)(4)(i)

A vertical ground wire which has been tested, found safe, and is connected to a power system multigrounded neutral or the grounded neutral of a power secondary system where there are at least three services connected;

1910.268(m)(4)(ii)

Communications cable sheath or shield and its supporting strand where the sheath or shield is:

..1910.268(m)(4)(ii)(A)

1910.268(m)(4)(ii)(A)

Bonded to an underground or buried cable which is connected to a central office ground, or

1910.268(m)(4)(ii)(B)

Bonded to an underground metallic piping system, or

1910.268(m)(4)(ii)(C)

Bonded to a power system multigrounded neutral or grounded neutral of a power secondary system which has at least three services connected;

1910.268(m)(4)(iii)

Guys which are bonded to the grounds specified in paragraphs (m)(4)(i) and (ii) of this section and which have continuity uninterrupted by an insulator; and

1910.268(m)(4)(iv)

If all of the preceding grounds are not available, arrays of driven ground rods where the resultant resistance to ground will be low enough to eliminate danger to personnel or permit prompt operation of protective devices.

..1910.268(m)(5)

1910.268(m)(5)

Attaching and removing temporary bonds. When attaching grounds (bonds), the first attachment shall be made to the protective ground. When removing bonds, the connection to the line or equipment shall be removed first. Insulating gloves shall be worn during these operations.

1910.268(m)(6)

Temporary grounding of suspension strand.

1910.268(m)(6)(i)

The suspension strand shall be grounded to the existing grounds listed in paragraph (m)(4) of this section when being placed on jointly used poles or during thunderstorm activity.

1910.268(m)(6)(ii)

Where power crossings are encountered on nonjoint lines, the strand shall be bonded to an existing ground listed in paragraph (m)(4) of this section as close as possible to the crossing. This bonding is not required where crossings are made on a common crossing pole unless there is an upward change in grade at the pole.

1910.268(m)(6)(iii)

Where roller-type bonds are used, they shall be restrained so as to avoid stressing the electrical connections.

1910.268(m)(6)(iv)

Bonds between the suspension strand and the existing ground shall be at least No. 6AWG copper.

1910.268(m)(6)(v)

Temporary bonds shall be left in place until the strand has been tensioned, dead-ended, and permanently grounded.

1910.268(m)(6)(vi)

The requirements of paragraphs (m)(6)(i) through (m)(6)(v) of this section do not apply to the installation of insulated strand.

1910.268(m)(7)

Antenna work-radio transmitting stations 3–30 MHZ.

1910.268(m)(7)(i)

Prior to grounding a radio transmitting station antenna, the employer shall insure that the rigger in charge:

..1910.268(m)(7)(i)(A)

1910.268(m)(7)(i)(A)

Prepares a danger tag signed with his signature,

1910.268(m)(7)(i)(B)

Requests the transmitting technician to shutdown the transmitter and to ground the antenna with its grounding switch,

1910.268(m)(7)(i)(C)

Is notified by the transmitting technician that the transmitter has been shutdown, and

1910.268(m)(7)(i)(D)

Tags the antenna ground switch personally in the presence of the transmitting technician after the antenna has been grounded by the transmitting technician.

1910.268(m)(7)(ii)

Power shall not be applied to the antenna, nor shall the grounding switch be opened under any circumstances while the tag is affixed.

1910.268(m)(7)(iii)(A)

Where no grounding switches are provided, grounding sticks shall be used, one on each side of line, and tags shall be placed on the grounding sticks, antenna switch, or plate power switch in a conspicuous place.

1910.268(m)(7)(iii)(B)

When necessary to further reduce excessive radio frequency pickup, ground sticks or short circuits shall be placed directly on the transmission lines near the transmitter in addition to the regular grounding switches.

1910.268(m)(7)(iii)(C)

In other cases, the antenna lines may be disconnected from ground and the transmitter to reduce pickup at the point in the field.

1910.268(m)(7)(iv)

All radio frequency line wires shall be tested for pickup with an insulated probe before they are handled either with bare hands or with metal tools.

1910.268(m)(7)(v)

The employer shall insure that the transmitting technician warn the riggers about adjacent lines which are, or may become energized.

1910.268(m)(7)(vi)

The employer shall insure that when antenna work has been completed, the rigger in charge of the job returns to the transmitter, notifies the transmitting technician in charge that work has been completed, and personally removes the tag from the antenna ground switch.

1910.268(n)

Overhead lines—

1910.268(n)(1)

Handling suspension strand.

..*1910.268(n)(1)(i)*
1910.268(n)(1)(i)

The employer shall insure that when handling cable suspension strand which is being installed on poles carrying exposed energized power conductors, employees shall wear insulating gloves and shall avoid body contact with the strand until after it has been tensioned, dead-ended and permanently grounded.

1910.268(n)(1)(ii)

The strand shall be restrained against upward movement during installation:

1910.268(n)(1)(ii)(A)

On joint-use poles, where there is an upward change in grade at the pole, and

1910.268(n)(1)(ii)(B)

On non-joint-use poles, where the line crosses under energized power conductors.

1910.268(n)(2)

Need for testing wood poles. Unless temporary guys or braces are attached, the following poles shall be tested in accordance with paragraph (n)(3) of this section and determined to be safe before employees are permitted to climb them:

1910.268(n)(2)(i)

Dead-end poles, except properly braced or guyed "Y" or "T" cable junction poles,

1910.268(n)(2)(ii)

Straight line poles which are not storm guyed and where adjacent span lengths exceed 165 feet,

1910.268(n)(2)(iii)

Poles at which there is a downward change in grade and which are not guyed or braced corner poles or cable junction poles,

1910.268(n)(2)(iv)

Poles which support only telephone drop wire, and

1910.268(n)(2)(v)

Poles which carry less than ten communication line wires. On joint use poles, one power line wire shall be considered as two communication wires for purposes of this paragraph (n)(2)(v).

1910.268(n)(3)

Methods for testing wood poles. One of the following methods or an equivalent method shall be used for testing wood poles:

1910.268(n)(3)(i)

Rap the pole sharply with a hammer weighing about 3 pounds, starting near the ground line and continuing upwards circumferentially around the pole to a height of approximately 6 feet. The hammer will produce a clear sound and rebound sharply when striking sound wood. Decay pockets will be indicated by a dull sound and/or a less pronounced hammer rebound. When decay pockets are indicated, the pole shall be considered unsafe. Also, prod the pole as near the ground line as possible using a pole prod or a screwdriver with a blade at least 5 inches long. If substantial decay is encountered, the pole shall be considered unsafe.

..1910.268(n)(3)(ii)

1910.268(n)(3)(ii)

Apply a horizontal force to the pole and attempt to rock it back and forth in a direction perpendicular to the line. Caution shall be exercised to avoid causing power wires to swing together. The force may be applied either by pushing with a pike pole or pulling with a rope. If the pole cracks during the test, it shall be considered unsafe.

1910.268(n)(4)

Unsafe poles or structures. Poles or structures determined to be unsafe by test or observation may not be climbed until made safe by guying, bracing or other adequate means. Poles determined to be unsafe to climb shall, until they are made safe, be tagged in a conspicuous place to alert and warn all employees of the unsafe condition.

1910.268(n)(5)

Test requirements for cable suspension strand.

1910.268(n)(5)(i)

Before attaching a splicing platform to a cable suspension strand, the strand shall be tested and determined to have strength sufficient to support the weight of the platform and the employee. Where the strand crosses above power wires or railroad tracks it may not be tested but shall be inspected in accordance with paragraph (n)(6) of this section.

1910.268(n)(5)(ii)

The following method or an equivalent method shall be used for testing the strength of the strand: A rope, at least three-eighths inch in diameter, shall be thrown over the strand. On joint lines, the rope shall be passed over the strand using tree pruner handles or a wire raising tool. If two employees are present, both shall grip the double rope and slowly transfer their entire weight to the rope and attempt to raise themselves off the ground. If only one employee is present, one end of the rope which has been passed over the strand shall be tied o the bumper of the truck, or other equally secure anchorage. The employee then shall grasp the other end of the rope and attempt to raise himself off the ground.

..1910.268(n)(6)

1910.268(n)(6)

Inspection of strand. Where strand passes over electric power wires or railroad tracks, it shall be inspected from an elevated working position at each pole supporting the span in question. The strand may not be used to support any splicing platform, scaffold or cable car, if any of the following conditions exist:

1910.268(n)(6)(i)

Corrosion so that no galvanizing can be detected,

1910.268(n)(6)(ii)

One or more wires of the strand are broken,

1910.268(n)(6)(iii)

Worn spots, or

1910.268(n)(6)(iv)

Burn marks such as those caused by contact with electric power wires.

1910.268(n)(7)

Outside work platforms. Unless adequate railings are provided, safety straps and body belts shall be used while working on elevated work platforms such as aerial splicing platforms, pole platforms, ladder platforms and terminal balconies.

1910.268(n)(8)

Other elevated locations. Safety straps and body belts shall be worn when working at elevated positions on poles, towers or similar structures, which do not have adequately guarded work areas.

..1910.268(n)(9)

1910.268(n)(9)

Installing and removing wire and cable. Before installing or removing wire or cable, the pole or structure shall be guyed, braced, or otherwise supported, as necessary, to prevent failure of the pole or structure.

1910.268(n)(10)

Avoiding contact with energized power conductors or equipment. When cranes, derricks, or other mechanized equipment are used for setting, moving, or removing poles, all necessary precautions shall be taken to avoid contact with energized power conductors or equipment.

1910.268(n)(11)

Handling poles near energized power conductors.

1910.268(n)(11)(i)

Joint use poles may not be set, moved, or removed where the nominal voltage of open electrical power conductors exceeds 34.5kV phase to phase (20kV to ground).

1910.268(n)(11)(ii)

Poles that are to be placed, moved or removed during heavy rains, sleet or wet snow in joint lines carrying more than 8.7kV phase to phase voltage (5kV to ground) shall be guarded or otherwise prevented from direct contact with over-head energized power conductors.

1910.268(n)(11)(iii)(A)

In joint lines where the power voltage is greater than 750 volts but less than 34.5kV phase to phase (20 kV to ground), wet poles being placed, moved or removed shall be insulated with either a rubber insulating blanket, a fiberglass box guide, or equivalent protective equipment.

1910.268(n)(11)(iii)(B)

In joint lines where the power voltage is greater than 8.7 kV phase to phase (5kV to ground) but less than 34.5kV phase to phase (20 kV to ground), dry poles being placed, moved, or removed shall be insulated with either a rubber insulating blanket, a fiberglass box guide, or equivalent protective equipment.

1910.268(n)(11)(iii)(C)

Where wet or dry poles are being removed, insulation of the pole is not required if the pole is cut off 2 feet or more below the lowest power wire and also cut off near the ground line.

1910.268(n)(11)(iv)

Insulating gloves shall be worn when handling the pole with either hands or tools, when there exists a possibility that the pole may contact a power conductor. Where the voltage to ground of the power conductor exceeds 15kV to ground, Class II gloves (as defined in ANSI J6.6-1971) shall be used. For voltages not exceeding 15kV to ground, insulating gloves shall have a breakdown voltage of at least 17kV.

1910.268(n)(11)(v)

The guard or insulating material used to protect the pole shall meet the appropriate 3 minute proof test voltage requirements contained in the ANSI J6.4-1971.

..1910.268(n)(11)(vi)

1910.268(n)(11)(vi)

When there exists a possibility of contact between the pole or the vehicle-mounted equipment used to handle the pole, and an energized power conductor, the following precautions shall be observed:

1910.268(n)(11)(vi)(A)

When on the vehicle which carries the derrick, avoid all contact with the ground, with persons standing on the ground, and with all grounded objects such as guys, tree limbs, or metal sign posts. To the extent feasible, remain on the vehicle as long as the possibility of contact exists.

1910.268(n)(11)(vi)(B)

When it is necessary to leave the vehicle, step onto an insulating blanket and break all contact with the vehicle before stepping off the blanket and onto the ground. As a last resort, if a blanket is not available, the employee may jump cleanly from the vehicle.

1910.268(n)(11)(vi)(C)

When it is necessary to enter the vehicle, first step onto an insulating blanket and break all contact with the ground, grounded objects and other persons before touching the truck or derrick.

1910.268(n)(12)

Working position on poles. Climbing and working are prohibited above the level of the lowest electric power conducter on the pole (exclusive of vertical runs and street light wiring), except:

1910.268(n)(12)(i)

Where communications facilities are attached above the electric power conductors, and a rigid fixed barrier is installed between the electric power facility and the communications facility, or

1910.268(n)(12)(ii)

Where the electric power conductors are cabled secondary service drops carrying less than 300 volts to ground and are attached 40 inches or more below the communications conductors or cables.

1910.268(n)(13)

Metal tapes and ropes.

1910.268(n)(13)(i)

Metal measuring tapes, metal measuring ropes, or tapes containing conductive strands may not be used when working near exposed energized parts.

1910.268(n)(13)(ii)

Where it is necessary to measure clearances from energized parts, only nonconductive devices shall be used.

1910.268(o)

Underground lines.   The provisions of this paragraph apply to the guarding of manholes and street openings, and to the ventilation and testing for gas in manholes and unvented vaults, where telecommunications field work is performed on or with underground lines.

1910.268(o)(1)

Guarding manholes and street openings.

1910.268(o)(1)(i)

When covers of manholes or vaults are removed, the opening shall be promptly guarded by a railing, temporary cover, or other suitable temporary barrier which is appropriate to prevent an accidental fall through the opening and to protect employees working in the manhole from foreign objects entering the manhole.

..1910.268(o)(1)(ii)

1910.268(o)(1)(ii)

While work is being performed in the manhole, a person with basic first aid training shall be immediately available to render assistance if there is cause for believing that a safety hazard exists, and if the requirements contained in paragraphs (d)(1) and (o)(1)(i) of this section do not adequately protect the employee(s). Examples of manhole worksite hazards which shall be considered to constitute a safety hazard include, but are not limited to:

1910.268(o)(1)(ii)(A)

Manhole worksites where safety hazards are created by traffic patterns that cannot be corrected by provisions of paragraph (d)(1) of this section.

1910.268(o)(1)(ii)(B)

Manhole worksites that are subject to unusual water hazards that cannot be abated by conventional means.

1910.268(o)(1)(ii)(C)

Manhole worksites that are occupied jointly with power utilities as described in paragraph (o)(3) of this section.

1910.268(o)(2)

Requirements prior to entering manholes and unvented vaults.

1910.268(o)(2)(i)

Before an employee enters a manhole, the following steps shall be taken:

1910.268(o)(2)(i)(A)

The internal atmosphere shall be tested for combustible gas and, except when continuous forced ventilation is provided, the atmosphere shall also be tested for oxygen deficiency.

1910.268(o)(2)(i)(B)

When unsafe conditions are detected by testing or other means, the work area shall be ventilated and otherwise made safe before entry.

..1910.268(o)(2)(ii)

1910.268(o)(2)(ii)

An adequate continuous supply of air shall be provided while work is performed in manholes under any of the following conditions:

1910.268(o)(2)(ii)(A)

Where combustible or explosive gas vapors have been initially detected and subsequently reduced to a safe level by ventilation,

1910.268(o)(2)(ii)(B)

Where organic solvents are used in the work procedure,

1910.268(o)(2)(ii)(C)

Where open flame torches are used in the work procedure,

1910.268(o)(2)(ii)(D)

Where the manhole is located in that portion of a public right of way open to vehicular traffic and/or exposed to a seepage of gas or gases, or

1910.268(o)(2)(ii)(E)

Where a toxic gas or oxygen deficiency is found.

1910.268(o)(2)(iii)(A)

The requirements of paragraphs (o)(2)(i) and (ii) of this section do not apply to work in central office cable vaults that are adequately ventilated.

1910.268(o)(2)(iii)(B)

The requirements of paragraphs (o)(2)(i) and (ii) of this section apply to work in unvented vaults.

1910.268(o)(3)

Joint power and telecommunication manholes. While work is being performed in a manhole occupied jointly by an electric utility and a telecommunication utility, an employee with basic first aid training shall be available in the immediate vicinity to render emergency assistance as may be required. The employee whose presence is required in the immediate vicinity for the purposes of rendering emergency assistance is not to be precluded from occasionally entering a manhole to provide assistance other than in an emergency. The requirement of this paragraph (o)(3) does not preclude a qualified employee, working alone, from entering for brief periods of time, a manhole where energized cables or equipment are in service, for the purpose of inspection, housekeeping, taking readings, or similar work if such work can be performed safely.

1910.268(o)(4)

Ladders. Ladders shall be used to enter and exit manholes exceeding 4 feet in depth.

1910.268(o)(5)

Flames. When open flames are used in manholes, the following precautions shall be taken to protect against the accumulation of combustible gas:

1910.268(o)(5)(i)

A test for combustible gas shall be made immediately before using the open flame device, and at least once per hour while using the device; and

1910.268(o)(5)(ii)

a fuel tank (e.g., acetylene) may not be in the manhole unless in actual use.

..1910.268(p)

1910.268(p)

Microwave transmission—

1910.268(p)(1)

Eye protection. Employers shall insure that employees do not look into an open waveguide which is connected to an energized source of microwave radiation.

1910.268(p)(2)

Hazardous area. Accessible areas associated with microwave communication systems where the electromagnetic radiation level exceeds the radiation protection guide given in § 1910.97 shall be posted as described in that section. The lower half of the warning symbol shall include the following:

Radiation in this area may exceed hazard limitations and special precautions are required. Obtain specific instruction before entering.

1910.268(p)(3)

Protective measures. When an employee works in an area where the electromagnetic radiation exceeds the radiation protection guide, the employer shall institute measures that insure that the employee's exposure is not greater than that permitted by the radiation guide. Such measures shall include, but not be limited to those of an administrative or engineering nature or those involving personal protective equipment.

1910.268(q)

Tree trimming—electrical hazards—

..1910.268(q)(1)

1910.268(q)(1)

General.

1910.268(q)(1)(i)

Employees engaged in pruning, trimming, removing, or clearing trees from lines shall be required to consider all overhead and underground electrical power conductors to be energized with potentially fatal voltages, never to be touched (contacted) either directly or indirectly.

1910.268(q)(1)(ii)

Employees engaged in line-clearing operations shall be instructed that:

1910.268(q)(1)(ii)(A)

A direct contact is made when any part of the body touches or contacts an energized conductor, or other energized electrical fixture or apparatus.

1910.268(q)(1)(ii)(B)

An indirect contact is made when any part of the body touches any object in contact with an energized electrical conductor, or other energized fixture or apparatus.

1910.268(q)(1)(ii)(C)

An indirect contact can be made through conductive tools, tree branches, trucks, equipment, or other objects, or as a result of communications wires, cables, fences, or guy wires being accidentally energized.

1910.268(q)(1)(ii)(D)

Electric shock will occur when an employee, by either direct or indirect contact with an energized conductor, energized tree limb, tool, equipment, or other object, provides a path for the flow of electricity to a grounded object or to the ground itself. Simultaneous contact with two energized conductors will also cause electric shock which may result in serious or fatal injury.

1910.268(q)(1)(iii)

Before any work is performed in proximity to energized conductors, the system operator/owner of the energized conductors shall be contacted to ascertain if he knows of any hazards associated with the conductors which may not be readily apparent. This rule does not apply when operations are performed by or on behalf of, the system operator/owner.

1910.268(q)(2)
Working in proximity to electrical hazards.

1910.268(q)(2)(i)

Employers shall ensure that a close inspection is made by the employee and by the foremen or supervisor in charge before climbing, entering, or working around any tree, to determine whether an electrical power conductor passes through the tree, or passes within reaching distance of an employee working in the tree. If any of these conditions exist either directly or indirectly, an electrical hazard shall be considered to exist unless the system operator/owner has caused the hazard to be removed by deenergizing the lines, or installing protective equipment.

..1910.268(q)(2)(ii)
1910.268(q)(2)(ii)

Only qualified employees or trainees, familiar with the special techniques and hazards involved in line clearance, shall be permitted to perform the work if it is found that an electrical hazard exists.

1910.268(q)(2)(iii)

During all tree working operations aloft where an electrical hazard of more than 750V exists, there shall be a second employee or trainee qualified in line clearance tree trimming within normal voice communication.

1910.268(q)(2)(iv)

Where tree work is performed by employees qualified in line-clearance tree trimming and trainees qualified in line-clearance tree trimming, the clearances from energized conductors given in Table R-3 shall apply.

Table R-3   Minimum Working Distances from Energized Conductors for Line-Clearance Tree Trimmers and Line-Clearance Tree-Trimmer Trainees

Voltage range (phase to phase)(kilovolts)	Minimum working distance
2.1 to 15.0	2 ft. 0 in.
15.1 to 35.0	2 ft. 4 in.
35.1 to 46.0	2 ft. 6 in.
46.1 to 72.5	3 ft. 0 in.
72.6 to 121.0	3 ft. 4 in.
138.0 to 145.0	3 ft. 6 in.
161.0 to 169.0	3 ft. 8 in.
230.0 to 242.0	5 ft. 0 in.
345.0 to 362.0	7 ft. 0 in.
500.0 to 552.0	11 ft. 0 in.
700.0 to 765.0	15 ft. 0 in.

1910.268(q)(2)(v)

Branches hanging on an energized conductor may only be removed using appropriately insulated equipment.

1910.268(q)(2)(vi)

Rubber footwear, including lineman's overshoes, shall not be considered as providing any measure of safety from electrical hazards.

1910.268(q)(2)(vii)

Ladders, platforms, and aerial devices, including insulated aerial devices, may not be brought in contact with an electrical conductor. Reliance shall not be placed on their dielectric capabilities.

..1910.268(q)(2)(viii)

1910.268(q)(2)(viii)

When an aerial lift device contacts an electrical conductor, the truck supporting the aerial lift device shall be considered as energized.

1910.268(q)(3)

Storm work and emergency conditions.

1910.268(q)(3)(i)

Since storm work and emergency conditions create special hazards, only authorized representatives of the electric utility system operator/owner and not telecommunication workers may perform tree work in these situations where energized electrical power conductors are involved.

1910.268(q)(3)(ii)

When an emergency condition develops due to tree operations, work shall be suspended and the system operator/owner shall be notified immediately.

1910.268(r)

Buried facilities—Communications lines and power lines in the same trench (Reserved)

1910.268(s)

Definitions—

1910.268(s)(1)

Aerial lifts. Aerial lifts include the following types of vehicle-mounted aerial devices used to elevate personnel to jobsites above ground:

1910.268(s)(1)(i)

Extensible boom platforms,

1910.268(s)(1)(ii)

Aerial ladders,

1910.268(s)(1)(iii)

Articulating boom platforms,

..1910.268(s)(1)(iv)
1910.268(s)(1)(iv)

Vertical towers,

1910.268(s)(1)(v)

A combination of any of the above defined in ANSI A92.2-1969, which is incorporated by reference as specified in § 1910.6. These devices are made of metal, wood, fiberglass reinforced plastic (FRP), or other material; are powered or manually operated; and are deemed to be aerial lifts whether or not they are capable of rotating about a substantially vertical axis.

1910.268(s)(2)

Aerial splicing platform.   This consists of a platform, approximately 3 ft. × 4 ft., used to perform aerial cable work. It is furnished with fiber or synthetic ropes for supporting the platform from aerial strand, detachable guy ropes for anchoring it, and a device for raising and lowering it with a handline.

1910.268(s)(3)

Aerial tent.   A small tent usually constructed of vinyl coated canvas which is usually supported by light metal or plastic tubing. It is designed to protect employees in inclement weather while working on ladders, aerial splicing platforms, or aerial devices.

1910.268(s)(4)

Alive or live (energized).   Electrically connected to a source of potential difference, or electrically charged so as to have a potential significantly different from that of the earth in the vicinity. The term "live" is sometimes used in the place of the term "current-carrying," where the intent is clear, to avoid repetition of the longer term.

1910.268(s)(5)

Barricade.   A physical obstruction such as tapes, cones, or "A" frame type wood and/or metal structure intended to warn and limit access to a work area.

1910.268(s)(6)

Barrier.   A physical obstruction which is intended to prevent contact with energized lines or equipment, or to prevent unauthorized access to work area.

1910.268(s)(7)

Bond.   An electrical connection from one conductive element to another for the purpose of minimizing potential differences or providing suitable conductivity for fault current or for mitigation of leakage current and electrolytic action.

1910.268(s)(8)

Cable. A conductor with insulation, or a stranded conductor with or without insulation and other coverings (single-conductor cable), or a combination of conductors insulated from one another (multiple-conductor cable).

1910.268(s)(9)

Cable sheath. A protective covering applied to cables.
Note: A cable sheath may consist of multiple layers of which one or more is conductive.

1910.268(s)(10)

Circuit. A conductor or system of conductors through which an electric current is intended to flow.

..1910.268(s)(11)

1910.268(s)(11)

Communication lines. The conductors and their supporting or containing structures for telephone, telegraph, railroad signal, data, clock, fire, police-alarm, community television antenna and other systems which are used for public or private signal or communication service, and which operate at potentials not exceeding 400 volts to ground or 750 volts between any two points of the circuit, and the transmitted power of which does not exceed 150 watts. When communications lines operate at less than 150 volts to ground, no limit is placed on the capacity of the system. Specifically designed communications cables may include communication circuits not complying with the preceding limitations, where such circuits are also used incidentally to supply power to communication equipment.

1910.268(s)(12)

Conductor. A material, usually in the form of a wire, cable, or bus bar, suitable for carrying an electric current.

1910.268(s)(13)

Effectively grounded. Intentionally connected to earth through a ground connection or connections of sufficiently low impedance and having sufficient current-carrying capacity to prevent the build-up of voltages which may result in undue hazard to connected equipment or to persons.

1910.268(s)(14)

Equipment. A general term which includes materials, fittings, devices, appliances, fixtures, apparatus, and similar items used as part of, or in connection with, a supply or communications installation.

1910.268(s)(15)

Ground (reference). That conductive body, usually earth, to which an electric potential is referenced.

1910.268(s)(16)

Ground (as a noun). A conductive connection, whether intentional or accidental, by which an electric circuit or equipment is connected to reference ground.

1910.268(s)(17)

Ground (as a verb). The connecting or establishment of a connection, whether by intention or accident, of an electric circuit or equipment to reference ground.

1910.268(s)(18)

Ground tent. A small tent usually constructed of vinyl coated canvas supported by a metal or plastic frame. Its purpose is to protect employees from inclement weather while working at buried cable pedestal sites or similar locations.

1910.268(s)(19)

Grounded conductor. A system or circuit conductor which is intentionally grounded.

..1910.268(s)(20)

1910.268(s)(20)

Grounded systems. A system of conductors in which at least one conductor or point (usually the middle wire, or the neutral point of transformer or generator windings) is intentionally grounded, either solidly or through a current-limiting device (not a current-interrupting device).

1910.268(s)(21)

Grounding electrode conductor. (Grounding conductor). A conductor used to connect equipment or the grounded circuit of a wiring system to a grounding electrode.

1910.268(s)(22)

Insulated. Separated from other conducting surfaces by a dielectric substance (including air space) offering a high resistance to the passage of current.

Note: When any object is said to be insulated, it is understood to be insulated in suitable manner for the conditions to which it is subjected. Otherwise, it is, within the purpose of these rules, uninsulated. Insulating coverings of conductors in one means of making the conductor insulated.

1910.268(s)(23)

Insulation (as applied to cable). That which is relied upon to insulate the conductor from other conductors or conducting parts or from ground.

1910.268(s)(24)

Joint use. The sharing of a common facility, such as a manhole, trench or pole, by two or more different kinds of utilities (e.g., power and telecommunications).

1910.268(s)(25)

Ladder platform. A device designed to facilitate working aloft from an extension ladder. A typical device consists of a platform (approximately 9" × 18") hinged to a welded pipe frame. The rear edge of the platform and the bottom cross-member of the frame are equipped with latches to lock the platform to ladder rungs.

1910.268(s)(26)

Ladder seat. A removable seat used to facilitate work at an elevated position on rolling ladders in telecommunication centers.

1910.268(s)(27)

Manhole. A subsurface enclosure which personnel may enter and which is used for the purpose of installing, operating, and maintaining submersible equipment and/or cable.

1910.268(s)(28)

Manhole platform. A platform consisting of separate planks which are laid across steel platform supports. The ends of the supports are engaged in the manhole cable racks.

1910.268(s)(29)

Microwave transmission. The act of communicating or signaling utilizing a frequency between 1 GHz (gigahertz) and 300 GHz inclusively.

..1910.268(s)(30)
1910.268(s)(30)

Nominal voltage. The nominal voltage of a system or circuit is the value assigned to a system or circuit of a given voltage class for the purpose of convenient designation. The actual voltage may vary above or below this value.

1910.268(s)(31)

Pole balcony or seat. A balcony or seat used as a support for workmen at pole-mounted equipment or terminal boxes. A typical device consists of a bolted assembly of steel details and a wooden platform. Steel braces run from the pole to the underside of the balcony. A guard rail (approximately 30" high) may be provided.

1910.268(s)(32)

Pole platform. A platform intended for use by a workman in splicing and maintenance operations in an elevated position adjacent to a pole. It consists of a platform equipped at one end with a hinged chain binder for securing the platform to a pole. A brace from the pole to the underside of the platform is also provided.

1910.268(s)(33)

Qualified employee. Any worker who by reason of his training and experience has demonstrated his ability to safely perform his duties.

1910.268(s)(34)

Qualified line-clearance tree trimmer. A tree worker who through related training and on-the-job experience is familiar with the special techniques and hazards involved in line clearance.

1910.268(s)(35)

Qualified line-clearance tree-trimmer trainee. Any worker regularly assigned to a line-clearance tree-trimming crew and undergoing on-the-job training who, in the course of such training, has demonstrated his ability to perform his duties safely at his level of training.

1910.268(s)(36)

System operator/owner. The person or organization that operates or controls the electrical conductors involved.

..1910.268(s)(37)

1910.268(s)(37)

Telecommunications center. An installation of communication equipment under the exclusive control of an organization providing telecommunications service, that is located outdoors or in a vault, chamber, or a building space used primarily for such installations.

Note: Telecommunication centers are facilities established, equipped and arranged in accordance with engineered plans for the purpose of providing telecommunications service. They may be located on premises owned or leased by the organization providing telecommunication service, or on the premises owned or leased by others. This definition includes switch rooms (whether electromechanical, electronic, or computer controlled), terminal rooms, power rooms, repeater rooms, transmitter and receiver rooms, switchboard operating rooms, cable vaults, and miscellaneous communications equipment rooms. Simulation rooms of telecommunication centers for training or developmental purposes are also included.

1910.268(s)(38)

Telecommunications derricks. Rotating or nonrotating derrick structures permanently mounted on vehicles for the purpose of lifting, lowering, or positioning hardware and materials used in telecommunications work.

1910.268(s)(39)

Telecommunication line truck. A truck used to transport men, tools, and material, and to serve as a traveling workshop for telecommunication installation and maintenance work. It is sometimes equipped with a boom and auxiliary equipment for setting poles, digging holes, and elevating material or men.

1910.268(s)(40)

Telecommunication service. The furnishing of a capability to signal or communicate at a distance by means such as telephone, telegraph, police and firealarm, community antenna television, or similar system, using wire, conventional cable, coaxial cable, wave guides, microwave transmission, or other similar means.

1910.268(s)(41)

Unvented vault. An enclosed vault in which the only openings are access openings.

1910.268(s)(42)

Vault. An enclosure above or below ground which personnel may enter, and which is used for the purpose of installing, operating, and/or maintaining equipment and/or cable which need not be of submersible design.

..1910.268(s)(43)

1910.268(s)(43)

Vented vault. An enclosure as described in paragraph(s) (42) of this section, with provision for air changes using exhaust flue stack(s) and low level air intake(s), operating on differentials of pressure and temperature providing for air flow.

1910.268(s)(44)

Voltage of an effectively grounded circuit. The voltage between any conductor and ground unless otherwise indicated.

1910.268(s)(45)

Voltage of a circuit not effectively grounded. The voltage between any two conductors. If one circuit is directly connected to and supplied from another circuit of higher voltage (as in the case of an autotransformer), both are considered as of the higher voltage, unless the circuit of lower voltage is effectively grounded, in which case its voltage is not determined by the circuit of higher voltage. Direct connection implies electric connection as distinguished from connection merely through electromagnetic or electrostatic induction.

[40 FR 13441, Mar. 26, 1975, as amended at 43 FR 49751, Oct. 24, 1978; 47 FR 14706, Apr. 6, 1982; 52 FR 36387, Sept. 28, 1987; 54 FR 24334, June 7, 1989; 61 FR 9227, March 7, 1996; 63 FR 33450, June 18, 1998; 67 FR 67965, Nov. 7, 2002; 69 FR 31882, June 8, 2004; 70 FR 1141, Jan. 5, 2005]

Reprinted from www.osha.gov

Occupational Safety & Health Administration
200 Constitution Avenue, NW
Washington, DC 20210

Appendix B

OSHA Standards Related to the NESC Work Rules

1910.269 (Includes Appendix A–E)

Electric Power Generation, Transmission, and Distribution
(Operation and Maintenance)
1910 Subpart R—Special Industries

OSHA 1910.269 Paragraph Titles:

(a) General
(b) Medical services and first aid
(c) Job briefing
(d) Hazardous energy control (lockout/tagout) procedures
(e) Enclosed spaces
(f) Excavations
(g) Personal protective equipment
(h) Ladders, platforms, step bolts, and manhole steps
(i) Hand and portable power tools
(j) Live-line tools
(k) Materials handling and storage
(l) Working on or near exposed energized parts
(m) Deenergizing lines and equipment for employee protection
(n) Grounding for the protection of employees
(o) Testing and test facilities
(p) Mechanical equipment
(q) Overhead lines
(r) Line-clearance tree trimming operations

(s) Communication facilities
(t) Underground electrical installations
(u) Substations
(v) Power generation
(w) Special conditions
(x) Definitions

Appendix A – Flow Charts
Appendix B – Working on Exposed Energized Parts
Appendix C – Protection from Step and Touch Potentials
Appendix D – Methods of Inspecting and Testing Wood Poles
Appendix E – Reference Documents

REGULATIONS (STANDARDS - 29 CFR)

ELECTRIC POWER GENERATION, TRANSMISSION, AND DISTRIBUTION. - 1910.269

- **Part Number:** 1910
- **Part Title:** Occupational Safety and Health Standards
- **Subpart:** R
- **Subpart Title:** Special Industries
- **Standard Number:** 1910.269
- **Title:** Electric Power Generation, Transmission, and Distribution.
- **Appendix:** A, B, C, D, E

1910.269(a)

"General."

1910.269(a)(1)

"Application."

1910.269(a)(1)(i)

This section covers the operation and maintenance of electric power generation, control, transformation, transmission, and distribution lines and equipment. These provisions apply to:

1910.269(a)(1)(i)(A)

Power generation, transmission, and distribution installations, including related equipment for the purpose of communication or metering, which are accessible only to qualified employees;

Note: The types of installations covered by this paragraph include the generation, transmission, and distribution installations of electric utilities, as well as equivalent installations of industrial establishments. Supplementary electric generating equipment that is used to supply a workplace for emergency, standby, or similar purposes only is covered under Subpart S of this Part. (See paragraph (a)(1)(ii)(B) of this section.)

1910.269(a)(1)(i)(B)

Other installations at an electric power generating station, as follows:

1910.269(a)(1)(i)(B)(1)

Fuel and ash handling and processing installations, such as coal conveyors,

1910.269(a)(1)(i)(B)(2)

Water and steam installations, such as penstocks, pipelines, and tanks, providing a source of energy for electric generators, and

1910.269(a)(1)(i)(B)(3)

Chlorine and hydrogen systems:

..1910.269(a)(1)(i)(C)

1910.269(a)(1)(i)(C)

Test sites where electrical testing involving temporary measurements associated with electric power generation, transmission, and distribution is performed in laboratories, in the field, in substations, and on lines, as opposed to metering, relaying, and routine line work;

1910.269(a)(1)(i)(D)

Work on or directly associated with the installations covered in paragraphs (a)(1)(i)(A) through (a)(1)(i)(C) of this section; and

1910.269(a)(1)(i)(E)

Line-clearance tree-trimming operations, as follows:

1910.269(a)(1)(i)(E)(1)

Entire 1910.269 of this Part, except paragraph (r)(1) of this section, applies to line-clearance tree-trimming operations performed by qualified employees (those who are knowledgeable in the construction and operation of electric power generation, transmission, or distribution equipment involved, along with the associated hazards).

1910.269(a)(1)(i)(E)(2)

Paragraphs (a)(2), (b), (c), (g), (k), (p), and (r) of this section apply to line-clearance tree-trimming operations performed by line-clearance tree trimmers who are not qualified employees.

1910.269(a)(1)(ii)

Notwithstanding paragraph (A)(1)(i) of this section, 1910.269 of this Part does not apply:

1910.269(a)(1)(ii)(A)

To construction work, as defined in 1910.12 of this Part; or

1910.269(a)(1)(ii)(B)

To electrical installations, electrical safety-related work practices, or electrical maintenance considerations covered by Subpart S of this Part.

Note 1: Work practices conforming to 1910.332 through 1910.335 of this Part are considered as complying with the electrical safety-related work practice requirements of this section identified in Table 1 of Appendix A-2 to this section, provided the work is being performed on a generation or distribution installation meeting 1910.303 through 1910.308 of this Part. This table also identifies provisions in this section that apply to work by qualified persons directly on or associated with installations of electric power generation, transmission, and distribution lines or equipment, regardless of compliance with 1910.332 through 1910.335 of this Part.

Note 2: Work practices performed by qualified persons and conforming to 1910.269 of this Part are considered as complying with 1910.333(c) and 1910.335 of this Part.

..1910.269(a)(1)(iii)

1910.269(a)(1)(iii)

This section applies in addition to all other applicable standards contained in this Part 1910. Specific references in this section to other sections of Part 1910 are provided for emphasis only.

1910.269(a)(2)

"Training."

1910.269(a)(2)(i)

Employees shall be trained in and familiar with the safety-related work practices, safety procedures, and other safety requirements in this section that pertain to their respective job assignments. Employees shall also be trained in and familiar with any other safety practices, including applicable emergency procedures (such as pole top and manhole rescue), that are not specifically addressed by this section but that are related to their work and are necessary for their safety.

1910.269(a)(2)(ii)

Qualified employees shall also be trained and competent in:

1910.269(a)(2)(ii)(A)

The skills and techniques necessary to distinguish exposed live parts from other parts of electric equipment,

1910.269(a)(2)(ii)(B)

The skills and techniques necessary to determine the nominal voltage of exposed live parts,

1910.269(a)(2)(ii)(C)

The minimum approach distances specified in this section corresponding to the voltages to which the qualified employee will be exposed, and

..1910.269(a)(2)(ii)(D)

1910.269(a)(2)(ii)(D)

The proper use of the special precautionary techniques, personal protective equipment, insulating and shielding materials, and insulated tools for working on or near exposed energized parts of electric equipment.

Note: For the purposes of this section, a person must have this training in order to be considered a qualified person.

1910.269(a)(2)(iii)

The employer shall determine, through regular supervision and through inspections conducted on at least an annual basis, that each employee is complying with the safety-related work practices required by this section.

1910.269(a)(2)(iv)

An employee shall receive additional training (or retraining) under any of the following conditions:

1910.269(a)(2)(iv)(A)

If the supervision and annual inspections required by paragraph (a)(2)(iii) of this section indicate that the employee is not complying with the safety-related work practices required by this section, or

1910.269(a)(2)(iv)(B)

If new technology, new types of equipment, or changes in procedures necessitate the use of safety-related work practices that are different from those which the employee would normally use, or

1910.269(a)(2)(iv)(C)

If he or she must employ safety-related work practices that are not normally used during his or her regular job duties.

Note: OSHA would consider tasks that are performed less often than once per year to necessitate retraining before the performance of the work practices involved.

1910.269(a)(2)(v)

The training required by paragraph (a)(2) of this section shall be of the classroom or on-the-job type.

..1910.269(a)(2)(vi)

1910.269(a)(2)(vi)

The training shall establish employee proficiency in the work practices required by this section and shall introduce the procedures necessary for compliance with this section.

1910.269(a)(2)(vii)

The employer shall certify that each employee has received the training required by paragraph (a)(2) of this section. This certification shall be made when the employee demonstrates proficiency in the work practices involved and shall be maintained for the duration of the employee's employment.

Note: Employment records that indicate that an employee has received the required training are an acceptable means of meeting this requirement.

1910.269(a)(3)

"Existing conditions." Existing conditions related to the safety of the work to be performed shall be determined before work on or near electric lines or equipment is started. Such conditions include, but are not limited to, the nominal voltages of lines and equipment, the maximum switching transient voltages, the presence of hazardous induced voltages, the presence and condition of protective grounds and equipment grounding conductors, the condition of poles, environmental conditions relative to safety, and the locations of circuits and equipment, including power and communication lines and fire protective signaling circuits.

1910.269(b)

"Medical services and first aid." The employer shall provide medical services and first aid as required in 1910.151 of this Part. In addition to the requirements of 1910.151 of this Part, the following requirements also apply:

..1910.269(b)(1)

1910.269(b)(1)

"Cardiopulmonary resuscitation and first aid training." When employees are performing work on or associated with exposed lines or equipment energized at 50 volts or more, persons trained in first aid including cardiopulmonary resuscitation (CPR) shall be available as follows:

1910.269(b)(1)(i)

For field work involving two or more employees at a work location, at least two trained persons shall be available. However, only one trained person need be available if all new employees are trained in first aid, including CPR, within 3 months of their hiring dates.

1910.269(b)(1)(ii)

For fixed work locations such as generating stations, the number of trained persons available shall be sufficient to ensure that each employee exposed to electric shock can be reached within 4 minutes by a trained person. However, where the existing number of employees is insufficient to meet this requirement (at a remote substation, for example), all employees at the work location shall be trained.

1910.269(b)(2)

"First aid supplies." First aid supplies required by 1910.151(b) of this Part shall be placed in weatherproof containers if the supplies could be exposed to the weather.

1910.269(b)(3)

"First aid kits." Each first aid kit shall be maintained, shall be readily available for use, and shall be inspected frequently enough to ensure that expended items are replaced but at least once per year.

..1910.269(c)

1910.269(c)

"Job briefing." The employer shall ensure that the employee in charge conducts a job briefing with the employees involved before they start each job. The briefing shall cover at least the following subjects: hazards associated with the job, work procedures involved, special precautions, energy source controls, and personal protective equipment requirements.

1910.269(c)(1)

"Number of briefings." If the work or operations to be performed during the work day or shift are repetitive and similar, at least one job briefing shall be conducted before the start of the first job of each day or shift. Additional job briefings shall be held if significant changes, which might affect the safety of the employees, occur during the course of the work.

1910.269(c)(2)

"Extent of briefing." A brief discussion is satisfactory if the work involved is routine and if the employee, by virtue of training and experience, can reasonably be expected to recognize and avoid the hazards involved in the job. A more extensive discussion shall be conducted:

1910.269(c)(2)(i)

If the work is complicated or particularly hazardous, or

1910.269(c)(2)(ii)

If the employee cannot be expected to recognize and avoid the hazards involved in the job.

Note: The briefing is always required to touch on all the subjects listed in the introductory text to paragraph (c) of this section.

1910.269(c)(3)

"Working alone." An employee working alone need not conduct a job briefing. However, the employer shall ensure that the tasks to be performed are planned as if a briefing were required.

..1910.269(d)

1910.269(d)

"Hazardous energy control (lockout/tagout) procedures."

1910.269(d)(1)

"Application." The provisions of paragraph (d) of this section apply to the use of lockout/tagout procedures for the control of energy sources in installations for the purpose of electric power generation, including related equipment for communication or metering. Locking and tagging procedures for the deenergizing of electric energy sources which are used exclusively for purposes of transmission and distribution are addressed by paragraph (m) of this section.

Note 1: Installations in electric power generation facilities that are not an integral part of, or inextricably commingled with, power generation processes or equipment are covered under 1910.147 and Subpart S of this Part.

Note 2: Lockout and tagging procedures that comply with paragraphs (c) through (f) of 1910.147 of this Part will also be deemed to comply with paragraph of (d) this section if the procedures address the hazards covered by paragraph (d) of this section.

1910.269(d)(2)

"General."

1910.269(d)(2)(i)

The employer shall establish a program consisting of energy control procedures, employee training, and periodic inspections to ensure that, before any employee performs any servicing or maintenance on a machine or equipment where the unexpected energizing, start up, or release of stored energy could occur and cause injury, the machine or equipment is isolated from the energy source and rendered inoperative.

1910.269(d)(2)(ii)

The employer's energy control program under paragraph (d)(2) of this section shall meet the following requirements:

1910.269(d)(2)(ii)(A)

If an energy isolating device is not capable of being locked out, the employer's program shall use a tagout system.

1910.269(d)(2)(ii)(B)

If an energy isolating device is capable of being locked out, the employer's program shall use lockout, unless the employer can demonstrate that the use of a tagout system will provide full employee protection as follows:

..1910.269(d)(2)(ii)(B)(1)
1910.269(d)(2)(ii)(B)(1)

When a tagout device is used on an energy isolating device which is capable of being locked out, the tagout device shall be attached at the same location that the lockout device would have been attached, and the employer shall demonstrate that the tagout program will provide a level of safety equivalent to that obtained by the use of a lockout program.

1910.269(d)(2)(ii)(B)(2)

In demonstrating that a level of safety is achieved in the tagout program equivalent to the level of safety obtained by the use of a lockout program, the employer shall demonstrate full compliance with all tagout-related provisions of this standard together with such additional elements as are necessary to provide the equivalent safety available from the use of a lockout device. Additional means to be considered as part of the demonstration of full employee protection shall include the implementation of additional safety measures such as the removal of an isolating circuit element, blocking of a controlling switch, opening of an extra disconnecting device, or the removal of a valve handle to reduce the likelihood of inadvertent energizing.

1910.269(d)(2)(ii)(C)

After November 1, 1994, whenever replacement or major repair, renovation, or modification of a machine or equipment is performed, and whenever new machines or equipment are installed, energy isolating devices for such machines or equipment shall be designed to accept a lockout device.

1910.269(d)(2)(iii)

Procedures shall be developed, documented, and used for the control of potentially hazardous energy covered by paragraph (d) of this section.

..1910.269(d)(2)(iv)

1910.269(d)(2)(iv)

The procedure shall clearly and specifically outline the scope, purpose, responsibility, authorization, rules, and techniques to be applied to the control of hazardous energy, and the measures to enforce compliance including, but not limited to, the following:

1910.269(d)(2)(iv)(A)

A specific statement of the intended use of this procedure;

1910.269(d)(2)(iv)(B)

Specific procedural steps for shutting down, isolating, blocking and securing machines or equipment to control hazardous energy;

1910.269(d)(2)(iv)(C)

Specific procedural steps for the placement, removal, and transfer of lockout devices or tagout devices and the responsibility for them; and

1910.269(d)(2)(iv)(D)

Specific requirements for testing a machine or equipment to determine and verify the effectiveness of lockout devices, tagout devices, and other energy control measures.

1910.269(d)(2)(v)

The employer shall conduct a periodic inspection of the energy control procedure at least annually to ensure that the procedure and the provisions of paragraph (d) of this section are being followed.

1910.269(d)(2)(v)(A)

The periodic inspection shall be performed by an authorized employee who is not using the energy control procedure being inspected.

..1910.269(d)(2)(v)(B)

1910.269(d)(2)(v)(B)

The periodic inspection shall be designed to identify and correct any deviations or inadequacies.

1910.269(d)(2)(v)(C)

If lockout is used for energy control, the periodic inspection shall include a review, between the inspector and each authorized employee, of that employee's responsibilities under the energy control procedure being inspected.

1910.269(d)(2)(v)(D)

Where tagout is used for energy control, the periodic inspection shall include a review, between the inspector and each authorized and affected employee, of that employee's responsibilities under the energy control procedure being inspected, and the elements set forth in paragraph (d)(2)(vii) of this section.

1910.269(d)(2)(v)(E)

The employer shall certify that the inspections required by paragraph (d)(2)(v) of this section have been accomplished. The certification shall identify the machine or equipment on which the energy control procedure was being used, the date of the inspection, the employees included in the inspection, and the person performing the inspection.

Note: If normal work schedule and operation records demonstrate adequate inspection activity and contain the required information, no additional certification is required.

1910.269(d)(2)(vi)

The employer shall provide training to ensure that the purpose and function of the energy control program are understood by employees and that the knowledge and skills required for the safe application, usage, and removal of energy controls are acquired by employees. The training shall include the following:

..1910.269(d)(2)(vi)(A)

1910.269(d)(2)(vi)(A)

Each authorized employee shall receive training in the recognition of applicable hazardous energy sources, the type and magnitude of energy available in the workplace, and in the methods and means necessary for energy isolation and control.

1910.269(d)(2)(vi)(B)

Each affected employee shall be instructed in the purpose and use of the energy control procedure.

1910.269(d)(2)(vi)(C)

All other employees whose work operations are or may be in an area where energy control procedures may be used shall be instructed about the procedures and about the prohibition relating to attempts to restart or reenergize machines or equipment that are locked out or tagged out.

1910.269(d)(2)(vii)

When tagout systems are used, employees shall also be trained in the following limitations of tags:

1910.269(d)(2)(vii)(A)

Tags are essentially warning devices affixed to energy isolating devices and do not provide the physical restraint on those devices that is provided by a lock.

1910.269(d)(2)(vii)(B)

When a tag is attached to an energy isolating means, it is not to be removed without authorization of the authorized person responsible for it, and it is never to be bypassed, ignored, or otherwise defeated.

..1910.269(d)(2)(vii)(C)

1910.269(d)(2)(vii)(C)

Tags must be legible and understandable by all authorized employees, affected employees, and all other employees whose work operations are or may be in the area, in order to be effective.

1910.269(d)(2)(vii)(D)

Tags and their means of attachment must be made of materials which will withstand the environmental conditions encountered in the workplace.

1910.269(d)(2)(vii)(E)

Tags may evoke a false sense of security, and their meaning needs to be understood as part of the overall energy control program.

1910.269(d)(2)(vii)(F)

Tags must be securely attached to energy isolating devices so that they cannot be inadvertently or accidentally detached during use.

1910.269(d)(2)(viii)

Retraining shall be provided by the employer as follows:

1910.269(d)(2)(viii)(A)

Retraining shall be provided for all authorized and affected employees whenever there is a change in their job assignments, a change in machines, equipment, or processes that present a new hazard or whenever there is a change in the energy control procedures.

1910.269(d)(2)(viii)(B)

Retraining shall also be conducted whenever a periodic inspection under paragraph (d)(2)(v) of this section reveals, or whenever the employer has reason to believe, that there are deviations from or inadequacies in an employee's knowledge or use of the energy control procedures.

..1910.269(d)(2)(viii)(C)

1910.269(d)(2)(viii)(C)

The retraining shall reestablish employee proficiency and shall introduce new or revised control methods and procedures, as necessary.

1910.269(d)(2)(ix)

The employer shall certify that employee training has been accomplished and is being kept up to date. The certification shall contain each employee's name and dates of training.

1910.269(d)(3)

"Protective materials and hardware."

1910.269(d)(3)(i)

Locks, tags, chains, wedges, key blocks, adapter pins, self-locking fasteners, or other hardware shall be provided by the employer for isolating, securing, or blocking of machines or equipment from energy sources.

1910.269(d)(3)(ii)

Lockout devices and tagout devices shall be singularly identified; shall be the only devices used for controlling energy; may not be used for other purposes; and shall meet the following requirements:

1910.269(d)(3)(ii)(A)

Lockout devices and tagout devices shall be capable of withstanding the environment to which they are exposed for the maximum period of time that exposure is expected.

..1910.269(d)(3)(ii)(A)(1)

1910.269(d)(3)(ii)(A)(1)

Tagout devices shall be constructed and printed so that exposure to weather conditions or wet and damp locations will not cause the tag to deteriorate or the message on the tag to become illegible.

1910.269(d)(3)(ii)(A)(2)

Tagout devices shall be so constructed as not to deteriorate when used in corrosive environments.

1910.269(d)(3)(ii)(B)

Lockout devices and tagout devices shall be standardized within the facility in at least one of the following criteria: color, shape, size. Additionally, in the case of tagout devices, print and format shall be standardized.

1910.269(d)(3)(ii)(C)

Lockout devices shall be substantial enough to prevent removal without the use of excessive force or unusual techniques, such as with the use of bolt cutters or metal cutting tools.

1910.269(d)(3)(ii)(D)

Tagout devices, including their means of attachment, shall be substantial enough to prevent inadvertent or accidental removal. Tagout device attachment means shall be of a non-reusable type, attachable by hand, self-locking, and non-releasable with a minimum unlocking strength of no less than 50 pounds and shall have the general design and basic characteristics of being at least equivalent to a one-piece, all-environment-tolerant nylon cable tie.

1910.269(d)(3)(ii)(E)

Each lockout device or tagout device shall include provisions for the identification of the employee applying the device.

..1910.269(d)(3)(ii)(F)

1910.269(d)(3)(ii)(F)

Tagout devices shall warn against hazardous conditions if the machine or equipment is energized and shall include a legend such as the following: Do Not Start, Do Not Open, Do Not Close, Do Not Energize, Do Not Operate.

Note: For specific provisions covering accident prevention tags, see 1910.145 of this Part.

1910.269(d)(4)

"Energy isolation." Lockout and tagout device application and removal may only be performed by the authorized employees who are performing the servicing or maintenance.

1910.269(d)(5)

"Notification." Affected employees shall be notified by the employer or authorized employee of the application and removal of lockout or tagout devices. Notification shall be given before the controls are applied and after they are removed from the machine or equipment.

Note: See also paragraph (d)(7) of this section, which requires that the second notification take place before the machine or equipment is reenergized.

1910.269(d)(6)

"Lockout/tagout application." The established procedures for the application of energy control (the lockout or tagout procedures) shall include the following elements and actions, and these procedures shall be performed in the following sequence:

1910.269(d)(6)(i)

Before an authorized or affected employee turns off a machine or equipment, the authorized employee shall have knowledge of the type and magnitude of the energy, the hazards of the energy to be controlled, and the method or means to control the energy.

..1910.269(d)(6)(ii)

1910.269(d)(6)(ii)

The machine or equipment shall be turned off or shut down using the procedures established for the machine or equipment. An orderly shutdown shall be used to avoid any additional or increased hazards to employees as a result of the equipment stoppage.

1910.269(d)(6)(iii)

All energy isolating devices that are needed to control the energy to the machine or equipment shall be physically located and operated in such a manner as to isolate the machine or equipment from energy sources.

1910.269(d)(6)(iv)

Lockout or tagout devices shall be affixed to each energy isolating device by authorized employees.

1910.269(d)(6)(iv)(A)

Lockout devices shall be attached in a manner that will hold the energy isolating devices in a "safe" or "off" position.

1910.269(d)(6)(iv)(B)

Tagout devices shall be affixed in such a manner as will clearly indicate that the operation or movement of energy isolating devices from the "safe" or "off" position is prohibited.

1910.269(d)(6)(iv)(B)(1)

Where tagout devices are used with energy isolating devices designed with the capability of being locked out, the tag attachment shall be fastened at the same point at which the lock would have been attached.

1910.269(d)(6)(iv)(B)(2)

Where a tag cannot be affixed directly to the energy isolating device, the tag shall be located as close as safely possible to the device, in a position that will be immediately obvious to anyone attempting to operate the device.

..1910.269(d)(6)(v)

1910.269(d)(6)(v)

Following the application of lockout or tagout devices to energy isolating devices, all potentially hazardous stored or residual energy shall be relieved, disconnected, restrained, or otherwise rendered safe.

1910.269(d)(6)(vi)

If there is a possibility of reaccumulation of stored energy to a hazardous level, verification of isolation shall be continued until the servicing or maintenance is completed or until the possibility of such accumulation no longer exists.

1910.269(d)(6)(vii)

Before starting work on machines or equipment that have been locked out or tagged out, the authorized employee shall verify that isolation and deenergizing of the machine or equipment have been accomplished. If normally energized parts will be exposed to contact by an employee while the machine or equipment is deenergized, a test shall be performed to ensure that these parts are deenergized.

1910.269(d)(7)

"Release from lockout/tagout." Before lockout or tagout devices are removed and energy is restored to the machine or equipment, procedures shall be followed and actions taken by the authorized employees to ensure the following:

1910.269(d)(7)(i)

The work area shall be inspected to ensure that nonessential items have been removed and that machine or equipment components are operationally intact.

1910.269(d)(7)(ii)

The work area shall be checked to ensure that all employees have been safely positioned or removed.

..1910.269(d)(7)(iii)

1910.269(d)(7)(iii)

After lockout or tagout devices have been removed and before a machine or equipment is started, affected employees shall be notified that the lockout or tagout devices have been removed.

1910.269(d)(7)(iv)

Each lockout or tagout device shall be removed from each energy isolating device by the authorized employee who applied the lockout or tagout device. However, if that employee is not available to remove it, the device may be removed under the direction of the employer, provided that specific procedures and training for such removal have been developed, documented, and incorporated into the employer's energy control program. The employer shall demonstrate that the specific procedure provides a degree of safety equivalent to that provided by the removal of the device by the authorized employee who applied it. The specific procedure shall include at least the following elements:

1910.269(d)(7)(iv)(A)

Verification by the employer that the authorized employee who applied the device is not at the facility;

1910.269(d)(7)(iv)(B)

Making all reasonable efforts to contact the authorized employee to inform him or her that his or her lockout or tagout device has been removed; and

1910.269(d)(7)(iv)(C)

Ensuring that the authorized employee has this knowledge before he or she resumes work at that facility.

..1910.269(d)(8)

1910.269(d)(8)

"Additional requirements."

1910.269(d)(8)(i)

If the lockout or tagout devices must be temporarily removed from energy isolating devices and the machine or equipment must be energized to test or position the machine, equipment, or component thereof, the following sequence of actions shall be followed:

1910.269(d)(8)(i)(A)

Clear the machine or equipment of tools and materials in accordance with paragraph (d)(7)(i) of this section;

1910.269(d)(8)(i)(B)

Remove employees from the machine or equipment area in accordance with paragraphs (d)(7)(ii) and (d)(7)(iii) of this section;

1910.269(d)(8)(i)(C)

Remove the lockout or tagout devices as specified in paragraph (d)(7)(iv) of this section;

1910.269(d)(8)(i)(D)

Energize and proceed with the testing or positioning; and

1910.269(d)(8)(i)(E)

Deenergize all systems and reapply energy control measures in accordance with paragraph (d)(6) of this section to continue the servicing or maintenance.

..1910.269(d)(8)(ii)

1910.269(d)(8)(ii)

When servicing or maintenance is performed by a crew, craft, department, or other group, they shall use a procedure which affords the employees a level of protection equivalent to that provided by the implementation of a personal lockout or tagout device. Group lockout or tagout devices shall be used in accordance with the procedures required by paragraphs (d)(2)(iii) and (d)(2)(iv) of this section including, but not limited to, the following specific requirements:

1910.269(d)(8)(ii)(A)

Primary responsibility shall be vested in an authorized employee for a set number of employees working under the protection of a group lockout or tagout device (such as an operations lock);

1910.269(d)(8)(ii)(B)

Provision shall be made for the authorized employee to ascertain the exposure status of all individual group members with regard to the lockout or tagout of the machine or equipment;

1910.269(d)(8)(ii)(C)

When more than one crew, craft, department, or other group is involved, assignment of overall job-associated lockout or tagout control responsibility shall be given to an authorized employee designated to coordinate affected work forces and ensure continuity of protection; and

1910.269(d)(8)(ii)(D)

Each authorized employee shall affix a personal lockout or tagout device to the group lockout device, group lockbox, or comparable mechanism when he or she begins work and shall remove those devices when he or she stops working on the machine or equipment being serviced or maintained.

..1910.269(d)(8)(iii)

1910.269(d)(8)(iii)

Procedures shall be used during shift or personnel changes to ensure the continuity of lockout or tagout protection, including provision for the orderly transfer of lockout or tagout device protection between off-going and on-coming employees, to minimize their exposure to hazards from the unexpected energizing or start-up of the machine or equipment or from the release of stored energy.

1910.269(d)(8)(iv)

Whenever outside servicing personnel are to be engaged in activities covered by paragraph (d) of this section, the on-site employer and the outside employer shall inform each other of their respective lockout or tagout procedures, and each employer shall ensure that his or her personnel understand and comply with restrictions and prohibitions of the energy control procedures being used.

1910.269(d)(8)(v)

If energy isolating devices are installed in a central location and are under the exclusive control of a system operator, the following requirements apply:

1910.269(d)(8)(v)(A)

The employer shall use a procedure that affords employees a level of protection equivalent to that provided by the implementation of a personal lockout or tagout device.

1910.269(d)(8)(v)(B)

The system operator shall place and remove lockout and tagout devices in place of the authorized employee under paragraphs (d)(4), (d)(6)(iv), and (d)(7)(iv) of this section.

..1910.269(d)(8)(v)(C)

1910.269(d)(8)(v)(C)

Provisions shall be made to identify the authorized employee who is responsible for (that is, being protected by) the lockout or tagout device, to transfer responsibility for lockout and tagout devices, and to ensure that an authorized employee requesting removal or transfer of a lockout or tagout device is the one responsible for it before the device is removed or transferred.

1910.269(e)

"Enclosed spaces." This paragraph covers enclosed spaces that may be entered by employees. It does not apply to vented vaults if a determination is made that the ventilation system is operating to protect employees before they enter the space. This paragraph applies to routine entry into enclosed spaces in lieu of the permit-space entry requirements contained in paragraphs (d) through (k) of 1910.146 of this Part. If, after the precautions given in paragraphs (e) and (t) of this section are taken, the hazards remaining in the enclosed space endanger the life of an entrant or could interfere with escape from the space, then entry into the enclosed space shall meet the permit-space entry requirements of paragraphs (d) through (k) of 1910.146 of this Part.

Note: Entries into enclosed spaces conducted in accordance with the permit-space entry requirements of paragraphs (d) through (k) of 1910.146 of this Part are considered as complying with paragraph (e) of this section.

1910.269(e)(1)

"Safe work practices." The employer shall ensure the use of safe work practices for entry into and work in enclosed spaces and for rescue of employees from such spaces.

1910.269(e)(2)

"Training." Employees who enter enclosed spaces or who serve as attendants shall be trained in the hazards of enclosed space entry, in enclosed space entry procedures, and in enclosed space rescue procedures.

..1910.269(e)(3)

1910.269(e)(3)

"Rescue equipment." Employers shall provide equipment to ensure the prompt and safe rescue of employees from the enclosed space.

1910.269(e)(4)

"Evaluation of potential hazards." Before any entrance cover to an enclosed space is removed, the employer shall determine whether it is safe to do so by checking for the presence of any atmospheric pressure or temperature differences and by evaluating whether there might be a hazardous atmosphere in the space. Any conditions making it unsafe to remove the cover shall be eliminated before the cover is removed.

Note: The evaluation called for in this paragraph may take the form of a check of the conditions expected to be in the enclosed space. For example, the cover could be checked to see if it is hot and, if it is fastened in place, could be loosened gradually to release any residual pressure. A determination must also be made of whether conditions at the site could cause a hazardous atmosphere, such as an oxygen deficient or flammable atmosphere, to develop within the space.

1910.269(e)(5)

"Removal of covers." When covers are removed from enclosed spaces, the opening shall be promptly guarded by a railing, temporary cover, or other barrier intended to prevent an accidental fall through the opening and to protect employees working in the space from objects entering the space.

1910.269(e)(6)

"Hazardous atmosphere." Employees may not enter any enclosed space while it contains a hazardous atmosphere, unless the entry conforms to the generic permit-required confined spaces standard in 1910.146 of this Part.

Note: The term "entry" is defined in 1910.146(b) of this Part.

..1910.269(e)(7)

1910.269(e)(7)

"Attendants." While work is being performed in the enclosed space, a person with first aid training meeting paragraph (b) of this section shall be immediately available outside the enclosed space to render emergency assistance if there is reason to believe that a hazard may exist in the space or if a hazard exists because of traffic patterns in the area of the opening used for entry. That person is not precluded from performing other duties outside the enclosed space if these duties do not distract the attendant from monitoring employees within the space.

Note: See paragraph (t)(3) of this section for additional requirements on attendants for work in manholes.

1910.269(e)(8)

"Calibration of test instruments." Test instruments used to monitor atmospheres in enclosed spaces shall be kept in calibration, with a minimum accuracy of + or − 10 percent.

1910.269(e)(9)

"Testing for oxygen deficiency." Before an employee enters an enclosed space, the internal atmosphere shall be tested for oxygen deficiency with a direct-reading meter or similar instrument, capable of collection and immediate analysis of data samples without the need for off-site evaluation. If continuous forced air ventilation is provided, testing is not required provided that the procedures used ensure that employees are not exposed to the hazards posed by oxygen deficiency.

1910.269(e)(10)

"Testing for flammable gases and vapors." Before an employee enters an enclosed space, the internal atmosphere shall be tested for flammable gases and vapors with a direct-reading meter or similar instrument capable of collection and immediate analysis of data samples without the need for off-site evaluation. This test shall be performed after the oxygen testing and ventilation required by paragraph (e)(9) of this section demonstrate that there is sufficient oxygen to ensure the accuracy of the test for flammability.

..1910.269(e)(11)

1910.269(e)(11)

"Ventilation and monitoring." If flammable gases or vapors are detected or if an oxygen deficiency is found, forced air ventilation shall be used to maintain oxygen at a safe level and to prevent a hazardous concentration of flammable gases and vapors from accumulating. A continuous monitoring program to ensure that no increase in flammable gas or vapor concentration occurs may be followed in lieu of ventilation, if flammable gases or vapors are detected at safe levels.

Note: See the definition of hazardous atmosphere for guidance in determining whether or not a given concentration of a substance is considered to be hazardous.

1910.269(e)(12)

"Specific ventilation requirements." If continuous forced air ventilation is used, it shall begin before entry is made and shall be maintained long enough to ensure that a safe atmosphere exists before employees are allowed to enter the work area. The forced air ventilation shall be so directed as to ventilate the immediate area where employees are present within the enclosed space and shall continue until all employees leave the enclosed space.

1910.269(e)(13)

"Air supply." The air supply for the continuous forced air ventilation shall be from a clean source and may not increase the hazards in the enclosed space.

1910.269(e)(14)

"Open flames." If open flames are used in enclosed spaces, a test for flammable gases and vapors shall be made immediately before the open flame device is used and at least once per hour while the device is used in the space. Testing shall be conducted more frequently if conditions present in the enclosed space indicate that once per hour is insufficient to detect hazardous accumulations of flammable gases or vapors.

Note: See the definition of hazardous atmosphere for guidance in determining whether or not a given concentration of a substance is considered to be hazardous.

..1910.269(f)

1910.269(f)

"Excavations." Excavation operations shall comply with Subpart P of Part 1926 of this chapter.

1910.269(g)

"Personal protective equipment."

1910.269(g)(1)

"General." Personal protective equipment shall meet the requirements of Subpart I of this Part.

1910.269(g)(2)

"Fall protection."

1910.269(g)(2)(i)

Personal fall arrest equipment shall meet the requirements of Subpart M of Part 1926 of this Chapter.

1910.269(g)(2)(ii)

Body belts and safety straps for work positioning shall meet the requirements of 1926.959 of this Chapter.

1910.269(g)(2)(iii)

Body belts, safety straps, lanyards, lifelines, and body harnesses shall be inspected before use each day to determine that the equipment is in safe working condition. Defective equipment may not be used.

1910.269(g)(2)(iv)

Lifelines shall be protected against being cut or abraded.

..1910.269(g)(2)(v)

1910.269(g)(2)(v)

Fall arrest equipment, work positioning equipment, or travel restricting equipment shall be used by employees working at elevated locations more than 4 feet (1.2 m) above the ground on poles, towers, or similar structures if other fall protection has not been provided. Fall protection equipment is not required to be used by a qualified employee climbing or changing location on poles, towers, or similar structures, unless conditions, such as, but not limited to, ice, high winds, the design of the structure (for example, no provision for holding on with hands), or the presence of contaminants on the structure, could cause the employee to lose his or her grip or footing.

Note 1: This paragraph applies to structures that support overhead electric power generation, transmission, and distribution lines and equipment. It does not apply to portions of buildings, such as loading docks, to electric equipment, such as transformers and capacitors, nor to aerial lifts. Requirements for fall protection associated with walking and working surfaces are contained in Subpart D of this Part; requirements for fall protection associated with aerial lifts are contained in 1910.67 of this Part.

Note 2: Employees undergoing training are not considered "qualified employees" for the purposes of this provision. Unqualified employees (including trainees) are required to use fall protection any time they are more than 4 feet (1.2 m) above the ground.

1910.269(g)(2)(vi)

The following requirements apply to personal fall arrest systems:

1910.269(g)(2)(vi)(A)

When stopping or arresting a fall, personal fall arrest systems shall limit the maximum arresting force on an employee to 900 pounds (4 kN) if used with a body belt.

1910.269(g)(2)(vi)(B)

When stopping or arresting a fall, personal fall arrest systems shall limit the maximum arresting force on an employee to 1800 pounds (8 kN) if used with a body harness.

1910.269(g)(2)(vi)(C)

Personal fall arrest systems shall be rigged such that an employee can neither free fall more than 6 feet (1.8 m) nor contact any lower level.

1910.269(g)(2)(vii)

If vertical lifelines or droplines are used, not more than one employee may be attached to any one lifeline.

1910.269(g)(2)(viii)

Snaphooks may not be connected to loops made in webbing-type lanyards.

1910.269(g)(2)(ix)

Snaphooks may not be connected to each other.

..1910.269(h)

1910.269(h)

"Ladders, platforms, step bolts, and manhole steps."

1910.269(h)(1)

"General." Requirements for ladders contained in Subpart D of this Part apply, except as specifically noted in paragraph (h)(2) of this section.

1910.269(h)(2)

"Special ladders and platforms." Portable ladders and platforms used on structures or conductors in conjunction with overhead line work need not meet paragraphs (d)(2)(i) and (d)(2)(iii) of 1910.25 of this Part or paragraph (c)(3)(iii) of 1910.26 of this Part. However, these ladders and platforms shall meet the following requirements:

1910.269(h)(2)(i)

Ladders and platforms shall be secured to prevent their becoming accidentally dislodged.

1910.269(h)(2)(ii)

Ladders and platforms may not be loaded in excess of the working loads for which they are designed.

1910.269(h)(2)(iii)

Ladders and platforms may be used only in applications for which they were designed.

1910.269(h)(2)(iv)

In the configurations in which they are used, ladders and platforms shall be capable of supporting without failure at least 2.5 times the maximum intended load.

..1910.269(h)(3)

1910.269(h)(3)

"Conductive ladders." Portable metal ladders and other portable conductive ladders may not be used near exposed energized lines or equipment. However, in specialized high-voltage work, conductive ladders shall be used where the employer can demonstrate that nonconductive ladders would present a greater hazard than conductive ladders.

1910.269(i)

"Hand and portable power tools."

1910.269(i)(1)

"General." Paragraph (i)(2) of this section applies to electric equipment connected by cord and plug. Paragraph (i)(3) of this section applies to portable and vehicle-mounted generators used to supply cord- and plug-connected equipment. Paragraph (i)(4) of this section applies to hydraulic and pneumatic tools.

1910.269(i)(2)

"Cord- and plug-connected equipment."

1910.269(i)(2)(i)

Cord- and plug-connected equipment supplied by premises wiring is covered by Subpart S of this Part.

1910.269(i)(2)(ii)

Any cord- and plug-connected equipment supplied by other than premises wiring shall comply with one of the following in lieu of 1910.243(a)(5) of this Part:

1910.269(i)(2)(ii)(A)

It shall be equipped with a cord containing an equipment grounding conductor connected to the tool frame and to a means for grounding the other end (however, this option may not be used where the introduction of the ground into the work environment increases the hazard to an employee); or

..1910.269(i)(2)(ii)(B)

1910.269(i)(2)(ii)(B)

It shall be of the double-insulated type conforming to Subpart S of this Part; or

1910.269(i)(2)(ii)(C)

It shall be connected to the power supply through an isolating transformer with an ungrounded secondary.

1910.269(i)(3)

"Portable and vehicle-mounted generators." Portable and vehicle-mounted generators used to supply cord- and plug-connected equipment shall meet the following requirements:

1910.269(i)(3)(i)

The generator may only supply equipment located on the generator or the vehicle and cord- and plug-connected equipment through receptacles mounted on the generator or the vehicle.

1910.269(i)(3)(ii)

The non-current-carrying metal parts of equipment and the equipment grounding conductor terminals of the receptacles shall be bonded to the generator frame.

1910.269(i)(3)(iii)

In the case of vehicle-mounted generators, the frame of the generator shall be bonded to the vehicle frame.

1910.269(i)(3)(iv)

Any neutral conductor shall be bonded to the generator frame.

1910.269(i)(4)

"Hydraulic and pneumatic tools."

1910.269(i)(4)(i)

Safe operating pressures for hydraulic and pneumatic tools, hoses, valves, pipes, filters, and fittings may not be exceeded.

Note: If any hazardous defects are present, no operating pressure would be safe, and the hydraulic or pneumatic equipment involved may not be used. In the absence of defects, the maximum rated operating pressure is the maximum safe pressure.

..1910.269(i)(4)(ii)

1910.269(i)(4)(ii)

A hydraulic or pneumatic tool used where it may contact exposed live parts shall be designed and maintained for such use.

1910.269(i)(4)(iii)

The hydraulic system supplying a hydraulic tool used where it may contact exposed live parts shall provide protection against loss of insulating value for the voltage involved due to the formation of a partial vacuum in the hydraulic line.

Note: Hydraulic lines without check valves having a separation of more than 35 feet (10.7 m) between the oil reservoir and the upper end of the hydraulic system promote the formation of a partial vacuum.

1910.269(i)(4)(iv)

A pneumatic tool used on energized electric lines or equipment or used where it may contact exposed live parts shall provide protection against the accumulation of moisture in the air supply.

1910.269(i)(4)(v)

Pressure shall be released before connections are broken, unless quick acting, self-closing connectors are used. Hoses may not be kinked.

1910.269(i)(4)(vi)

Employees may not use any part of their bodies to locate or attempt to stop a hydraulic leak.

1910.269(j)

"Live-line tools."

1910.269(j)(1)

"Design of tools." Live-line tool rods, tubes, and poles shall be designed and constructed to withstand the following minimum tests:

..1910.269(j)(1)(i)

1910.269(j)(1)(i)

100,000 volts per foot (3281 volts per centimeter) of length for 5 minutes if the tool is made of fiberglass-reinforced plastic (FRP), or

1910.269(j)(1)(ii)

75,000 volts per foot (2461 volts per centimeter) of length for 3 minutes if the tool is made of wood, or

1910.269(j)(1)(iii)

Other tests that the employer can demonstrate are equivalent.

Note: Live-line tools using rod and tube that meet ASTM F711-89, Standard Specification for Fiberglass-Reinforced Plastic (FRP) Rod and Tube Used in Live-Line Tools, conform to paragraph (j)(1)(i) of this section.

1910.269(j)(2)

"Condition of tools."

1910.269(j)(2)(i)

Each live-line tool shall be wiped clean and visually inspected for defects before use each day.

1910.269(j)(2)(ii)

If any defect or contamination that could adversely affect the insulating qualities or mechanical integrity of the live-line tool is present after wiping, the tool shall be removed from service and examined and tested according to paragraph (j)(2)(iii) of this section before being returned to service.

1910.269(j)(2)(iii)

Live-line tools used for primary employee protection shall be removed from service every 2 years and whenever required under paragraph (j)(2)(ii) of this section for examination, cleaning, repair, and testing as follows:

1910.269(j)(2)(iii)(A)

Each tool shall be thoroughly examined for defects.

..1910.269(j)(2)(iii)(B)
1910.269(j)(2)(iii)(B)

If a defect or contamination that could adversely affect the insulating qualities or mechanical integrity of the live-line tool is found, the tool shall be repaired and refinished or shall be permanently removed from service. If no such defect or contamination is found, the tool shall be cleaned and waxed.

1910.269(j)(2)(iii)(C)

The tool shall be tested in accordance with paragraphs (j)(2)(iii)(D) and (j)(2)(iii)(E) of this section under the following conditions:

1910.269(j)(2)(iii)(C)(1)

After the tool has been repaired or refinished; and

1910.269(j)(2)(iii)(C)(2)

After the examination if repair or refinishing is not performed, unless the tool is made of FRP rod or foam-filled FRP tube and the employer can demonstrate that the tool has no defects that could cause it to fail in use.

1910.269(j)(2)(iii)(D)

The test method used shall be designed to verify the tool's integrity along its entire working length and, if the tool is made of fiberglass-reinforced plastic, its integrity under wet conditions.

1910.269(j)(2)(iii)(E)

The voltage applied during the tests shall be as follows:

1910.269(j)(2)(iii)(E)(1)

75,000 volts per foot (2461 volts per centimeter) of length for 1 minute if the tool is made of fiberglass, or

1910.269(j)(2)(iii)(E)(2)

50,000 volts per foot (1640 volts per centimeter) of length for 1 minute if the tool is made of wood, or

1910.269(j)(2)(iii)(E)(3)

Other tests that the employer can demonstrate are equivalent.

Note: Guidelines for the examination, cleaning, repairing, and in-service testing of live-line tools are contained in the Institute of Electrical and Electronics Engineers Guide for In-Service Maintenance and Electrical Testing of Live-Line Tools, IEEE Std. 978-1984.

..1910.269(k)

1910.269(k)

"Materials handling and storage."

1910.269(k)(1)

"General." Material handling and storage shall conform to the requirements of Subpart N of this Part.

1910.269(k)(2)

"Materials storage near energized lines or equipment."

1910.269(k)(2)(i)

In areas not restricted to qualified persons only, materials or equipment may not be stored closer to energized lines or exposed energized parts of equipment than the following distances plus an amount providing for the maximum sag and side swing of all conductors and providing for the height and movement of material handling equipment:

1910.269(k)(2)(i)(A)

For lines and equipment energized at 50 kV or less, the distance is 10 feet (305 cm).

1910.269(k)(2)(i)(B)

For lines and equipment energized at more than 50 kV, the distance is 10 feet (305 cm) plus 4 inches (10 cm) for every 10 kV over 50 kV.

1910.269(k)(2)(ii)

In areas restricted to qualified employees, material may not be stored within the working space about energized lines or equipment.

Note: Requirements for the size of the working space are contained in paragraphs (u)(1) and (v)(3) of this section.

1910.269(l)

"Working on or near exposed energized parts." This paragraph applies to work on exposed live parts, or near enough to them, to expose the employee to any hazard they present.

..1910.269(l)(1)

1910.269(l)(1)

"General." Only qualified employees may work on or with exposed energized lines or parts of equipment. Only qualified employees may work in areas containing unguarded, uninsulated energized lines or parts of equipment operating at 50 volts or more. Electric lines and equipment shall be considered and treated as energized unless the provisions of paragraph (d) or paragraph (m) of this section have been followed.

1910.269(l)(1)(i)

Except as provided in paragraph (l)(1)(ii) of this section, at least two employees shall be present while the following types of work are being performed:

1910.269(l)(1)(i)(A)

Installation, removal, or repair of lines that are energized at more than 600 volts,

1910.269(l)(1)(i)(B)

Installation, removal, or repair of deenergized lines if an employee is exposed to contact with other parts energized at more than 600 volts,

1910.269(l)(1)(i)(C)

Installation, removal, or repair of equipment, such as transformers, capacitors, and regulators, if an employee is exposed to contact with parts energized at more than 600 volts,

1910.269(l)(1)(i)(D)

Work involving the use of mechanical equipment, other than insulated aerial lifts, near parts energized at more than 600 volts, and

1910.269(l)(1)(i)(E)

Other work that exposes an employee to electrical hazards greater than or equal to those posed by operations that are specifically listed in paragraphs (l)(1)(i)(A) through (l)(1)(i)(D) of this section.

..1910.269(l)(1)(ii)

1910.269(l)(1)(ii)

Paragraph (l)(1)(i) of this section does not apply to the following operations:

1910.269(l)(1)(ii)(A)

Routine switching of circuits, if the employer can demonstrate that conditions at the site allow this work to be performed safely,

1910.269(l)(1)(ii)(B)

Work performed with live-line tools if the employee is positioned so that he or she is neither within reach of nor otherwise exposed to contact with energized parts, and

1910.269(l)(1)(ii)(C)

Emergency repairs to the extent necessary to safeguard the general public.

1910.269(l)(2)

"Minimum approach distances." The employer shall ensure that no employee approaches or takes any conductive object closer to exposed energized parts than set forth in Table R-6 through Table R-10, unless:

1910.269(l)(2)(i)

The employee is insulated from the energized part (insulating gloves or insulating gloves and sleeves worn in accordance with paragraph (l)(3) of this section are considered insulation of the employee only with regard to the energized part upon which work is being performed), or

..1910.269(l)(2)(ii)

1910.269(l)(2)(ii)

The energized part is insulated from the employee and from any other conductive object at a different potential, or

1910.269(l)(2)(iii)

The employee is insulated from any other exposed conductive object, as during live-line bare-hand work.

Note: Paragraphs (u)(5)(i) and (v)(5)(i) and of this section contain requirements for the guarding and isolation of live parts. Parts of electric circuits that meet these two provisions are not considered as "exposed" unless a guard is removed or an employee enters the space intended to provide isolation from the live parts.

1910.269(l)(3)

"Type of insulation." If the employee is to be insulated from energized parts by the use of insulating gloves (under paragraph (l)(2)(i) of this section), insulating sleeves shall also be used. However, insulating sleeves need not be used under the following conditions:

1910.269(l)(3)(i)

If exposed energized parts on which work is not being performed are insulated from the employee and

1910.269(l)(3)(ii)

If such insulation is placed from a position not exposing the employee's upper arm to contact with other energized parts.

1910.269(l)(4)

"Working position." The employer shall ensure that each employee, to the extent that other safety-related conditions at the worksite permit, works in a position from which a slip or shock will not bring the employee's body into contact with exposed, uninsulated parts energized at a potential different from the employee.

1910.269(l)(5)

"Making connections." The employer shall ensure that connections are made as follows:

..1910.269(l)(5)(i)

1910.269(l)(5)(i)

In connecting deenergized equipment or lines to an energized circuit by means of a conducting wire or device, an employee shall first attach the wire to the deenergized part;

1910.269(l)(5)(ii)

When disconnecting equipment or lines from an energized circuit by means of a conducting wire or device, an employee shall remove the source end first; and

1910.269(l)(5)(iii)

When lines or equipment are connected to or disconnected from energized circuits, loose conductors shall be kept away from exposed energized parts.

1910.269(l)(6)

"Apparel."

1910.269(l)(6)(i)

When work is performed within reaching distance of exposed energized parts of equipment, the employer shall ensure that each employee removes or renders nonconductive all exposed conductive articles, such as key or watch chains, rings, or wrist watches or bands, unless such articles do not increase the hazards associated with contact with the energized parts.

1910.269(l)(6)(ii)

The employer shall train each employee who is exposed to the hazards of flames or electric arcs in the hazards involved.

1910.269(l)(6)(iii)

The employer shall ensure that each employee who is exposed to the hazards of flames or electric arcs does not wear clothing that, when exposed to flames or electric arcs, could increase the extent of injury that would be sustained by the employee.

Note: Clothing made from the following types of fabrics, either alone or in blends, is prohibited by this paragraph, unless the employer can demonstrate that the fabric has been treated to withstand the conditions that may be encountered or that the clothing is worn in such a manner as to eliminate the hazard involved: acetate, nylon, polyester, rayon.

..1910.269(l)(7)

1910.269(l)(7)

"Fuse handling." When fuses must be installed or removed with one or both terminals energized at more than 300 volts or with exposed parts energized at more than 50 volts, the employer shall ensure that tools or gloves rated for the voltage are used. When expulsion-type fuses are installed with one or both terminals energized at more than 300 volts, the employer shall ensure that each employee wears eye protection meeting the requirements of Subpart I of this Part, uses a tool rated for the voltage, and is clear of the exhaust path of the fuse barrel.

1910.269(l)(8)

"Covered (noninsulated) conductors." The requirements of this section which pertain to the hazards of exposed live parts also apply when work is performed in the proximity of covered (noninsulated) wires.

1910.269(l)(9)

"Noncurrent-carrying metal parts." Noncurrent-carrying metal parts of equipment or devices, such as transformer cases and circuit breaker housings, shall be treated as energized at the highest voltage to which they are exposed, unless the employer inspects the installation and determines that these parts are grounded before work is performed.

1910.269(l)(10)

"Opening circuits under load." Devices used to open circuits under load conditions shall be designed to interrupt the current involved.

Table R-6. AC Live-Line Work Minimum Approach Distance

Nominal voltage in kilovolts phase to phase	Distance			
	Phase to ground exposure		Phase to phase exposure	
	(ft-in)	(m)	(ft-in)	(m)
0.05 to 1.0	(4)	(4)	(4)	(4)
1.1 to 15.0	2-1	0.64	2-2	0.66
15.1 to 36.0	2-4	0.72	2-7	0.77
36.1 to 46.0	2-7	0.77	2-10	0.85
46.1 to 72.5	3-0	0.90	3-6	1.05
72.6 to 121	3-2	0.95	4-3	1.29
138 to 145	3-7	1.09	4-11	1.50
161 to 169	4-0	1.22	5-8	1.71
230 to 242	5-3	1.59	7-6	2.27
345 to 362	8-6	2.59	12-6	3.80
500 to 550	11-3	3.42	18-1	5.50
765 to 800	14-11	4.53	26-0	7.91

Footnote(1) These distances take into consideration the highest switching surge an employee will be exposed to on any system with air as the insulating medium and the maximum voltages shown.

Footnote(2) The clear live-line tool distance shall equal or exceed the values for the indicated voltage ranges.

Footnote(3) See Appendix B to this section for information on how the minimum approach distances listed in the tables were derived.

Footnote(4) Avoid contact.

Table R-7. AC Live-Line Work Minimum Approach Distance with Overvoltage Factor Phase-to-Ground Exposure

Maximum anticipated per-unit transient overvoltage	Distance in feet-inches						
	Maximum phase-to-phase voltage in kilovolts						
	121	145	169	242	362	552	800
1.5	....	....	....	....	....	6-0	9-8
1.6	....	....	....	....	....	6-6	10-8
1.7	....	....	....	....	....	7-0	11-8
1.8	....	....	....	....	....	7-7	12-8
1.9	....	....	....	....	....	8-1	13-9
2.0	2-5	2-9	3-0	3-10	5-3	8-9	14-11
2.1	2-6	2-10	3-2	4-0	5-5	9-4	....
2.2	2-7	2-11	3-3	4-1	5-9	9-11	....
2.3	2-8	3-0	3-4	4-3	6-1	10-6	....
2.4	2-9	3-1	3-5	4-5	6-4	11-3	....
2.5	2-9	3-2	3-6	4-6	6-8	....	....
2.6	2-10	3-3	3-8	4-8	7-1	....	....
2.7	2-11	3-4	3-9	4-10	7-5	....	....
2.8	3-0	3-5	3-10	4-11	7-9	....	....
2.9	3-1	3-6	3-11	5-1	8-2	....	....
3.0	3-2	3-7	4-0	5-3	8-6	....	....

Note 1: The distance specified in this table may be applied only where the maximum anticipated per-unit transient overvoltage has been determined by engineering analysis and has been supplied by the employer. Table R-6 applies otherwise.

Note 2: The distances specified in this table are the air, bare-hand, and live-line tool distances.

Note 3: See Appendix B to this section for information on how the minimum approach distances listed in the tables were derived and on how to calculate revised minimum approach distances based on the control of transient overvoltages.

Table R-8. AC Live-Line Work Minimum Approach Distance with Overvoltage Factor
Phase-to-Phase Exposure

Maximum anticipated per-unit transient overvoltage	Distance in feet-inches						
	Maximum phase-to-phase voltage in kilovolts						
	121	145	169	242	362	552	800
1.5	....	....	....	....	....	7-4	12-1
1.6	....	....	....	....	....	8-9	14-6
1.7	....	....	....	....	....	10-2	17-2
1.8	....	....	....	....	....	11-7	19-11
1.9	....	....	....	....	....	13-2	22-11
2.0	3-7	4-1	4-8	6-1	8-7	14-10	26-0
2.1	3-7	4-2	4-9	6-3	8-10	15-7	....
2.2	3-8	4-3	4-10	6-4	9-2	16-4	....
2.3	3-9	4-4	4-11	6-6	9-6	17-2	....
2.4	3-10	4-5	5-0	6-7	9-11	18-1	....
2.5	3-11	4-6	5-2	6-9	10-4	....	....
2.6	4-0	4-7	5-3	6-11	10-9	....	....
2.7	4-1	4-8	5-4	7-0	11-2	....	....
2.8	4-1	4-9	5-5	7-2	11-7	....	....
2.9	4-2	4-10	5-6	7-4	12-1	....	....
3.0	4-3	4-11	5-8	7-6	12-6	....	....

Note 1: The distance specified in this table may be applied only where the maximum anticipated per-unit transient overvoltage has been determined by engineering analysis and has been supplied by the employer. Table R-6 applies otherwise.

Note 2: The distances specified in this table are the air, bare-hand, and live-line tool distances.

Note 3: See Appendix B to this section for information on how the minimum approach distances listed in the tables were derived and on how to calculate revised minimum approach distances based on the control of transient overvoltages.

Table R-9. DC Live-Line Work Minimum Approach Distance with Overvoltage Factor

Maximum anticipated per-unit transient overvoltage	Distance in feet-inches				
	Maximum line-to-ground voltage in kilovolts				
	250	400	500	600	750
1.5 or lower	3-8	5-3	6-9	8-7	11-10
1.6	3-10	5-7	7-4	9-5	13-1
1.7	4-1	6-0	7-11	10-3	14-4
1.8	4-3	6-5	8-7	11-2	15-9

Note 1: The distances specified in this table may be applied only where the maximum anticipated per-unit transient overvoltage has been determined by engineering analysis and has been supplied by the employer. However, if the transient overvoltage factor is not known, a factor of 1.8 shall be assumed.

Note 2: The distances specified in this table are the air, bare-hand, and live-line tool distances.

Table R-10. Altitude Correction Factor

Altitude				Correction factor	
ft	ft	m	m		
3000	10000	900	3000	1.00	1.20
4000	12000	1200	3600	1.02	1.25
5000	14000	1500	4200	1.05	1.30
6000	16000	1800	4800	1.08	1.35
7000	18000	2100	5400	1.11	1.39
8000	20000	2400	6000	1.14	1.44
9000		2700		1.17	

Note: If the work is performed at elevations greater than 3000 ft (900 m) above mean sea level, the minimum approach distance shall be determined by multiplying the distances in Table R-6 through Table R-9 by the correction factor corresponding to the altitude at which work is performed.

1910.269(m)

"Deenergizing lines and equipment for employee protection."

..1910.269(m)(1)

1910.269(m)(1)

"Application." Paragraph (m) of this section applies to the deenergizing of transmission and distribution lines and equipment for the purpose of protecting employees. Control of hazardous energy sources used in the generation of electric energy is covered in paragraph (d) of this section. Conductors and parts of electric equipment that have been deenergized under procedures other than those required by paragraph (d) or (m) of this section, as applicable, shall be treated as energized.

1910.269(m)(2)

"General."

1910.269(m)(2)(i)

If a system operator is in charge of the lines or equipment and their means of disconnection, all of the requirements of paragraph (m)(3) of this section shall be observed, in the order given.

1910.269(m)(2)(ii)

If no system operator is in charge of the lines or equipment and their means of disconnection, one employee in the crew shall be designated as being in charge of the clearance. All of the requirements of paragraph (m)(3) of this section apply, in the order given, except as provided in paragraph (m)(2)(iii) of this section. The employee in charge of the clearance shall take the place of the system operator, as necessary.

1910.269(m)(2)(iii)

If only one crew will be working on the lines or equipment and if the means of disconnection is accessible and visible to and under the sole control of the employee in charge of the clearance, paragraphs (m)(3)(i), (m)(3)(iii), (m)(3)(iv), (m)(3)(viii), and (m)(3)(xii) of this section do not apply. Additionally, tags required by the remaining provisions of paragraph (m)(3) of this section need not be used.

..1910.269(m)(2)(iv)

1910.269(m)(2)(iv)

Any disconnecting means that are accessible to persons outside the employer's control (for example, the general public) shall be rendered inoperable while they are open for the purpose of protecting employees.

1910.269(m)(3)

"Deenergizing lines and equipment."

1910.269(m)(3)(i)

A designated employee shall make a request of the system operator to have the particular section of line or equipment deenergized. The designated employee becomes the employee in charge (as this term is used in paragraph (m)(3) of this section) and is responsible for the clearance.

1910.269(m)(3)(ii)

All switches, disconnectors, jumpers, taps, and other means through which known sources of electric energy may be supplied to the particular lines and equipment to be deenergized shall be opened. Such means shall be rendered inoperable, unless its design does not so permit, and tagged to indicate that employees are at work.

1910.269(m)(3)(iii)

Automatically and remotely controlled switches that could cause the opened disconnecting means to close shall also be tagged at the point of control. The automatic or remote control feature shall be rendered inoperable, unless its design does not so permit.

1910.269(m)(3)(iv)

Tags shall prohibit operation of the disconnecting means and shall indicate that employees are at work.

..1910.269(m)(3)(v)

1910.269(m)(3)(v)

After the applicable requirements in paragraphs (m)(3)(i) through (m)(3)(iv) of this section have been followed and the employee in charge of the work has been given a clearance by the system operator, the lines and equipment to be worked shall be tested to ensure that they are deenergized.

1910.269(m)(3)(vi)

Protective grounds shall be installed as required by paragraph (n) of this section.

1910.269(m)(3)(vii)

After the applicable requirements of paragraphs (m)(3)(i) through (m)(3)(vi) of this section have been followed, the lines and equipment involved may be worked as deenergized.

1910.269(m)(3)(viii)

If two or more independent crews will be working on the same lines or equipment, each crew shall independently comply with the requirements in paragraph (m)(3) of this section.

1910.269(m)(3)(ix)

To transfer the clearance, the employee in charge (or, if the employee in charge is forced to leave the worksite due to illness or other emergency, the employee's supervisor) shall inform the system operator; employees in the crew shall be informed of the transfer; and the new employee in charge shall be responsible for the clearance.

1910.269(m)(3)(x)

To release a clearance, the employee in charge shall:

1910.269(m)(3)(x)(A)

Notify employees under his or her direction that the clearance is to be released;

..*1910.269(m)(3)(x)(B)*
1910.269(m)(3)(x)(B)

Determine that all employees in the crew are clear of the lines and equipment;

1910.269(m)(3)(x)(C)

Determine that all protective grounds installed by the crew have been removed; and

1910.269(m)(3)(x)(D)

Report this information to the system operator and release the clearance.

1910.269(m)(3)(xi)

The person releasing a clearance shall be the same person that requested the clearance, unless responsibility has been transferred under paragraph (m)(3)(ix) of this section.

1910.269(m)(3)(xii)

Tags may not be removed unless the associated clearance has been released under paragraph (m)(3)(x) of this section.

1910.269(m)(3)(xiii)

Only after all protective grounds have been removed, after all crews working on the lines or equipment have released their clearances, after all employees are clear of the lines and equipment, and after all protective tags have been removed from a given point of disconnection, may action be initiated to reenergize the lines or equipment at that point of disconnection.

..1910.269(n)

1910.269(n)

"Grounding for the protection of employees."

1910.269(n)(1)

"Application." Paragraph (n) of this section applies to the grounding of transmission and distribution lines and equipment for the purpose of protecting employees. Paragraph (n)(4) of this section also applies to the protective grounding of other equipment as required elsewhere in this section.

1910.269(n)(2)

"General." For the employee to work lines or equipment as deenergized, the lines or equipment shall be deenergized under the provisions of paragraph (m) of this section and shall be grounded as specified in paragraphs (n)(3) through (n)(9) of this section. However, if the employer can demonstrate that installation of a ground is impracticable or that the conditions resulting from the installation of a ground would present greater hazards than working without grounds, the lines and equipment may be treated as deenergized provided all of the following conditions are met:

1910.269(n)(2)(i)

The lines and equipment have been deenergized under the provisions of paragraph (m) of this section.

1910.269(n)(2)(ii)

There is no possibility of contact with another energized source.

1910.269(n)(2)(iii)

The hazard of induced voltage is not present.

1910.269(n)(3)

"Equipotential zone." Temporary protective grounds shall be placed at such locations and arranged in such a manner as to prevent each employee from being exposed to hazardous differences in electrical potential.

..1910.269(n)(4)

1910.269(n)(4)

"Protective grounding equipment."

1910.269(n)(4)(i)

Protective grounding equipment shall be capable of conducting the maximum fault current that could flow at the point of grounding for the time necessary to clear the fault. This equipment shall have an ampacity greater than or equal to that of No. 2 AWG copper.

Note: Guidelines for protective grounding equipment are contained in American Society for Testing and Materials Standard Specifications for Temporary Grounding Systems to be Used on De-Energized Electric Power Lines and Equipment, ASTM F855-1990.

1910.269(n)(4)(ii)

Protective grounds shall have an impedance low enough to cause immediate operation of protective devices in case of accidental energizing of the lines or equipment.

1910.269(n)(5)

"Testing." Before any ground is installed, lines and equipment shall be tested and found absent of nominal voltage, unless a previously installed ground is present.

1910.269(n)(6)

"Order of connection." When a ground is to be attached to a line or to equipment, the ground-end connection shall be attached first, and then the other end shall be attached by means of a live-line tool.

1910.269(n)(7)

"Order of removal." When a ground is to be removed, the grounding device shall be removed from the line or equipment using a live-line tool before the ground-end connection is removed.

1910.269(n)(8)

"Additional precautions." When work is performed on a cable at a location remote from the cable terminal, the cable may not be grounded at the cable terminal if there is a possibility of hazardous transfer of potential should a fault occur.

..1910.269(n)(9)

1910.269(n)(9)

"Removal of grounds for test." Grounds may be removed temporarily during tests. During the test procedure, the employer shall ensure that each employee uses insulating equipment and is isolated from any hazards involved, and the employer shall institute any additional measures as may be necessary to protect each exposed employee in case the previously grounded lines and equipment become energized.

1910.269(o)

"Testing and test facilities."

1910.269(o)(1)

"Application." Paragraph (o) of this section provides for safe work practices for high-voltage and high-power testing performed in laboratories, shops, and substations, and in the field and on electric transmission and distribution lines and equipment. It applies only to testing involving interim measurements utilizing high voltage, high power, or combinations of both, and not to testing involving continuous measurements as in routine metering, relaying, and normal line work.

Note: Routine inspection and maintenance measurements made by qualified employees are considered to be routine line work and are not included in the scope of paragraph (o) of this section, as long as the hazards related to the use of intrinsic high-voltage or high-power sources require only the normal precautions associated with routine operation and maintenance work required in the other paragraphs of this section. Two typical examples of such excluded test work procedures are "phasing-out" testing and testing for a "no-voltage" condition.

1910.269(o)(2)

"General requirements."

1910.269(o)(2)(i)

The employer shall establish and enforce work practices for the protection of each worker from the hazards of high-voltage or high-power testing at all test areas, temporary and permanent. Such work practices shall include, as a minimum, test area guarding, grounding, and the safe use of measuring and control circuits. A means providing for periodic safety checks of field test areas shall also be included. (See paragraph (o)(6) of this section.)

1910.269(o)(2)(ii)

Employees shall be trained in safe work practices upon their initial assignment to the test area, with periodic reviews and updates provided as required by paragraph (a)(2) of this section.

..1910.269(o)(3)

1910.269(o)(3)

"Guarding of test areas."

1910.269(o)(3)(i)

Permanent test areas shall be guarded by walls, fences, or barriers designed to keep employees out of the test areas.

1910.269(o)(3)(ii)

In field testing, or at a temporary test site where permanent fences and gates are not provided, one of the following means shall be used to prevent unauthorized employees from entering:

1910.269(o)(3)(ii)(A)

The test area shall be guarded by the use of distinctively colored safety tape that is supported approximately waist high and to which safety signs are attached,

1910.269(o)(3)(ii)(B)

The test area shall be guarded by a barrier or barricade that limits access to the test area to a degree equivalent, physically and visually, to the barricade specified in paragraph (o)(3)(ii)(A) of this section, or

1910.269(o)(3)(ii)(C)

The test area shall be guarded by one or more test observers stationed so that the entire area can be monitored.

1910.269(o)(3)(iii)

The barriers required by paragraph (o)(3)(ii) of this section shall be removed when the protection they provide is no longer needed.

..1910.269(o)(3)(iv)

1910.269(o)(3)(iv)

Guarding shall be provided within test areas to control access to test equipment or to apparatus under test that may become energized as part of the testing by either direct or inductive coupling, in order to prevent accidental employee contact with energized parts.

1910.269(o)(4)

"Grounding practices."

1910.269(o)(4)(i)

The employer shall establish and implement safe grounding practices for the test facility.

1910.269(o)(4)(i)(A)

All conductive parts accessible to the test operator during the time the equipment is operating at high voltage shall be maintained at ground potential except for portions of the equipment that are isolated from the test operator by guarding.

1910.269(o)(4)(i)(B)

Wherever ungrounded terminals of test equipment or apparatus under test may be present, they shall be treated as energized until determined by tests to be deenergized.

1910.269(o)(4)(ii)

Visible grounds shall be applied, either automatically or manually with properly insulated tools, to the high-voltage circuits after they are deenergized and before work is performed on the circuit or item or apparatus under test. Common ground connections shall be solidly connected to the test equipment and the apparatus under test.

..1910.269(o)(4)(iii)

1910.269(o)(4)(iii)

In high-power testing, an isolated ground-return conductor system shall be provided so that no intentional passage of current, with its attendant voltage rise, can occur in the ground grid or in the earth. However, an isolated ground-return conductor need not be provided if the employer can demonstrate that both the following conditions are met:

1910.269(o)(4)(iii)(A)

An isolated ground-return conductor cannot be provided due to the distance of the test site from the electric energy source, and

1910.269(o)(4)(iii)(B)

Employees are protected from any hazardous step and touch potentials that may develop during the test.

Note: See Appendix C to this section for information on measures that can be taken to protect employees from hazardous step and touch potentials.

1910.269(o)(4)(iv)

In tests in which grounding of test equipment by means of the equipment grounding conductor located in the equipment power cord cannot be used due to increased hazards to test personnel or the prevention of satisfactory measurements, a ground that the employer can demonstrate affords equivalent safety shall be provided, and the safety ground shall be clearly indicated in the test set-up.

1910.269(o)(4)(v)

When the test area is entered after equipment is deenergized, a ground shall be placed on the high-voltage terminal and any other exposed terminals.

1910.269(o)(4)(v)(A)

High capacitance equipment or apparatus shall be discharged through a resistor rated for the available energy.

1910.269(o)(4)(v)(B)

A direct ground shall be applied to the exposed terminals when the stored energy drops to a level at which it is safe to do so.

..1910.269(o)(4)(vi)

1910.269(o)(4)(vi)

If a test trailer or test vehicle is used in field testing, its chassis shall be grounded. Protection against hazardous touch potentials with respect to the vehicle, instrument panels, and other conductive parts accessible to employees shall be provided by bonding, insulation, or isolation.

1910.269(o)(5)

"Control and measuring circuits."

1910.269(o)(5)(i)

Control wiring, meter connections, test leads and cables may not be run from a test area unless they are contained in a grounded metallic sheath and terminated in a grounded metallic enclosure or unless other precautions are taken that the employer can demonstrate as ensuring equivalent safety.

1910.269(o)(5)(ii)

Meters and other instruments with accessible terminals or parts shall be isolated from test personnel to protect against hazards arising from such terminals and parts becoming energized during testing. If this isolation is provided by locating test equipment in metal compartments with viewing windows, interlocks shall be provided to interrupt the power supply if the compartment cover is opened.

1910.269(o)(5)(iii)

The routing and connections of temporary wiring shall be made secure against damage, accidental interruptions and other hazards. To the maximum extent possible, signal, control, ground, and power cables shall be kept separate.

..1910.269(o)(5)(iv)

1910.269(o)(5)(iv)

If employees will be present in the test area during testing, a test observer shall be present. The test observer shall be capable of implementing the immediate deenergizing of test circuits for safety purposes.

1910.269(o)(6)

"Safety check."

1910.269(o)(6)(i)

Safety practices governing employee work at temporary or field test areas shall provide for a routine check of such test areas for safety at the beginning of each series of tests.

1910.269(o)(6)(ii)

The test operator in charge shall conduct these routine safety checks before each series of tests and shall verify at least the following conditions:

1910.269(o)(6)(ii)(A)

That barriers and guards are in workable condition and are properly placed to isolate hazardous areas;

1910.269(o)(6)(ii)(B)

That system test status signals, if used, are in operable condition;

1910.269(o)(6)(ii)(C)

That test power disconnects are clearly marked and readily available in an emergency;

1910.269(o)(6)(ii)(D)

That ground connections are clearly identifiable;

1910.269(o)(6)(ii)(E)

That personal protective equipment is provided and used as required by Subpart I of this Part and by this section; and

1910.269(o)(6)(ii)(F)

That signal, ground, and power cables are properly separated.

..1910.269(p)

1910.269(p)

"Mechanical equipment."

1910.269(p)(1)

"General requirements."

1910.269(p)(1)(i)

The critical safety components of mechanical elevating and rotating equipment shall receive a thorough visual inspection before use on each shift.

Note: Critical safety components of mechanical elevating and rotating equipment are components whose failure would result in a free fall or free rotation of the boom.

1910.269(p)(1)(ii)

No vehicular equipment having an obstructed view to the rear may be operated on off-highway jobsites where any employee is exposed to the hazards created by the moving vehicle, unless:

1910.269(p)(1)(ii)(A)

The vehicle has a reverse signal alarm audible above the surrounding noise level, or

1910.269(p)(1)(ii)(B)

The vehicle is backed up only when a designated employee signals that it is safe to do so.

1910.269(p)(1)(iii)

The operator of an electric line truck may not leave his or her position at the controls while a load is suspended, unless the employer can demonstrate that no employee (including the operator) might be endangered.

1910.269(p)(1)(iv)

Rubber-tired, self-propelled scrapers, rubber-tired front-end loaders, rubber-tired dozers, wheel-type agricultural and industrial tractors, crawler-type tractors, crawler-type loaders, and motor graders, with or without attachments, shall have roll-over protective structures that meet the requirements of Subpart W of Part 1926 of this chapter.

..1910.269(p)(2)

1910.269(p)(2)

"Outriggers."

1910.269(p)(2)(i)

Vehicular equipment, if provided with outriggers, shall be operated with the outriggers extended and firmly set as necessary for the stability of the specific configuration of the equipment. Outriggers may not be extended or retracted outside of clear view of the operator unless all employees are outside the range of possible equipment motion.

1910.269(p)(2)(ii)

If the work area or the terrain precludes the use of outriggers, the equipment may be operated only within its maximum load ratings for the particular configuration of the equipment without outriggers.

1910.269(p)(3)

"Applied loads." Mechanical equipment used to lift or move lines or other material shall be used within its maximum load rating and other design limitations for the conditions under which the work is being performed.

1910.269(p)(4)

"Operations near energized lines or equipment."

1910.269(p)(4)(i)

Mechanical equipment shall be operated so that the minimum approach distances of Table R-6 through Table R-10 are maintained from exposed energized lines and equipment. However, the insulated portion of an aerial lift operated by a qualified employee in the lift is exempt from this requirement.

..1910.269(p)(4)(ii)

1910.269(p)(4)(ii)

A designated employee other than the equipment operator shall observe the approach distance to exposed lines and equipment and give timely warnings before the minimum approach distance required by paragraph (p)(4)(i) is reached, unless the employer can demonstrate that the operator can accurately determine that the minimum approach distance is being maintained.

1910.269(p)(4)(iii)

If, during operation of the mechanical equipment, the equipment could become energized, the operation shall also comply with at least one of paragraphs (p)(4)(iii)(A) through (p)(4)(iii)(C) of this section.

1910.269(p)(4)(iii)(A)

The energized lines exposed to contact shall be covered with insulating protective material that will withstand the type of contact that might be made during the operation.

1910.269(p)(4)(iii)(B)

The equipment shall be insulated for the voltage involved. The equipment shall be positioned so that its uninsulated portions cannot approach the lines or equipment any closer than the minimum approach distances specified in Table R-6 through Table R-10.

1910.269(p)(4)(iii)(C)

Each employee shall be protected from hazards that might arise from equipment contact with the energized lines. The measures used shall ensure that employees will not be exposed to hazardous differences in potential. Unless the employer can demonstrate that the methods in use protect each employee from the hazards that might arise if the equipment contacts the energized line, the measures used shall include all of the following techniques:

..1910.269(p)(4)(iii)(C)(1)

1910.269(p)(4)(iii)(C)(1)

Using the best available ground to minimize the time the lines remain energized,

1910.269(p)(4)(iii)(C)(2)

Bonding equipment together to minimize potential differences,

1910.269(p)(4)(iii)(C)(3)

Providing ground mats to extend areas of equipotential, and

1910.269(p)(4)(iii)(C)(4)

Employing insulating protective equipment or barricades to guard against any remaining hazardous potential differences.

Note: Appendix C to this section contains information on hazardous step and touch potentials and on methods of protecting employees from hazards resulting from such potentials.

1910.269(q)

"Overhead lines." This paragraph provides additional requirements for work performed on or near overhead lines and equipment.

1910.269(q)(1)

"General."

1910.269(q)(1)(i)

Before elevated structures, such as poles or towers, are subjected to such stresses as climbing or the installation or removal of equipment may impose, the employer shall ascertain that the structures are capable of sustaining the additional or unbalanced stresses. If the pole or other structure cannot withstand the loads which will be imposed, it shall be braced or otherwise supported so as to prevent failure.

Note: Appendix D to this section contains test methods that can be used in ascertaining whether a wood pole is capable of sustaining the forces that would be imposed by an employee climbing the pole. This paragraph also requires the employer to ascertain that the pole can sustain all other forces that will be imposed by the work to be performed.

1910.269(q)(1)(ii)

When poles are set, moved, or removed near exposed energized overhead conductors, the pole may not contact the conductors.

..1910.269(q)(1)(iii)
1910.269(q)(1)(iii)

When a pole is set, moved, or removed near an exposed energized overhead conductor, the employer shall ensure that each employee wears electrical protective equipment or uses insulated devices when handling the pole and that no employee contacts the pole with uninsulated parts of his or her body.

1910.269(q)(1)(iv)

To protect employees from falling into holes into which poles are to be placed, the holes shall be attended by employees or physically guarded whenever anyone is working nearby.

1910.269(q)(2)

"Installing and removing overhead lines." The following provisions apply to the installation and removal of overhead conductors or cable.

1910.269(q)(2)(i)

The employer shall use the tension stringing method, barriers, or other equivalent measures to minimize the possibility that conductors and cables being installed or removed will contact energized power lines or equipment.

1910.269(q)(2)(ii)

The protective measures required by paragraph (p)(4)(iii) of this section for mechanical equipment shall also be provided for conductors, cables, and pulling and tensioning equipment when the conductor or cable is being installed or removed close enough to energized conductors that any of the following failures could energize the pulling or tensioning equipment or the wire or cable being installed or removed:

1910.269(q)(2)(ii)(A)

Failure of the pulling or tensioning equipment,

..1910.269(q)(2)(ii)(B)
1910.269(q)(2)(ii)(B)

Failure of the wire or cable being pulled, or

1910.269(q)(2)(ii)(C)

Failure of the previously installed lines or equipment.

1910.269(q)(2)(iii)

If the conductors being installed or removed cross over energized conductors in excess of 600 volts and if the design of the circuit-interrupting devices protecting the lines so permits, the automatic-reclosing feature of these devices shall be made inoperative.

1910.269(q)(2)(iv)

Before lines are installed parallel to existing energized lines, the employer shall make a determination of the approximate voltage to be induced in the new lines, or work shall proceed on the assumption that the induced voltage is hazardous. Unless the employer can demonstrate that the lines being installed are not subject to the induction of a hazardous voltage or unless the lines are treated as energized, the following requirements also apply:

1910.269(q)(2)(iv)(A)

Each bare conductor shall be grounded in increments so that no point along the conductor is more than 2 miles (3.22 km) from a ground.

1910.269(q)(2)(iv)(B)

The grounds required in paragraph (q)(2)(iv)(A) of this section shall be left in place until the conductor installation is completed between dead ends.

..1910.269(q)(2)(iv)(C)
1910.269(q)(2)(iv)(C)

The grounds required in paragraph (q)(2)(iv)(A) of this section shall be removed as the last phase of aerial cleanup.

1910.269(q)(2)(iv)(D)

If employees are working on bare conductors, grounds shall also be installed at each location where these employees are working, and grounds shall be installed at all open dead-end or catch-off points or the next adjacent structure.

1910.269(q)(2)(iv)(E)

If two bare conductors are to be spliced, the conductors shall be bonded and grounded before being spliced.

1910.269(q)(2)(v)

Reel handling equipment, including pulling and tensioning devices, shall be in safe operating condition and shall be leveled and aligned.

1910.269(q)(2)(vi)

Load ratings of stringing lines, pulling lines, conductor grips, load-bearing hardware and accessories, rigging, and hoists may not be exceeded.

1910.269(q)(2)(vii)

Pulling lines and accessories shall be repaired or replaced when defective.

1910.269(q)(2)(viii)

Conductor grips may not be used on wire rope, unless the grip is specifically designed for this application.

..1910.269(q)(2)(ix)
1910.269(q)(2)(ix)

Reliable communications, through two-way radios or other equivalent means, shall be maintained between the reel tender and the pulling rig operator.

1910.269(q)(2)(x)

The pulling rig may only be operated when it is safe to do so.

Note: Examples of unsafe conditions include employees in locations prohibited by paragraph (q)(2)(xi) of this section, conductor and pulling line hang-ups, and slipping of the conductor grip.

1910.269(q)(2)(xi)

While the conductor or pulling line is being pulled (in motion) with a power-driven device, employees are not permitted directly under overhead operations or on the cross arm, except as necessary to guide the stringing sock or board over or through the stringing sheave.

1910.269(q)(3)

"Live-line bare-hand work." In addition to other applicable provisions contained in this section, the following requirements apply to live-line bare-hand work:

1910.269(q)(3)(i)

Before using or supervising the use of the live-line bare-hand technique on energized circuits, employees shall be trained in the technique and in the safety requirements of paragraph (q)(3) of this section. Employees shall receive refresher training as required by paragraph (a)(2) of this section.

1910.269(q)(3)(ii)

Before any employee uses the live-line bare-hand technique on energized high-voltage conductors or parts, the following information shall be ascertained:

..1910.269(q)(3)(ii)(A)

1910.269(q)(3)(ii)(A)

The nominal voltage rating of the circuit on which the work is to be performed,

1910.269(q)(3)(ii)(B)

The minimum approach distances to ground of lines and other energized parts on which work is to be performed, and

1910.269(q)(3)(ii)(C)

The voltage limitations of equipment to be used.

1910.269(q)(3)(iii)

The insulated equipment, insulated tools, and aerial devices and platforms used shall be designed, tested, and intended for live-line bare-hand work. Tools and equipment shall be kept clean and dry while they are in use.

1910.269(q)(3)(iv)

The automatic-reclosing feature of circuit-interrupting devices protecting the lines shall be made inoperative, if the design of the devices permits.

1910.269(q)(3)(v)

Work may not be performed when adverse weather conditions would make the work hazardous even after the work practices required by this section are employed. Additionally, work may not be performed when winds reduce the phase-to-phase or phase-to-ground minimum approach distances at the work location below that specified in paragraph (q)(3)(xiii) of this section, unless the grounded objects and other lines and equipment are covered by insulating guards.

Note: Thunderstorms in the immediate vicinity, high winds, snow storms, and ice storms are examples of adverse weather conditions that are presumed to make live-line bare-hand work too hazardous to perform safely.

..1910.269(q)(3)(vi)

1910.269(q)(3)(vi)

A conductive bucket liner or other conductive device shall be provided for bonding the insulated aerial device to the energized line or equipment.

1910.269(q)(3)(vi)(A)

The employee shall be connected to the bucket liner or other conductive device by the use of conductive shoes, leg clips, or other means.

1910.269(q)(3)(vi)(B)

Where differences in potentials at the worksite pose a hazard to employees, electrostatic shielding designed for the voltage being worked shall be provided.

1910.269(q)(3)(vii)

Before the employee contacts the energized part, the conductive bucket liner or other conductive device shall be bonded to the energized conductor by means of a positive connection. This connection shall remain attached to the energized conductor until the work on the energized circuit is completed.

1910.269(q)(3)(viii)

Aerial lifts to be used for live-line bare-hand work shall have dual controls (lower and upper) as follows:

1910.269(q)(3)(viii)(A)

The upper controls shall be within easy reach of the employee in the bucket. On a two-bucket-type lift, access to the controls shall be within easy reach from either bucket.

..1910.269(q)(3)(viii)(B)

1910.269(q)(3)(viii)(B)

The lower set of controls shall be located near the base of the boom, and they shall be so designed that they can override operation of the equipment at any time.

1910.269(q)(3)(ix)

Lower (ground-level) lift controls may not be operated with an employee in the lift, except in case of emergency.

1910.269(q)(3)(x)

Before employees are elevated into the work position, all controls (ground level and bucket) shall be checked to determine that they are in proper working condition.

1910.269(q)(3)(xi)

Before the boom of an aerial lift is elevated, the body of the truck shall be grounded, or the body of the truck shall be barricaded and treated as energized.

1910.269(q)(3)(xii)

A boom-current test shall be made before work is started each day, each time during the day when higher voltage is encountered, and when changed conditions indicate a need for an additional test. This test shall consist of placing the bucket in contact with an energized source equal to the voltage to be encountered for a minimum of 3 minutes. The leakage current may not exceed 1 microampere per kilovolt of nominal phase-to-ground voltage. Work from the aerial lift shall be immediately suspended upon indication of a malfunction in the equipment.

..1910.269(q)(3)(xiii)

1910.269(q)(3)(xiii)

The minimum approach distances specified in Table R-6 through Table R-10 shall be maintained from all grounded objects and from lines and equipment at a potential different from that to which the live-line bare-hand equipment is bonded, unless such grounded objects and other lines and equipment are covered by insulating guards.

1910.269(q)(3)(xiv)

While an employee is approaching, leaving, or bonding to an energized circuit, the minimum approach distances in Table R-6 through Table R-10 shall be maintained between the employee and any grounded parts, including the lower boom and portions of the truck.

1910.269(q)(3)(xv)

While the bucket is positioned alongside an energized bushing or insulator string, the phase-to-ground minimum approach distances of Table R-6 through Table R-10 shall be maintained between all parts of the bucket and the grounded end of the bushing or insulator string or any other grounded surface.

1910.269(q)(3)(xvi)

Hand lines may not be used between the bucket and the boom or between the bucket and the ground. However, non-conductive-type hand lines may be used from conductor to ground if not supported from the bucket. Ropes used for live-line bare-hand work may not be used for other purposes.

1910.269(q)(3)(xvii)

Uninsulated equipment or material may not be passed between a pole or structure and an aerial lift while an employee working from the bucket is bonded to an energized part.

..1910.269(q)(3)(xviii)

1910.269(q)(3)(xviii)

A minimum approach distance table reflecting the minimum approach distances listed in Table R-6 through Table R-10 shall be printed on a plate of durable non-conductive material. This table shall be mounted so as to be visible to the operator of the boom.

1910.269(q)(3)(xix)

A non-conductive measuring device shall be readily accessible to assist employees in maintaining the required minimum approach distance.

1910.269(q)(4)

"Towers and structures." The following requirements apply to work performed on towers or other structures which support overhead lines.

1910.269(q)(4)(i)

The employer shall ensure that no employee is under a tower or structure while work is in progress, except where the employer can demonstrate that such a working position is necessary to assist employees working above.

1910.269(q)(4)(ii)

Tag lines or other similar devices shall be used to maintain control of tower sections being raised or positioned, unless the employer can demonstrate that the use of such devices would create a greater hazard.

1910.269(q)(4)(iii)

The loadline may not be detached from a member or section until the load is safely secured.

1910.269(q)(4)(iv)

Except during emergency restoration procedures, work shall be discontinued when adverse weather conditions would make the work hazardous in spite of the work practices required by this section.

Note: Thunderstorms in the immediate vicinity, high winds, snow storms, and ice storms are examples of adverse weather conditions that are presumed to make this work too hazardous to perform, except under emergency conditions.

..1910.269(r)

1910.269(r)

"Line-clearance tree trimming operations." This paragraph provides additional requirements for line-clearance tree-trimming operations and for equipment used in these operations.

1910.269(r)(1)

"Electrical hazards." This paragraph does not apply to qualified employees.

1910.269(r)(1)(i)

Before an employee climbs, enters, or works around any tree, a determination shall be made of the nominal voltage of electric power lines posing a hazard to employees. However, a determination of the maximum nominal voltage to which an employee will be exposed may be made instead, if all lines are considered as energized at this maximum voltage.

1910.269(r)(1)(ii)

There shall be a second line-clearance tree trimmer within normal (that is, unassisted) voice communication under any of the following conditions:

1910.269(r)(1)(ii)(A)

If a line-clearance tree trimmer is to approach more closely than 10 feet (305 cm) any conductor or electric apparatus energized at more than 750 volts or

1910.269(r)(1)(ii)(B)

If branches or limbs being removed are closer to lines energized at more than 750 volts than the distances listed in Table R-6, Table R-9, and Table R-10 or

1910.269(r)(1)(ii)(C)

If roping is necessary to remove branches or limbs from such conductors or apparatus.

..1910.269(r)(1)(iii)

1910.269(r)(1)(iii)

Line-clearance tree trimmers shall maintain the minimum approach distances from energized conductors given in Table R-6, Table R-9, and Table R-10.

1910.269(r)(1)(iv)

Branches that are contacting exposed energized conductors or equipment or that are within the distances specified in Table R-6, Table R-9, and Table R-10 may be removed only through the use of insulating equipment.

Note: A tool constructed of a material that the employer can demonstrate has insulating qualities meeting paragraph (j)(1) of this section is considered as insulated under this paragraph if the tool is clean and dry.

1910.269(r)(1)(v)

Ladders, platforms, and aerial devices may not be brought closer to an energized part than the distances listed in Table R-6, Table R-9, and Table R-10.

1910.269(r)(1)(vi)

Line-clearance tree-trimming work may not be performed when adverse weather conditions make the work hazardous in spite of the work practices required by this section. Each employee performing line-clearance tree trimming work in the aftermath of a storm or under similar emergency conditions shall be trained in the special hazards related to this type of work.

Note: Thunderstorms in the immediate vicinity, high winds, snow storms, and ice storms are examples of adverse weather conditions that are presumed to make line-clearance tree trimming work too hazardous to perform safely.

1910.269(r)(2)

"Brush chippers."

1910.269(r)(2)(i)

Brush chippers shall be equipped with a locking device in the ignition system.

1910.269(r)(2)(ii)

Access panels for maintenance and adjustment of the chipper blades and associated drive train shall be in place and secure during operation of the equipment.

..1910.269(r)(2)(iii)

1910.269(r)(2)(iii)

Brush chippers not equipped with a mechanical infeed system shall be equipped with an infeed hopper of length sufficient to prevent employees from contacting the blades or knives of the machine during operation.

1910.269(r)(2)(iv)

Trailer chippers detached from trucks shall be chocked or otherwise secured.

1910.269(r)(2)(v)

Each employee in the immediate area of an operating chipper feed table shall wear personal protective equipment as required by Subpart I of this Part.

1910.269(r)(3)

"Sprayers and related equipment."

1910.269(r)(3)(i)

Walking and working surfaces of sprayers and related equipment shall be covered with slip-resistant material. If slipping hazards cannot be eliminated, slip-resistant footwear or handrails and stair rails meeting the requirements of Subpart D may be used instead of slip-resistant material.

1910.269(r)(3)(ii)

Equipment on which employees stand to spray while the vehicle is in motion shall be equipped with guardrails around the working area. The guardrail shall be constructed in accordance with Subpart D of this Part.

1910.269(r)(4)

"Stump cutters."

1910.269(r)(4)(i)

Stump cutters shall be equipped with enclosures or guards to protect employees.

..1910.269(r)(4)(ii)

1910.269(r)(4)(ii)

Each employee in the immediate area of stump grinding operations (including the stump cutter operator) shall wear personal protective equipment as required by Subpart I of this Part.

1910.269(r)(5)

"Gasoline-engine power saws." Gasoline-engine power saw operations shall meet the requirements of 1910.266(e) and the following:

1910.269(r)(5)(i)

Each power saw weighing more than 15 pounds (6.8 kilograms, service weight) that is used in trees shall be supported by a separate line, except when work is performed from an aerial lift and except during topping or removing operations where no supporting limb will be available.

1910.269(r)(5)(ii)

Each power saw shall be equipped with a control that will return the saw to idling speed when released.

1910.269(r)(5)(iii)

Each power saw shall be equipped with a clutch and shall be so adjusted that the clutch will not engage the chain drive at idling speed.

1910.269(r)(5)(iv)

A power saw shall be started on the ground or where it is otherwise firmly supported. Drop starting of saws over 15 pounds (6.8 kg) is permitted outside of the bucket of an aerial lift only if the area below the lift is clear of personnel.

..1910.269(r)(5)(v)

1910.269(r)(5)(v)

A power saw engine may be started and operated only when all employees other than the operator are clear of the saw.

1910.269(r)(5)(vi)

A power saw may not be running when the saw is being carried up into a tree by an employee.

1910.269(r)(5)(vii)

Power saw engines shall be stopped for all cleaning, refueling, adjustments, and repairs to the saw or motor, except as the manufacturer's servicing procedures require otherwise.

1910.269(r)(6)

"Backpack power units for use in pruning and clearing."

1910.269(r)(6)(i)

While a backpack power unit is running, no one other than the operator may be within 10 feet (305 cm) of the cutting head of a brush saw.

1910.269(r)(6)(ii)

A backpack power unit shall be equipped with a quick shutoff switch readily accessible to the operator.

1910.269(r)(6)(iii)

Backpack power unit engines shall be stopped for all cleaning, refueling, adjustments, and repairs to the saw or motor, except as the manufacturer's servicing procedures require otherwise.

..1910.269(r)(7)

1910.269(r)(7)

"Rope."

1910.269(r)(7)(i)

Climbing ropes shall be used by employees working aloft in trees. These ropes shall have a minimum diameter of 0.5 inch (1.2 cm) with a minimum breaking strength of 2300 pounds (10.2 kN). Synthetic rope shall have elasticity of not more than 7 percent.

1910.269(r)(7)(ii)

Rope shall be inspected before each use and, if unsafe (for example, because of damage or defect), may not be used.

1910.269(r)(7)(iii)

Rope shall be stored away from cutting edges and sharp tools. Rope contact with corrosive chemicals, gas, and oil shall be avoided.

1910.269(r)(7)(iv)

When stored, rope shall be coiled and piled, or shall be suspended, so that air can circulate through the coils.

1910.269(r)(7)(v)

Rope ends shall be secured to prevent their unraveling.

1910.269(r)(7)(vi)

Climbing rope may not be spliced to effect repair.

1910.269(r)(7)(vii)

A rope that is wet, that is contaminated to the extent that its insulating capacity is impaired, or that is otherwise not considered to be insulated for the voltage involved may not be used near exposed energized lines.

1910.269(r)(8)

"Fall protection." Each employee shall be tied in with a climbing rope and safety saddle when the employee is working above the ground in a tree, unless he or she is ascending into the tree.

..1910.269(s)

1910.269(s)

"Communication facilities."

1910.269(s)(1)

"Microwave transmission."

1910.269(s)(1)(i)

The employer shall ensure that no employee looks into an open waveguide or antenna that is connected to an energized microwave source.

1910.269(s)(1)(ii)

If the electromagnetic radiation level within an accessible area associated with microwave communications systems exceeds the radiation protection guide given in 1910.97(a)(2) of this Part, the area shall be posted with the warning symbol described in 1910.97(a)(3) of this Part. The lower half of the warning symbol shall include the following statements or ones that the employer can demonstrate are equivalent:

Radiation in this area may exceed hazard limitations and special precautions are required. Obtain specific instruction before entering.

1910.269(s)(1)(iii)

When an employee works in an area where the electromagnetic radiation could exceed the radiation protection guide, the employer shall institute measures that ensure that the employee's exposure is not greater than that permitted by that guide. Such measures may include administrative and engineering controls and personal protective equipment.

1910.269(s)(2)

"Power line carrier." Power line carrier work, including work on equipment used for coupling carrier current to power line conductors, shall be performed in accordance with the requirements of this section pertaining to work on energized lines.

1910.269(t)

"Underground electrical installations." This paragraph provides additional requirements for work on underground electrical installations.

..1910.269(t)(1)
1910.269(t)(1)

"Access." A ladder or other climbing device shall be used to enter and exit a manhole or subsurface vault exceeding 4 feet (122 cm) in depth. No employee may climb into or out of a manhole or vault by stepping on cables or hangers.

1910.269(t)(2)

"Lowering equipment into manholes." Equipment used to lower materials and tools into manholes or vaults shall be capable of supporting the weight to be lowered and shall be checked for defects before use. Before tools or material are lowered into the opening for a manhole or vault, each employee working in the manhole or vault shall be clear of the area directly under the opening.

1910.269(t)(3)

"Attendants for manholes."

1910.269(t)(3)(i)

While work is being performed in a manhole containing energized electric equipment, an employee with first aid and CPR training meeting paragraph (b)(1) of this section shall be available on the surface in the immediate vicinity to render emergency assistance.

1910.269(t)(3)(ii)

Occasionally, the employee on the surface may briefly enter a manhole to provide assistance, other than emergency.

Note 1: An attendant may also be required under paragraph (e)(7) of this section. One person may serve to fulfill both requirements. However, attendants required under paragraph (e)(7) of this section are not permitted to enter the manhole.

Note 2: Employees entering manholes containing unguarded, uninsulated energized lines or parts of electric equipment operating at 50 volts or more are required to be qualified under paragraph (l)(1) of this section.

1910.269(t)(3)(iii)

For the purpose of inspection, housekeeping, taking readings, or similar work, an employee working alone may enter, for brief periods of time, a manhole where energized cables or equipment are in service, if the employer can demonstrate that the employee will be protected from all electrical hazards.

..1910.269(t)(3)(iv)

1910.269(t)(3)(iv)

Reliable communications, through two-way radios or other equivalent means, shall be maintained among all employees involved in the job.

1910.269(t)(4)

"Duct rods." If duct rods are used, they shall be installed in the direction presenting the least hazard to employees. An employee shall be stationed at the far end of the duct line being rodded to ensure that the required minimum approach distances are maintained.

1910.269(t)(5)

"Multiple cables." When multiple cables are present in a work area, the cable to be worked shall be identified by electrical means, unless its identity is obvious by reason of distinctive appearance or location or by other readily apparent means of identification. Cables other than the one being worked shall be protected from damage.

1910.269(t)(6)

"Moving cables." Energized cables that are to be moved shall be inspected for defects.

1910.269(t)(7)

"Defective cables." Where a cable in a manhole has one or more abnormalities that could lead to or be an indication of an impending fault, the defective cable shall be deenergized before any employee may work in the manhole, except when service load conditions and a lack of feasible alternatives require that the cable remain energized. In that case, employees may enter the manhole provided they are protected from the possible effects of a failure by shields or other devices that are capable of containing the adverse effects of a fault in the joint.

Note: Abnormalities such as oil or compound leaking from cable or joints, broken cable sheaths or joint sleeves, hot localized surface temperatures of cables or joints, or joints that are swollen beyond normal tolerance are presumed to lead to or be an indication of an impending fault.

..1910.269(t)(8)

1910.269(t)(8)

"Sheath continuity." When work is performed on buried cable or on cable in manholes, metallic sheath continuity shall be maintained or the cable sheath shall be treated as energized.

1910.269(u)

"Substations." This paragraph provides additional requirements for substations and for work performed in them.

1910.269(u)(1)

"Access and working space." Sufficient access and working space shall be provided and maintained about electric equipment to permit ready and safe operation and maintenance of such equipment.

Note: Guidelines for the dimensions of access and working space about electric equipment in substations are contained in American National Standard - National Electrical Safety Code, ANSI C2-1987. Installations meeting the ANSI provisions comply with paragraph (u)(1) of this section. An installation that does not conform to this ANSI standard will, nonetheless, be considered as complying with paragraph (u)(1) of this section if the employer can demonstrate that the installation provides ready and safe access based on the following evidence:
1. That the installation conforms to the edition of ANSI C2 that was in effect at the time the installation was made,
2. That the configuration of the installation enables employees to maintain the minimum approach distances required by paragraph (l)(2) of this section while they are working on exposed, energized parts, and
3. That the precautions taken when work is performed on the installation provide protection equivalent to the protection that would be provided by access and working space meeting ANSI C2-1987.

1910.269(u)(2)

"Draw-out-type circuit breakers." When draw-out-type circuit breakers are removed or inserted, the breaker shall be in the open position. The control circuit shall also be rendered inoperative, if the design of the equipment permits.

1910.269(u)(3)

"Substation fences." Conductive fences around substations shall be grounded. When a substation fence is expanded or a section is removed, fence grounding continuity shall be maintained, and bonding shall be used to prevent electrical discontinuity.

1910.269(u)(4)

"Guarding of rooms containing electric supply equipment."

1910.269(u)(4)(i)

Rooms and spaces in which electric supply lines or equipment are installed shall meet the requirements of paragraphs (u)(4)(ii) through (u)(4)(v) of this section under the following conditions:

..1910.269(u)(4)(i)(A)

1910.269(u)(4)(i)(A)

If exposed live parts operating at 50 to 150 volts to ground are located within 8 feet of the ground or other working surface inside the room or space,

1910.269(u)(4)(i)(B)

If live parts operating at 151 to 600 volts and located within 8 feet of the ground or other working surface inside the room or space are guarded only by location, as permitted under paragraph (u)(5)(i) of this section, or

1910.269(u)(4)(i)(C)

If live parts operating at more than 600 volts are located within the room or space, unless:

1910.269(u)(4)(i)(C)(1)

The live parts are enclosed within grounded, metal-enclosed equipment whose only openings are designed so that foreign objects inserted in these openings will be deflected from energized parts, or

1910.269(u)(4)(i)(C)(2)

The live parts are installed at a height above ground and any other working surface that provides protection at the voltage to which they are energized corresponding to the protection provided by an 8-foot height at 50 volts.

1910.269(u)(4)(ii)

The rooms and spaces shall be so enclosed within fences, screens, partitions, or walls as to minimize the possibility that unqualified persons will enter.

1910.269(u)(4)(iii)

Signs warning unqualified persons to keep out shall be displayed at entrances to the rooms and spaces.

..1910.269(u)(4)(iv)

1910.269(u)(4)(iv)

Entrances to rooms and spaces that are not under the observation of an attendant shall be kept locked.

1910.269(u)(4)(v)

Unqualified persons may not enter the rooms or spaces while the electric supply lines or equipment are energized.

1910.269(u)(5)

"Guarding of energized parts."

1910.269(u)(5)(i)

Guards shall be provided around all live parts operating at more than 150 volts to ground without an insulating covering, unless the location of the live parts gives sufficient horizontal or vertical or a combination of these clearances to minimize the possibility of accidental employee contact.

Note: Guidelines for the dimensions of clearance distances about electric equipment in substations are contained in American National Standard - National Electrical Safety Code, ANSI C2-1987. Installations meeting the ANSI provisions comply with paragraph (u)(5)(i) of this section. An installation that does not conform to this ANSI standard will, nonetheless, be considered as complying with paragraph (u)(5)(i) of this section if the employer can demonstrate that the installation provides sufficient clearance based on the following evidence:

1. That the installation conforms to the edition of ANSI C2 that was in effect at the time the installation was made,
2. That each employee is isolated from energized parts at the point of closest approach, and
3. That the precautions taken when work is performed on the installation provide protection equivalent to the protection that would be provided by horizontal and vertical clearances meeting ANSI C2-1987.

1910.269(u)(5)(ii)

Except for fuse replacement and other necessary access by qualified persons, the guarding of energized parts within a compartment shall be maintained during operation and maintenance functions to prevent accidental contact with energized parts and to prevent tools or other equipment from being dropped on energized parts.

1910.269(u)(5)(iii)

When guards are removed from energized equipment, barriers shall be installed around the work area to prevent employees who are not working on the equipment, but who are in the area, from contacting the exposed live parts.

..1910.269(u)(6)

1910.269(u)(6)

"Substation entry."

1910.269(u)(6)(i)

Upon entering an attended substation, each employee other than those regularly working in the station shall report his or her presence to the employee in charge in order to receive information on special system conditions affecting employee safety.

1910.269(u)(6)(ii)

The job briefing required by paragraph (c) of this section shall cover such additional subjects as the location of energized equipment in or adjacent to the work area and the limits of any deenergized work area.

1910.269(v)

"Power generation." This paragraph provides additional requirements and related work practices for power generating plants.

1910.269(v)(1)

"Interlocks and other safety devices."

1910.269(v)(1)(i)

Interlocks and other safety devices shall be maintained in a safe, operable condition.

1910.269(v)(1)(ii)

No interlock or other safety device may be modified to defeat its function, except for test, repair, or adjustment of the device.

1910.269(v)(2)

"Changing brushes." Before exciter or generator brushes are changed while the generator is in service, the exciter or generator field shall be checked to determine whether a ground condition exists. The brushes may not be changed while the generator is energized if a ground condition exists.

1910.269(v)(3)

"Access and working space." Sufficient access and working space shall be provided and maintained about electric equipment to permit ready and safe operation and maintenance of such equipment.

Note: Guidelines for the dimensions of access and working space about electric equipment in generating stations are contained in American National Standard - National Electrical Safety Code, ANSI C2-1987. Installations meeting the ANSI provisions comply with paragraph (v)(3) of this section. An installation that does not conform to this ANSI standard will, nonetheless, be considered as complying with paragraph (v)(3) of this section if the employer can demonstrate that the installation provides ready and safe access based on the following evidence:

1. That the installation conforms to the edition of ANSI C2 that was in effect at the time the installation was made,
2. That the configuration of the installation enables employees to maintain the minimum approach distances required by paragraph (l)(2) of this section while they are working on exposed, energized parts, and
3. That the precautions taken when work is performed on the installation provide protection equivalent to the protection that would be provided by access and working space meeting ANSI C2-1987.

..1910.269(v)(4)

1910.269(v)(4)

"Guarding of rooms containing electric supply equipment."

1910.269(v)(4)(i)

Rooms and spaces in which electric supply lines or equipment are installed shall meet the requirements of paragraphs (v)(4)(ii) through (v)(4)(v) of this section under the following conditions:

1910.269(v)(4)(i)(A)

If exposed live parts operating at 50 to 150 volts to ground are located within 8 feet of the ground or other working surface inside the room or space,

1910.269(v)(4)(i)(B)

If live parts operating at 151 to 600 volts and located within 8 feet of the ground or other working surface inside the room or space are guarded only by location, as permitted under paragraph (v)(5)(i) of this section, or

1910.269(v)(4)(i)(C)

If live parts operating at more than 600 volts are located within the room or space, unless:

1910.269(v)(4)(i)(C)(1)

The live parts are enclosed within grounded, metal-enclosed equipment whose only openings are designed so that foreign objects inserted in these openings will be deflected from energized parts, or

1910.269(v)(4)(i)(C)(2)

The live parts are installed at a height above ground and any other working surface that provides protection at the voltage to which they are energized corresponding to the protection provided by an 8-foot height at 50 volts.

..1910.269(v)(4)(ii)

1910.269(v)(4)(ii)

The rooms and spaces shall be so enclosed within fences, screens, partitions, or walls as to minimize the possibility that unqualified persons will enter.

1910.269(v)(4)(iii)

Signs warning unqualified persons to keep out shall be displayed at entrances to the rooms and spaces.

1910.269(v)(4)(iv)

Entrances to rooms and spaces that are not under the observation of an attendant shall be kept locked.

1910.269(v)(4)(v)

Unqualified persons may not enter the rooms or spaces while the electric supply lines or equipment are energized.

1910.269(v)(5)

"Guarding of energized parts."

1910.269(v)(5)(i)

Guards shall be provided around all live parts operating at more than 150 volts to ground without an insulating covering, unless the location of the live parts gives sufficient horizontal or vertical or a combination of these clearances to minimize the possibility of accidental employee contact.

Note: Guidelines for the dimensions of clearance distances about electric equipment in generating stations are contained in American National Standard - National Electrical Safety Code, ANSI C2-1987. Installations meeting the ANSI provisions comply with paragraph (v)(5)(i) of this section. An installation that does not conform to this ANSI standard will, nonetheless, be considered as complying with paragraph (v)(5)(i) of this section if the employer can demonstrate that the installation provides sufficient clearance based on the following evidence:

1. That the installation conforms to the edition of ANSI C2 that was in effect at the time the installation was made,
2. That each employee is isolated from energized parts at the point of closest approach, and
3. That the precautions taken when work is performed on the installation provide protection equivalent to the protection that would be provided by horizontal and vertical clearances meeting ANSI C2-1987.

1910.269(v)(5)(ii)

Except for fuse replacement or other necessary access by qualified persons, the guarding of energized parts within a compartment shall be maintained during operation and maintenance functions to prevent accidental contact with energized parts and to prevent tools or other equipment from being dropped on energized parts.

..1910.269(v)(5)(iii)

1910.269(v)(5)(iii)

When guards are removed from energized equipment, barriers shall be installed around the work area to prevent employees who are not working on the equipment, but who are in the area, from contacting the exposed live parts.

1910.269(v)(6)

"Water or steam spaces." The following requirements apply to work in water and steam spaces associated with boilers:

1910.269(v)(6)(i)

A designated employee shall inspect conditions before work is permitted and after its completion. Eye protection, or full face protection if necessary, shall be worn at all times when condenser, heater, or boiler tubes are being cleaned.

1910.269(v)(6)(ii)

Where it is necessary for employees to work near tube ends during cleaning, shielding shall be installed at the tube ends.

1910.269(v)(7)

"Chemical cleaning of boilers and pressure vessels." The following requirements apply to chemical cleaning of boilers and pressure vessels:

1910.269(v)(7)(i)

Areas where chemical cleaning is in progress shall be cordoned off to restrict access during cleaning. If flammable liquids, gases, or vapors or combustible materials will be used or might be produced during the cleaning process, the following requirements also apply:

1910.269(v)(7)(i)(A)

The area shall be posted with signs restricting entry and warning of the hazards of fire and explosion; and

..1910.269(v)(7)(i)(B)
1910.269(v)(7)(i)(B)

Smoking, welding, and other possible ignition sources are prohibited in these restricted areas.

1910.269(v)(7)(ii)

The number of personnel in the restricted area shall be limited to those necessary to accomplish the task safely.

1910.269(v)(7)(iii)

There shall be ready access to water or showers for emergency use.

Note: See 1910.141 of this Part for requirements that apply to the water supply and to washing facilities.

1910.269(v)(7)(iv)

Employees in restricted areas shall wear protective equipment meeting the requirements of Subpart I of this Part and including, but not limited to, protective clothing, boots, goggles, and gloves.

1910.269(v)(8)

"Chlorine systems."

1910.269(v)(8)(i)

Chlorine system enclosures shall be posted with signs restricting entry and warning of the hazard to health and the hazards of fire and explosion.

Note: See Subpart Z of this Part for requirements necessary to protect the health of employees from the effects of chlorine.

1910.269(v)(8)(ii)

Only designated employees may enter the restricted area. Additionally, the number of personnel shall be limited to those necessary to accomplish the task safely.

1910.269(v)(8)(iii)

Emergency repair kits shall be available near the shelter or enclosure to allow for the prompt repair of leaks in chlorine lines, equipment, or containers.

..1910.269(v)(8)(iv)

1910.269(v)(8)(iv)

Before repair procedures are started, chlorine tanks, pipes, and equipment shall be purged with dry air and isolated from other sources of chlorine.

1910.269(v)(8)(v)

The employer shall ensure that chlorine is not mixed with materials that would react with the chlorine in a dangerously exothermic or other hazardous manner.

1910.269(v)(9)

"Boilers."

1910.269(v)(9)(i)

Before internal furnace or ash hopper repair work is started, overhead areas shall be inspected for possible falling objects. If the hazard of falling objects exists, overhead protection such as planking or nets shall be provided.

1910.269(v)(9)(ii)

When opening an operating boiler door, employees shall stand clear of the opening of the door to avoid the heat blast and gases which may escape from the boiler.

1910.269(v)(10)

"Turbine generators."

1910.269(v)(10)(i)

Smoking and other ignition sources are prohibited near hydrogen or hydrogen sealing systems, and signs warning of the danger of explosion and fire shall be posted.

1910.269(v)(10)(ii)

Excessive hydrogen makeup or abnormal loss of pressure shall be considered as an emergency and shall be corrected immediately.

..1910.269(v)(10)(iii)

1910.269(v)(10)(iii)

A sufficient quantity of inert gas shall be available to purge the hydrogen from the largest generator.

1910.269(v)(11)

"Coal and ash handling."

1910.269(v)(11)(i)

Only designated persons may operate railroad equipment.

1910.269(v)(11)(ii)

Before a locomotive or locomotive crane is moved, a warning shall be given to employees in the area.

1910.269(v)(11)(iii)

Employees engaged in switching or dumping cars may not use their feet to line up drawheads.

1910.269(v)(11)(iv)

Drawheads and knuckles may not be shifted while locomotives or cars are in motion.

1910.269(v)(11)(v)

When a railroad car is stopped for unloading, the car shall be secured from displacement that could endanger employees.

1910.269(v)(11)(vi)

An emergency means of stopping dump operations shall be provided at railcar dumps.

1910.269(v)(11)(vii)

The employer shall ensure that employees who work in coal- or ash-handling conveyor areas are trained and knowledgeable in conveyor operation and in the requirements of paragraphs (v)(11)(viii) through (v)(11)(xii) of this section.

..1910.269(v)(11)(viii)

1910.269(v)(11)(viii)

Employees may not ride a coal- or ash-handling conveyor belt at any time. Employees may not cross over the conveyor belt, except at walkways, unless the conveyor's energy source has been deenergized and has been locked out or tagged in accordance with paragraph (d) of this section.

1910.269(v)(11)(ix)

A conveyor that could cause injury when started may not be started until personnel in the area are alerted by a signal or by a designated person that the conveyor is about to start.

1910.269(v)(11)(x)

If a conveyor that could cause injury when started is automatically controlled or is controlled from a remote location, an audible device shall be provided that sounds an alarm that will be recognized by each employee as a warning that the conveyor will start and that can be clearly heard at all points along the conveyor where personnel may be present. The warning device shall be actuated by the device starting the conveyor and shall continue for a period of time before the conveyor starts that is long enough to allow employees to move clear of the conveyor system. A visual warning may be used in place of the audible device if the employer can demonstrate that it will provide an equally effective warning in the particular circumstances involved.

Exception: If the employer can demonstrate that the system's function would be seriously hindered by the required time delay, warning signs may be provided in place of the audible warning device. If the system was installed before January 31, 1995, warning signs may be provided in place of the audible warning device until such time as the conveyor or its control system is rebuilt or rewired. These warning signs shall be clear, concise, and legible and shall indicate that conveyors and allied equipment may be started at any time, that danger exists, and that personnel must keep clear. These warning signs shall be provided along the conveyor at areas not guarded by position or location.

..1910.269(v)(11)(xi)

1910.269(v)(11)(xi)

Remotely and automatically controlled conveyors, and conveyors that have operating stations which are not manned or which are beyond voice and visual contact from drive areas, loading areas, transfer points, and other locations on the conveyor path not guarded by location, position, or guards shall be furnished with emergency stop buttons, pull cords, limit switches, or similar emergency stop devices. However, if the employer can demonstrate that the design, function, and operation of the conveyor do not expose an employee to hazards, an emergency stop device is not required.

1910.269(v)(11)(xi)(A)

Emergency stop devices shall be easily identifiable in the immediate vicinity of such locations.

1910.269(v)(11)(xi)(B)

An emergency stop device shall act directly on the control of the conveyor involved and may not depend on the stopping of any other equipment.

1910.269(v)(11)(xi)(C)

Emergency stop devices shall be installed so that they cannot be overridden from other locations.

1910.269(v)(11)(xii)

Where coal-handling operations may produce a combustible atmosphere from fuel sources or from flammable gases or dust, sources of ignition shall be eliminated or safely controlled to prevent ignition of the combustible atmosphere.

Note: Locations that are hazardous because of the presence of combustible dust are classified as Class II hazardous locations. See 1910.307 of this Part.

1910.269(v)(11)(xiii)

An employee may not work on or beneath overhanging coal in coal bunkers, coal silos, or coal storage areas, unless the employee is protected from all hazards posed by shifting coal.

..1910.269(v)(11)(xiv)

1910.269(v)(11)(xiv)

An employee entering a bunker or silo to dislodge the contents shall wear a body harness with lifeline attached. The lifeline shall be secured to a fixed support outside the bunker and shall be attended at all times by an employee located outside the bunker or facility.

1910.269(v)(12)

"Hydroplants and equipment." Employees working on or close to water gates, valves, intakes, forebays, flumes, or other locations where increased or decreased water flow or levels may pose a significant hazard shall be warned and shall vacate such dangerous areas before water flow changes are made.

1910.269(w)

"Special conditions."

1910.269(w)(1)

"Capacitors." The following additional requirements apply to work on capacitors and on lines connected to capacitors.

Note: See paragraphs (m) and (n) of this section for requirements pertaining to the deenergizing and grounding of capacitor installations.

1910.269(w)(1)(i)

Before employees work on capacitors, the capacitors shall be disconnected from energized sources and, after a wait of at least 5 minutes from the time of disconnection, short-circuited.

1910.269(w)(1)(ii)

Before the units are handled, each unit in series-parallel capacitor banks shall be short-circuited between all terminals and the capacitor case or its rack. If the cases of capacitors are on ungrounded substation racks, the racks shall be bonded to ground.

1910.269(w)(1)(iii)

Any line to which capacitors are connected shall be short-circuited before it is considered deenergized.

..1910.269(w)(2)

1910.269(w)(2)

"Current transformer secondaries." The secondary of a current transformer may not be opened while the transformer is energized. If the primary of the current transformer cannot be deenergized before work is performed on an instrument, a relay, or other section of a current transformer secondary circuit, the circuit shall be bridged so that the current transformer secondary will not be opened.

1910.269(w)(3)

"Series streetlighting."

1910.269(w)(3)(i)

If the open-circuit voltage exceeds 600 volts, the series streetlighting circuit shall be worked in accordance with paragraph (q) or (t) of this section, as appropriate.

1910.269(w)(3)(ii)

A series loop may only be opened after the streetlighting transformer has been deenergized and isolated from the source of supply or after the loop is bridged to avoid an open-circuit condition.

1910.269(w)(4)

"Illumination." Sufficient illumination shall be provided to enable the employee to perform the work safely.

1910.269(w)(5)

"Protection against drowning."

1910.269(w)(5)(i)

Whenever an employee may be pulled or pushed or may fall into water where the danger of drowning exists, the employee shall be provided with and shall use U.S. Coast Guard approved personal flotation devices.

1910.269(w)(5)(ii)

Each personal flotation device shall be maintained in safe condition and shall be inspected frequently enough to ensure that it does not have rot, mildew, water saturation, or any other condition that could render the device unsuitable for use.

..1910.269(w)(5)(iii)

1910.269(w)(5)(iii)

An employee may cross streams or other bodies of water only if a safe means of passage, such as a bridge, is provided.

1910.269(w)(6)

"Employee protection in public work areas."

1910.269(w)(6)(i)

Traffic control signs and traffic control devices used for the protection of employees shall meet the requirements of 1926.200(g)(2) of this Chapter.

1910.269(w)(6)(ii)

Before work is begun in the vicinity of vehicular or pedestrian traffic that may endanger employees, warning signs or flags and other traffic control devices shall be placed in conspicuous locations to alert and channel approaching traffic.

1910.269(w)(6)(iii)

Where additional employee protection is necessary, barricades shall be used.

1910.269(w)(6)(iv)

Excavated areas shall be protected with barricades.

1910.269(w)(6)(v)

At night, warning lights shall be prominently displayed.

..1910.269(w)(7)

1910.269(w)(7)

"Backfeed." If there is a possibility of voltage backfeed from sources of cogeneration or from the secondary system (for example, backfeed from more than one energized phase feeding a common load), the requirements of paragraph (l) of this section apply if the lines or equipment are to be worked as energized, and the requirements of paragraphs (m) and (n) of this section apply if the lines or equipment are to be worked as deenergized.

1910.269(w)(8)

"Lasers." Laser equipment shall be installed, adjusted, and operated in accordance with 1926.54 of this Chapter.

1910.269(w)(9)

"Hydraulic fluids." Hydraulic fluids used for the insulated sections of equipment shall provide insulation for the voltage involved.

1910.269(x)

"Definitions."

"Affected employee." An employee whose job requires him or her to operate or use a machine or equipment on which servicing or maintenance is being performed under lockout or tagout, or whose job requires him or her to work in an area in which such servicing or maintenance is being performed.

"Attendant." An employee assigned to remain immediately outside the entrance to an enclosed or other space to render assistance as needed to employees inside the space.

"Authorized employee." An employee who locks out or tags out machines or equipment in order to perform servicing or maintenance on that machine or equipment. An affected employee becomes an authorized employee when that employee's duties include performing servicing or maintenance covered under this section.

"Automatic circuit recloser." A self-controlled device for interrupting and reclosing an alternating current circuit with a predetermined sequence of opening and reclosing followed by resetting, hold-closed, or lockout operation.

"Barricade." A physical obstruction such as tapes, cones, or A-frame type wood or metal structures intended to provide a warning about and to limit access to a hazardous area.

"Barrier." A physical obstruction which is intended to prevent contact with energized lines or equipment or to prevent unauthorized access to a work area.

"Bond." The electrical interconnection of conductive parts designed to maintain a common electrical potential.

"Bus." A conductor or a group of conductors that serve as a common connection for two or more circuits.

"Bushing." An insulating structure, including a through conductor or providing a passageway for such a conductor, with provision for mounting on a barrier, conducting or otherwise, for the purposes of insulating the conductor from the barrier and conducting current from one side of the barrier to the other.

"Cable." A conductor with insulation, or a stranded conductor with or without insulation and other coverings (single-conductor cable), or a combination of conductors insulated from one another (multiple-conductor cable).

"Cable sheath." A conductive protective covering applied to cables.

Note: A cable sheath may consist of multiple layers of which one or more is conductive.

"Circuit." A conductor or system of conductors through which an electric current is intended to flow.

"Clearance (between objects)." The clear distance between two objects measured surface to surface.

"Clearance (for work)." Authorization to perform specified work or permission to enter a restricted area.

"Communication lines. (See Lines, communication.)"

"Conductor." A material, usually in the form of a wire, cable, or bus bar, used for carrying an electric current.

"Covered conductor." A conductor covered with a dielectric having no rated insulating strength or having a rated insulating strength less than the voltage of the circuit in which the conductor is used.

"Current-carrying part." A conducting part intended to be connected in an electric circuit to a source of voltage. Non-current-carrying parts are those not intended to be so connected.

"Deenergized." Free from any electrical connection to a source of potential difference and from electric charge; not having a potential different from that of the earth.

Note: The term is used only with reference to current-carrying parts, which are sometimes energized (alive).

"Designated employee (designated person)." An employee (or person) who is designated by the employer to perform specific duties under the terms of this section and who is knowledgeable in the construction and operation of the equipment and the hazards involved.

"Electric line truck." A truck used to transport personnel, tools, and material for electric supply line work.

"Electric supply equipment." Equipment that produces, modifies, regulates, controls, or safeguards a supply of electric energy.

"Electric supply lines. (See Lines, electric supply.)"

"Electric utility." An organization responsible for the installation, operation, or maintenance of an electric supply system.

"Enclosed space." A working space, such as a manhole, vault, tunnel, or shaft, that has a limited means of egress or entry, that is designed for periodic employee entry under normal operating conditions, and that under normal conditions does not contain a hazardous atmosphere, but that may contain a hazardous atmosphere under abnormal conditions.

Note: Spaces that are enclosed but not designed for employee entry under normal operating conditions are not considered to be enclosed spaces for the purposes of this section. Similarly, spaces that are enclosed and that are expected to contain a hazardous atmosphere are not considered to be enclosed spaces for the purposes of this section. Such spaces meet the definition of permit spaces in 1910.146 of this Part, and entry into them must be performed in accordance with that standard.

"Energized (alive, live)." Electrically connected to a source of potential difference, or electrically charged so as to have a potential significantly different from that of earth in the vicinity.

"Energy isolating device." A physical device that prevents the transmission or release of energy, including, but not limited to, the following: a manually operated electric circuit breaker, a disconnect switch, a manually operated switch, a slide gate, a slip blind, a line valve, blocks, and any similar device with a visible indication of the position of the device. (Push buttons, selector switches, and other control-circuit-type devices are not energy isolating devices.)

"Energy source." Any electrical, mechanical, hydraulic, pneumatic, chemical, nuclear, thermal, or other energy source that could cause injury to personnel.

"Equipment (electric)." A general term including material, fittings, devices, appliances, fixtures, apparatus, and the like used as part of or in connection with an electrical installation.

"Exposed." Not isolated or guarded.

"Ground." A conducting connection, whether intentional or accidental, between an electric circuit or equipment and the earth, or to some conducting body that serves in place of the earth.

"Grounded." Connected to earth or to some conducting body that serves in place of the earth.

"Guarded." Covered, fenced, enclosed, or otherwise protected, by means of suitable covers or casings, barrier rails or screens, mats, or platforms, designed to minimize the possibility, under normal conditions, of dangerous approach or accidental contact by persons or objects.

Note: Wires which are insulated, but not otherwise protected, are not considered as guarded.

"Hazardous atmosphere" means an atmosphere that may expose employees to the risk of death, incapacitation, impairment of ability to self-rescue (that is, escape unaided from an enclosed space), injury, or acute illness from one or more of the following causes:

1910.269(x)(1)

Flammable gas, vapor, or mist in excess of 10 percent of its lower flammable limit (LFL);

1910.269(x)(2)

Airborne combustible dust at a concentration that meets or exceeds its LFL;

Note: This concentration may be approximated as a condition in which the dust obscures vision at a distance of 5 feet (1.52 m) or less.

1910.269(x)(3)

Atmospheric oxygen concentration below 19.5 percent or above 23.5 percent;

1910.269(x)(4)

Atmospheric concentration of any substance for which a dose or a permissible exposure limit is published in Subpart G, "Occupational Health and Environmental Control", or in Subpart Z, "Toxic and Hazardous Substances," of this Part and which could result in employee exposure in excess of its dose or permissible exposure limit;

Note: An atmospheric concentration of any substance that is not capable of causing death, incapacitation, impairment of ability to self-rescue, injury, or acute illness due to its health effects is not covered by this provision.

..1910.269(x)(5)

1910.269(x)(5)

Any other atmospheric condition that is immediately dangerous to life or health.

Note: For air contaminants for which OSHA has not determined a dose or permissible exposure limit, other sources of information, such as Material Safety Data Sheets that comply with the Hazard Communication Standard, 1910.1200 of this Part, published information, and internal documents can provide guidance in establishing acceptable atmospheric conditions.

"High-power tests." Tests in which fault currents, load currents, magnetizing currents, and line-dropping currents are used to test equipment, either at the equipment's rated voltage or at lower voltages.

"High-voltage tests." Tests in which voltages of approximately 1000 volts are used as a practical minimum and in which the voltage source has sufficient energy to cause injury.

"High wind." A wind of such velocity that the following hazards would be present:

1. An employee would be exposed to being blown from elevated locations, or
2. An employee or material handling equipment could lose control of material being handled, or
3. An employee would be exposed to other hazards not controlled by the standard involved.

Note: Winds exceeding 40 miles per hour (64.4 kilometers per hour), or 30 miles per hour (48.3 kilometers per hour) if material handling is involved, are normally considered as meeting this criteria unless precautions are taken to protect employees from the hazardous effects of the wind.

"Immediately dangerous to life or health (IDLH)." means any condition that poses an immediate or delayed threat to life or that would cause irreversible adverse health effects or that would interfere with an individual's ability to escape unaided from a permit space.

Note: Some materials—hydrogen fluoride gas and cadmium vapor, for example—may produce immediate transient effects that, even if severe, may pass without medical attention, but are followed by sudden, possibly fatal collapse 12–72 hours after exposure. The victim "feels normal" from recovery from transient effects until collapse. Such materials in hazardous quantities are considered to be "immediately" dangerous to life or health.

"Insulated." Separated from other conducting surfaces by a dielectric (including air space) offering a high resistance to the passage of current.

Note: When any object is said to be insulated, it is understood to be insulated for the conditions to which it is normally subjected. Otherwise, it is, within the purpose of this section, uninsulated.

"Insulation (cable)." That which is relied upon to insulate the conductor from other conductors or conducting parts or from ground.

"Line-clearance tree trimmer." An employee who, through related training or on-the-job experience or both, is familiar with the special techniques and hazards involved in line-clearance tree trimming.

Note 1: An employee who is regularly assigned to a line-clearance tree-trimming crew and who is undergoing on-the-job training and who, in the course of such training, has demonstrated an ability to perform duties safely at his or her level of training and who is under the direct supervision of a line-clearance tree trimmer is considered to be a line-clearance tree trimmer for the performance of those duties.

Note 2: A line-clearance tree trimmer is not considered to be a "qualified employee" under this section unless he or she has the training required for a qualified employee under paragraph (a)(2)(ii) of this section. However, under the electrical safety-related work practices standard in Subpart S of this Part, a line-clearance tree trimmer is considered to be a "qualified employee". Tree trimming performed by such "qualified employees" is not subject to the electrical safety-related work practice requirements contained in 1910.331 through 1910.335 of this Part. (See also the note following 1910.332(b)(3) of this Part for information regarding the training an employee must have to be considered a qualified employee under 1910.331 through 1910.335 of this part.)

"Line-clearance tree trimming." The pruning, trimming, repairing, maintaining, removing, or clearing of trees or the cutting of brush that is within 10 feet (305 cm) of electric supply lines and equipment.

"Lines. 1. Communication lines." The conductors and their supporting or containing structures which are used for public or private signal or communication service, and which operate at potentials not exceeding 400 volts to ground or 750 volts between any two points of the circuit, and the transmitted power of which does not exceed 150 watts. If the lines are operating at less than 150 volts, no limit is placed on the transmitted power of the system. Under certain conditions, communication cables may include communication circuits exceeding these limitations where such circuits are also used to supply power solely to communication equipment.

Note: Telephone, telegraph, railroad signal, data, clock, fire, police alarm, cable television, and other systems conforming to this definition are included. Lines used for signaling purposes, but not included under this definition, are considered as electric supply lines of the same voltage.

2. "Electric supply lines". Conductors used to transmit electric energy and their necessary supporting or containing structures. Signal lines of more than 400 volts are always supply lines within this section, and those of less than 400 volts are considered as supply lines, if so run and operated throughout.

"Manhole." A subsurface enclosure which personnel may enter and which is used for the purpose of installing, operating, and maintaining submersible equipment or cable.

"Manhole steps." A series of steps individually attached to or set into the walls of a manhole structure.

"Minimum approach distance." The closest distance an employee is permitted to approach an energized or a grounded object.

"Qualified employee (qualified person)." One knowledgeable in the construction and operation of the electric power generation, transmission, and distribution equipment involved, along with the associated hazards.

Note 1: An employee must have the training required by paragraph (a)(2)(ii) of this section in order to be considered a qualified employee.

Note 2: Except under paragraph (g)(2)(v) of this section, an employee who is undergoing on-the-job training and who, in the course of such training, has demonstrated an ability to perform duties safely at his or her level of training and who is under the direct supervision of a qualified person is considered to be a qualified person for the performance of those duties.

"Step bolt." A bolt or rung attached at intervals along a structural member and used for foot placement during climbing or standing.

"Switch." A device for opening and closing or for changing the connection of a circuit. In this section, a switch is understood to be manually operable, unless otherwise stated.

"System operator." A qualified person designated to operate the system or its parts.

"Vault." An enclosure, above or below ground, which personnel may enter and which is used for the purpose of installing, operating, or maintaining equipment or cable.

"Vented vault." A vault that has provision for air changes using exhaust flue stacks and low level air intakes operating on differentials of pressure and temperature providing for airflow which precludes a hazardous atmosphere from developing.

"Voltage." The effective (rms) potential difference between any two conductors or between a conductor and ground. Voltages are expressed in nominal values unless otherwise indicated. The nominal voltage of a system or circuit is the value assigned to a system or circuit of a given voltage class for the purpose of convenient designation. The operating voltage of the system may vary above or below this value.

[59 FR 40672, Aug. 9, 1994; 59 FR 51672, Oct. 12, 1994]

Reprinted from www.osha.gov

Occupational Safety & Health Administration
200 Constitution Avenue, NW
Washington, DC 20210

REGULATIONS (STANDARDS - 29 CFR)

FLOW CHARTS. - 1910.269APPA

- **Part Number:** 1910
- **Part Title:** Occupational Safety and Health Standards
- **Subpart:** R
- **Subpart Title:** Special Industries
- **Standard Number:** 1910.269 App A
- **Title:** Flow Charts.

Appendix A to § 1910.269—Flow Charts

This appendix presents information, in the form of flow charts, that illustrates the scope and application of 1910.269. This appendix addresses the interface between 1910.269 and Subpart S of this Part (Electrical), between 1910.269 and 1910.146 of this Part (Permit-required confined spaces), and between 1910.269 and 1910.147 of this Part (The control of hazardous energy (lockout/tagout)). These flow charts provide guidance for employers trying to implement the requirements of 1910.269 in combination with other General Industry Standards contained in Part 1910.

Appendix A-1 to §1910.269—Application of §1910.269 and Subpart S of this Part to Electrical Installations.

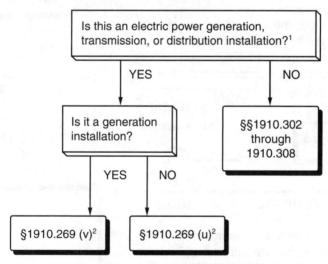

[1] Electrical installation design requirements only. See Appendix 1B for electrical safety-related work practices. Supplementary electric generating equipment that is used to supply a workplace for emergency, standby, or similar purposes only is not considered to be an electric power generation installation.

[2] See Table 1 of Appendix A-2 for requirements that can be met through compliance with Subpart S.

Appendix A-2 to §1910.269—Application of §1910.269 and Subpart S of this Part to Electrical Safety-Related Work Practices.

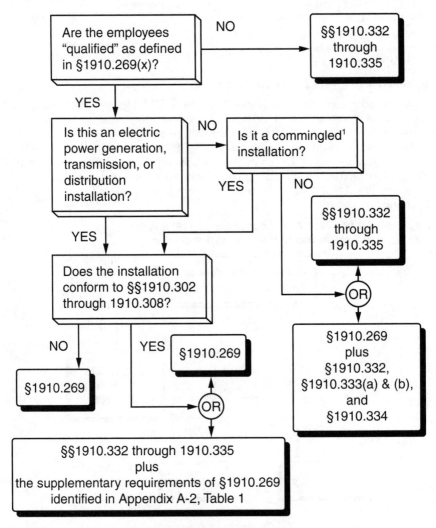

[1] Commingled to the extent that the electric power generation, transmission, or distribution installation poses the greater hazard.

Table 1.—Electrical Safety-Related Work Practices in §1910.269

Compliance with subpart S is considered as compliance with § 1910.269[1]	Paragraphs that apply regardless of compliance with subpart S
(d) electric shock hazards only	(a)(2)[2] and (a)(3)[2].
(h)(3)	(b)[2].
(i)(2)	(c)[2].
(k)	(d), other than electric shock hazards.
(i)(1) through (i)(4), (i)(6)(i), and (i)(8) through (i)(10)	(e).
(m)	(f).
(p)(4)	(g).
(s)(2)	(h)(1) and (h)(2).
(u)(1) and (u)(3) through (u)(5)	(i)(3)[2] and (i)(4)[2].
(v)(3) through (v)(5)	(i)[2].
(w)(1) and (w)(7)	(i)(5)[2], (i)(6)(ii)[2], (i)(6)(iii)[2], and (i)(7)[2].
	(n)[2].
	(o)[2].
	(p)(1) through (p)(3).
	(q)[2].
	(r)[2].
	(s)(1).
	(i)[2].
	(u)(2)[2] and (u)(6)[2].
	(v)(1), (v)(2)[2], and (v)(6) through (v)(12).
	(w)(2) through (w)(6)[2], (w)(8), and (w)(9)[2].

[1]If the electrical installation meets the requirements of §§ 1910.303 through 1910.308 of this part, then the electrical installation and any associated electrical safety-related work practices conforming to §§ 1910.332 through 1910.335 of this part are considered to comply with these provisions of § 1910.269 of this part.
[2]These provisions include electrical safety requirements that must be met regardless of compliance with subpart S of this part.

Appendix A-3 to §1910.269—Application of §1910.269 and Subpart S of this Part to Tree-Trimming Operations.

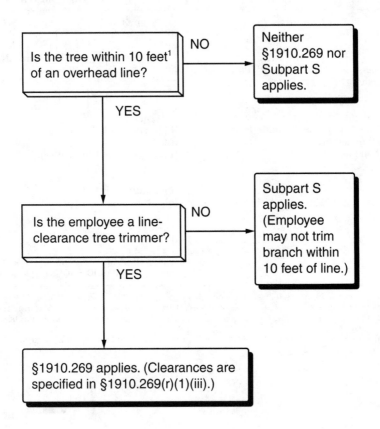

Is the tree within 10 feet[1] of an overhead line?

NO → Neither §1910.269 nor Subpart S applies.

YES

Is the employee a line-clearance tree trimmer?

NO → Subpart S applies. (Employee may not trim branch within 10 feet of line.)

YES

§1910.269 applies. (Clearances are specified in §1910.269(r)(1)(iii).)

[1] 10 feet plus 4 inches for every 10 kilovolts over 50 kilovolts.

Appendix A-4 to §1910.269—Application of §§1910.147, 1910.269 and 1910.333 to Hazardous Energy Control Procedures (Lockout/Tagout).

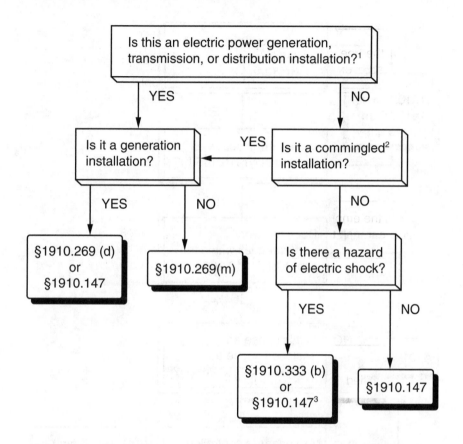

[1] If the installation conforms to §§1910.303 through 1910.308, the lockout and tagging procedures of 1910.333(b) may be followed for electric shock hazards.

[2] Commingled to the extent that the electric power generation, transmission, or distribution installation poses the greater hazard.

[3] §1910.333(b)(2)(iii)(D) and (b)(2)(iv)(B) still apply.

Appendix A-5 to §1910.269—Application of §§1910.146 and 1910.269 to Permit-Required Confined Spaces.

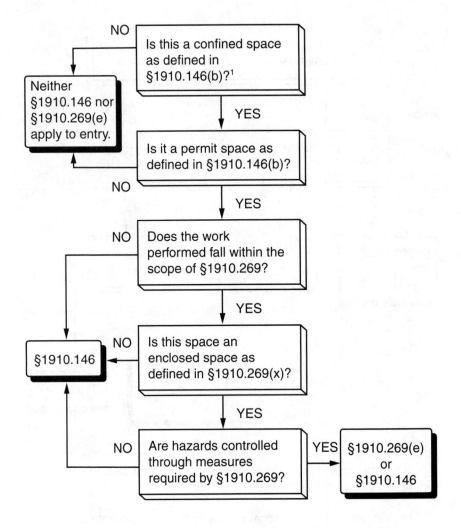

[1] See §1910.146(c) for general non-entry requirements that apply to all confined spaces.

Reprinted from www.osha.gov

Occupational Safety & Health Administration
200 Constitution Avenue, NW
Washington, DC 20210

REGULATIONS (STANDARDS - 29 CFR)

WORKING ON EXPOSED ENERGIZED PARTS. - 1910.269APPB
- **Part Number:** 1910
- **Part Title:** Occupational Safety and Health Standards
- **Subpart:** R
- **Subpart Title:** Special Industries
- **Standard Number:** 1910.269 App B
- **Title:** Working on Exposed Energized Parts.

Appendix B to §1910.269—Working on Exposed Energized Parts

I. Introduction

Electric transmission and distribution line installations have been designed to meet National Electrical Safety Code (NESC), ANSI C2, requirements and to provide the level of line outage performance required by system reliability criteria. Transmission and distribution lines are also designed to withstand the maximum overvoltages expected to be impressed on the system. Such overvoltages can be caused by such conditions as switching surges, faults, or lightning. Insulator design and lengths and the clearances to structural parts (which, for low voltage through extra-high voltage, or EHV, facilities, are generally based on the performance of the line as a result of contamination of the insulation or during storms) have, over the years, come closer to the minimum approach distances used by workers (which are generally based on non-storm conditions). Thus, as minimum approach (working) distances and structural distances (clearances) converge, it is increasingly important that basic considerations for establishing safe approach distances for performing work be understood by the designers and the operating and maintenance personnel involved.

The information in this Appendix will assist employers in complying with the minimum approach distance requirements contained in paragraphs (l)(2) and (q)(3) of this section. The technical criteria and methodology presented herein is mandatory for employers using reduced minimum approach distances as permitted in Table R-7 and Table R-8. This Appendix is intended to provide essential background information and technical criteria for the development or modification, if possible, of the safe minimum approach distances for electric transmission and distribution live-line work. The development of these safe distances must be undertaken by persons knowledgeable in the techniques discussed in this appendix and competent in the field of electric transmission and distribution system design.

II. General

A. Definitions

The following definitions from 1910.269(x) relate to work on or near transmission and distribution lines and equipment and the electrical hazards they present.
Exposed.   Not isolated or guarded.
Guarded.   Covered, fenced, enclosed, or otherwise protected, by means of suitable covers or casings, barrier rails or screens, mats, or platforms, designed

to minimize the possibility, under normal conditions, of dangerous approach or accidental contact by persons or objects.

NOTE: Wires which are insulated, but not otherwise protected, are not considered as guarded.

Insulated. Separated from other conducting surfaces by a dielectric (including air space) offering a high resistance to the passage of current.

NOTE: When any object is said to be insulated, it is understood to be insulated for the conditions to which it is normally subjected. Otherwise, it is, within the purpose of this section, uninsulated.

B. Installations Energized at 50 to 300 Volts

The hazards posed by installations energized at 50 to 300 volts are the same as those found in many other workplaces. That is not to say that there is no hazard, but the complexity of electrical protection required does not compare to that required for high voltage systems. The employee must avoid contact with the exposed parts, and the protective equipment used (such as rubber insulating gloves) must provide insulation for the voltages involved.

C. Exposed Energized Parts over 300 Volts AC

Table R-6, Table R-7, and Table R-8 of 1910.269 provide safe approach and working distances in the vicinity of energized electric apparatus so that work can be done safely without risk of electrical flashover.

The working distances must withstand the maximum transient overvoltage that can reach the work site under the working conditions and practices in use. Normal system design may provide or include a means to control transient overvoltages, or temporary devices may be employed to achieve the same result. The use of technically correct practices or procedures to control overvoltages (for example, portable gaps or preventing the automatic control from initiating breaker reclosing) enables line design and operation to be based on reduced transient overvoltage values. Technical information for U.S. electrical systems indicates that current design provides for the following maximum transient overvoltage values (usually produced by switching surges): 362 kV and less—3.0 per unit; 552 kV—2.4 per unit; 800 kV—2.0 per unit.

Additional discussion of maximum transient overvoltages can be found in paragraph IV.A.2, later in this Appendix.

III. Determination of the Electrical Component of Minimum Approach Distances

A. Voltages of 1.1 kV to 72.5 kV

For voltages of 1.1 kV to 72.5 kV, the electrical component of minimum approach distances is based on American National Standards Institute (ANSI)/American Institute of Electrical Engineers (AIEE) Standard No.4, March 1943, Tables III and IV. (AIEE is the predecessor technical society to the Institute of Electrical and Electronic Engineers (IEEE).) These distances are calculated by the following formula:

Equation (1) - For voltages of 1.1 kV to 72.5 kV

$$D = (V_{max} \times pu/124)^{1.63}$$

where D = Electrical component of the minimum approach distance in air
in feet
V(max) = Maximum rated line-to-ground rms voltage in kV
pu = Maximum transient overvoltage factor in per unit

Source: AIEE Standard No. 4, 1943.

This formula has been used to generate Table 1.

Table 1. AC Energized Line-Work Phase-to-Ground Electrical Component of the Minimum Approach Distance - 1.1 to 72.5 kV

Maximum anticipated per-unit transient overvoltage	Phase to phase voltage			
	15,000	36,000	46,000	72,500
3.0	0.08	0.33	0.49	1.03

NOTE: The distances given (in feet) are for air as the insulating medium and provide no additional clearance for inadvertent movement.

B. Voltages of 72.6 kV to 800 kV

For voltages of 72.6 kV to 800 kV, the electrical component of minimum approach distances is based on ANSI/IEEE Standard 516-1987, "IEEE Guide for Maintenance Methods on Energized Power Lines." This standard gives the electrical component of the minimum approach distance based on power frequency rod-gap data, supplemented with transient overvoltage information and a saturation factor for high voltages. The distances listed in ANSI/IEEE Standard 516 have been calculated according to the following formula:

Equation (2) - For voltages of 72.6 kV to 800 kV

$$D = (C + a)pu\ V(MAX)$$

where D = Electrical component of the minimum approach distance in air in feet
C = 0.01 to take care of correction factors associated with the variation of gap sparkover with voltage
a = A factor relating to the saturation of air at voltages of 345 kV or higher
pu = Maximum anticipated transient overvoltage, in per unit (p.u.)
V(MAX) = Maximum rms system line-to-ground voltage in kilovolts—it should be the "actual" maximum, or the normal highest voltage for the range (for example, 10 percent above the nominal voltage)
SOURCE: Formula developed from ANSI/IEEE Standard No. 516, 1987.

This formula is used to calculate the electrical component of the minimum approach distances in air and is used in the development of Table 2 and Table 3.

Table 2.   AC Energized Line-Work Phase-to-Ground Electrical Component of the Minimum Approach Distance - 121 to 242 kV

Maximum anticipated per-unit transient overvoltage	Phase to phase voltage			
	121,000	145,000	169,000	242,000
2.0	1.40	1.70	2.00	2.80
2.1	1.47	1.79	2.10	2.94
2.2	1.54	1.87	2.20	3.08
2.3	1.61	1.96	2.30	3.22
2.4	1.68	2.04	2.40	3.35
2.5	1.75	2.13	2.50	3.50
2.6	1.82	2.21	2.60	3.64
2.7	1.89	2.30	2.70	3.76
2.8	1.96	2.38	2.80	3.92
2.9	2.03	2.47	2.90	4.05
3.0	2.10	2.55	3.00	4.29

NOTE: The distances given (in feet) are for air as the insulating medium and provide no additional clearance for inadvertent movement.

Table 3.   AC Energized Line-Work Phase-to-Ground Electrical Component of the Minimum Approach Distance - 362 to 800 kV

Maximum anticipated per-unit transient overvoltage	Phase to phase voltage		
	362,000	552,000	800,000
1.5		4.97	8.66
1.6		5.46	9.60
1.7		5.98	10.60
1.8		6.51	11.64
1.9		7.08	12.73
2.0	4.20	7.68	13.86
2.1	4.41	8.27	
2.2	4.70	8.87	
2.3	5.01	9.49	
2.4	5.34	10.21	
2.5	5.67		
2.6	6.01		
2.7	6.36		
2.8	6.73		
2.9	7.10		
3.0	7.48		

NOTE: The distances given (in feet) are for air as the insulating medium and provide no additional clearance for inadvertent movement.

C. Provisions for Inadvertent Movement

The minimum approach distances (working distances) must include an "adder" to compensate for the inadvertent movement of the worker relative to an energized part or the movement of the part relative to the worker. A certain allowance must be made to account for this possible inadvertent movement and to provide the worker with a comfortable and safe zone in which to work. A distance for inadvertent movement (called the "ergonomic component of the minimum approach distance") must be added to the electrical component to determine the total safe minimum approach distances used in live-line work.

One approach that can be used to estimate the ergonomic component of the minimum approach distance is response time-distance analysis. When this technique is used, the total response time to a hazardous incident is estimated and converted to distance travelled. For example, the driver of a car takes a given amount of time to respond to a "stimulus" and stop the vehicle. The elapsed time involved results in a distance being travelled before the car comes to a complete stop. This distance is dependent on the speed of the car at the time the stimulus appears.

In the case of live-line work, the employee must first perceive that he or she is approaching the danger zone. Then, the worker responds to the danger and must decelerate and stop all motion toward the energized part. During the time it takes to stop, a distance will have been traversed. It is this distance that must be added to the electrical component of the minimum approach distance to obtain the total safe minimum approach distance.

At voltages below 72.5 kV, the electrical component of the minimum approach distance is smaller than the ergonomic component. At 72.5 kV the electrical component is only a little more than 1 foot. An ergonomic component of the minimum approach distance is needed that will provide for all the worker's unexpected movements. The usual live-line work method for these voltages is the use of rubber insulating equipment, frequently rubber gloves. The energized object needs to be far enough away to provide the worker's face with a safe approach distance, as his or her hands and arms are insulated. In this case, 2 feet has been accepted as a sufficient and practical value.

For voltages between 72.6 and 800 kV, there is a change in the work practices employed during energized line work. Generally, live-line tools (hot sticks) are employed to perform work while equipment is energized. These tools, by design, keep the energized part at a constant distance from the employee and thus maintain the appropriate minimum approach distance automatically.

The length of the ergonomic component of the minimum approach distance is also influenced by the location of the worker and by the nature of the work. In these higher voltage ranges, the employees use work methods that more tightly control their movements than when the workers perform rubber glove work. The worker is farther from energized line or equipment and needs to be more precise in his or her movements just to perform the work.

For these reasons, a smaller ergonomic component of the minimum approach distance is needed, and a distance of 1 foot has been selected for voltages between 72.6 and 800 kV.

Table 4 summarizes the ergonomic component of the minimum approach distance for the two voltage ranges.

Table 4.   Ergonomic Component of Minimum Approach Distance

Voltage range (kV)	Distance (feet)
1.1 to 72.5	2.0
72.6 to 800	1.0

NOTE: This distance must be added to the electrical component of the minimum approach distance to obtain the full minimum approach distance.

D. Bare-Hand Live-Line Minimum Approach Distances

Calculating the strength of phase-to-phase transient overvoltages is compli-cated by the varying time displacement between overvoltages on parallel con-ductors (electrodes) and by the varying ratio between the positive and negative voltages on the two electrodes. The time displacement causes the maximum voltage between phases to be less than the sum of the phase-to-ground voltages. The International Electrotechnical Commission (IEC) Technical Committee 28, Working Group 2, has developed the following formula for determining the phase-to-phase maximum transient overvoltage, based on the per unit (p.u.) of the system nominal voltage phase-to-ground crest:

$$pu(p) = pu(g) + 1.6.$$

where pu(g) = p.u. phase-to-ground maximum transient overvoltage
      pu(p) = p.u. phase-to-phase maximum transient overvoltage

This value of maximum anticipated transient overvoltage must be used in Equation (2) to calculate the phase-to-phase minimum approach distances for live-line bare-hand work.

E. Compiling the Minimum Approach Distance Tables

For each voltage involved, the distance in Table 4 in this appendix has been added to the distance in Table 1, Table 2 or Table 3 in this appendix to deter-mine the resulting minimum approach distances in Table R-6, Table R-7, and Table R-8 in 1910.269.

F. Miscellaneous Correction Factors

The strength of an air gap is influenced by the changes in the air medium that forms the insulation. A brief discussion of each factor follows, with a summary at the end.
1. Dielectric strength of air.   The dielectric strength of air in a uniform electric field at standard atmospheric conditions is approximately 31 kV (crest) per cm at 60 Hz. The disruptive gradient is affected by the air pressure, temperature, and humidity, by the shape, dimensions, and separation of the electrodes, and by the characteristics of the applied voltage (wave shape).
2. Atmospheric effect.   Flashover for a given air gap is inhibited by an increase in the density (humidity) of the air. The empirically determined electrical strength of a given gap is normally applicable at standard atmospheric condi-tions (20 deg. C, 101.3 kPa, 11 g/cm^3 humidity).

The combination of temperature and air pressure that gives the lowest gap flashover voltage is high temperature and low pressure. These are conditions not likely to occur simultaneously. Low air pressure is generally associated with high humidity, and this causes increased electrical strength. An average air pres-sure is more likely to be associated with low humidity. Hot and dry working conditions are thus normally associated with reduced electrical strength.

The electrical component of the minimum approach distances in Table 1, Table 2, and Table 3 has been calculated using the maximum transient overvoltages to determine withstand voltages at standard atmospheric conditions.

3. Altitude. The electrical strength of an air gap is reduced at high altitude, due principally to the reduced air pressure. An increase of 3 percent per 300 meters in the minimum approach distance for altitudes above 900 meters is required. Table R-10 of 1910.269 presents this information in tabular form.

Summary. After taking all these correction factors into account and after considering their interrelationships relative to the air gap insulation strength and the conditions under which live work is performed, one finds that only a correction for altitude need be made. An elevation of 900 meters is established as the base elevation, and the values of the electrical component of the minimum approach distances has been derived with this correction factor in mind. Thus, the values used for elevations below 900 meters are conservative without any change; corrections have to be made only above this base elevation.

IV. Determination of Reduced Minimum Approach Distances

A. Factors Affecting Voltage Stress at the Work Site

1. System voltage (nominal). The nominal system voltage range sets the absolute lower limit for the minimum approach distance. The highest value within the range, as given in the relevant table, is selected and used as a reference for per unit calculations.

2. Transient overvoltages. Transient overvoltages may be generated on an electrical system by the operation of switches or breakers, by the occurrence of a fault on the line or circuit being worked or on an adjacent circuit, and by similar activities. Most of the overvoltages are caused by switching, and the term "switching surge" is often used to refer generically to all types of overvoltages. However, each overvoltage has an associated transient voltage wave shape. The wave shape arriving at the site and its magnitude vary considerably.

The information used in the development of the minimum approach distances takes into consideration the most common wave shapes; thus, the required minimum approach distances are appropriate for any transient overvoltage level usually found on electric power generation, transmission, and distribution systems. The values of the per unit (p.u.) voltage relative to the nominal maximum voltage are used in the calculation of these distances.

3. Typical magnitude of overvoltages. The magnitude of typical transient overvoltages is given in Table 5.

4. Standard deviation—air-gap withstand. For each air gap length, and under the same atmospheric conditions, there is a statistical variation in the breakdown voltage. The probability of the breakdown voltage is assumed to have a normal (Gaussian) distribution. The standard deviation of this distribution varies with the wave shape, gap geometry, and atmospheric conditions. The withstand voltage of the air gap used in calculating the electrical component of the minimum approach distance has been set at three standard deviations (3 sigma[1]) below the critical flashover voltage. (The critical flashover voltage is the crest value of the impulse wave that, under specified conditions, causes flashover on 50 percent of the applications. An impulse wave of three standard

---

[1]Sigma is the symbol for standard deviation.

deviations below this value, that is, the withstand voltage, has a probability of flashover of approximately 1 in 1000.)

Table 5.   Magnitude of Typical Transient Overvoltages

Cause	Magnitude (per unit)
Energized 200 mile line without closing resistors	3.5
Energized 200 mile line with one step closing resistor	2.1
Energized 200 mile line with multi-step resistor	2.5
Reclosed with trapped charge one step resistor	2.2
Opening surge with single restrike	3.0
Fault initiation unfaulted phase	2.1
Fault initiation adjacent circuit	2.5
Fault clearing	1.7 - 1.9

SOURCE: ANSI/IEEE Standard No. 516, 1987.

5. Broken Insulators.   Tests have shown that the insulation strength of an insulator string with broken skirts is reduced. Broken units may have lost up to 70% of their withstand capacity. Because the insulating capability of a broken unit cannot be determined without testing it, damaged units in an insulator are usually considered to have no insulating value. Additionally, the overall insulating strength of a string with broken units may be further reduced in the presence of a live-line tool alongside it. The number of good units that must be present in a string is based on the maximum overvoltage possible at the worksite.

B. Minimum Approach Distances Based on Known Maximum Anticipated Per-Unit Transient Overvoltages

1. Reduction of the minimum approach distance for AC systems.   When the transient overvoltage values are known and supplied by the employer, Table R-7 and Table R-8 of 1910.269 allow the minimum approach distances from energized parts to be reduced. In order to determine what this maximum overvoltage is, the employer must undertake an engineering analysis of the system. As a result of this engineering study, the employer must provide new live work procedures, reflecting the new minimum approach distances, the conditions and limitations of application of the new minimum approach distances, and the specific practices to be used when these procedures are implemented.

2. Calculation of reduced approach distance values.   The following method of calculating reduced minimum approach distances is based on ANSI/IEEE Standard 516:

Step 1. Determine the maximum voltage (with respect to a given nominal voltage range) for the energized part.

Step 2. Determine the maximum transient overvoltage (normally a switching surge) that can be present at the work site during work operation.

Step 3. Determine the technique to be used to control the maximum transient overvoltage. (See paragraphs IV.C and IV.D of this appendix.) Determine the maximum voltage that can exist at the work site with that form of control in place and with a confidence level of 3 sigma. This voltage is considered to be the withstand voltage for the purpose of calculating the appropriate minimum approach distance.

Step 4. Specify in detail the control technique to be used, and direct its implementation during the course of the work.

Step 5. Using the new value of transient overvoltage in per unit (p.u.), determine the required phase-to-ground minimum approach distance from Table R-7 or Table R-8 of 1910.269.

C. Methods of Controlling Possible Transient Overvoltage Stress Found on a System

1. Introduction.   There are several means of controlling overvoltages that occur on transmission systems. First, the operation of circuit breakers or other switching devices may be modified to reduce switching transient overvoltages. Second, the overvoltage itself may be forcibly held to an acceptable level by means of installation of surge arresters at the specific location to be protected. Third, the transmission system may be changed to minimize the effect of switching operations.

2. Operation of circuit breakers.[2]   The maximum transient overvoltage that can reach the work site is often due to switching on the line on which work is being performed. If the automatic-reclosing is removed during energized line work so that the line will not be re-energized after being opened for any reason, the maximum switching surge overvoltage is then limited to the larger of the opening surge or the greatest possible fault-generated surge, provided that the devices (for example, insertion resistors) are operable and will function to limit the transient overvoltage. It is essential that the operating ability of such devices be assured when they are employed to limit the overvoltage level. If it is prudent not to remove the reclosing feature (because of system operating conditions), other methods of controlling the switching surge level may be necessary.

Transient surges on an adjacent line, particularly for double circuit construction, may cause a significant overvoltage on the line on which work is being performed. The coupling to adjacent lines must be accounted for when minimum approach distances are calculated based on the maximum transient overvoltage.

3. Surge arresters.   The use of modern surge arresters has permitted a reduction in the basic impulse-insulation levels of much transmission system equipment. The primary function of early arresters was to protect the system insulation from the effects of lightning. Modern arresters not only dissipate lightning-caused transients, but may also control many other system transients that may be caused by switching or faults.

It is possible to use properly designed arresters to control transient overvoltages along a transmission line and thereby reduce the requisite length of the insulator string. On the other hand, if the installation of arresters has not been used to reduce the length of the insulator string, it may be used to reduce the minimum approach distance instead.[3]

---

[2]The detailed design of a circuit interrupter, such as the design of the contacts, of resistor insertion, and of breaker timing control, are beyond the scope of this appendix. These features are routinely provided as part of the design for the system. Only features that can limit the maximum switching transient overvoltage on a system are discussed in this appendix.

4. Switching Restrictions. Another form of overvoltage control is the establishment of switching restrictions, under which breakers are not permitted to be operated until certain system conditions are satisfied. Restriction of switching is achieved by the use of a tagging system, similar to that used for a "permit," except that the common term used for this activity is a "hold-off" or "restriction." These terms are used to indicate that operation is not prevented, but only modified during the live-work activity.

D. Minimum Approach Distance Based on Control of Voltage Stress (Overvoltages) at the Work Site.

Reduced minimum approach distances can be calculated as follows:

1. First Method—Determining the reduced minimum approach distance from a given withstand voltage.[4]

Step 1. Select the appropriate withstand voltage for the protective gap based on system requirements and an acceptable probability of actual gap flashover.

Step 2. Determine a gap distance that provides a withstand voltage[5] greater than or equal to the one selected in the first step.[6]

Step 3. Using 110 percent of the gap's critical flashover voltage, determine the electrical component of the minimum approach distance from Equation (2) or Table 6, which is a tabulation of distance vs. withstand voltage based on Equation (2).

Step 4. Add the 1-foot ergonomic component to obtain the total minimum approach distance to be maintained by the employee.

2. Second Method—Determining the necessary protective gap length from a desired (reduced) minimum approach distance.

Step 1. Determine the desired minimum approach distance for the employee. Subtract the 1-foot ergonomic component of the minimum approach distance.

Step 2. Using this distance, calculate the air gap withstand voltage from Equation (2). Alternatively, find the voltage corresponding to the distance in Table 6.[7]

Step 3. Select a protective gap distance corresponding to a critical flashover voltage that, when multiplied by 110 percent, is less than or equal to the withstand voltage from Step 2.

---

[3]Surge arrestor application is beyond the scope of this appendix. However, if the arrestor is installed near the work site, the application would be similar to protective gaps as discussed in paragraph IV.D. of this appendix.

[4]Since a given rod gap of a given configuration corresponds to a certain withstand voltage, this method can also be used to determine the minimum approach distance for a known gap.

[5]The withstand voltage for the gap is equal to 85 percent of its critical flashover voltage.

[6]Switch steps 1 and 2 if the length of the protective gap is known. The withstand voltage must then be checked to ensure that it provides an acceptable probability of gap flashover. In general, it should be at least 1.25 times the maximum crest operating voltage.

[7]Since the value of the saturation factor, a, in Equation (2) is dependent on the maximum voltage, several iterative computations may be necessary to determine the correct withstand voltage using the equation. A graph of withstand voltage vs. distance is given in ANSI/IEEE Std. 516, 1987. This graph could also be used to determine the appropriate withstand voltage for the minimum approach distance involved.

Step 4. Calculate the withstand voltage of the protective gap (85 percent of the critical flashover voltage) to ensure that it provides an acceptable risk of flashover during the time the gap is installed.

Table 6. Withstand Distances for Transient Overvoltages

Crest voltage (kV)	Withstand distance (in feet) air gap
100	0.71
150	1.06
200	1.41
250	1.77
300	2.12
350	2.47
400	2.83
450	3.18
500	3.54
550	3.89
600	4.24
650	4.60
700	5.17
750	5.73
800	6.31
850	6.91
900	7.57
950	8.23
1000	8.94
1050	9.65
1100	10.42
1150	11.18
1200	12.05
1250	12.90
1300	13.79
1350	14.70
1400	15.64
1450	16.61
1500	17.61
1550	18.63

SOURCE: Calculations are based on Equation (2).
NOTE: The air gap is based on the 60-Hz rod-gap withstand distance.

3. Sample protective gap calculations.

Problem 1: Work is to be performed on a 500-kV transmission line that is subject to transient overvoltages of 2.4 p.u. The maximum operating voltage of the line is 552 kV. Determine the length of the protective gap that will provide the minimum practical safe approach distance. Also, determine what that minimum approach distance is.

Step 1. Calculate the smallest practical maximum transient overvoltage (1.25 times the crest line-to-ground voltage)[8]

$$552 \text{ kV} \times (\sqrt{2}/\sqrt{3}) \times 1.25 = 563 \text{ kV}$$

[8]To eliminate unwanted flashovers due to minor system disturbances, it is desirable to have the crest withstand voltage no lower than 1.25 p.u.

This will be the withstand voltage of the protective gap.

Step 2. Using test data for a particular protective gap, select a gap that has a critical flashover voltage greater than or equal to:

$$563 \text{ kV}/0.85 = 662 \text{ kV}.$$

For example, if a protective gap with a 4.0-foot spacing tested to a critical flashover voltage of 665 kV, crest, select this gap spacing.

Step 3. This protective gap corresponds to a 110 percent of critical flashover voltage value of:

$$665 \text{ kV} \times 1.10 = 732 \text{ kV}.$$

This corresponds to the withstand voltage of the electrical component of the minimum approach distance.

Step 4. Using this voltage in Equation (2) results in an electrical component of the minimum approach distance of:

$$D = (0.01 + 0.0006) \times (552 \text{ kV}/\sqrt{3}) = 5.5 \text{ ft.}$$

Step 5. Add 1 foot to the distance calculated in step 4, resulting in a total minimum approach distance of 6.5 feet.

Problem 2: For a line operating at a maximum voltage of 552 kV subject to a maximum transient overvoltage of 2.4 p.u., find a protective gap distance that will permit the use of a 9.0-foot minimum approach distance. (A minimum approach distance of 11 feet, 3 inches is normally required.)

Step 1. The electrical component of the minimum approach distance is 8.0 feet (9.0-1.0).

Step 2. From Table 6, select the withstand voltage corresponding to a distance of 8.0 feet. By interpolation:

$$900 \text{ kV} + \left[50 \times \frac{(8.00 - 7.57)}{(8.23 - 7.57)} \right] = 933 \text{ kV}.$$

Step 3. The voltage calculated in Step 2 corresponds to 110 percent of the critical flashover voltage of the gap that should be employed. Using test data for a particular protective gap, select a gap that has a critical flashover voltage less than or equal to:

$$D = (0.01 + 0.0006) \times 732 \text{ kV divided by square root of 2}$$

For example, if a protective gap with a 5.8-foot spacing tested to a critical flashover voltage of 820 kV, crest, select this gap spacing.

Step 4. The withstand voltage of this protective gap would be:

$$820 \text{ kV} \times 0.85 = 697 \text{ kV.}$$

The maximum operating crest voltage would be:

$$552 \text{ kV} \times (\sqrt{2}/\sqrt{3}) = 449 \text{ kV}$$

The crest withstand voltage of the protective gap in per unit is thus

$$697 \text{ kV} + 449 \text{ kV} = 1.55 \text{ p.u.}$$

If this is acceptable, the protective gap could be installed with a 5.8-foot spacing, and the minimum approach distance could then be reduced to 9.0 feet.

4. *Comments and variations.* The 1-foot ergonomic component of the minimum approach distance must be added to the electrical component of the minimum approach distance calculated under paragraph IV.D of this appendix. The calculations may be varied by starting with the protective gap distance or by starting with the minimum approach distance.

E. Location of Protective Gaps

1. Installation of the protective gap on a structure adjacent to the work site is an acceptable practice, as this does not significantly reduce the protection afforded by the gap.
2. Gaps installed at terminal stations of lines or circuits provide a given level of protection. The level may not, however, extend throughout the length of the line to the worksite. The use of gaps at terminal stations must be studied in depth. The use of substation terminal gaps raises the possibility that separate surges could enter the line at opposite ends, each with low enough magnitude to pass the terminal gaps without flashover. When voltage surges are initiated simultaneously at each end of a line and travel toward each other, the total voltage on the line at the point where they meet is the arithmetic sum of the two surges. A gap that is installed within 0.5 mile of the work site will protect against such intersecting waves. Engineering studies of a particular line or system may indicate that adequate protection can be provided by even more distant gaps.
3. If protective gaps are used at the work site, the work site impulse insulation strength is established by the gap setting. Lightning strikes as much as 6 miles away from the worksite may cause a voltage surge greater than the insulation withstand voltage, and a gap flashover may occur. The flashover will not occur between the employee and the line, but across the protective gap instead.
4. There are two reasons to disable the automatic-reclosing feature of circuit-interrupting devices while employees are performing live-line maintenance:
   - To prevent the reenergizing of a circuit faulted by actions of a worker, which could possibly create a hazard or compound injuries or damage produced by the original fault;
   - To prevent any transient overvoltage caused by the switching surge that would occur if the circuit were reenergized.

However, due to system stability considerations, it may not always be feasible to disable the automatic-reclosing feature.

Reprinted from www.osha.gov

Occupational Safety & Health Administration
200 Constitution Avenue, NW
Washington, DC 20210

REGULATIONS (STANDARDS - 29 CFR)

PROTECTION FROM STEP AND TOUCH POTENTIALS. - 1910.269APPC
- **Part Number:** 1910
- **Part Title:** Occupational Safety and Health Standards
- **Subpart:** R
- **Subpart Title:** Special Industries
- **Standard Number:** 1910.269 App C
- **Title:** Protection from Step and Touch Potentials.

Appendix C to §1910.269—Protection From Step and Touch Potentials

I. Introduction

When a ground fault occurs on a power line, voltage is impressed on the "grounded" object faulting the line. The voltage to which this object rises depends largely on the voltage on the line, on the impedance of the faulted conductor, and on the impedance to "true," or "absolute," ground represented by the object. If the object causing the fault represents a relatively large impedance, the voltage impressed on it is essentially the phase-to-ground system voltage. However, even faults to well grounded transmission towers or substation structures can result in hazardous voltages.[1] The degree of the hazard depends upon the magnitude of the fault current and the time of exposure.

II. Voltage-Gradient Distribution

A. Voltage-Gradient Distribution Curve

The dissipation of voltage from a grounding electrode (or from the grounded end of an energized grounded object) is called the ground potential gradient. Voltage drops associated with this dissipation of voltage are called ground potentials. Figure 1 is a typical voltage-gradient distribution curve (assuming a uniform soil texture). This graph shows that voltage decreases rapidly with increasing distance from the grounding electrode.

B. Step and Touch Potentials

"Step potential" is the voltage between the feet of a person standing near an energized grounded object. It is equal to the difference in voltage, given by the voltage distribution curve, between two points at different distances from the "electrode." A person could be at risk of injury during a fault simply by standing near the grounding point.

"Touch potential" is the voltage between the energized object and the feet of a person in contact with the object. It is equal to the difference in voltage between the object (which is at a distance of 0 feet) and a point some distance away. It should be noted that the touch potential could be nearly the full voltage across the grounded object if that object is grounded at a point remote from the place

---

[1]This appendix provides information primarily with respect to employee protection from contact between equipment being used and an energized power line. The information presented is also relevant to ground faults to transmission towers and substation structures; however, grounding systems for these structures should be designed to minimize the step and touch potentials involved.

where the person is in contact with it. For example, a crane that was grounded to the system neutral and that contacted an energized line would expose any person in contact with the crane or its uninsulated load line to a touch potential nearly equal to the full fault voltage.

Step and touch potentials are illustrated in Figure 2.

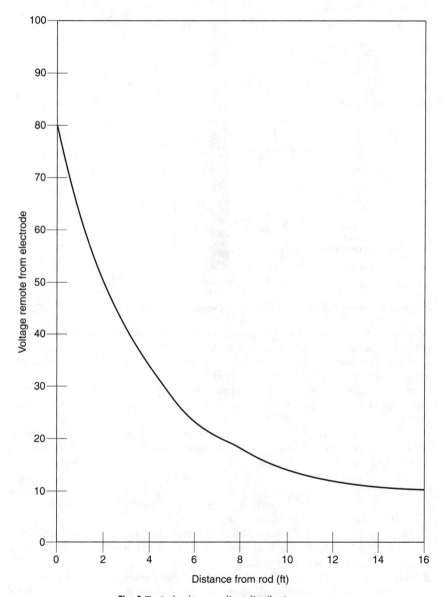

Fig. 1 Typical voltage-gradient distribution curve.

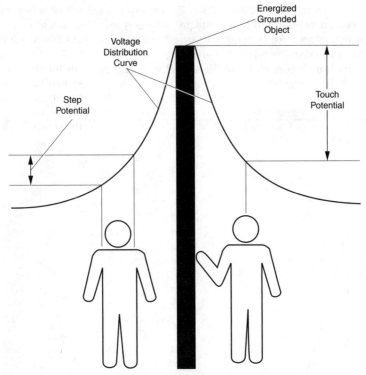

Fig. 2 Step and touch potentials.

C. Protection from the Hazards of Ground-Potential Gradients.

An engineering analysis of the power system under fault conditions can be used to determine whether or not hazardous step and touch voltages will develop. The result of this analysis can ascertain the need for protective measures and can guide the selection of appropriate precautions.

Several methods may be used to protect employees from hazardous ground-potential gradients, including equipotential zones, insulating equipment, and restricted work areas.

1. The creation of an equipotential zone will protect a worker standing within it from hazardous step and touch potentials. (See Figure 3.) Such a zone can be produced through the use of a metal mat connected to the grounded object. In some cases, a grounding grid can be used to equalize the voltage within the grid. Equipotential zones will not, however, protect employees who are either wholly or partially outside the protected area. Bonding conductive objects in the immediate work area can also be used to minimize the potential between the objects and between each object and ground. (Bonding an object outside the work area can increase the touch potential to that object in some cases, however.)

2. The use of insulating equipment, such as rubber gloves, can protect employees handling grounded equipment and conductors from hazardous touch potentials. The insulating equipment must be rated for the highest voltage

that can be impressed on the grounded objects under fault conditions (rather than for the full system voltage).

3. Restricting employees from areas where hazardous step or touch potentials could arise can protect employees not directly involved in the operation being performed. Employees on the ground in the vicinity of transmission structures should be kept at a distance where step voltages would be insufficient to cause injury. Employees should not handle grounded conductors or equipment likely to become energized to hazardous voltages unless the employees are within an equipotential zone or are protected by insulating equipment.

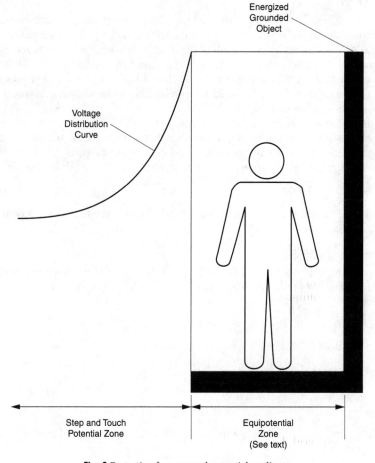

Fig. 3 Protection from ground-potential gradients.

Reprinted from www.osha.gov

Occupational Safety & Health Administration
200 Constitution Avenue, NW
Washington, DC 20210

REGULATIONS (STANDARDS - 29 CFR)

METHODS OF INSPECTING AND TESTING WOOD POLES. - 1910.269APPD

- **Part Number:** 1910
- **Part Title:** Occupational Safety and Health Standards
- **Subpart:** R
- **Subpart Title:** Special Industries
- **Standard Number:** 1910.269 App D
- **Title:** Methods of Inspecting and Testing Wood Poles.

I. Introduction

When work is to be performed on a wood pole, it is important to determine the condition of the pole before it is climbed. The weight of the employee, the weight of equipment being installed, and other working stresses (such as the removal or retensioning of conductors) can lead to the failure of a defective pole or one that is not designed to handle the additional stresses.[1] For these reasons, it is essential that an inspection and test of the condition of a wood pole be performed before it is climbed.

If the pole is found to be unsafe to climb or to work from, it must be secured so that it does not fail while an employee is on it. The pole can be secured by a line truck boom, by ropes or guys, or by lashing a new pole alongside it. If a new one is lashed alongside the defective pole, work should be performed from the new one.

II. Inspection of Wood Poles

Wood poles should be inspected by a qualified employee for the following conditions:[2]

A. General Condition

The pole should be inspected for buckling at the ground line and for an unusual angle with respect to the ground. Buckling and odd angles may indicate that the pole has rotted or is broken.

B. Cracks

The pole should be inspected for cracks. Horizontal cracks perpendicular to the grain of the wood may weaken the pole. Vertical ones, although not considered to be a sign of a defective pole, can pose a hazard to the climber, and the employee should keep his or her gaffs away from them while climbing.

C. Holes

Hollow spots and woodpecker holes can reduce the strength of a wood pole.

---

[1] A properly guyed pole in good condition should, at a minimum, be able to handle the weight of an employee climbing it.

[2] The presence of any of these conditions is an indication that the pole may not be safe to climb or to work from. The employee performing the inspection must be qualified to make a determination as to whether or not it is safe to perform the work without taking additional precautions.

D. Shell Rot and Decay

Rotting and decay are cutout hazards and are possible indications of the age and internal condition of the pole.

E. Knots

One large knot or several smaller ones at the same height on the pole may be evidence of a weak point on the pole.

F. Depth of Setting

Evidence of the existence of a former ground line substantially above the existing ground level may be an indication that the pole is no longer buried to a sufficient extent.

G. Soil Conditions

Soft, wet, or loose soil may not support any changes of stress on the pole.

H. Burn Marks

Burning from transformer failures or conductor faults could damage the pole so that it cannot withstand mechanical stress changes.

III. Testing of Wood Poles

The following tests, which have been taken from 1910.268(n)(3), are recognized as acceptable methods of testing wood poles:

A. Hammer Test

Rap the pole sharply with a hammer weighing about 3 pounds, starting near the ground line and continuing upwards circumferentially around the pole to a height of approximately 6 feet. The hammer will produce a clear sound and rebound sharply when striking sound wood. Decay pockets will be indicated by a dull sound or a less pronounced hammer rebound. Also, prod the pole as near the ground line as possible using a pole prod or a screwdriver with a blade at least 5 inches long. If substantial decay is encountered, the pole is considered unsafe.

B. Rocking Test

Apply a horizontal force to the pole and attempt to rock it back and forth in a direction perpendicular to the line. Caution must be exercised to avoid causing power lines to swing together. The force may be applied either by pushing with a pike pole or pulling with a rope. If the pole cracks during the test, it shall be considered unsafe.

Reprinted from www.osha.gov

Occupational Safety & Health Administration
200 Constitution Avenue, NW
Washington, DC 20210

REGULATIONS (STANDARDS - 29 CFR)

REFERENCE DOCUMENTS. - 1910.269APPE

- **Part Number:** 1910
- **Part Title:** Occupational Safety and Health Standards
- **Subpart:** R
- **Subpart Title:** Special Industries
- **Standard Number:** 1910.269 App E
- **Title:** Reference Documents.

The references contained in this appendix provide information that can be helpful in understanding and complying with the requirements contained in 1910.269. The national consensus standards referenced in this appendix contain detailed specifications that employers may follow in complying with the more performance-oriented requirements of OSHA's final rule. Except as specifically noted in 1910.269, however, compliance with the national consensus standards is not a substitute for compliance with the provisions of the OSHA standard.

ANSI/SIA A92.2-1990, American National Standard for Vehicle-Mounted Elevating and Rotating Aerial Devices.

ANSI C2-1993, National Electrical Safety Code.

ANSI Z133.1-1988, American National Standard Safety Requirements for Pruning, Trimming, Repairing, Maintaining, and Removing Trees, and for Cutting Brush.

ANSI/ASME B20.1-1990, Safety Standard for Conveyors and Related Equipment.

ANSI/IEEE Std. 4-1978 (Fifth Printing), IEEE Standard Techniques for High-Voltage Testing.

ANSI/IEEE Std. 100-1988, IEEE Standard Dictionary of Electrical and Electronic Terms.

ANSI/IEEE Std. 516-1987, IEEE Guide for Maintenance Methods on Energized Power-Lines.

ANSI/IEEE Std. 935-1989, IEEE Guide on Terminology for Tools and Equipment to Be Used in Live Line Working.

ANSI/IEEE Std. 957-1987, IEEE Guide for Cleaning Insulators.

ANSI/IEEE Std. 978-1984 (R1991), IEEE Guide for In-Service Maintenance and Electrical Testing of Live-Line Tools.

ASTM D 120-87, Specification for Rubber Insulating Gloves.

ASTM D 149-92, Test Method for Dielectric Breakdown Voltage and Dielectric Strength of Solid Electrical Insulating Materials at Commercial Power Frequencies.

ASTM D 178-93, Specification for Rubber Insulating Matting.

ASTM D 1048-93, Specification for Rubber Insulating Blankets.

ASTM D 1049-93, Specification for Rubber Insulating Covers.

ASTM D 1050-90, Specification for Rubber Insulating Line Hose.

ASTM D 1051-87, Specification for Rubber Insulating Sleeves.

ASTM F 478-92, Specification for In-Service Care of Insulating Line Hose and Covers.

ASTM F 479-93, Specification for In-Service Care of Insulating Blankets.

ASTM F 496-93B, Specification for In-Service Care of Insulating Gloves and Sleeves.

ASTM F 711-89, Specification for Fiberglass-Reinforced Plastic (FRP) Rod and Tube Used in Live Line Tools.

ASTM F 712-88, Test Methods for Electrically Insulating Plastic Guard Equipment for Protection of Workers.

ASTM F 819-83a (1988), Definitions of Terms Relating to Electrical Protective Equipment for Workers.

ASTM F 855-90, Specifications for Temporary Grounding Systems to Be Used on De-Energized Electric Power Lines and Equipment.

ASTM F 887-91a, Specifications for Personal Climbing Equipment.

ASTM F 914-91, Test Method for Acoustic Emission for Insulated Aerial Personnel Devices.

ASTM F 968-93, Specification for Electrically Insulating Plastic Guard Equipment for Protection of Workers.

ASTM F 1116-88, Test Method for Determining Dielectric Strength of Overshoe Footwear.

ASTM F 1117-87, Specification for Dielectric Overshoe Footwear.

ASTM F 1236-89, Guide for Visual Inspection of Electrical Protective Rubber Products.

ASTM F 1505-94, Standard Specification for Insulated and Insulating Hand Tools.

ASTM F 1506-94, Standard Performance Specification for Textile Materials for Wearing Apparel for Use by Electrical Workers Exposed to Momentary Electric Arc and Related Thermal Hazards.

IEEE Std. 62-1978, IEEE Guide for Field Testing Power Apparatus Insulation.

IEEE Std. 524-1992, IEEE Guide to the Installation of Overhead Transmission Line Conductors.

IEEE Std. 1048-1990, IEEE Guide for Protective Grounding of Power Lines.

IEEE Std. 1067-1990, IEEE Guide for the In-Service Use, Care, Maintenance, and Testing of Conductive Clothing for Use on Voltages up to 765 kV AC.

[59 FR 4437, Jan. 31, 1994; 59 FR 33658, June 30, 1994; 59 FR 40729, Aug. 9, 1994]

Reprinted from www.osha.gov

Occupational Safety & Health Administration
200 Constitution Avenue, NW
Washington, DC 20210

Appendix B

OSHA Standards Related to the NESC Work Rules

1926.950 through 1926.960

Power Transmission and Distribution (Construction)
1926 Subpart V—Power Transmission and Distribution

OSHA 1926.950–1926.960 Paragraph Titles:

OSHA 1926 Subpart V—Authority for 1926 Subpart V
Authority

OSHA 1926.950—General Requirements
(a) Application
(b) Initial inspections, tests, or determinations
(c) Clearances
(d) Deenergizing lines and equipment
(e) Emergency procedures and first aid
(f) Night work
(g) Work near and over water
(h) Sanitation facilities
(i) Hydraulic fluids

OSHA 1926.951—Tools and Protective Equipment
(a) Protective equipment
(b) Personal climbing equipment
(c) Ladders
(d) Live-line tools
(e) Measuring tapes or measuring ropes
(f) Handtools

OSHA 1926.952—Mechanical Equipment
(a) General
(b) Aerial lifts

(c) Derrick trucks, cranes and other lifting equipment

OSHA 1926.953—Material Handling
(a) Unloading
(b) Pole hauling
(c) Storage
(d) Tag line
(e) Oil filled equipment
(f) Framing
(g) Attaching the load

OSHA 1926.954—Grounding for Protection of Employees
(a) General
(b) New construction
(c) Communication conductors
(d) Voltage testing
(e) Attaching grounds
(f) Grounds shall be placed...
(g) Testing without grounds
(h) Grounding electrode
(i) Grounding to tower
(j) Ground lead

OSHA 1926.955—Overhead Lines
(a) Overhead lines
(b) Metal tower construction
(c) Stringing or removing deenergized conductors
(d) Stringing adjacent to energized lines
(e) Live-line bare-hand work

OSHA 1926.956—Underground Lines
(a) Guarding and ventilating street opening...
(b) Work in manholes
(c) Trenching and excavating

OSHA 1926.957—Construction in Energized Substations
(a) Work near energized equipment facilities
(b) Deenergized equipment or lines
(c) Barricades and barriers
(d) Control panels
(e) Mechanized equipment
(f) Storage
(g) Substation fences
(h) Footing excavation

OSHA 1926.958—External Load Helicopters
In all operations performed using a rotorcraft...

OSHA 1926.959—Lineman's Body Belts, Safety Straps, and Lanyards
(a) General requirements
(b) Specific requirements

OSHA 1926.960—Definitions Applicable to this Part
Definitions

REGULATIONS (STANDARDS - 29 CFR)

GENERAL REQUIREMENTS. - 1926.950

- **Part Number:** 1926
- **Part Title:** Safety and Health Regulations for Construction
- **Subpart:** V
- **Subpart Title:** Power Transmission and Distribution
- **Standard Number:** 1926.950
- **Title:** General requirements.

1926.950(a)

Application. The occupational safety and health standards contained in this Subpart V shall apply to the construction of electric transmission and distribution lines and equipment.

1926.950(a)(1)

As used in this Subpart V the term "construction" includes the erection of new electric transmission and distribution lines and equipment, and the alteration, conversion, and improvement of existing electric transmission and distribution lines and equipment.

1926.950(a)(2)

Existing electric transmission and distribution lines and electrical equipment need not be modified to conform to the requirements of applicable standards in this Subpart V, until such work as described in paragraph (a)(1) of this section is to be performed on such lines or equipment.

1926.950(a)(3)

The standards set forth in this Subpart V provide minimum requirements for safety and health. Employers may require adherence to additional standards which are not in conflict with the standards contained in this Subpart V.

..1926.950(b)

1926.950(b)

Initial inspections, tests, or determinations.

1926.950(b)(1)

Existing conditions shall be determined before starting work, by an inspection or a test. Such conditions shall include, but not be limited to, energized lines and equipment, conditions of poles, and the location of circuits and equipment, including power and communication lines, CATV and fire alarm circuits.

1926.950(b)(2)

Electric equipment and lines shall be considered energized until determined to be deenergized by tests or other appropriate methods or means.

1926.950(b)(3)

Operating voltage of equipment and lines shall be determined before working on or near energized parts.

1926.950(c)

Clearances. The provisions of paragraph (c) (1) or (2) of this section shall be observed.

1926.950(c)(1)

No employee shall be permitted to approach or take any conductive object without an approved insulating handle closer to exposed energized parts than shown in Table V-1, unless:

1926.950(c)(1)(i)

The employee is insulated or guarded from the energized part (gloves or gloves with sleeves rated for the voltage involved shall be considered insulation of the employee from the energized part), or

1926.950(c)(1)(ii)

The energized part is insulated or guarded from him and any other conductive object at a different potential, or

..1926.950(c)(1)(iii)

1926.950(c)(1)(iii)

The employee is isolated, insulated, or guarded from any other conductive object(s), as during live-line bare-hand work.

1926.950(c)(2)

1926.950(c)(2)(i)

The minimum working distance and minimum clear hot stick distances stated in Table V-1 shall not be violated. The minimum clear hot stick distance is that for the use of live-line tools held by linemen when performing live-line work.

1926.950(c)(2)(ii)

Conductor support tools, such as link sticks, strain carriers, and insulator cradles, may be used: Provided, That the clear insulation is at least as long as the insulator string or the minimum distance specified in Table V-1 for the operating voltage.

1926.950(d)

Deenergizing lines and equipment.

1926.950(d)(1)

When deenergizing lines and equipment operated in excess of 600 volts, and the means of disconnecting from electric energy is not visibly open or visibly locked out, the provisions of paragraphs (d)(1) (i) through (vii) of this section shall be complied with:

1926.950(d)(1)(i)

The particular section of line or equipment to be deenergized shall be clearly identified, and it shall be isolated from all sources of voltage.

Voltage range (phase to phase) (kilovolt)	Minimum working and clear hot stick distance
2.1 to 15	2 ft. 0 in.
15.1 to 35	2 ft. 4 in.
35.1 to 46	2 ft. 6 in.
46.1 to 72.5	3 ft. 0 in.
72.6 to 121	3 ft. 4 in.
138 to 145	3 ft. 6 in.
161 to 169	3 ft. 8 in.
230 to 242	5 ft. 0 in.
345 to 362	7 ft. 0 in.[1]
500 to 552	11 ft. 0 in.[1]
700 to 765	15 ft. 0 in.[1]

[1]NOTE: For 345-362 kV, 500-552 kV, and 700-765 kV, minimum clear hot stick distance may be reduced provided that such distances are not less than the shortest distance between the energized part and the grounded surface.

1926.950(d)(1)(ii)

Notification and assurance from the designated employee shall be obtained that:

..1926.950(d)(1)(ii)(a)

1926.950(d)(1)(ii)(a)

All switches and disconnectors through which electric energy may be supplied to the particular section of line or equipment to be worked have been deenergized;

1926.950(d)(1)(ii)(b)

All switches and disconnectors are plainly tagged indicating that men are at work;

1926.950(d)(1)(ii)(c)

And that where design of such switches and disconnectors permits, they have been rendered inoperable.

1926.950(d)(1)(iii)

After all designated switches and disconnectors have been opened, rendered inoperable, and tagged, visual inspection or tests shall be conducted to insure that equipment or lines have been deenergized.

1926.950(d)(1)(iv)

Protective grounds shall be applied on the disconnected lines or equipment to be worked on.

1926.950(d)(1)(v)

Guards or barriers shall be erected as necessary to adjacent energized lines.

1926.950(d)(1)(vi)

When more than one independent crew requires the same line or equipment to be deenergized, a prominent tag for each such independent crew shall be placed on the line or equipment by the designated employee in charge.

..1926.950(d)(1)(vii)

1926.950(d)(1)(vii)

Upon completion of work on deenergized lines or equipment, each designated employee in charge shall determine that all employees in his crew are clear, that protective grounds installed by his crew have been removed, and he shall report to the designated authority that all tags protecting his crew may be removed.

1926.950(d)(2)

When a crew working on a line or equipment can clearly see that the means of disconnecting from electric energy are visibly open or visibly locked-out, the provisions of paragraphs (d)(i), and (ii) of this section shall apply:

1926.950(d)(2)(i)

Guards or barriers shall be erected as necessary to adjacent energized lines.

1926.950(d)(2)(ii)

Upon completion of work on deenergized lines or equipment, each designated employee in charge shall determine that all employees in his crew are clear, that protective grounds installed by his crew have been removed, and he shall report to the designated authority that all tags protecting his crew may be removed.

1926.950(e)

Emergency procedures and first aid.

1926.950(e)(1)

The employer shall provide training or require that his employees are knowledgeable and proficient in:

1926.950(e)(1)(i)

Procedures involving emergency situations, and

1926.950(e)(1)(ii)

First-aid fundamentals including resuscitation.

..1926.950(e)(2)

1926.950(e)(2)

In lieu of paragraph (e)(1) of this section the employer may comply with the provisions of 1926.50(c) regarding first-aid requirements.

1926.950(f)

Night work. When working at night, spotlights or portable lights for emergency lighting shall be provided as needed to perform the work safely.

1926.950(g)

Work near and over water. When crews are engaged in work over or near water and when danger of drowning exists, suitable protection shall be provided as stated in 1926.104, or 1926.105, or 1926.106.

1926.950(h)

Sanitation facilities. The requirements of 1926.51 of Subpart D of this part shall be complied with for sanitation facilities.

1926.950(i)

Hydraulic fluids. All hydraulic fluids used for the insulated sections of derrick trucks, aerial lifts, and hydraulic tools which are used on or around energized lines and equipment shall be of the insulating type. The requirements for fire resistant fluids of 1926.302(d)(1) do not apply to hydraulic tools covered by this paragraph.

Reprinted from www.osha.gov

Occupational Safety & Health Administration
200 Constitution Avenue, NW
Washington, DC 20210

REGULATIONS (STANDARDS - 29 CFR)

TOOLS AND PROTECTIVE EQUIPMENT. - 1926.951

- **Part Number:** 1926
- **Part Title:** Safety and Health Regulations for Construction
- **Subpart:** V
- **Subpart Title:** Power Transmission and Distribution
- **Standard Number:** 1926.951
- **Title:** Tools and protective equipment.

1926.951(a)

Protective equipment.

1926.951(a)(1)

1926.951(a)(1)(i)

Rubber protective equipment shall be in accordance with the provisions of the American National Standards Institute (ANSI), ANSI J6 series, as follows:

Item	Standard
Rubber insulating gloves	J6.6-1971.
Rubber matting for use around electric apparatus	J6.7-1935 (R1971).
Rubber insulating blankets	J6.4-1971.
Rubber insulating hoods	J6.2-1950 (R1971).
Rubber insulating line hose	J6.1-1950 (R1971).
Rubber insulating sleeves	J6.5-1971.

1926.951(a)(1)(ii)

Rubber protective equipment shall be visually inspected prior to use.

1926.951(a)(1)(iii)

In addition, an "air" test shall be performed for rubber gloves prior to use.

1926.951(a)(1)(iv)

Protective equipment of material other than rubber shall provide equal or better electrical and mechanical protection.

1926.951(a)(2)

Protective hats shall be in accordance with the provisions of ANSI Z89.2-1971 Industrial Protective Helmets for Electrical Workers, Class B, and shall be worn at the jobsite by employees who are exposed to the hazards of falling objects, electric shock, or burns.

..1926.951(b)

1926.951(b)

Personal climbing equipment.

1926.951(b)(1)

Body belts with straps or lanyards shall be worn to protect employees working at elevated locations on poles, towers, or other structures except where such use creates a greater hazard to the safety of the employees, in which case other safeguards shall be employed.

1926.951(b)(2)

Body belts and safety straps shall meet the requirements of 1926.959. In addition to being used as an employee safeguarding item, body belts with approved tool loops may be used for the purpose of holding tools. Body belts shall be free from additional metal hooks and tool loops other than those permitted in 1926.959.

1926.951(b)(3)

Body belts and straps shall be inspected before use each day to determine that they are in safe working condition.

1926.951(b)(4)
1926.951(b)(4)(i)

Lifelines and lanyards shall comply with the provisions of 1926.502.

1926.951(b)(4)(ii)

Safety lines are not intended to be subjected to shock loading and are used for emergency rescue such as lowering a man to the ground. Such safety lines shall be a minimum of one-half-inch diameter and three or four strand first-grade manila or its equivalent in strength (2,650 lb.) and durability.

1926.951(b)(5)

Defective ropes shall be replaced.

..1926.951(c)
1926.951(c)

Ladders.

1926.951(c)(1)

Portable metal or conductive ladders shall not be used near energized lines or equipment except as may be necessary in specialized work such as in high voltage substations where nonconductive ladders might present a greater hazard than conductive ladders. Conductive or metal ladders shall be prominently marked as conductive and all necessary precautions shall be taken when used in specialized work.

1926.951(c)(2)

Hook or other type ladders used in structures shall be positively secured to prevent the ladder from being accidentally displaced.

1926.951(d)

Live-line tools.

1926.951(d)(1)

Only live-line tool poles having a manufacturer's certification to withstand the following minimum tests shall be used:

1926.951(d)(1)(i)

100,000 volts per foot of length for 5 minutes when the tool is made of fiberglass; or

1926.951(d)(1)(ii)

75,000 volts per foot of length for 3 minutes when the tool is made of wood; or

1926.951(d)(1)(iii)

Other tests equivalent to paragraph (d) (i) or (ii) of this section as appropriate.

1926.951(d)(2)

All live-line tools shall be visually inspected before use each day. Tools to be used shall be wiped clean and if any hazardous defects are indicated such tools shall be removed from service.

..1926.951(e)

1926.951(e)

Measuring tapes or measuring ropes. Measuring tapes or measuring ropes which are metal or contain conductive strands shall not be used when working on or near energized parts.

1926.951(f)

Handtools.

1926.951(f)(1)

Switches for all powered hand tools shall comply with 1926.300(d).

1926.951(f)(2)

All portable electric handtools shall:

1926.951(f)(2)(i)

Be equipped with three-wire cord having the ground wire permanently connected to the tool frame and means for grounding the other end; or

1926.951(f)(2)(ii)

Be of the double insulated type and permanently labeled as "Double Insulated"; or

1926.951(f)(2)(iii)

Be connected to the power supply by means of an isolating transformer, or other isolated power supply.

1926.951(f)(3)

All hydraulic tools which are used on or around energized lines or equipment shall use nonconducting hoses having adequate strength for the normal operating pressures. It should be noted that the provisions of 1926.302(d)(2) shall also apply.

1926.951(f)(4)

All pneumatic tools which are used on or around energized lines or equipment shall:

..1926.951(f)(4)(i)

1926.951(f)(4)(i)

Have nonconducting hoses having adequate strength for the normal operating pressures, and

1926.951(f)(4)(ii)

Have an accumulator on the compressor to collect moisture.

[59 FR 40730, Aug. 9, 1994]

Reprinted from www.osha.gov

Occupational Safety & Health Administration
200 Constitution Avenue, NW
Washington, DC 20210

REGULATIONS (STANDARDS - 29 CFR)

MECHANICAL EQUIPMENT. - 1926.952

- **Part Number:** 1926
- **Part Title:** Safety and Health Regulations for Construction
- **Subpart:** V
- **Subpart Title:** Power Transmission and Distribution
- **Standard Number:** 1926.952
- **Title:** Mechanical equipment.

1926.952(a)

General.

1926.952(a)(1)

Visual inspections shall be made of the equipment to determine that it is in good condition each day the equipment is to be used.

1926.952(a)(2)

Tests shall be made at the beginning of each shift during which the equipment is to be used to determine that the brakes and operating systems are in proper working condition.

1926.952(a)(3)

No employer shall use any motor vehicle equipment having an obstructed view to the rear unless:

1926.952(a)(3)(i)

The vehicle has a reverse signal alarm audible above the surrounding noise level or:

1926.952(a)(3)(ii)

The vehicle is backed up only when an observer signals that it is safe to do so.

1926.952(b)

Aerial lifts.

1926.952(b)(1)

The provisions of 1926.556, Subpart N of this part, shall apply to the utilization of aerial lifts.

..1926.952(b)(2)
1926.952(b)(2)

When working near energized lines or equipment, aerial lift trucks shall be grounded or barricaded and considered as energized equipment, or the aerial lift truck shall be insulated for the work being performed.

1926.952(b)(3)

Equipment or material shall not be passed between a pole or structure and an aerial lift while an employee working from the basket is within reaching distance of energized conductors or equipment that are not covered with insulating protective equipment.

1926.952(c)

Derrick trucks, cranes and other lifting equipment.

1926.952(c)(1)

All derrick trucks, cranes and other lifting equipment shall comply with Subpart N and O of this part except:

1926.952(c)(1)(i)

As stated in 1926.550(a)(15) (i) and (ii) relating to clearance (for clearances in this subpart see Table V-1) and

1926.952(c)(1)(ii)

Derrick truck (electric line trucks) shall not be required to comply with 1926.550(a)(7)(vi), (a)(17), (b)(2), and (e).

1926.952(c)(2)

With the exception of equipment certified for work on the proper voltage, mechanical equipment shall not be operated closer to any energized line or equipment than the clearances set forth in 1926.950(c) unless:

1926.952(c)(2)(i)

An insulated barrier is installed between the energized part and the mechanical equipment, or

..1926.952(c)(2)(ii)

1926.952(c)(2)(ii)

The mechanical equipment is grounded, or

1926.952(c)(2)(iii)

The mechanical equipment is insulated, or

1926.952(c)(2)(iv)

The mechanical equipment is considered as energized.

Reprinted from www.osha.gov

Occupational Safety & Health Administration
200 Constitution Avenue, NW
Washington, DC 20210

REGULATIONS (STANDARDS - 29 CFR)

MATERIAL HANDLING. - 1926.953

- **Part Number:** 1926
- **Part Title:** Safety and Health Regulations for Construction
- **Subpart:** V
- **Subpart Title:** Power Transmission and Distribution
- **Standard Number:** 1926.953
- **Title:** Material handling.

1926.953(a)

Unloading. Prior to unloading steel, poles, cross arms and similar material, the load shall be thoroughly examined to ascertain if the load has shifted, binders or stakes have broken or the load is otherwise hazardous to employees.

1926.953(b)

Pole hauling.

1926.953(b)(1)

During pole hauling operations, all loads shall be secured to prevent displacement and a red flag shall be displayed at the trailing end of the longest pole.

1926.953(b)(2)

Precautions shall be exercised to prevent blocking of roadways or endangering other traffic.

1926.953(b)(3)

When hauling poles during the hours of darkness, illuminated warning devices shall be attached to the trailing end of the longest pole.

1926.953(c)

Storage.

1926.953(c)(1)

No materials or equipment shall be stored under energized bus, energized lines, or near energized equipment, if it is practical to store them elsewhere.

..1926.953(c)(2)

1926.953(c)(2)

When materials or equipment are stored under energized lines or near energized equipment, applicable clearances shall be maintained as stated in Table V-1; and extraordinary caution shall be exercised when moving materials near such energized equipment.

1926.953(d)

Tag line. Where hazards to employees exist tag lines or other suitable devices shall be used to control loads being handled by hoisting equipment.

1926.953(e)

Oil filled equipment. During construction or repair of oil filled equipment the oil may be stored in temporary containers other than those required in 1926.152, such as pillow tanks.

1926.953(f)

Framing. During framing operations, employees shall not work under a pole or a structure suspended by a crane, A-frame or similar equipment unless the pole or structure is adequately supported.

1926.953(g)

Attaching the load. The hoist rope shall not be wrapped around the load. This provision shall not apply to electric construction crews when setting or removing poles.

Reprinted from www.osha.gov

Occupational Safety & Health Administration
200 Constitution Avenue, NW
Washington, DC 20210

REGULATIONS (STANDARDS - 29 CFR)

GROUNDING FOR PROTECTION OF EMPLOYEES. - 1926.954

- **Part Number:** 1926
- **Part Title:** Safety and Health Regulations for Construction
- **Subpart:** V
- **Subpart Title:** Power Transmission and Distribution
- **Standard Number:** 1926.954
- **Title:** Grounding for protection of employees.

1926.954(a)

General. All conductors and equipment shall be treated as energized until tested or otherwise determined to be deenergized or until grounded.

1926.954(b)

New construction. New lines or equipment may be considered deenergized and worked as such where:

1926.954(b)(1)

The lines or equipment are grounded, or

1926.954(b)(2)

The hazard of induced voltages is not present, and adequate clearances or other means are implemented to prevent contact with energized lines or equipment and the new lines or equipment.

1926.954(c)

Communication conductors. Bare wire communication conductors on power poles or structures shall be treated as energized lines unless protected by insulating materials.

1926.954(d)

Voltage testing. Deenergized conductors and equipment which are to be grounded shall be tested for voltage. Results of this voltage test shall determine the subsequent procedures as required in 1926.950(d).

..1926.954(e)

1926.954(e)

Attaching grounds.

1926.954(e)(1)

When attaching grounds, the ground end shall be attached first, and the other end shall be attached and removed by means of insulated tools or other suitable devices.

1926.954(e)(2)

When removing grounds, the grounding device shall first be removed from the line or equipment using insulating tools or other suitable devices.

1926.954(f)

Grounds shall be placed between work location and all sources of energy and as close as practicable to the work location, or grounds shall be placed at the work location. If work is to be performed at more than one location in a line section, the line section must be grounded and short circuited at one location in the line section and the conductor to be worked on shall be grounded at each work location. The minimum distance shown in Table V-1 shall be maintained from ungrounded conductors at the work location. Where the making of a ground is impracticable, or the conditions resulting therefrom would be more hazardous than working on the lines or equipment without grounding, the grounds may be omitted and the line or equipment worked as energized.

1926.954(g)

Testing without grounds. Grounds may be temporarily removed only when necessary for test purposes and extreme caution shall be exercised during the test procedures.

..1926.954(h)

1926.954(h)

Grounding electrode. When grounding electrodes are utilized, such electrodes shall have a resistance to ground low enough to remove the danger of harm to personnel or permit prompt operation of protective devices.

1926.954(i)

Grounding to tower. Grounding to tower shall be made with a tower clamp capable of conducting the anticipated fault current.

1926.954(j)

Ground lead. A ground lead, to be attached to either a tower ground or driven ground, shall be capable of conducting the anticipated fault current and shall have a minimum conductance of No. 2 AWG copper.

Reprinted from www.osha.gov

Occupational Safety & Health Administration
200 Constitution Avenue, NW
Washington, DC 20210

REGULATIONS (STANDARDS - 29 CFR)

OVERHEAD LINES. - 1926.955

- **Part Number:** 1926
- **Part Title:** Safety and Health Regulations for Construction
- **Subpart:** V
- **Subpart Title:** Power Transmission and Distribution
- **Standard Number:** 1926.955
- **Title:** Overhead lines.

1926.955(a)

Overhead lines.

1926.955(a)(1)

When working on or with overhead lines the provisions of paragraphs (a) (2) through (8) of this section shall be complied with in addition to other applicable provisions of this subpart.

1926.955(a)(2)

Prior to climbing poles, ladders, scaffolds, or other elevated structures, an inspection shall be made to determine that the structures are capable of sustaining the additional or unbalanced stresses to which they will be subjected.

1926.955(a)(3)

Where poles or structures may be unsafe for climbing, they shall not be climbed until made safe by guying, bracing, or other adequate means.

1926.955(a)(4)

Before installing or removing wire or cable, strains to which poles and structures will be subjected shall be considered and necessary action taken to prevent failure of supporting structures.

..1926.955(a)(5)

1926.955(a)(5)

1926.955(a)(5)(i)

When setting, moving, or removing poles using cranes, derricks, gin poles, A-frames, or other mechanized equipment near energized lines or equipment, precautions shall be taken to avoid contact with energized lines or equipment, except in bare-hand live-line work, or where barriers or protective devices are used.

1926.955(a)(5)(ii)

Equipment and machinery operating adjacent to energized lines or equipment shall comply with 1926.952(c)(2).

1926.955(a)(6)

1926.955(a)(6)(i)

Unless using suitable protective equipment for the voltage involved, employees standing on the ground shall avoid contacting equipment or machinery working adjacent to energized lines or equipment.

1926.955(a)(6)(ii)

Lifting equipment shall be bonded to an effective ground or it shall be considered energized and barricaded when utilized near energized equipment or lines.

1926.955(a)(7)

Pole holes shall not be left unattended or unguarded in areas where employees are currently working.

1926.955(a)(8)

Tag lines shall be of a nonconductive type when used near energized lines.

..1926.955(b)

1926.955(b)

Metal tower construction.

1926.955(b)(1)

When working in unstable material the excavation for pad- or pile-type footings in excess of 5 feet deep shall be either sloped to the angle of repose as required in 1926.652 or shored if entry is required. Ladders shall be provided for access to pad- or pile-type footing excavations in excess of 4 feet.

1926.955(b)(2)

When working in unstable material provision shall be made for cleaning out auger-type footings without requiring an employee to enter the footing unless shoring is used to protect the employee.

1926.955(b)(3)

1926.955(b)(3)(i)

A designated employee shall be used in directing mobile equipment adjacent to footing excavations.

1926.955(b)(3)(ii)

No one shall be permitted to remain in the footing while equipment is being spotted for placement.

1926.955(b)(3)(iii)

Where necessary to assure the stability of mobile equipment the location of use for such equipment shall be graded and leveled.

1926.955(b)(4)

1926.955(b)(4)(i)

Tower assembly shall be carried out with a minimum exposure of employees to falling objects when working at two or more levels on a tower.

1926.955(b)(4)(ii)

Guy lines shall be used as necessary to maintain sections or parts of sections in position and to reduce the possibility of tipping.

..1926.955(b)(4)(iii)

1926.955(b)(4)(iii)

Members and sections being assembled shall be adequately supported.

1926.955(b)(5)

When assembling and erecting towers the provisions of paragraphs (b)(5)(i), (ii) and (iii) of this section shall be complied with:

1926.955(b)(5)(i)

The construction of transmission towers and the erecting of poles, hoisting machinery, site preparation machinery, and other types of construction machinery shall conform to the applicable requirements of this part.

1926.955(b)(5)(ii)

No one shall be permitted under a tower which is in the process of erection or assembly, except as may be required to guide and secure the section being set.

1926.955(b)(5)(iii)

When erecting towers using hoisting equipment adjacent to energized transmission lines, the lines shall be deenergized when practical. If the lines are not deenergized, extraordinary caution shall be exercised to maintain the minimum clearance distances required by 1926.950(c), including Table V-1.

1926.955(b)(6)

1926.955(b)(6)(i)

Erection cranes shall be set on firm level foundations and when the cranes are so equipped outriggers shall be used.

..1926.955(b)(6)(ii)

1926.955(b)(6)(ii)

Tag lines shall be utilized to maintain control of tower sections being raised and positioned, except where the use of such lines would create a greater hazard.

1926.955(b)(6)(iii)

The loadline shall not be detached from a tower section until the section is adequately secured.

1926.955(b)(6)(iv)

Except during emergency restoration procedures erection shall be discontinued in the event of high wind or other adverse weather conditions which would make the work hazardous.

1926.955(b)(6)(v)

Equipment and rigging shall be regularly inspected and maintained in safe operating condition.

1926.955(b)(7)

Adequate traffic control shall be maintained when crossing highways and railways with equipment as required by the provisions of 1926.200(g)(1) and (2).

1926.955(b)(8)

A designated employee shall be utilized to determine that required clearance is maintained in moving equipment under or near energized lines.

1926.955(c)

Stringing or removing deenergized conductors.

1926.955(c)(1)

When stringing or removing deenergized conductors, the provisions of paragraphs (c)(2) through (12) of this section shall be complied with.

..1926.955(c)(2)

1926.955(c)(2)

Prior to stringing operations a briefing shall be held setting forth the plan of operation and specifying the type of equipment to be used, grounding devices and procedures to be followed, crossover methods to be employed, and the clearance authorization required.

1926.955(c)(3)

Where there is a possibility of the conductor accidentally contacting an energized circuit or receiving a dangerous induced voltage buildup, to further protect the employee from the hazards of the conductor, the conductor being installed or removed shall be grounded or provisions made to insulate or isolate the employee.

1926.955(c)(4)

1926.955(c)(4)(i)

If the existing line is deenergized, proper clearance authorization shall be secured and the line grounded on both sides of the crossover or, the line being strung or removed shall be considered and worked as energized.

1926.955(c)(4)(ii)

When crossing over energized conductors in excess of 600 volts, rope nets or guard structures shall be installed unless provision is made to isolate or insulate the workman or the energized conductor. Where practical the automatic reclosing feature of the circuit interrupting device shall be made inoperative. In addition, the line being strung shall be grounded on either side of the crossover or considered and worked as energized.

1926.955(c)(5)

Conductors being strung in or removed shall be kept under positive control by the use of adequate tension reels, guard structures, tielines, or other means to prevent accidental contact with energized circuits.

..1926.955(c)(6)

1926.955(c)(6)

Guard structure members shall be sound and of adequate dimension and strength, and adequately supported.

1926.955(c)(7)

1926.955(c)(7)(i)

Catch-off anchors, rigging, and hoists shall be of ample capacity to prevent loss of the lines.

1926.955(c)(7)(ii)

The manufacturer's load rating shall not be exceeded for stringing lines, pulling lines, sock connections, and all load-bearing hardware and accessories.

1926.955(c)(7)(iii)

Pulling lines and accessories shall be inspected regularly and replaced or repaired when damaged or when dependability is doubtful. The provisions of 1926.251(c)(4)(ii) (concerning splices) shall not apply.

1926.955(c)(8)

Conductor grips shall not be used on wire rope unless designed for this application.

1926.955(c)(9)

While the conductor or pulling line is being pulled (in motion) employees shall not be permitted directly under overhead operations, nor shall any employee be permitted on the crossarm.

..1926.955(c)(10)

1926.955(c)(10)

A transmission clipping crew shall have a minimum of two structures clipped in between the crew and the conductor being sagged. When working on bare conductors, clipping and tying crews shall work between grounds at all times. The grounds shall remain intact until the conductors are clipped in, except on dead end structures.

1926.955(c)(11)

1926.955(c)(11)(i)

Except during emergency restoration procedures, work from structures shall be discontinued when adverse weather (such as high wind or ice on structures) makes the work hazardous.

1926.955(c)(11)(ii)

Stringing and clipping operations shall be discontinued during the progress of an electrical storm in the immediate vicinity.

1926.955(c)(12)

1926.955(c)(12)(i)

Reel handling equipment, including pulling and braking machines, shall have ample capacity, operate smoothly, and be leveled and aligned in accordance with the manufacturer's operating instructions.

1926.955(c)(12)(ii)

Reliable communications between the reel tender and pulling rig operator shall be provided.

1926.955(c)(12)(iii)

Each pull shall be snubbed or dead ended at both ends before subsequent pulls.

..1926.955(d)

1926.955(d)

Stringing adjacent to energized lines.

1926.955(d)(1)

Prior to stringing parallel to an existing energized transmission line a competent determination shall be made to ascertain whether dangerous induced voltage buildups will occur, particularly during switching and ground fault conditions. When there is a possibility that such dangerous induced voltage may exist the employer shall comply with the provisions of paragraphs (d) (2) through (9) of this section in addition to the provisions of paragraph (c) of this 1926.955, unless the line is worked as energized.

1926.955(d)(2)

When stringing adjacent to energized lines the tension stringing method or other methods which preclude unintentional contact between the lines being pulled and any employee shall be used.

1926.955(d)(3)

All pulling and tensioning equipment shall be isolated, insulated, or effectively grounded.

1926.955(d)(4)

A ground shall be installed between the tensioning reel setup and the first structure in order to ground each bare conductor, subconductor, and overhead ground conductor during stringing operations.

1926.955(d)(5)

During stringing operations, each bare conductor, subconductor, and overhead ground conductor shall be grounded at the first tower adjacent to both the tensioning and pulling setup and in increments so that no point is more than 2 miles from a ground.

1926.955(d)(5)(i)

The grounds shall be left in place until conductor installation is completed.

1926.955(d)(5)(ii)

Such grounds shall be removed as the last phase of aerial cleanup.

1926.955(d)(5)(iii)

Except for moving type grounds, the grounds shall be placed and removed with a hot stick.

..1926.955(d)(6)

1926.955(d)(6)

Conductors, subconductors, and overhead ground conductors shall be grounded at all dead-end or catch-off points.

1926.955(d)(7)

A ground shall be located at each side and within 10 feet of working areas where conductors, subconductors, or overhead ground conductors are being spliced at ground level. The two ends to be spliced shall be bonded to each other. It is recommended that splicing be carried out on either an insulated platform or on a conductive metallic grounding mat bonded to both grounds. When a grounding mat is used, it is recommended that the grounding mat be roped off and an insulated walkway provided for access to the mat.

1926.955(d)(8)

1926.955(d)(8)(i)

All conductors, subconductors, and overhead ground conductors shall be bonded to the tower at any isolated tower where it may be necessary to complete work on the transmission line.

1926.955(d)(8)(ii)

Work on dead-end towers shall require grounding on all deenergized lines.

1926.955(d)(8)(iii)

Grounds may be removed as soon as the work is completed: Provided, That the line is not left open circuited at the isolated tower at which work is being completed.

..1926.955(d)(9)

1926.955(d)(9)

When performing work from the structures, clipping crews and all others working on conductors, subconductors, or overhead ground conductors shall be protected by individual grounds installed at every work location.

1926.955(e)

Live-line bare-hand work. In addition to any other applicable standards contained elsewhere in this subpart all live-line bare-hand work shall be performed in accordance with the following requirements:

1926.955(e)(1)

Employees shall be instructed and trained in the live-line bare-hand technique and the safety requirements pertinent thereto before being permitted to use the technique on energized circuits.

1926.955(e)(2)

Before using the live-line bare-hand technique on energized high-voltage conductors or parts, a check shall be made of:

1926.955(e)(2)(i)

The voltage rating of the circuit on which the work is to be performed;

1926.955(e)(2)(ii)

The clearances to ground of lines and other energized parts on which work is to be performed; and

1926.955(e)(2)(iii)

The voltage limitations of the aerial-lift equipment intended to be used.

1926.955(e)(3)

Only equipment designed, tested, and intended for live-line bare-hand work shall be used.

..*1926.955(e)(4)*

1926.955(e)(4)

All work shall be personally supervised by a person trained and qualified to perform live-line bare-hand work.

1926.955(e)(5)

The automatic reclosing feature of circuit interrupting devices shall be made inoperative where practical before working on any energized line or equipment.

1926.955(e)(6)

Work shall not be performed during the progress of an electrical storm in the immediate vicinity.

1926.955(e)(7)

A conductive bucket liner or other suitable conductive device shall be provided for bonding the insulated aerial device to the energized line or equipment.

1926.955(e)(7)(i)

The employee shall be connected to the bucket liner by use of conductive shoes, leg clips, or other suitable means.

1926.955(e)(7)(ii)

Where necessary, adequate electrostatic shielding for the voltage being worked or conductive clothing shall be provided.

1926.955(e)(8)

Only tools and equipment intended for live-line bare-hand work shall be used, and such tools and equipment shall be kept clean and dry.

..1926.955(e)(9)

1926.955(e)(9)

Before the boom is elevated, the outriggers on the aerial truck shall be extended and adjusted to stabilize the truck and the body of the truck shall be bonded to an effective ground, or barricaded and considered as energized equipment.

1926.955(e)(10)

Before moving the aerial lift into the work position, all controls (ground level and bucket) shall be checked and tested to determine that they are in proper working condition.

1926.955(e)(11)

Arm current tests shall be made before starting work each day, each time during the day when higher voltage is going to be worked and when changed conditions indicate a need for additional tests. Aerial buckets used for bare-hand live-line work shall be subjected to an arm current test. This test shall consist of placing the bucket in contact with an energized source equal to the voltage to be worked upon for a minimum time of three (3) minutes, the leakage current shall not exceed 1 microampere per kilovolt of nominal line-to-line voltage. Work operations shall be suspended immediately upon any indication of a malfunction in the equipment.

1926.955(e)(12)

All aerial lifts to be used for live-line bare-hand work shall have dual controls (lower and upper) as required by paragraph (e)(12)(i) and (ii) of this section.

1926.955(e)(12)(i)

The upper controls shall be within easy reach of the employee in the basket. If a two basket type lift is used access to the controls shall be within easy reach from either basket.

..1926.955(e)(12)(ii)

1926.955(e)(12)(ii)

The lower set of controls shall be located near base of the boom that will permit over-ride operation of equipment at any time.

1926.955(e)(13)

Ground level lift control shall not be operated unless permission has been obtained from the employee in lift, except in case of emergency.

1926.955(e)(14)

Before the employee contacts the energized part to be worked on, the conductive bucket liner shall be bonded to the energized conductor by means of a positive connection which shall remain attached to the energized conductor until the work on the energized circuit is completed.

1926.955(e)(15)

The minimum clearance distances for live-line bare-hand work shall be as specified in Table V-2. These minimum clearance distances shall be maintained from all grounded objects and from lines and equipment at a different potential than that to which the insulated aerial device is bonded unless such grounded objects or other lines and equipment are covered by insulated guards. These distances shall be maintained when approaching, leaving, and when bonded to the energized circuit.

TABLE V-2 Minimum Clearance Distances for Live-Line Bare-Hand Work (Alternating Current)

Voltage range (phase to phase) kilovolts	Distance in feet and inches for maximum voltage	
	Phase to ground	Phase to phase
2.1 to 15	2′0″	2′0″
15.1 to 35	2′4″	2′4″
35.1 to 46	2′6″	2′6″
46.1 to 72.5	3′0″	3′0″
72.6 to 121	3′4″	4′6″
138 to 145	3′6″	5′0″
161 to 169	3′8″	5′6″
230 to 242	5′0″	8′4″
345 to 362	(1)7′0″	(1)13′4″
500 to 552	(1)11′0″	(1)20′0″
700 to 765	(1)15′0″	(1)31′0″

[1]For 345-362kV, 500-552kV, and 700-765kV, the minimum clearance distance may be reduced provided the distances are not made less than the shortest distance between the energized part and the grounded surface.

1926.955(e)(16)

When approaching, leaving, or bonding to an energized circuit the minimum distances in Table V-2 shall be maintained between all parts of the insulated boom assembly and any grounded parts (including the lower arm or portions of the truck).

..1926.955(e)(17)

1926.955(e)(17)

When positioning the bucket alongside an energized bushing or insulator string, the minimum line-to-ground clearances of Table V-2 must be maintained between all parts of the bucket and the grounded end of the bushing or insulator string.

1926.955(e)(18)

1926.955(e)(18)(i)

The use of handlines between buckets, booms, and the ground is prohibited.

1926.955(e)(18)(ii)

No conductive materials over 36 inches long shall be placed in the bucket, except for appropriate length jumpers, armor rods, and tools.

1926.955(e)(18)(iii)

Nonconductive-type handlines may be used from line to ground when not supported from the bucket.

1926.955(e)(19)

The bucket and upper insulated boom shall not be overstressed by attempting to lift or support weights in excess of the manufacturer's rating.

1926.955(e)(20)

1926.955(e)(20)(i)

A minimum clearance table (as shown in table V-2) shall be printed on a plate of durable nonconductive material, and mounted in the buckets or its vicinity so as to be visible to the operator of the boom.

1926.955(e)(20)(ii)

It is recommended that insulated measuring sticks be used to verify clearance distances.

Reprinted from www.osha.gov

Occupational Safety & Health Administration
200 Constitution Avenue, NW
Washington, DC 20210

REGULATIONS (STANDARDS - 29 CFR)

UNDERGROUND LINES. - 1926.956

- **Part Number:** 1926
- **Part Title:** Safety and Health Regulations for Construction
- **Subpart:** V
- **Subpart Title:** Power Transmission and Distribution
- **Standard Number:** 1926.956
- **Title:** Underground lines.

1926.956(a)

Guarding and ventilating street opening used for access to underground lines or equipment.

1926.956(a)(1)

Appropriate warning signs shall be promptly placed when covers of manholes, handholes, or vaults are removed. What is an appropriate warning sign is dependent upon the nature and location of the hazards involved.

1926.956(a)(2)

Before an employee enters a street opening, such as a manhole or an unvented vault, it shall be promptly protected with a barrier, temporary cover, or other suitable guard.

1926.956(a)(3)

When work is to be performed in a manhole or unvented vault:

1926.956(a)(3)(i)

No entry shall be permitted unless forced ventilation is provided or the atmosphere is found to be safe by testing for oxygen deficiency and the presence of explosive gases or fumes;

1926.956(a)(3)(ii)

Where unsafe conditions are detected, by testing or other means, the work area shall be ventilated and otherwise made safe before entry;

1926.956(a)(3)(iii)

Provisions shall be made for an adequate continuous supply of air.

..1926.956(b)
1926.956(b)

Work in manholes.

1926.956(b)(1)

While work is being performed in manholes, an employee shall be available in the immediate vicinity to render emergency assistance as may be required. This shall not preclude the employee in the immediate vicinity from occasionally entering a manhole to provide assistance, other than emergency. This requirement does not preclude a qualified employee, working alone, from entering for brief periods of time, a manhole where energized cables or equipment are in service, for the purpose of inspection, housekeeping, taking readings, or similar work if such work can be performed safely.

1926.956(b)(2)

When open flames must be used or smoking is permitted in manholes, extra precautions shall be taken to provide adequate ventilation.

1926.956(b)(3)

Before using open flames in a manhole or excavation in an area where combustible gases or liquids may be present, such as near a gasoline service station, the atmosphere of the manhole or excavation shall be tested and found safe or cleared of the combustible gases or liquids.

1926.956(c)

Trenching and excavating.

1926.956(c)(1)

During excavation or trenching, in order to prevent the exposure of employees to the hazards created by damage to dangerous underground facilities, efforts shall be made to determine the location of such facilities and work conducted in a manner designed to avoid damage.

1926.956(c)(2)

Trenching and excavation operations shall comply with 1926.651 and 1926.652.

1926.956(c)(3)

When underground facilities are exposed (electric, gas, water, telephone, etc.) they shall be protected as necessary to avoid damage.

1926.956(c)(4)

Where multiple cables exist in an excavation, cables other than the one being worked on shall be protected as necessary.

..1926.956(c)(5)
1926.956(c)(5)

When multiple cables exist in an excavation, the cable to be worked on shall be identified by electrical means unless its identity is obvious by reason of distinctive appearance.

1926.956(c)(6)

Before cutting into a cable or opening a splice, the cable shall be identified and verified to be the proper cable.

1926.956(c)(7)

When working on buried cable or on cable in manholes, metallic sheath continuity shall be maintained by bonding across the opening or by equivalent means.

Reprinted from www.osha.gov

Occupational Safety & Health Administration
200 Constitution Avenue, NW
Washington, DC 20210

REGULATIONS (STANDARDS - 29 CFR)

CONSTRUCTION IN ENERGIZED SUBSTATIONS. - 1926.957

- **Part Number:** 1926
- **Part Title:** Safety and Health Regulations for Construction
- **Subpart:** V
- **Subpart Title:** Power Transmission and Distribution
- **Standard Number:** 1926.957
- **Title:** Construction in energized substations.

1926.957(a)

Work near energized equipment facilities.

1926.957(a)(1)

When construction work is performed in an energized substation, authorization shall be obtained from the designated, authorized person before work is started.

1926.957(a)(2)

When work is to be done in an energized substation, the following shall be determined:

1926.957(a)(2)(i)

What facilities are energized, and

1926.957(a)(2)(ii)

What protective equipment and precautions are necessary for the safety of personnel.

1926.957(a)(3)

Extraordinary caution shall be exercised in the handling of busbars, tower steel, materials, and equipment in the vicinity of energized facilities. The requirements set forth in 1926.950(c), shall be complied with.

1926.957(b)

Deenergized equipment or lines. When it is necessary to deenergize equipment or lines for protection of employees, the requirements of 1926.950(d) shall be complied with.

..1926.957(c)

1926.957(c)

Barricades and barriers.

1926.957(c)(1)

Barricades or barriers shall be installed to prevent accidental contact with energized lines or equipment.

1926.957(c)(2)

Where appropriate, signs indicating the hazard shall be posted near the barricade or barrier. These signs shall comply with 1926.200.

1926.957(d)

Control panels.

1926.957(d)(1)

Work on or adjacent to energized control panels shall be performed by designated employees.

1926.957(d)(2)

Precaution shall be taken to prevent accidental operation of relays or other protective devices due to jarring, vibration, or improper wiring.

1926.957(e)

Mechanized equipment.

1926.957(e)(1)

Use of vehicles, gin poles, cranes, and other equipment in restricted or hazardous areas shall at all times be controlled by designated employees.

1926.957(e)(2)

All mobile cranes and derricks shall be effectively grounded when being moved or operated in close proximity to energized lines or equipment, or the equipment shall be considered energized.

1926.957(e)(3)

Fenders shall not be required for lowboys used for transporting large electrical equipment, transformers, or breakers.

..1926.957(f)

1926.957(f)

Storage. The storage requirements of 1926.953(c) shall be complied with.

1926.957(g)

Substation fences.

1926.957(g)(1)

When a substation fence must be expanded or removed for construction purposes, a temporary fence affording similar protection when the site is unattended, shall be provided. Adequate interconnection with ground shall be maintained between temporary fence and permanent fence.

1926.957(g)(2)

All gates to all unattended substations shall be locked, except when work is in progress.

1926.957(h)

Footing excavation.

1926.957(h)(1)

Excavation for auger, pad and piling type footings for structures and towers shall require the same precautions as for metal tower construction (see 1926.955(b)(1)).

1926.957(h)(2)

No employee shall be permitted to enter an unsupported auger-type excavation in unstable material for any purpose. Necessary clean-out in such cases shall be accomplished without entry.

Reprinted from www.osha.gov

Occupational Safety & Health Administration
200 Constitution Avenue, NW
Washington, DC 20210

REGULATIONS (STANDARDS - 29 CFR)

EXTERNAL LOAD HELICOPTERS. - 1926.958

- **Part Number:** 1926
- **Part Title:** Safety and Health Regulations for Construction
- **Subpart:** V
- **Subpart Title:** Power Transmission and Distribution
- **Standard Number:** 1926.958
- **Title:** External load helicopters.

In all operations performed using a rotorcraft for moving or placing external loads, the provisions of 1926.551 of Subpart N of this part shall be complied with.

Reprinted from www.osha.gov

Occupational Safety & Health Administration
200 Constitution Avenue, NW
Washington, DC 20210

REGULATIONS (STANDARDS - 29 CFR)

LINEMAN'S BODY BELTS, SAFETY STRAPS, AND LANYARDS. - 1926.959

- **Part Number:** 1926
- **Part Title:** Safety and Health Regulations for Construction
- **Subpart:** V
- **Subpart Title:** Power Transmission and Distribution
- **Standard Number:** 1926.959
- **Title:** Lineman's body belts, safety straps, and lanyards.

1926.959(a)

General requirements. The requirements of paragraphs (a) and (b) of this section shall be complied with for all lineman's body belts, safety straps and lanyards acquired for use after the effective date of this subpart.

1926.959(a)(1)

Hardware for lineman's body belts, safety straps, and lanyards shall be drop forged or pressed steel and have a corrosive resistive finish tested to American Society for Testing and Materials B117-64 (50-hour test). Surfaces shall be smooth and free of sharp edges.

1926.959(a)(2)

All buckles shall withstand a 2,000-pound tensile test with a maximum permanent deformation no greater than one sixty-fourth inch.

1926.959(a)(3)

D rings shall withstand a 5,000-pound tensile test without failure. Failure of a D ring shall be considered cracking or breaking.

1926.959(a)(4)

Snaphooks shall withstand a 5,000-pound tensile test without failure. Failure of a snaphook shall be distortion sufficient to release the keeper.

..1926.959(b)

1926.959(b)

Specific requirements.

1926.959(b)(1)

1926.959(b)(1)(i)

All fabric used for safety straps shall withstand an A.C. dielectric test of not less than 25,000 volts per foot "dry" for 3 minutes, without visible deterioration.

1926.959(b)(1)(ii)

All fabric and leather used shall be tested for leakage current and shall not exceed 1 milliampere when a potention of 3,000 volts is applied to the electrodes positioned 12 inches apart.

1926.959(b)(1)(iii)

Direct current tests may be permitted in lieu of alternating current tests.

1926.959(b)(2)

The cushion part of the body belt shall:

1926.959(b)(2)(i)

Contain no exposed rivets on the inside;

1926.959(b)(2)(ii)

Be at least three (3) inches in width;

1926.959(b)(2)(iii)

Be at least five thirty-seconds (5/32) inch thick, if made of leather; and

1926.959(b)(2)(iv)

Have pocket tabs that extended at least 1 1/2 inches down and three (3) inches back of the inside of circle of each D ring for riveting on plier or tool pockets. On shifting D belts, this measurement for pocket tabs shall be taken when the D ring section is centered.

..*1926.959(b)(3)*

1926.959(b)(3)

A maximum of four (4) tool loops shall be so situated on the body belt that four (4) inches of the body belt in the center of the back, measuring from D ring to D ring, shall be free of tool loops, and any other attachments.

1926.959(b)(4)

Suitable copper, steel, or equivalent liners shall be used around bar of D rings to prevent wear between these members and the leather or fabric enclosing them.

1926.959(b)(5)

All stitching shall be of a minimum 42-pound weight nylon or equivalent thread and shall be lock stitched. Stitching parallel to an edge shall not be less than three-sixteenths (3/16) inch from edge of narrowest member caught by the thread. The use of cross stitching on leather is prohibited.

1926.959(b)(6)

The keeper of snaphooks shall have a spring tension that will not allow the keeper to begin to open with a weight of 2 1/2 pounds or less, but the keeper of snaphooks shall begin to open with a weight of four (4) pounds, when the weight is supported on the keeper against the end of the nose.

1926.959(b)(7)

Testing of lineman's safety straps, body belts and lanyards shall be in accordance with the following procedure:

1926.959(b)(7)(i)

Attach one end of the safety strap or lanyard to a rigid support, the other end shall be attached to a 250-pound canvas bag of sand:

..1926.959(b)(7)(ii)

1926.959(b)(7)(ii)

Allow the 250-pound canvas bag of sand to free fall 4 feet for (safety strap test) and 6 feet for (lanyard test); in each case stopping the fall of the 250-pound bag:

1926.959(b)(7)(iii)

Failure of the strap or lanyard shall be indicated by any breakage, or slippage sufficient to permit the bag to fall free of the strap or lanyard. The entire "body belt assembly" shall be tested using one D ring. A safety strap or lanyard shall be used that is capable of passing the "impact loading test" and attached as required in paragraph (b)(7)(i) of this section. The body belt shall be secured to the 250-pound bag of sand at a point to simulate the waist of a man and allowed to drop as stated in paragraph (b)(7)(ii) of this section. Failure of the body belt shall be indicated by any breakage, or slippage sufficient to permit the bag to fall free of the body belt.

Reprinted from www.osha.gov

Occupational Safety & Health Administration
200 Constitution Avenue, NW
Washington, DC 20210

REGULATIONS (STANDARDS - 29 CFR)

DEFINITIONS APPLICABLE TO THIS SUBPART. - 1926.960

- **Part Number:** 1926
- **Part Title:** Safety and Health Regulations for Construction
- **Subpart:** V
- **Subpart Title:** Power Transmission and Distribution
- **Standard Number:** 1926.960
- **Title:** Definitions applicable to this subpart.

1926.960(a)

Alive or live (energized). The term means electrically connected to a source of potential difference, or electrically charged so as to have a potential significantly different from that of the earth in the vicinity. The term "live" is sometimes used in place of the term "current-carrying," where the intent is clear, to avoid repetition of the longer term.

1926.960(b)

Automatic circuit recloser. The term means a self-controlled device for automatically interrupting and reclosing an alternating current circuit with a predetermined sequence of opening and reclosing followed by resetting, hold closed, or lockout operation.

1926.960(c)

Barrier. The term means a physical obstruction which is intended to prevent contact with energized lines or equipment.

1926.960(d)

Barricade. The term means a physical obstruction such as tapes, screens, or cones intended to warn and limit access to a hazardous area.

1926.960(e)

Bond. The term means an electrical connection from one conductive element to another for the purpose of minimizing potential differences or providing suitable conductivity for fault current or for mitigation of leakage current and electrolytic action.

..1926.960(f)

1926.960(f)

Bushing. The term means an insulating structure including a through conductor, or providing a passageway for such a conductor, with provision for mounting on a barrier, conducting or otherwise, for the purpose of insulating the conductor from the barrier and conducting current from one side of the barrier to the other.

1926.960(g)

Cable. The term means a conductor with insulation, or a stranded conductor with or without insulation and other coverings (single-conductor cable) or a combination of conductors insulated from one another (multiple-conductor cable).

1926.960(h)

Cable sheath. The term means a protective covering applied to cables.

NOTE: A cable sheath may consist of multiple layers of which one or more is conductive.

1926.960(i)

Circuit. The term means a conductor or system of conductors through which an electric current is intended to flow.

1926.960(j)

Communication lines. The term means the conductors and their supporting or containing structures which are used for public or private signal or communication service, and which operate at potentials not exceeding 400 volts to ground or 750 volts between any two points of the circuit, and the transmitted power of which does not exceed 150 watts. When operating at less than 150 volts no limit is placed on the capacity of the system.

NOTE: Telephone, telegraph, railroad signal, data, clock, fire, police-alarm, community television antenna, and other systems conforming with the above are included. Lines used for signaling purposes, but not included under the above definition, are considered as supply lines of the same voltage and are to be so run.

1926.960(k)

Conductor. The term means a material, usually in the form of a wire, cable, or bus bar suitable for carrying an electric current.

..1926.960(l)

1926.960(l)

Conductor shielding. The term means an envelope which encloses the conductor of a cable and provides an equipotential surface in contact with the cable insulation.

1926.960(m)

Current-carrying part. The term means a conducting part intended to be connected in an electric circuit to a source of voltage. Non-current-carrying parts are those not intended to be so connected.

1926.960(n)

Dead (deenergized). The term means free from any electrical connection to a source of potential difference and from electrical charges: Not having a potential difference from that of earth.

NOTE: The term is used only with reference to current-carrying parts which are sometimes alive (energized).

1926.960(o)

Designated employee. The term means a qualified person delegated to perform specific duties under the conditions existing.

1926.960(p)

Effectively grounded. The term means intentionally connected to earth through a ground connection or connections of sufficiently low impedance and having sufficient current-carrying capacity to prevent the buildup of voltages which may result in undue hazard to connected equipment or to persons.

1926.960(q)

Electric line trucks. The term means a truck used to transport men, tools, and material, and to serve as a traveling workshop for electric power line construction and maintenance work. It is sometimes equipped with a boom and auxiliary equipment for setting poles, digging holes, and elevating material or men.

..1926.960(r)

1926.960(r)

Enclosed. The term means surrounded by a case, cage, or fence, which will protect the contained equipment and prevent accidental contact of a person with live parts.

1926.960(s)

Equipment. This is a general term which includes fittings, devices, appliances, fixtures, apparatus, and the like, used as part of, or in connection with, an electrical power transmission and distribution system, or communication systems.

1926.960(t)

Exposed. The term means not isolated or guarded.

1926.960(u)

Electric supply lines. The term means those conductors used to transmit electric energy and their necessary supporting or containing structures. Signal lines of more than 400 volts to ground are always supply lines within the meaning of the rules, and those of less than 400 volts to ground may be considered as supply lines, if so run and operated throughout.

1926.960(v)

Guarded. The term means protected by personnel, covered, fenced, or enclosed by means of suitable casings, barrier rails, screens, mats, platforms, or other suitable devices in accordance with standard barricading techniques designed to prevent dangerous approach or contact by persons or objects.

NOTE: Wires, which are insulated but not otherwise protected, are not considered as guarded.

1926.960(w)

Ground. (Reference). The term means that conductive body, usually earth, to which an electric potential is referenced.

..1926.960(x)

1926.960(x)

Ground (as a noun). The term means a conductive connection whether intentional or accidental, by which an electric circuit or equipment is connected to reference ground.

1926.960(y)

Ground (as a verb). The term means the connecting or establishment of a connection, whether by intention or accident of an electric circuit or equipment to reference ground.

1926.960(z)

Grounding electrode (ground electrode). The term grounding electrode means a conductor embedded in the earth, used for maintaining ground potential on conductors connected to it, and for dissipating into the earth current conducted to it.

1926.960(aa)

Grounding electrode resistance. The term means the resistance of the grounding electrode to earth.

1926.960(bb)

Grounding electrode conductor (grounding conductor). The term means a conductor used to connect equipment or the grounded circuit of a wiring system to a grounding electrode.

1926.960(cc)

Grounded conductor. The term means a system or circuit conductor which is intentionally grounded.

..1926.960(dd)

1926.960(dd)

Grounded system. The term means a system of conductors in which at least one conductor or point (usually the middle wire, or neutral point of transformer or generator windings) is intentionally grounded, either solidly or through a current-limiting device (not a current-interrupting device).

1926.960(ee)

Hotline tools and ropes. The term means those tools and ropes which are especially designed for work on energized high voltage lines and equipment. Insulated aerial equipment especially designed for work on energized high voltage lines and equipment shall be considered hot line.

1926.960(ff)

Insulated. The term means separated from other conducting surfaces by a dielectric substance (including air space) offering a high resistance to the passage of current.

NOTE: When any object is said to be insulated, it is understood to be insulated in suitable manner for the conditions to which it is subjected. Otherwise, it is within the purpose of this subpart, uninsulated. Insulating covering of conductors is one means of making the conductor insulated.

1926.960(gg)

Insulation (as applied to cable). The term means that which is relied upon to insulate the conductor from other conductors or conducting parts or from ground.

1926.960(hh)

Insulation shielding. The term means an envelope which encloses the insulation of a cable and provides an equipotential surface in contact with cable insulation.

1926.960(ii)

Isolated. The term means an object that is not readily accessible to persons unless special means of access are used.

1926.960(jj)

Manhole. The term means a subsurface enclosure which personnel may enter and which is used for the purpose of installing, operating, and maintaining equipment and/or cable.

..1926.960(kk)
1926.960(kk)

Pulling tension. The term means the longitudinal force exerted on a cable during installation.

1926.960(ll)

Qualified person. The term means a person who by reason of experience or training is familiar with the operation to be performed and the hazards involved.

1926.960(mm)

Switch. The term means a device for opening and closing or changing the connection of a circuit. In these rules, a switch is understood to be manually operable, unless otherwise stated.

1926.960(nn)

Tag. The term means a system or method of identifying circuits, systems or equipment for the purpose of alerting persons that the circuit, system or equipment is being worked on.

1926.960(oo)

Unstable material. The term means earth material, other than running, that because of its nature or the influence of related conditions, cannot be depended upon to remain in place without extra support, such as would be furnished by a system of shoring.

1926.960(pp)

Vault. The term means an enclosure above or below ground which personnel may enter and is used for the purpose of installing, operating, and/or maintaining equipment and/or cable.

..1926.960(qq)

1926.960(qq)

Voltage. The term means the effective (rms) potential difference between any two conductors or between a conductor and ground. Voltages are expressed in nominal values. The nominal voltage of a system or circuit is the value assigned to a system or circuit of a given voltage class for the purpose of convenient designation. The operating voltage of the system may vary above or below this value.

1926.960(rr)

Voltage of an effectively grounded circuit. The term means the voltage between any conductor and ground unless otherwise indicated.

1926.960(ss)

Voltage of a circuit not effectively grounded. The term means the voltage between any two conductors. If one circuit is directly connected to and supplied from another circuit of higher voltage (as in the case of an autotransformer), both are considered as of the higher voltage, unless the circuit of lower voltage is effectively grounded, in which case its voltage is not determined by the circuit of higher voltage. Direct connection implies electric connection as distinguished from connection merely through electromagnetic or electrostatic induction.

Reprinted from www.osha.gov

Occupational Safety & Health Administration
200 Constitution Avenue, NW
Washington, DC 20210

Index